Intelligent Systems Reference Library

Volume 97

Series editors

Janusz Kacprzyk, Polish Academy of Sciences, Warsaw, Poland
e-mail: kacprzyk@ibspan.waw.pl

Lakhmi C. Jain, Bournemouth University, Poole, UK and
University of South Australia, Adelaide, Australia
e-mail: Lakhmi.Jain@unisa.edu.au

About this Series

The aim of this series is to publish a Reference Library, including novel advances and developments in all aspects of Intelligent Systems in an easily accessible and well structured form. The series includes reference works, handbooks, compendia, textbooks, well-structured monographs, dictionaries, and encyclopedias. It contains well integrated knowledge and current information in the field of Intelligent Systems. The series covers the theory, applications, and design methods of Intelligent Systems. Virtually all disciplines such as engineering, computer science, avionics, business, e-commerce, environment, healthcare, physics and life science are included.

More information about this series at http://www.springer.com/series/8578

Cengiz Kahraman · Seda Yanık
Editors

Intelligent Decision Making in Quality Management

Theory and Applications

 Springer

Editors
Cengiz Kahraman
Industrial Engineering Department
Istanbul Technical University
Istanbul
Turkey

Seda Yanık
Industrial Engineering Department
Istanbul Technical University
Istanbul
Turkey

ISSN 1868-4394   ISSN 1868-4408   (electronic)
Intelligent Systems Reference Library
ISBN 978-3-319-24497-6   ISBN 978-3-319-24499-0   (eBook)
DOI 10.1007/978-3-319-24499-0

Library of Congress Control Number: 2015955831

Springer Cham Heidelberg New York Dordrecht London

Printed on acid-free paper

Springer International Publishing AG Switzerland is part of Springer Science+Business Media (www.springer.com)

I dedicate this book to my grandfathers Hüseyin Kahraman and Hüseyin Çelikkol and my grandmothers Medine Kahraman and Hayriye Çelikkol.

Prof. Cengiz Kahraman

I dedicate this book to my parents Sırma Yanık and Hulusi Yanık and my brother Emre Yanık.

Dr. Seda Yanık

Preface

The objective of quality management is to maintain a desired level of excellence in service and production systems. Quality management is composed of management activities and functions involved in determination of product and service quality policy and its implementation through quality planning and assurance, as well as quality control and quality improvement.

In recent years, there has been a significant increase in the interest for designing intelligent systems to address complex decision systems. Intelligent decision-making deals with solving complex problems in engineering, social sciences, computer sciences, etc., based on artificial intelligence. The aim of employing intelligent systems is to incorporate the uncertainties in real-world problems that cannot be handled by classical approaches. The types of uncertainties that intelligent techniques can handle are due to imprecise, vague, unreliable, contradictory, and incomplete data.

Intelligent techniques, in addition to the classical techniques, have improved the ability to solve such complex real-world problems in uncertain environments. Intelligent techniques include fuzzy and rough sets, neural networks, support vector machines, genetic algorithms, particle swarm optimization, simulated annealing, and Tabu search.

The nature of most problems in quality management is complex as they include many variables and parameters to model them. Besides, the data related to these problems are commonly vague, imprecise, or incomplete. Moreover, the data are represented by linguistic terms rather than exact numerical values in some cases. In such cases, intelligent decision-making techniques can be efficiently used for solving these types of quality management problems.

The contents of this book have been constituted by considering classical quality management books. The aim of the book is to gather intelligent applications suitable to this content. The complex problems of quality management such as process control, reliability analysis, and software and service quality have been tackled with using intelligent techniques such as neural networks, genetic algorithms, or fuzzy logic. Fifteen chapters have been invited from various countries, namely Turkey,

USA, India, France, Sweden, Hungary, China, Singapore, Spain, Italy, Croatia, and Iran.

Chapter 1 reviews the intelligent decision-making literature in order to reveal their usage in quality problems. It first classifies the intelligent techniques and then presents graphical illustrations to show the status of these techniques in the solutions of quality problems. These graphs display the publishing frequencies of the intelligent quality management papers with respect to their countries, universities, journals, authors, and document types.

Chapter 2 presents the fuzzy control charts for variables. It includes the development of fuzzy Shewhart control charts, fuzzy EWMA control charts, and unnatural pattern analyses under fuzziness.

Chapter 3 presents the fuzzy control charts for attributes, namely p-chart, np-chart, u-chart, and c-chart. Numerical examples are given for each type of these charts.

Chapter 4 considers the economical design of EWMA zone control charts for set of machines operating under Jidoka Production System (JPS). It provides an extensive literature review of intelligent systems in quality control deductively. It starts with an overview of quality control charts, then, reviews charts designed for special purposes such as EWMA, CUSUM, and zone control charts. Finally, it develops economical design and intelligent applications of EWMA control charts

Chapter 5 includes the categorization of the most essential works on fuzzy process capability indices in four main categories: Lee et al. method and its extensions, Parchami et al. method and its extensions, Kaya and Kahraman method and its extensions, and Yongting method and its extensions.

Chapter 6 presents a methodology for increasing the performance as well as productivity of the system by utilizing uncertain data. For this, an optimization problem is formulated by considering reliability, availability, and maintainability parameters as an objective function. The conflicting nature between the objectives is resolved by defining their nonlinear fuzzy goals and then aggregate by using a product aggregator operator.

Chapter 7 reviews acceptance sampling plans with intelligent techniques for solving important as well as fairly complex problems. Then, it introduces acceptance sampling under fuzziness in detail. Finally, multi-objective mathematical models for fuzzy single and fuzzy double acceptance sampling plans with illustrative examples are proposed.

Chapter 8 focuses on the most recent applications of experimental design related to heuristic optimization, fuzzy approach, and artificial intelligence with a special emphasis on the optimal experimental design and optimality criteria. The area of optimal experimentation, which deals with the calculation of the best scheme of measurements, is explained.

Chapter 9 shows how intelligent techniques can be used to design data-driven tools that are able to support the organization to continuously improve the effectiveness of their production according to the Plan–Do–Check–Act (PDCA) methodology. The chapter focuses on the application of data mining and multivariate statistical tools for process monitoring and quality control. Classical multivariate tools such as PLS and PCA are presented along with their nonlinear variants.

Chapter 10 provides a comprehensive review of the multicriteria approaches proposed for failure mode and effects analysis under uncertainty and offer a brief tutorial for those who are interested in these approaches.

Chapter 11 presents research on intelligent techniques for QFD with regard to four aspects, namely, determination of importance weights of customer requirements, modeling of functional relationships in QFD, determination of importance weights of engineering characteristics, and target value setting of engineering characteristics. A fuzzy analytic hierarchy process with an extent analysis approach together with a chaos-based fuzzy regression approach proposes a novel fuzzy group decision-making method, and an inexact genetic algorithm is proposed in this chapter.

Chapter 12 extends the understanding of what performance measures can be applied to processes in order to gain useful information and the emerging application of artificial neural networks to handle concurrent multiple feedback loops.

Chapter 13 employs the Taguchi method coupled with intelligent techniques on the fleet control of automated guided vehicles in a flexible manufacturing setting. Particularly, details and illustrations of combining the Taguchi method with a fuzzy system and a radial basis neural network are provided. In the simulation study, the adaptive fuzzy rules are formulated to base the decision-making process and the Taguchi method is applied to fine tune the rules for optimal performance.

Chapter 14 focuses on modeling and analysis of quality of service (QoS) tradeoffs of a software architecture based on optimization models. A particular emphasis is given to two aspects of this problem: (i) the mathematical foundations of QoS tradeoffs and their dependencies on the static and dynamic aspects of a software architecture, and (ii) the automation of architectural decisions driven by optimization models for QoS tradeoffs.

Chapter 15 describes the application of back-propagation neural networks (BPNN) in an extended importance–performance analysis (IPA) framework with the goal of discovering key areas of quality improvements. The value of the extended BPNN-based IPA is demonstrated using an empirical case example of airport service quality.

This book will provide a useful resource of ideas, techniques, and methods for the research on the theory and applications of intelligent techniques in quality management. Finally, we thank all the authors whose contributions and efforts made the publication of this book possible. We are grateful to the referees for their valuable and highly appreciated works contributed to select the high quality of chapters published in this book.

Cengiz Kahraman
Seda Yanık

Contents

Chapter 1
Intelligent Decision Making Techniques in Quality Management: A Literature Review

Cengiz Kahraman and Seda Yanık

Abstract Intelligent techniques present optimum or suboptimal solutions to complex problems, which cannot be solved by the classical mathematical programming techniques. The aim of this chapter is to review the intelligent decision making literature in order to reveal their usage in quality problems. We first classify the intelligent techniques and then present graphical illustrations to show the status of these techniques in the solutions of quality problems. These graphs display the publishing frequencies of the intelligent quality management papers with respect to their countries, universities, journals, authors, types (whether it is a conference paper, book chapter, journal 1 paper, etc.)

Keywords Intelligent techniques · Quality management · Quality control · Tabu search · Fuzzy sets · Swarm optimization · Genetic algorithm · Ant colony optimization · Neural networks · Simulated annealing

1.1 Introduction

Traditionally quality is defined as the fitness for use. As the marketplace evolved, a modern view of quality stated that quality is inversely proportional to variability (Montgomery 2012). Quality control and improvement efforts aim to control the variability in order to ensure a continuous specific quality level. To this end, one of the most effective tools is statistical process control (SPC) which uses the approach of probability and statistics. The main aim of SPC is to monitor and minimize process variations. Statistical process control (SPC) is very useful in maintaining an acceptable and stable level of some quality characteristics (Guh 2003). SPC helps to first draw conclusions about the populations (or processes) by statistical inference process.

C. Kahraman · S. Yanık (✉)
Department of Industrial Engineering, Istanbul Technical University,
34367 Macka Istanbul, Turkey
e-mail: sedayanik@itu.edu.tr

© Springer International Publishing Switzerland 2016
C. Kahraman and S. Yanık (eds.), *Intelligent Decision Making in Quality Management*, Intelligent Systems Reference Library 97,
DOI 10.1007/978-3-319-24499-0_1

Then, we decide diagnosing if the process is the deviated and take corrective actions if necessary.

The tools of quality management help us to draw the reasoning in order to make decisions for maintaining the quality. Intelligent decision making is defined as the computer-based artifacts performing human decision making task. Two main aspects of decision-making, diagnosis and look-ahead, may be adopted by the intelligent techniques such as expert systems, case-based reasoning, fuzzy set, rough set theories, neural networks (NNs) and optimization/evolutionary algorithms (Pomerol 1997). Intelligent systems either observe how people make the decision in the task at hand and reproduce the process in the machine or help to represent knowledge and reasoning (Pomerol 1997; Simon 1969, Newell and Simon 1972).

The advance of computer integrated manufacturing allowed automatic implementation of quality control tasks. The advanced data-collection systems (e.g. the machine vision system and scanning laser system) increased the rate and number of data input while decreasing the data collection costs. Quality problems commonly involve multivariate data that are not easy to model and/or optimize. Hence, computer-coded logic and algorithms are developed for the data analysis and decision making in quality control. Intelligent methods are extensively used for decision making in quality control together with data collection and analysis. Knowledge discovery with quality data has been achieved by various data mining techniques such association, clustering. On the other hand, decision-making tasks such as the description of product and process quality, prediction of quality, classification of quality and parameter optimization are achieved by intelligent techniques (Köksal et al. 2011). Fuzzy logic is used to capture the uncertainty and imprecision for the description of product and process quality. Rule-based experts systems provide a logical, symbolic approach for reasoning while neural networks use numeric and associative processing to mimic models of biological systems (Guh 2003). Both techniques help to predict or classify the quality. Evolutionary algorithms imitate the evolution process for the optimization of parameters in the quality control applications.

The aim of this chapter is to classify the intelligent techniques and to review their usage in the solution of complex quality problems. It provides an excellent review which summarizes the present status of quality problems solved by intelligent techniques.

The rest of the chapter is organized as follows. Section 1.2 classifies intelligent techniques. Section 1.3 presents a classification of literature of intelligent techniques in quality problems. Section 1.4 gives the results of literature review by some graphical illustrations. Section 1.5 concludes the chapter with future directions.

1.2   Intelligent Techniques

In this section, we introduce the intelligent decision making techniques which are used in the quality management literature.

1.2.1 Particle Swarm Optimization

Particle swarm optimization method is inspired from the social behaviour of bio-logical swarm systems such as the movement of organisms in a bird flock or fish school. PSO method was developed originally by Kennedy and Eberhart (1995). It is a population-based computational method which achieves optimization by iter-atively improving the candidate solutions. Particles, which are candidate solutions, form a population. The particles of the population are located in the search space according to the particle's position and velocity and the current optimum particles. The particles communicate either directly or indirectly with one another for search directions. As a result, the swarm is directed to the best solution. PSO has been used as an effective metaheuristic technique for various problem types of different applications. Some applications of the use of PSO method for different areas are reviewed as follows.

Image compression is an important tool which allows effective resource use. However, the quality of the compressed image should be assessed accurately. Optimal quantization tables which determine the compression ratio and the quality of the decoded images are selected using PSO method (Ma and Zhang 2013). A multi-objective model which also presents different trade-offs between image compression and quality is developed in their study. Real-time self tuning of autonomous microgrid operations of typical distributed generation units is needed for optimal power control strategy.

The parameter design problem for the lighting performance of a specific type of LED includes the settings of the geometric parameters and the refractive properties of the materials. Hsu (2012) proposed a hybrid approach for the selection of optimal design parameters using genetic programming (GP), Taguchi quality loss functions, and particle swarm optimization. The methodology helps to identify the key quality characteristics of a LED and outperforms the traditional Taguchi method in solving this multi-response parameter design problem.

Particle Swarm Optimization (PSO) is applied for real-time self-tuning of the power control parameters such as voltage and frequency regulation, and power sharing by Al-Saedi et al. (2012). The developed controller is shown to be effective to improve the quality of power supply of the microgrid.

The fuzzy logic and particle swarm optimization (PSO) method are also employed for the power quality by Hooshmand and Enshaee (2010). The single and combined power quality disturbances are aimed to be detected and classified using the proposed approach. The signals are used to identify the power quality distur-bances which are derived from the Fourier and wavelet transforms of the signal. Fuzzy rule-based system which is oriented with a PSO algorithm is developed to classify the type of the disturbances. The PSO algorithm is used to provide opti-mized values for the parameters of the membership functions of the fuzzy rule-based systems used for detection and classification.

In construction applications, quality is closely interrelated with the time and cost. The objective is in such settings is minimizing the cost and time while maximizing the quality. The optimal combination of construction methods is chosen by fuzzy-multi-objective particle swarm optimization in the study of Zhang and Xing (2010). The imprecise or vague data related to quality is represented by fuzzy numbers. The proposed methodology presents the solution for time–cost–quality trade off problem of selected construction methods.

Shirani et al. (2015) introduced a hybrid algorithm, specifically designed to work with optimized decision tree with particle swarm optimization (PSO-DT), for the prediction of Soil physical quality indicators (i.e., air capacity, AC; plant-available water capacity, PAWC; and relative field capacity, RFC). The potential power of using the PSO-DT algorithm in setting up a framework for identifying the most determinant parameters affecting the physical quality of agricultural soils in a semiarid region of Iran is also investigated.

1.2.2   Genetic Algorithms

Genetic Algorithms (GAs) are heuristic procedures that use the principles of evolutionary algorithms. The methodology of Genetic algorithms have been developed by Holland (1975) and applied extensively to various types of optimization problems. GAs are inspired from the biological process of natural selection and the survival of the fittest. A pool of solutions defined as a population of chromosomes and a search process is achieved by generations of crossovers. Improvement is aimed to be obtained by selecting the competitive chromosomes that weed out poor solutions and carry over the genetic material to the offspring. At each iteration, the competitive solutions are recombined with other solutions to obtain hopefully better solutions in terms of objective function value or the "fitness" value. The resulting better solutions are then used to replace inferior solutions in the population. For further details on Genetic Algorithms, the interested reader is referred to the study by Reeves (2003).

Genetic algorithms have been mainly used for the calibration of model parameters which aim commonly to predict or classify quality and to search for the optimal design of quality detection and monitoring systems. Examples of studies from the literature are reviewed as follows.

The efficiency of electrical devices is highly related to harmonic occurrence in the inverters. Linear equations which are functions of the switching angles are needed to be solved to eliminate harmonics. Tutkun (2010) developed a hybrid genetic algorithm method based on the refinement of the genetic algorithms results through the Newton–Raphson method to simultaneously solve such non-linear equations.

In power systems, maintaining the quality is important and various methods have been studied to achieve power quality. One of the most common power disturbances is due to voltage sag which is a decrease in voltage or current at the power frequency for short durations. To obtain information related to voltage sags, power quality monitoring system are implemented in power supply networks. GAs have been used to find the optimal number and location of monitored sites to minimize the number of monitors and to reduce monitoring costs without missing essential voltage sag information (Gupta et al. 2014). Kazemi et al. (2013) have offered to use GAs to determine the optimal number and placement of power quality monitors (PQMs) in power systems. Specifically, a GA was developed to evaluate the optimum number of allocated monitors, which is defined as the minimum difference between the Mallow's Cp value and the number of variables used in the multivariable regression model for estimating the unmonitored buses. Mallow's Cp is a statistical criterion for selecting among many alternative subset regressions.

Selection of optimal number and location of quality monitoring sites is also a problem of water quality assurance. Park et al. (2006) used a genetic algorithm (GA) and a geographic information system (GIS) for the design of an effective water quality monitoring network in a large river system. Fitness functions were defined with five criteria: representativeness of a river system, compliance with water quality standards, supervision of water use, surveillance of pollution sources and examination of water quality changes. GIS data was used for obtaining the fitness levels.

Ng and Perera (2003) developed a GA to optimise model parameters of river water quality models. Then, Pelletier et al. (2006) has used a GA to find the combination of kinetic rate parameters and constants that results in a best fit for a model application compared with observed data. They modelled the relation between kinetics and the water quality in streams and rivers. Preis and Ostfeld (2008) also aimed to predict flow and water quality load in watersheds. They have used extensive data and employed a data driven modelling approach. The methodology included a coupled model tree–genetic algorithm scheme. The model tree predicted flow and water quality constituents while the genetic algorithm was employed for calibrating the model tree parameters.

Advances in the electronic systems and machine learning allow the uses of devices used for the detection of quality. Shi et al. (2013) used GAs to select and optimize the effective sensors of electronic nose which is aimed to contribute the modeling of production areas and tree species.

Artificial intelligence techniques are commonly used to detect quality. An application on the classification of the cotton yarn quality is presented by Amin (2013). A hybrid technique involving Artificial Neural Network (ANN) and genetic algorithm (GA) is developed. GA is used to find the optimal values of the input chromosomes (input attributes of the ANN) which maximize the nonlinear exponential function of the output node of ANN. Rules for classification are derived using the optimum chromosomes.

Chou et al. (2010) proposes a virtual metrology (VM) system for real-time quality measurement of wafers and detection of the performance degradation of machines in manufacturing of semiconductor and thin-film transistor liquid crystal display. Support vector machines model is used for detection and a GA is developed for the training/learning of support vector machine (SVM) model.

An application for the identification of materials at mines is presented by Chatterjee and Bhattacherjee (2011). They developed an image analysis-based method which efficiently and cost effectively determines the quality parameters of material at mines. A GA was designed to reduce the dimensions of the image features effectively. Then the features are modelled using neural networks against the actual grade values of the samples generated by chemical analysis.

Parameters related to manufacturing process are selected to for achieving high quality for more than one quality characteristics. Su and Chiang (2003) applied a neural–genetic algorithm to select these parameters. the neural network is used to formulate a fitness function for predicting the value of the response based on the parameter settings. GA then takes the fitness function from the trained neural network to search for the optimal parameter combination.

Castellini et al. (2015) presented an adaptive illumination system for image quality enhancement in vision-based quality control systems. In particular, a spatial modulation of illumination intensity was proposed in order to improve image quality, thus compensating for different target scattering properties, local reflections and fluctuations of ambient light.

1.2.3 Fuzzy Sets

Fuzzy sets are the basic concept supporting the fuzzy set theory. The main research fields in the fuzzy set theory are fuzzy sets, fuzzy logic, and fuzzy measure. Fuzzy reasoning or approximate reasoning is an application of fuzzy logic to knowledge processing. Fuzzy control is an application of fuzzy reasoning to control devices. One feature of FSs is the ability to realize a complex nonlinear input–output relation as a synthesis of multiple simple input–output relations.

The fuzzy set theory has been used in several intelligent technologies by today ranging from control, automation technology, robotics, image processing, pattern recognition, medical diagnosis etc. Some examples are

- A major application area is automotive industry. Fuzzy control has been applied to control automatic transmission system, suspension system, engine system, climate system and antilock brake system.
- Another example is washing machines that adjust their washing strategy based on sensed dirt level, fabric type, load size and water level.
- Fuzzy logic has been used to enhance processing of digital image and signals. Autofocus, auto-zoom, auto-white balancing and auto-exposure systems of cameras.
- Electrophotography process of photocopying machines has been improved. The image quality has been improved by better toner supply control based on fuzzy control.
- Some other successful applications are hand written language recognition and voice recognition.

Fuzzy logic and fuzzy set theory have been successfully applied to handle imperfect, vague, and imprecise information. Nevertheless, to handle vague and imprecise information whereby two or more sources of vagueness appear simultaneously, the modeling tools of ordinary fuzzy sets are limited. For this reason, different generalizations and extensions of fuzzy sets have been introduced (Rodriguez et al. 2012): Type-2 fuzzy sets, nonstationary fuzzy sets, intuitionistic fuzzy sets, fuzzy multisets, and hesitant fuzzy sets.

Fuzzy logic has been extensively used in quality measurement and control problems in the literature. Some of the recently published works are as follows: Yuen (2014) proposes a hybrid framework of Fuzzy Cognitive Network Process, Aggregative Grading Clustering, and Quality Function Deployment (F-CNP-AGC-QFD) for the criteria evaluation and analysis in QFD. The fuzzy QFD enables rating flexibility for the expert judgment to handle uncertainty. The Fuzzy Cognitive Network Process (FCNP) is used for the evaluation of the criteria weights.

An et al. (2014) introduce a fuzzy rough set to perform attribute reduction. Then, an attribute recognition theoretical model and entropy method are combined to assess water quality in the Harbin reach of the Songhuajiang River in China. A dataset consisting of ten parameters is collected from January to October in 2012. Fuzzy rough set is applied to reduce the ten parameters to four parameters.

Li et al. (2014) study the problem of multiple attribute decision making in which the decision making information values are triangular fuzzy numbers and a relative entropy decision making method for software quality evaluation is proposed. Then, according to the concept of the relative entropy, the relative closeness degree is defined to determine the ranking order of all alternatives by calculating the relative entropy to both the fuzzy positive-ideal solution (FPIS) and fuzzy negative-ideal solution (FNIS) simultaneously.

Ghorbani et al. (2014) provides a new method to categorize and select distributors for supply chain management. After determining criteria according to the service quality dimensions as a novel innovation, the fuzzy adaptive resonance theory (ART) algorithm is utilized to categorize distributors according to their similarity. Then, AHP and fuzzy TOPSIS are utilized to arrange distributors in their relative category.

Hsu (2015) integrated the fuzzy analytic network process and fuzzy VIKOR method in a fuzzy multi-criteria decision-making model to provide a complete process to diagnose managerial strategies to reduce customer gaps in service quality efficiently.

Wei et al. (2015) improved fuzzy comprehensive evaluation (FCE) by importing trustworthy degree to it and proposed an automatic hotel service quality assessment method using the improved FCE, which can automatically get more trustworthy evaluation from a large amount of less trustworthy online comments. Then, the causal relations among evaluation indexes were mined from online comments to build the fuzzy cognitive map for the hotel service quality, which was useful to unfold the problematic areas of hotel service quality, and recommended more economical solutions for improving the service quality.

1.2.4   Ant Colony Optimization

Ant Colony Optimization (ACO) is a metaheuristic approach for solving hard combinatorial optimization problems. Ant colony optimization (ACO) algorithm based on the foraging behaviour of ants has been first introduced by Dorigo and Gambardella (1997). The basic idea of ACO is to imitate the cooperative behavior of ant colonies. When searching for food, ants initially explore the area surrounding their nest in a random manner. As soon as an ant finds a food source, it evaluates it and carries some food back to the nest. During the return trip, the ant deposits a pheromone trail on the ground. The pheromone deposited, the amount of which may depend on the quantity and quality of the food, guides other ants to the food source (Socha and Dorigo 2008). Quantity of pheromone on the arc is decreased in time due to evaporating. Each ant decides to a path or way according to the quantity of pheromone which has been leaved by other ants. More pheromone trail consists in short path than long path. Because the ants drop pheromones every time they bring food, shorter paths are more likely to be stronger, hence optimizing the solution. The first ACO algorithm developed was the ant system (AS) (Dorigo 1992), and since then several improvement of the AS have been devised (Gambardella and Dorigo 1995, 1996; Stützle and Hoos 2000).

Bhaskara Murthy and Prabhakar Rao (2015) focused on the application of ACO with optimized link state routing protocol to improve quality of service in Mobile Ad Hoc Network, a dynamic multi-hop wireless network. Simulation results show that the proposed routing enhances the performance of the network.

1.2.5   Bee Colony Optimization

Artificial bee colony (ABC) algorithm was proposed by Karaboga (2005). Bee Colony Optimization (BCO) algorithm imitates the procedure of collective food search of honeybees. The initial search for the food is executed by a group of bees which inform their remaining bees in the hive about the location quantity and the quality of the food they have explored. A bee carrying out random search is called a scout. Moreover, the scout bees which will lead the followers also try to attract follower bees from the hive by a dance behaviour named as waggle dance. During the waggle dance, the quantity of the food is also given to the followers. Besides, it is known that the quality food is an important factor for strong commitment among the bees. The foraging bees under the lead of the explorer bee leave the hive and collect the food in the explored area. The collected food is returned back to the hive. As the bees collect the food, they return back to the hive to store the food. Then, those bees may choose one of the following options to go through: (i) it may continue to collect food at the same location under its previous leader; (ii) it may choose to build up its own team and try to attract followers to join its team (iii) they may separate from the leader bee and become an uncommitted bee. The exploration of new areas and food collection processes continuously take place.

The bee colony optimization algorithms are a newly developed swarm intelligence technique. Its application has mainly focused on job shop scheduling, location and transportation modelling as well as control theory (Davidović et al. 2011; Taheri et al. 2013; Ngamroo 2012). This newly proposed method provides a potential approach for the quality related intelligent decision making. Chen et al. (2015) firstly carried out a sensitivity analysis of a water quality model using the Monte Carlo method. Then, two hybrid swarm intelligence algorithms were proposed to identify the parameters of the model based on the artificial bee colony and quantum-behaved particle swarm algorithms. One hybrid strategy is to use sequential framework, and the other is to use parallel adaptive cooperative evolving. The results of sensitivity analysis reveal that the average velocity and area of the river section are well identified, and the longitudinal dispersion coefficient is difficult to identify.

Chen et al. (2015) firstly carried out a sensitivity analysis of a water quality model using the Monte Carlo method. Then, two hybrid swarm intelligence algorithms were proposed to identify the parameters of the model based on the artificial bee colony and quantum-behaved particle swarm algorithms. One hybrid strategy is to use sequential framework, and the other is to use parallel adaptive cooperative evolving. The results of sensitivity analysis reveal that the average velocity and area of the river section are well identified, and the longitudinal dispersion coefficient is difficult to identify.

1.2.6 Neural Networks

Neural networks are computational models which are inspired by the connected neurons of the nervous system. The network structure takes the inputs, then weighs and transforms them by predetermined functions, finally determines the output(s) through neurons. Using the principles of human brain processes, artificial neural networks achieve learning from experiences and present the use of these experiences via parallel processing units (i.e. neurons). The learning process takes place in the network and stored as weights among the connections of the neurons.

ANNs are commonly used for machine learning, data classification, generalization, feature extraction, optimization, data completion and pattern recognition. The fundamental property of ANNs is to process data and make decisions using the weights acquired from the learning phase.

Salehi et al. (2012) proposes a model consisting of two models which are effective in recognition of unnatural control chart patterns. The first model is a support vector machine (SVM)-classifier which recognizes the mean and variance shift. The second model consists of two neural networks for mean and variance to detect the magnitude of the shifts. Ebrahimzadeh et al. (2011) develops an intelligent method for recognition of the common types of control chart pattern.

Similarly the method includes two modules: clustering module uses a combination of the modified imperialist competitive algorithm (MICA) and the K-means algorithm whereas classifier module includes a neural network for determining the pattern type. Cheng and Cheng (2011) aims to recognize the bivariate process variance shifts using neural networks. They have explored the networks design factors window size, number of training examples, sample size, training algorithm with respect to the performance of the neural network, in terms of the ARL and run length distribution. Wu and Yu (2010) propose a network ensemble model to identify the mean and variance shifts in correlated processes and show that this model performs better than single NNs. Hosseinifard et al. (2011) proposes to use artificial neural networks to detect and classify the shifts in linear profiles which are defined as relation between a response variable and one or more explanatory variables. Velo et al. (2013) compared the performance of alkalinity level prediction of cruise ships using ANN and multilinear regression. Then the alkalinity estimation is used for quality control of measurements. Lopez-Lineros et al. (2014) developed a non-linear autoregressive neural network for the quality control of raw river stage data. Kesharaju et al. (2014) develop a ultrasonic sensor based neural network to identify defects in ceramic components. Neural network approach is used for the classification of defects.

Kadiyala and Kumar (2015) presented a methodology that combines the use of univariate time series and back propagation neural network (widely used ANN) methods in the development and evaluation of IAQ models for the monitored contaminants of carbon dioxide and carbon monoxide inside a public transportation bus using available software.

1.2.7 Simulated Annealing

Simulated annealing (SA) methods are the methods proposed for the problem of finding, numerically, a point of the global minimum of a function defined on a subset of a k-dimensional Euclidean space. The motivation of the methods lies in the physical process of annealing, in which a solid is heated to a liquid state and, when cooled sufficiently slowly, takes up the configuration with minimal inner energy. Metropolis et al. (1953) described this process mathematically. SA uses this mathematical description for the minimization of other functions than the energy. The first results published by Kirpatrick et al. (1983), German and German (1984), Cerny (1985),

SA algorithm is a technique to find a good solution of an optimization problem using a random variation of the current solution. A worse variation is accepted as the new solution with a probability that decreases as the computation proceeds. The slower the cooling schedule, or rate of decrease, the more likely the algorithm is to find an optimal or near-optimal solution (Xinchao 2011).

1.2.8 Tabu Search

The word tabu (or taboo) comes from Tongan, a language of Polynesia, where it was used by the aborigines of Tonga island to indicate things that cannot be touched because they are sacred. According to Webster's Dictionary, the word now also means "a prohibition imposed by social custom as a protective measure" or of something "banned as constituting a risk."

Difficulty in optimization problems encountered in practical settings such as telecommunications, logistics, financial planning, transportation and production has motivated in development of optimization techniques. Tabu search (TS) is a higher level heuristic algorithm for solving combinatorial optimization problems. It is an iterative improvement procedure that starts form an initial solution and attempts to determine a better solution.

1.2.9 Swarm Intelligence

Social insects work without supervision. In fact, their teamwork is largely self-organized, and coordination arises from the different interactions among individuals in the colony. Although these interactions might be primitive (one ant merely following the trail left by another, for instance), taken together they result in efficient solutions to difficult problems (such as finding the shortest route to a food source among myriad possible paths). The collective behaviour that emerges from a group of social insects has been dubbed swarm intelligence (Bonabeau and Meyer, 2001). SI indicates a recent computational and behavioural metaphor for solving distributed problems that originally took its inspiration from the biological examples provided by social insects (ants, termites, bees, wasps) and by swarming, flocking, herding behaviours in vertebrates.

1.2.10 Differential Evolution

Differential evolution (DE) is introduced by Storn and Price in 1996. DE is known as population-based optimisation algorithm similar to GAs using similar operators; crossover, mutation and selection. According to Karaboğa and Ökdem (2004), the main difference in constructing better solutions is that genetic algorithms rely on crossover while DE relies on mutation operation. This main operation is based on the differences of randomly sampled pairs of solutions in the population. DE algorithm uses mutation operation as a search mechanism and selection operation to direct the search toward the prospective regions in the search space. In addition to this, the DE algorithm uses a non-uniform crossover which can take child vector parameters from one parent more often than it does from others. By using the components of the existing population members to construct trial vectors, the recombination (crossover) operator efficiently shuffles information about successful combinations, enabling the search for a better solution space. An optimization task consisting of D parameters can be represented by a D-dimensional vector. In DE, a population of NP solution vectors is randomly created at the start. This population is successfully improved by applying mutation, crossover and selection operators.

Zhang et al. (2015) proposed a new method for water quality evaluation integrated self-adaptive differential evolution algorithm and extreme learning machine namely SADEELM algorithm to overcome the limitation of extreme learning machine, which not only can solve the problem of complicated non-linear relationship between influencing factors and the grade of water quality, but also can well perform in water quality evaluation.

1.3 Some Quality Problem Areas Solved by Intelligent Techniques

Table 1.1 presents a classification of more than 50 papers from the literature including the used intelligent methods, aims of the studies, and the application areas of the related methods. As it can be seen from Table 1.1, Either only one of the intelligent techniques or integrated intelligent techniques are used for the solutions of quality problems.

1.4 Graphical Analyses of Literature Review

In this section, the results of literature review are given by some graphical illustrations. Figure 1.1 illustrates the publication frequencies of papers with respect to the years, indicating a positive trend. The years 2012 and 2013 have the largest frequencies that intelligent techniques are used for quality problems in the papers.

Figure 1.2 shows the journals most publishing papers on quality problems with intelligent techniques. Applied Mechanics and Materials take the first rank in the recent years.

Figure 1.3 ranks the universities that *intelligent quality* based papers come from. Technische Universitat Wien and Chongqing University takes the first and second ranks, respectively.

Figure 1.4 ranks the countries that *intelligent quality* based papers come from. China and US take the first and second ranks, respectively.

Table 1.1 Examples of quality problem areas solved by intelligent techniques

Authors	Method	Aim	Problem area
Ma and Zhang (2013)	PSO	Quality prediction/classification	Power control
Al-Saedi et al. (2012)	PSO	Design parameter optimization	Power control
Hooshmand and Enshaee (2010)	PSO-fuzzy logic	Quality prediction/classification	Power control
Zhang and Xing (2010)	PSO-fuzzy logic	Quality prediction/classification	Construction
Hsu (2012)	PSO-Genetic programming	Design parameter optimization	Lighting performance of LEDs
Tutkun (2010)	GA	Quality prediction/classification	Electrical devices
Gupta et al. (2014)	GA	optimal number and location of quality monitoring	Power systems
Kazemi et al. (2013)	GA	optimal number and location of quality monitoring	Power systems
Park et al. (2006)	GA	optimal number and location of quality monitoring	Water quality assurance
Ng and Perera (2003)	GA	Design parameter optimization	Water quality assurance
Pelletier et al. (2006)	GA	Design parameter optimization	Water quality assurance
Preis and Ostfeld (2008)	GA	Quality prediction/classification	Water quality assurance
Shi et al. (2013)	GA	Quality prediction/classification	Electronic nose used for detecting tree production Areas
Amin (2013)	ANN-GA	Quality prediction/classification	Cotton yarn quality
Chou et al. (2010)	SVM-GA	Quality prediction/classification	Manufacturing of semiconductor and thin-film transistor liquid crystal display
Chatterjee and Bhattacherjee (2011)	GA-ANN	Quality prediction/classification	Mining
Su and Chiang (2003)	GA-ANN	Design parameter optimization	Multivariate quality at manufacturing
Lopez-Lineros et al. (2014)	ANN	Quality prediction/classification	River stage data validation
Salehi et al. (2012)	ANN-SVM	Pattern recognition	Multivariate process control
Cheng and Cheng (2011)	ANN	Quality prediction/classification	Multivariate process control

(continued)

Table 1.1 (continued)

Authors	Method	Aim	Problem area
Ebrahimzadeh et al. (2011)	ANN	Pattern recognition	Multivariate process control
Wu and Yu (2010)	ANN	Pattern recognition	Multivariate process control
Kesharaju et al. (2014)	ANN	Quality prediction/classification	Ceramics
Hosseinifard et al. (2011)	ANN	Quality prediction/classification	Multivariate process control
Velo et al. (2013)	ANN	Quality prediction/classification	
Yuen (2014)	Fuzzy logic	Cloud software product development	Quality function deployment
An et al. (2014)	Fuzzy logic	Attribute reduction	Water quality assessment
Li et al. (2014)	Fuzzy logic	Relative entropy decision making	Software quality evaluation
Ghorbani et al. (2014)	Fuzzy logic	Distributor categorization	Supply chain management
Valavi and Pramod (2015)	Fuzzy logic	Determination of weightages	Maintenance quality function deployment
Liu et al. (2015)	Fuzzy logic	Service quality analysis	Certification and inspection industry
Azar and Vybihal (2011)	ACO	Optimization of service quality prediction accuracy	Software quality evaluation
Neagoe et al. (2010)	ACO	Wine quality assessment	Data mining
Dhurandher et al. (2009)	ACO	Optimization of quality of service security	Wireless sensor networks
Ning and Wang (2009)	ACO	Construction quality maximization	Construction project management
Tong et al. (2012)	SA	Laser cutting quality control	Laser applications
Abdullah and Othman (2014)	SA	Optimization of quality of service	Job scheduling
Kulkarni and Babu (2005)	SA	Product quality optimization	Continuous casting system
Soliman et al. (2004)	SA	Power systems quality analysis	Harmonics and frequency evaluation
Garcia-Martinez et al. (2012)	TS	Voltage minimization	Network expansion
Rahim and Shakil (2011)	TS	Optimization of quality control parameters	Economic production and Quality control

(continued)

Table 1.1 (continued)

Authors	Method	Aim	Problem area
Mukherjee and Ray (2007)	TS	Grinding process optimization	Process functional approximation
Umapathi and Ramaraj (2014)	SI	Quality of service improvement	Wireless mobile hosts
Zhang et al. (2014)	SI	Swarm intelligence optimization	Quality assessment
Machado et al. (2013)	SI	Texture analysis	Material quality assessment
Goudarzi (2012)	SI	Solution improvement for delivering digital video	Multi-hop wireless networks
Sagar Reddy and Varadarajan (2011)	SI	Quality of service improvement	Mobile communication systems
Lv et al. (2014)	DE	Water quality prediction	Regional ecological and water management
Sathya Narayanan and Suribabu (2014)	DE	Multi-objective time-cost-quality optimization	Construction project
Biswal et al. (2012)	DE	Classification of power quality data	Automatic disturbance pattern classification
Zheng et al. (2009)	DE	Urban living quality examination	Urbanization and income growth

Fig. 1.1 Publication frequencies with respect to the years

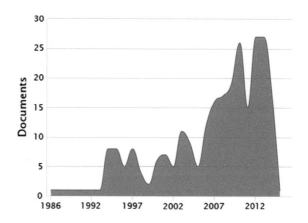

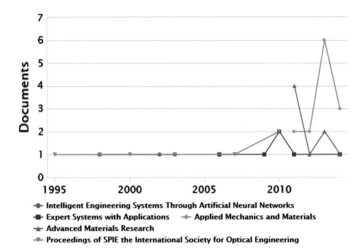

Fig. 1.2 The journals most publishing *intelligent quality* based papers

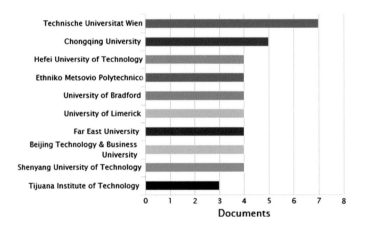

Fig. 1.3 The universities most publishing *intelligent quality* based papers

Figure 1.5 shows the distribution of the published intelligent quality based papers by their document types. Conference papers and journal papers take the first and second ranks, respectively.

Figure 1.6 shows the distribution of intelligent quality papers by their subject areas. Engineering and computer sciences take the first and second ranks, respectively.

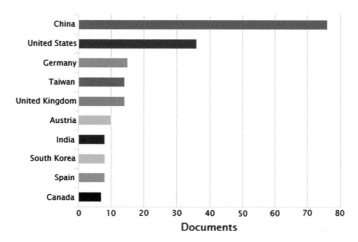

Fig. 1.4 *Intelligent quality* based papers by the countries

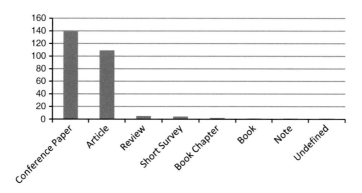

Fig. 1.5 Distribution of intelligent quality based papers by their document types

1.5 Conclusion and Future Trends

Complex quality problems are hard to solve by using classical optimization techniques which guarantee to find an optimal solution and to prove its optimality. Instead, intelligent techniques that may sacrifice the guarantee of finding optimal solutions for the sake of getting good solutions in a limited time have been introduced in this chapter. The quality problems solved by intelligent techniques in the literature are mostly related to power control and systems, water quality assurance, multivariate process control, software quality, and wireless networks. There is a significant increasing trend in the number of papers on quality problems using intelligence techniques after the year 2010. The subject area with the largest

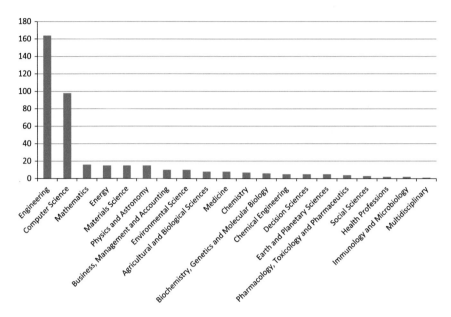

Fig. 1.6 The distribution of intelligent quality papers by their subject areas

frequency using intelligent techniques for quality problems is engineering sciences. Computer sciences then follow it. It can be concluded that researchers seem to more deal with intelligent techniques in the future as the complexity of the quality problems increases.

References

Abdullah, M., Othman, M.: Simulated annealing approach to cost-based multi-quality of service job scheduling in cloud computing enviroment. Am. J. Appl. Sci. **11**(6), 72–87 (2014)

Al-Saedi, W., Lachowicz, S.W., Habibi, D., Bass, O.: Power quality enhancement in autonomous microgrid operation using Particle Swarm Optimization. Int. J. Electr. Power Energy Syst. **42**(1), 139–149 (2012)

Amin, A.E.: A novel classification model for cotton yarn quality based on trained neural network using genetic algorithm. Knowl. Based Syst. **39**, 124–132 (2013)

An, Y., Zou, Z., Li, R.: Water quality assessment in the Harbin reach of the Songhuajiang River (China) based on a fuzzy rough set and an attribute recognition theoretical model. Int. J. Environ. Res. Public Health **11**(4), 3507–3520 (2014)

Azar, D., Vybihal, J.: An ant colony optimization algorithm to improve software quality prediction models: case of class stability. Inf. Softw. Technol. **53**(4), 388–393 (2011)

Bhaskara Murthy, M.V.H., Prabhakar Rao, B.: Ant colony based OLSR for improved quality of service for multimedia traffic. Int. J. Appl. Eng. Res. **10**(6), 15695–15710 (2015)

Biswal, B., Behera, H.S., Bisoi, R., Dash, P.K.: Classification of power quality data using decision tree and chemotactic differential evolution based fuzzy clustering. Swarm Evol. Comput. **4**, 12–24 (2012)

Bonabeou, E., Meyer, C. (Eds.).: Swarm intelligence: a whole new way to think about business. Harward Bus. Rev. (2001)

Castellini, P., Cecchini, S., Stroppa, L., Paone, N.: Optimization of spatial light distribution through genetic algorithms for vision systems applied to quality control. Meas. Sci. Technol. **26**(2), 025401 (2015)

Cerny, V.: A thermodynamical approach to the traveling salesman problem: an efficient simulation algorithm. J. Optim. Theory Appl. **45**, 41–51 (1985)

Chatterjee, S., Bhattacherjee, A.: Genetic algorithms for feature selection of image analysis-based quality monitoring model: an application to an iron mine. Eng. Appl. Artif. Intell. **24**(5), 786–795 (2011)

Chen, G., Wang, J., Li, R.: Parameter identification for a water quality model using two hybrid swarm intelligence algorithms. Soft Comput. 11 pp., (2015) (article in press)

Cheng, C.S., Cheng H.P.: Using neural networks to detect the bivariate process variance shifts pattern. Comput. Ind. Eng. **60**(2), 269–278 (2011)

Chou, P.-H., Wu, M.-J., Chen, K.-K.: Integrating support vector machine and genetic algorithm to implement dynamic wafer quality prediction system. Expert Syst. Appl. **37**(6), 4413–4424 (2010)

Davidović, T., Ramljak, D., Šelmić, M., Teodorović, D.: Bee colony optimization for the p-center problem. Comput. Oper. Res. **38**(10), 1367–1376 (2011)

Dhurandher, S.K., Misra, S., Obaidat, M.S., Gupta, N.: An Ant colony optimization approach for reputation and quality-of-service- based security in wireless sensor networks. Secur. Commun. Networks **2**(2), 215–224 (2009)

Dorigo, M.: Optimization, Learning and Natural Algorithms. Unpublished Doctoral Dissertation. University of Politecnico di Milano, Italy (1992)

Dorigo, M., Gambardella, L.M.: Ant colony system: a cooperative learning approach to the traveling salesman problem. IEEE Trans. Evol. Comput. **1**, 53–66 (1997)

Ebrahimzadeh, A., Addeh, J., Rahmani, Z.: Control chart pattern recognition using K-MICA clustering and neural networks. ISA Trans. **51**(1), 111–119 (2011)

Gambardella, L.M., Dorigo, M.: Ant-Q: a reinforcement learning approach to the travelling salesman problem. In: Proceedings of the Twelfth International Conference on Machine Learning. California, USA (1995)

Gambardella, L.M., Dorigo, M.: Solving symmetric and asymmetric TSPs by ant colonies. In: Proceedings of the IEEE Conference on Evolutionary Computation, pp. 622–627. Nagoya, Japan (1996)

Garcia-Martinez, S., Espinosa-Juarez, E., Rico-Melgoza, J.J.: Application of Tabu search for transmission expansion planning considering power quality aspects. In: CCE 2012—9th International Conference on Electrical Engineering, Computing Science and Automatic Control, Mexico City, Mexico, 26–28 Sept 2012

German, S., German, D.: Stochastic relaxation, Gibbs distributions, and the Bayesian restoration of images. IEEE Proc. Pattern Anal. Mach. Intell. **6**(6), 721–741 (1984)

Ghorbani, M., Arabzad, S.M., Tavakkoli-Moghaddam, R.: Service quality-based distributor selection problem: A hybrid approach using fuzzy ART and AHP-FTOPSIS. Int. J. Prod. Qual. Manag. **13**(2), 157–177 (2014)

Goudarzi, P.: Scalable video transmission over multi-hop wireless networks with enhanced quality of experience using swarm intelligence. Sig. Process. Image Commun. **27**(7), 722–736 (2012)

Guh, R.S.: Integrating artificial intelligence into on-line statistical process control. Qual. Reliab. Eng. Int. **19**(1), 1–20 (2003)

Gupta, N., Swarnkar, A., Niazi, K.R.: Distribution network reconfiguration for power quality and reliability improvement using Genetic Algorithms. Int. J. Electr. Power Energy Syst. **54**, 664–671 (2014)

Holland, J.H. (ed.): Adaptation in Natural and Artificial Systems: An Introductory Analysis with Applications to Biology, Control, and Artificial Intelligence. University of Michigan Press, Ann Arbor, MI (1975)

Hooshmand, R., Enshaee, A.: Detection and classification of single and combined power quality disturbances using fuzzy systems oriented by particle swarm optimization algorithm. Electr. Power Syst. Res. **80**(12), 1552–1561 (2010)

Hosseinifard, S.Z., Abdollahian, M., Zeephongsekul, P.: Application of artificial neural networks in linear profile monitoring. Expert Syst. Appl. **38**(5), 4920–4928 (2011)

Hsu, C.-M.: Improving the lighting performance of a 3535 packaged hi-power LED using genetic programming, quality loss functions and particle swarm optimization. Appl. Soft Comput. **12**(9), 2933–2947 (2012)

Hsu, W.: A fuzzy multiple-criteria decision-making system for analyzing gaps of service quality. Int. J. Fuzzy Syst. **17**(2), 256–267 (2015)

Kadiyala, A., Kumar, A.: Multivariate time series based back propagation neural network modeling of air quality inside a public transportation bus using available software. Environ. Prog. Sustain. Energ. **34**(5), 1259–1266 (2015)

Karaboğa, D., Ökdem, S.: A simple and global optimization algorithm for engineering problems: differential evolution algorithm. Turk. J. Electron. Eng. **12**(1) (2004)

Karaboğa, D.: An idea based on honeybee swarm for numerical optimization. Technical Report TR06, Erciyes University (2005)

Kazemi, A., Mohamed, H., Shareef, H.Zayandehroodi: Optimal power quality monitor placement using genetic algorithm and Mallow's Cp. Int. J. Electr. Power Energy Syst. **53**, 564–575 (2013)

Kennedy, J., Eberhart, R.C.: Particle swarm optimization. In: Proceedings of the IEEE international conference on neural networks IV, 1942–1948 (1995)

Kesharaju, M., Nagarajah, R., Zhang, T., Crouch, I.: Ultrasonic sensor based defect detection and characterisation of ceramics. Ultrasonics **54**(1), 312–317 (2014)

Kirpatrick, S., Gelat Jr, C.D., Vecchi, M.P.: Optimization by simulated annealing. Science **220**, 671–680 (1983)

Köksal, G., Batmaz, İ., Testik, M.C.: A review of data mining applications for quality improvement in manufacturing industry. Expert Syst. Appl. **38**(10), 13448–13467 (2011)

Kulkarni, M.S., Babu, A.S.: Managing quality in continuous casting process using product quality model and simulated annealing. J. Mater. Process. Technol. **166**(2), 294–306 (2005)

Li, Q., Zhao, X., Lin, R., Chen, B.: Relative entropy method for fuzzy multiple attribute decision making and its application to software quality evaluation. J. Intell. Fuzzy Syst. **26**(4), 1687–1693 (2014)

Liu, R., Cui, L., Zeng, G., Wu, H., Wang, C., Yan, S., Yan, B.: Applying the fuzzy SERVQUAL method to measure the service quality in certification and inspection industry. Appl. Soft Comput. J. **26**, 508–512 (2015)

López-Lineros, M., Estévez, J., Giráldez, J.V., Madueño, A.: A new quality control procedure based on non-linear autoregressive neural network for validating raw river stage data. J. Hydrol. **510**(14), 103–109 (2014)

Lv, J., Zou, W., Wang, X.: Water quality prediction using support vector machine with differential evolution optimization. ICIC Expr. Lett., Part B: Appl. **5**(3), 763–768 (2014)

Ma, H., Zhang, Q.: Research on cultural-based multi-objective particle swarm optimization in image compression quality assessment. Opt.—Int. J. Light Electron Opt. **124**(10), 957–961 (2013)

Machado, B.B., Gonçalves, W.N., Bruno, O.M.: Material quality assessment of silk nanofibers based on swarm intelligence. J. Phys.: Conf. Ser. **410**(1) (2013)

Metropolis, N., Rosenbluth, A., Rosenbluth, M., Teller, A., Teller, E.: Equation of state calculations by fast computing machines. J. Chem. Phys. **21**(6), 1087–1092 (1953)

Montgomery, D.C.: Statistical Quality Control 7th Ed., Wiley, New York (2012)

Mukherjee, I., Ray, P.K.: Multi-response grinding process functional approximation and its influence on solution quality of a modified tabu search. In: Proceedings of IEEM 2007: 2007 IEEE International Conference on Industrial Engineering and Engineering Management, pp. 837–841 (2007)

Neagoe, V.-E., Neghina, C.-E., Neghina, M.: Ant colony optimization for logistic regression and its application to wine quality assessment. In: International Conference on Mathematical Models for Engineering Science—Proceedings, MMES'10; Puerto de la Cruz, Tenerife, Spain, pp. 195–200. 30 Nov–2 Dec 2010

Newell, A., Simon, H.A.: Human problem solving. Prentice-Hall, Englewood Cliffs, NJ (1972)

Ng, A.W.M., Perera, B.J.C.: Selection of genetic algorithm operators for river water quality model calibration. Eng. Appl. Artif. Intell. **16**(5–6), 529–541 (2003)

Ngamroo, I.: Application of electrolyzer to alleviate power fluctuation in a stand alone microgrid based on an optimal fuzzy PID. Int. J. Electr. Power Energy Syst. **43**(1), 969–976 (2012)

Ning, X., Wang, L.-G.: Construction quality-cost trade-off using the pareto-based ant colony optimization algorithm. In: Proceedings—International Conference on Management and Service Science, MASS 2009, International Conference on Management and Service Science, Wuhan, China, 20–22 Sept 2009

Park, S.-Y., Choi, J.H., Wang, S., Park, S.S.: Design of a water quality monitoring network in a large river system using the genetic algorithm. Ecol. Model. **199**(3), 289–297 (2006)

Pelletier, G.J., Chapra, S.C., Tao, H.: QUAL2Kw—A framework for modeling water quality in streams and rivers using a genetic algorithm for calibration. Environ. Model Softw. **21**(3), 419–425 (2006)

Preis, A., Ostfeld, A.: A coupled model tree–genetic algorithm scheme for flow and water quality predictions in watersheds. J. Hydrol. **349**(3–4), 364–375 (2008)

Pomerol, J.C.: Artificial intelligence and human decision making. Eur. J. Oper. Res. **99**(1), 3–25 (1997)

Rahim, A., Shakil, M.: A tabu search algorithm for determining the economic design parameters of an integrated production planning, quality control and preventive maintenance policy. Int. J. Ind. Syst. Eng. **7**(4), 477–497 (2011)

Reeves, C.R.: Genetic alorithms. In: Glover, F., Kochenberge, G.A. (eds.) Handbook of Metaheuristics, pp. 55–82. Kluwer Academic, Boston (2003)

Rodriguez, R.M., Martinez, L., Herrera, F.: Hesitant fuzzy linguistic term sets for decision making. IEEE Trans. Fuzzy Syst. **20**(1), 109–119 (2012)

Sagar Reddy, K.S., Varadarajan, S.: Increasing quality of service using swarm intelligence technique through bandwidth reservation scheme in 4G mobile communication systems, In: International Conference on Sustainable Energy and Intelligent Systems, SEISCON 2011, Issue (583), pp. 616–621. IET Conference Publications, Chennai, India, 20–22 July 2011

Salehi, M., Kazemzadeh, R.B., Salmasnia, A.: On line detection of mean and variance shift using neural networks and support vector machine in multivariate processes. Appl. Soft Comput. **12**(9), 2973–2984 (2012)

Sathya Narayanan, A., Suribabu, C.R.: Multi-objective optimization of construction project time-cost-quality trade-off using differential evolution algorithm. Jordan J. Civil Eng. **8**(4), 375–392 (2014)

Shi, B., Lei Zhao, L., Zhi, R., Xi, X.: Optimization of electronic nose sensor array by genetic algorithms in Xihu-Longjing Tea quality analysis. Math. Comput. Model. **58**(3–4), 752–758 (2013)

Shirani, H., Habibi, M., Besalatpour, A.A., Esfandiarpour, I.: Determining the features influencing physical quality of calcareous soils in a semiarid region of Iran using a hybrid PSO-DT algorithm. Geoderma **259–260**, 1–11 (2015)

Simon, H.A.: The Sciences of the Artificial. MIT Press, Cambridge, MA (1969)

Socha, K., Dorigo, M.: Ant colony optimization for continuous domains. Eur. J. Oper. Res. **185**, 1155–1173 (2008)

Soliman, S.A, Mantaway, A.H., El-Hawary, M.E.: Simulated annealing optimization algorithm for power systems quality analysis. Int. J. Electr. Power Energy Syst. **26**(1), 31–36 (2004)

Stützle, T., Hoos, H.: MAX-MIN ant system. Future Gener. Comput. Syst. **16**(8), 889–904 (2000)

Su, C.T., Chiang, T.L.: Optimizing the IC wire bonding process using a neural networks/genetic algorithms approach. J. Intell. Manuf. **14**(2), 229–238 (2003)

Taheri, J., Lee, Y.C., Zomaya, A.Y., Siegel, H.J.: A Bee Colony based optimization approach for simultaneous job scheduling and data replication in grid environments. Comput. Oper. Res. **40**(6), 1564–1578 (2013)

Tong, G., Xu, H., Yu, H.: Control model of laser cutting quality based on simulated annealing and neural network. Appl. Mech. Mater. **148–149**, 206–211 (2012)

Tutkun, N.: Improved power quality in a single-phase PWM inverter voltage with bipolar notches through the hybrid genetic algorithms. Expert Syst. Appl. **37**(8), 5614–5620 (2010)

Umapathi, N., Ramaraj, N.: Swarm intelligence based dynamic source routing for improved quality of service. J. Theor. Appl. Inf. Technol. **61**(3), 604–608 (2014)

Valavi, D.G., Pramod, V.R.: A hybrid fuzzy MCDM approach to maintenance Quality Function Deployment. Decis. Sci. Lett. **4**(1), 97–108 (2015)

Velo, A., Péreza, F.F., Tanhuab, T., Gilcotoa, M., Ríosa, A.F., Key, R.M.: Total alkalinity estimation using MLR and neural network techniques. J. Mar. Syst. **111–112**, 11–18 (2013)

Wei, X., Luo, X., Li, Q., Zhang, J., Xu, Z.: Online comment-based hotel quality automatic assessment using improved fuzzy comprehensive evaluation and fuzzy cognitive map. IEEE Trans. Fuzzy Syst. **23**(1), 72–84 (2015)

Wu, B., Yu, J.: A neural network ensemble model for on-line monitoring of process mean and variance shifts in correlated processes. Expert Syst. Appl. **37**(6), 4058–4065 (2010)

Xinchao, Z.: Simulated annealing algorithm with adaptive neighborhood. Appl. Soft Comput. **11**, 1827–1836 (2011)

Yuen, K.K.F.: A hybrid fuzzy quality function deployment framework using cognitive network process and aggregative grading clustering: An application to cloud software product development. Neurocomputing **142**, 95–106 (2014)

Zhang, H., Xing, F.: Fuzzy-multi-objective particle swarm optimization for time–cost–quality tradeoff in construction. Autom. Constr. **19**(8), 1067–1075 (2010)

Zhang, Y., Cai, Z., Gong, W., Wang, X.: Self-adaptive differential evolution extreme learning machine and its application in water quality evaluation. J. Comput. Inf. Syst. **11**(4), 1443–1451 (2015)

Zhang, Z., Wang, G.-G., Zou, K., Zhang, J.: A solution quality assessment method for swarm intelligence optimization algorithms. Sci. World J. **183809** (2014)

Zheng, S., Fu, Y., Liu, H.: Demand for urban quality of living in China: evolution in compensating land-rent and wage-rate differentials. J. Real Estate Financ. Econ. **38**(3), 194–213 (2009)

Chapter 2
Intelligent Process Control Using Control Charts—I: Control Charts for Variables

Murat Gülbay and Cengiz Kahraman

Abstract Shewhart's control charts are used when you have enough and exact observed data. In case of incomplete and vague data, they can be still used by the help of the fuzzy set theory. In this chapter, we develop the fuzzy control charts for variables, which are namely $\overline{X}$ and R and $\overline{X}$ and S charts. Triangular fuzzy numbers have been used in the development of these charts. Unnatural patterns have been examined under fuzziness. Besides, fuzzy EWMA charts have been also developed in this chapter. For each fuzzy case, we present a numerical example.

Keywords Shewhart's control charts · EWMA control charts · Fuzzy sets · Triangular fuzzy numbers · Unnatural pattern

2.1 Introduction

Process control is an engineering discipline dealing with maintaining the output of a specific process, generally called a quality characteristic, within a desired range. Type of processes using the process control can be categorized into three main groups which are discrete, batch, and continuous processes. Applications having elements of both discrete, batch and continuous process control are often called hybrid applications.

A process may either be classified as "in control" or "out of control". The boundaries for these classifications are set by calculating mean, standard deviation, and range from a set of process data randomly collected when it is under stable operation. Based on the statistical methods, analytical decision-making tools which

M. Gülbay (✉)
Technical Sciences Vocational School, Gaziantep University, Gaziantep, Turkey
e-mail: gulbay@gantep.edu.tr

C. Kahraman
Department of Industrial Engineering, Istanbul Technical University,
34367 Maçka Istanbul, Turkey

© Springer International Publishing Switzerland 2016
C. Kahraman and S. Yanık (eds.), *Intelligent Decision Making in Quality Management*, Intelligent Systems Reference Library 97,
DOI 10.1007/978-3-319-24499-0_2

allow practitioners to measure, monitor, and control the process behavior working normally or not, are called "Statistical Process Control (SPC)". The most successful SPC tool is control charts, originally developed by Walter Shewhart in the early 1920s. Comparing with boundaries of a stable process with a graphical display, they enable online data tracing and abnormal conditions warning, which are an essential tool for continuous quality control. Basically, the control charts are the graphical display of a quality characteristic that has been measured or computed from a sample versus the sample number or time to monitor and show how the process is performing and how the capabilities are affected by changes to the process. This information is then used to make quality improvements. The control charts attempt to distinguish between two types of process variation that impede peak performance. These variations are as follows:

- Common cause variation, which is intrinsic to the process and will always be present.
- Special cause variation, which stems from external sources indicating that the process has assignable situation(s).

Based on the monitored quality characteristics in numerical or in "conforming" or "nonconforming" measurements, the control charts are categorized into two main groups, variables and attributes. This chapter deals with the control charts for variables. The most commonly used control charts for variables use the mean ($\bar{x}, \mu$), range (R), and standard deviations (σ, s) in terms of paired $\bar{X}$ and R charts, paired $\bar{X}$ and s charts, and moving average charts.

2.2 Classical Shewhart Control Charts for Variables

2.2.1 $\bar{X}$ and R Control Charts

Many quality characteristics can be expressed in terms of a precise numerical measurement. One of the efficient ways of determining whether the process is in control or not is checking the process mean and process variability. $\bar{X}$ Charts are used to control the process mean while R charts are used to control the process variability. In general, they are paired and interpreted by looking both of the control charts. When the sample size is constant and relatively small, say $n \leq 10$, the usage of $\bar{X}$ and R charts advantageous.

2.2.1.1 Control Limits for $\bar{X}$ and R Control Charts

Suppose that a quality characteristic "X" is normally distributed with the parameters of μ and σ both known. For a sample size of n ($X_1, X_2, ..., X_n$), the average of the sample is

$$\overline{X} = \frac{X_1 + X_2 + \cdots + X_n}{n} \qquad (2.1)$$

and it is known that $\bar{x}$ is normally distributed with mean μ and standard deviation $\sigma_{\bar{x}} = \sigma/\sqrt{n}$. The probability is $1 - \alpha$ that any sample mean will fall between

$$\mu + z_{\alpha/2}\sigma_{\bar{x}} = \mu + z_{\alpha/2}\frac{\sigma}{\sqrt{n}} \quad and \quad \mu - z_{\alpha/2}\sigma_{\bar{x}} = \mu - z_{\alpha/2}\frac{\sigma}{\sqrt{n}} \qquad (2.2)$$

If μ and σ are known, Eq. 2.2 can be used as upper and lower control limits on a control chart for sample means. It is customary to replace $z_{\alpha/2}$ by 3, so that three-sigma limits are employed.

In practice, we do not know μ and σ and estimate them from preliminary samples or subgroups usually based on at least 20–25 samples taken when the process is thought to be in control. If m samples are available, each containing n observations on the quality characteristic. Let $\overline{X}_1, \overline{X}_2, \ldots, \overline{X}_m$ be the average of each sample. Then, the best estimator of the process average μ is the grand average, and would be used as the center line of the $\overline{X}$ chart.

$$\overline{\overline{X}} = \frac{\overline{X}_1 + \overline{X}_2 + \cdots + \overline{X}_m}{m} \qquad (2.3)$$

The range of a sample (R) is the difference between the largest and smallest observations and the average range $(\overline{R})$ can be written as given in Eqs. 2.4 and 2.5, respectively.

$$R_i = X_{i,max} - X_{i,min} \quad i = 1, 2, \ldots, m \qquad (2.4)$$

$$\overline{R} = \frac{R_1 + R_2 + \cdots + R_m}{m} \qquad (2.5)$$

The formulas for constructing the control limits on the $\overline{X}$ and R charts are tabulated in Table 2.1. Development of these equations can be found in Montgomery (2001).

The constants A_2, D_3, and D_4 depend on the sample (observation) size and are tabulated for various sample sizes in Appendix A.

These initial set of control limits is usually treated as trial limits and subject to subsequent revision. The past hypothesis that is the process is thought to be in

Table 2.1 Control limits for $\overline{X}$ and R charts		$\overline{X}$ chart	R chart
	Center Line (CL)	$\overline{\overline{X}}$	$\overline{R}$
	Lower Control Limit (LCL)	$\overline{\overline{X}} - A_2\overline{R}$	$D_3\overline{R}$
	Upper Control Limit (UCL)	$\overline{\overline{X}} + A_2\overline{R}$	$D_4\overline{R}$

control when samples are takes should be checked. If one or more of the samples plot out of control, the hypothesis is rejected and trial control limits should be revised. This can be done by examining the out of control points, and looking for assignable causes. If an assignable cause is found, the point is eliminated and control limits are recalculated based on the remaining samples. Recalculated control limits are called revised control limits. This revision process is continued until all points plot in control, and the final limits are adapted to the process as the control chart limits.

2.2.1.2 A Numerical Example

In a packaging process, 25 samples of size of 4 are taken in order to control the process mean and deviation. Data obtained from the packaging process is shown in Table 2.2. Let's construct the $\overline{X}$ and R chart.

Table 2.2 Data for the example

Sample number	Observations				$\overline{X}_i$	R_i
	I	II	III	IV		
1	51.98	49.21	49.73	50.16	50.27	2.77
2	50.94	50.28	50.77	51.40	50.85	1.13
3	50.87	51.67	49.89	52.68	51.28	2.79
4	47.15	46.25	48.05	49.91	47.84	3.66
5	48.97	52.20	49.86	52.46	50.87	3.49
6	50.43	51.08	52.99	50.41	51.23	2.58
7	48.51	51.18	52.02	51.09	50.70	3.51
8	50.65	52.73	51.65	52.86	51.97	2.22
9	51.70	50.93	50.80	48.43	50.46	3.27
10	52.77	52.70	48.01	52.93	51.60	4.93
11	48.36	52.59	49.70	51.55	50.55	4.23
12	49.15	51.07	48.33	49.94	49.62	2.73
13	52.07	48.51	48.90	51.15	50.16	3.56
14	52.24	51.01	51.15	52.74	51.79	1.73
15	52.19	48.85	52.28	49.34	50.67	3.43
16	52.53	49.63	51.25	51.15	51.14	2.90
17	48.16	52.89	52.84	50.86	51.19	4.73
18	52.36	48.84	52.88	48.22	50.58	4.66
19	49.00	51.83	49.48	51.67	50.49	2.83
20	52.69	49.86	51.27	52.28	51.52	2.84
21	51.88	48.09	50.64	49.61	50.05	3.79
22	48.33	49.81	51.88	48.23	49.56	3.65
23	48.81	50.90	48.84	52.12	50.17	3.32
24	50.68	49.19	51.66	50.71	50.56	2.47
25	51.21	51.25	50.83	52.34	51.41	1.50
				Average	50.661	3.148

For the sample size of 4, the constants of A_2, D_3, and D_4 are 0.729, 0, and 2.282, respectively. Mean and range of each subgroup are determined by using Eqs. 2.3 and 2.4, and also shown in Table 2.2.

Trial control limits for the given process to construct $\overline{X}$ and R charts are as follows.

For $\overline{X}$ chart

$$CL = \overline{\overline{X}} = 50.661$$
$$UCL = \overline{\overline{X}} + A_2 \overline{R} = 50.661 + 0.729 \times 3.148 = 52.956$$
$$LCL = \overline{\overline{X}} - A_2 \overline{R} = 50.661 - 0.729 \times 3.148 = 48.366$$

For R chart

$$CL = \overline{R} = 3.148$$
$$UCL = D_4 \overline{R} = 2.282 \times 3.148 = 7.184$$
$$LCL = D_3 \overline{R} = 0 \times 3.148 = 0$$

By looking the $\overline{X}_i$'s of the 25 samples, it can be clearly seen that sample 4 plot out of control which requires for calculation of the revised control limits. Eliminating sample 4, revised control limits can be calculated by taking remaining 24 samples into consideration as shown in Table 2.3.

Revised control limits for the given process to construct $\overline{X}$ and R charts are as follows.

For $\overline{X}$ chart

$$CL = \overline{\overline{X}} = 50.779$$
$$UCL = \overline{\overline{X}} + A_2 \overline{R} = 50.779 + 0.729 \times 3.127 = 53.059$$
$$LCL = \overline{\overline{X}} - A_2 \overline{R} = 50.779 - 0.729 \times 3.127 = 48.499$$

For R chart

$$CL = \overline{R} = 3.127$$
$$UCL = D_4 \overline{R} = 2.282 \times 3.127 = 7.184$$
$$LCL = D_3 \overline{R} = 0 \times 3.127 = 0$$

Since all points plot in control, these limits can be set as the control limits to construct $\overline{X}$ and R charts as given in Figs. 2.1 and 2.2.

Table 2.3 Data for the calculation of the revised control limits

Sample number	Observations				$\overline{X}_i$	R_i
	I	II	III	IV		
1	51.98	49.21	49.73	50.16	50.27	2.77
2	50.94	50.28	50.77	51.40	50.85	1.13
3	50.87	51.67	49.89	52.68	51.28	2.79
5	48.97	52.20	49.86	52.46	50.87	3.49
6	50.43	51.08	52.99	50.41	51.23	2.58
7	48.51	51.18	52.02	51.09	50.70	3.51
8	50.65	52.73	51.65	52.86	51.97	2.22
9	51.70	50.93	50.80	48.43	50.46	3.27
10	52.77	52.70	48.01	52.93	51.60	4.93
11	48.36	52.59	49.70	51.55	50.55	4.23
12	49.15	51.07	48.33	49.94	49.62	2.73
13	52.07	48.51	48.90	51.15	50.16	3.56
14	52.24	51.01	51.15	52.74	51.79	1.73
15	52.19	48.85	52.28	49.34	50.67	3.43
16	52.53	49.63	51.25	51.15	51.14	2.90
17	48.16	52.89	52.84	50.86	51.19	4.73
18	52.36	48.84	52.88	48.22	50.58	4.66
19	49.00	51.83	49.48	51.67	50.49	2.83
20	52.69	49.86	51.27	52.28	51.52	2.84
21	51.88	48.09	50.64	49.61	50.05	3.79
22	48.33	49.81	51.88	48.23	49.56	3.65
23	48.81	50.90	48.84	52.12	50.17	3.32
24	50.68	49.19	51.66	50.71	50.56	2.47
25	51.21	51.25	50.83	52.34	51.41	1.50
				Average	50.779	3.127

Fig. 2.1 $\overline{X}$ chart

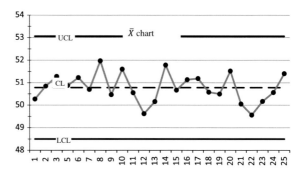

Fig. 2.2 *R* chart

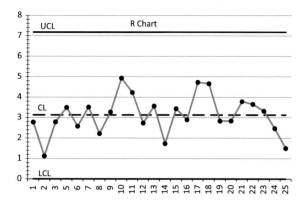

2.2.2 $\overline{X}$ and S Control Charts

When the sample size is variable and relatively large, say $n > 10$, the usage of $\overline{X}$ and s charts is advantageous.

2.2.2.1 Control Limits for $\overline{X}$ and S Control Charts

For the cases where a standard value is known and given for σ, control limits for the S chart can be determined as follows:

$$CL = c_4\sigma \tag{2.6}$$

$$UCL = B_6\sigma \tag{2.7}$$

$$LCL = B_5\sigma \tag{2.8}$$

where the constants c4, B5, and B6 depend on the sample (observation) size and are tabulated for various sample sizes in Appendix A.

When σ is unknown, we may write the parameters of the $\overline{X}$ and S control chart as given in the equations shown in Table 2.4.

where,

Table 2.4 Control limits for $\overline{X}$ and R charts

	$\overline{X}$ chart	s chart
CL	$\overline{\overline{X}}$	$\bar{s}$
LCL	$\overline{\overline{X}} - A_3\bar{s}$	$B_3\bar{s}$
UCL	$\overline{\overline{X}} + A_3\bar{s}$	$B_4\bar{s}$

$$s = \sqrt{\frac{\sum_{i=1}^{n}(X_i - \overline{X})^2}{n-1}} \qquad (2.9)$$

$$\bar{s} = \frac{s_1 + s_2 + \cdots + s_n}{n} \qquad (2.10)$$

The constants c_4, A_3, B_3, B_4, B_5 and B_6 depend on the sample (observation) size and are tabulated for various sample sizes in Appendix A.

If the sample size is variable, a weighted average approach is used in calculating $\bar{x}$ and $\bar{s}$ as given below. In this case, since the sample size differs, the constants c4, A_3, B_3, B_4, B_5 and B_6 for each subgroup will be different. Hence, upper and lower control limits for each subgroup will also change.

$$\overline{\overline{X}} = \frac{\sum_{i=1}^{m} n_i \overline{X}_i}{\sum_{i=1}^{m} n_i} \qquad (2.11)$$

$$\bar{s} = \left[\frac{\sum_{i=1}^{m}(n_i - 1)s_i^2}{\sum_{i=1}^{m}(n_i - m)} \right]^{1/2} \qquad (2.12)$$

2.2.2.2 A Numerical Example

Consider the example given in Sect. 2.2.1.2. Calculation of the $\bar{x}_i$ and s_i for each subgroup are tabulated in Table 2.5.

For the sample size of 4, the constants of A_3, B_3, and B_4 are 1.628, 0, and 2.266, respectively. Trial control limits for the given process to construct $\overline{X}$ and S charts are as follows.

For $\overline{X}$ chart

$$CL = \overline{\overline{X}} = 50.661$$
$$UCL = \overline{\overline{X}} + A_3\bar{s} = 50.661 + 1.628 \times 1.452 = 53.025$$
$$LCL = \overline{\overline{X}} - A_3\bar{s} = 50.661 - 1.628 \times 1.452 = 48.297$$

For s chart

$$CL = \bar{s} = 1.452$$
$$CL = B_4\bar{s} = 2.266 \times 1.452 = 3.290$$
$$LCL = B_3\bar{s} = 0 \times 1.452 = 0$$

By looking the $\overline{X}_i$'s of the 25 samples, it can be clearly seen that sample 4 plot out of control which requires for calculation of the revised control limits.

Table 2.5 Data for the $\overline{X}$ and S chart

Sample number	Observations				$\overline{X}_i$	s_i
	I	II	III	IV		
1	51.98	49.21	49.73	50.16	50.27	1.21
2	50.94	50.28	50.77	51.40	50.85	0.46
3	50.87	51.67	49.89	52.68	51.28	1.18
4	47.15	46.25	48.05	49.91	47.84	1.56
5	48.97	52.20	49.86	52.46	50.87	1.73
6	50.43	51.08	52.99	50.41	51.23	1.21
7	48.51	51.18	52.02	51.09	50.70	1.52
8	50.65	52.73	51.65	52.86	51.97	1.04
9	51.70	50.93	50.80	48.43	50.46	1.41
10	52.77	52.70	48.01	52.93	51.60	2.40
11	48.36	52.59	49.70	51.55	50.55	1.89
12	49.15	51.07	48.33	49.94	49.62	1.16
13	52.07	48.51	48.90	51.15	50.16	1.73
14	52.24	51.01	51.15	52.74	51.79	0.84
15	52.19	48.85	52.28	49.34	50.67	1.83
16	52.53	49.63	51.25	51.15	51.14	1.19
17	48.16	52.89	52.84	50.86	51.19	2.23
18	52.36	48.84	52.88	48.22	50.58	2.38
19	49.00	51.83	49.48	51.67	50.49	1.46
20	52.69	49.86	51.27	52.28	51.52	1.26
21	51.88	48.09	50.64	49.61	50.05	1.60
22	48.33	49.81	51.88	48.23	49.56	1.71
23	48.81	50.90	48.84	52.12	50.17	1.63
24	50.68	49.19	51.66	50.71	50.56	1.02
25	51.21	51.25	50.83	52.34	51.41	0.65
				Average	50.661	1.452

Eliminating sample 4, revised control limits can be calculated by taking remaining 24 samples into consideration as shown in Table 2.6.

Revised control limits for the given process to construct $\overline{X}$ and S charts are as follows.

Table 2.6 Data for the calculation of the revised control limits

Sample number	Observations				$\overline{X}_i$	s_i
	I	II	III	IV		
1	51.98	49.21	49.73	50.16	50.27	1.21
2	50.94	50.28	50.77	51.40	50.85	0.46
3	50.87	51.67	49.89	52.68	51.28	1.18
5	48.97	52.20	49.86	52.46	50.87	1.73
6	50.43	51.08	52.99	50.41	51.23	1.21
7	48.51	51.18	52.02	51.09	50.70	1.52
8	50.65	52.73	51.65	52.86	51.97	1.04
9	51.70	50.93	50.80	48.43	50.46	1.41
10	52.77	52.70	48.01	52.93	51.60	2.40
11	48.36	52.59	49.70	51.55	50.55	1.89
12	49.15	51.07	48.33	49.94	49.62	1.16
13	52.07	48.51	48.90	51.15	50.16	1.73
14	52.24	51.01	51.15	52.74	51.79	0.84
15	52.19	48.85	52.28	49.34	50.67	1.83
16	52.53	49.63	51.25	51.15	51.14	1.19
17	48.16	52.89	52.84	50.86	51.19	2.23
18	52.36	48.84	52.88	48.22	50.58	2.38
19	49.00	51.83	49.48	51.67	50.49	1.46
20	52.69	49.86	51.27	52.28	51.52	1.26
21	51.88	48.09	50.64	49.61	50.05	1.60
22	48.33	49.81	51.88	48.23	49.56	1.71
23	48.81	50.90	48.84	52.12	50.17	1.63
24	50.68	49.19	51.66	50.71	50.56	1.02
25	51.21	51.25	50.83	52.34	51.41	0.65
				Average	50.779	1.448

For $\overline{X}$ chart

$$CL = \overline{\overline{X}} = 50.779$$
$$UCL = \overline{\overline{X}} + A_3\overline{s} = 50.779 + 1.628 \times 1.448 = 53.136$$
$$LCL = \overline{\overline{X}} - A_3\overline{s} = 50.779 - 1.628 \times 1.448 = 48.422$$

For s chart

$$CL = \overline{s} = 1.448$$
$$UCL = B_4\overline{s} = 2.266 \times 1.448 = 3.281$$
$$LCL = B_3\overline{s} = 0 \times 1.448 = 0$$

Fig. 2.3 $\overline{X}$ chart

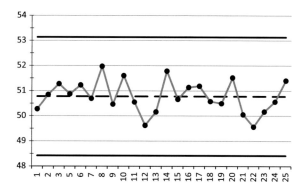

Fig. 2.4 S chart

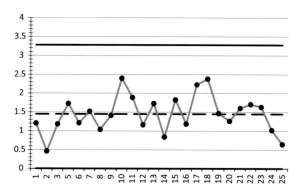

Since all points plot in control, these limits can be set as the control limits to construct $\overline{X}$ and S charts as given in Figs. 2.3 and 2.4.

2.3 Moving Average (MA) Control Charts

In the cases where data are collected slowly over a period of time, or data are expensive to collect, moving average (MA) control charts are beneficial. The MA charts can help bringing trends to light more rapidly than conventional charts. However, run tests are not valid, since the adjacent points on the MA charts are not independent. As another disadvantage, there is a tendency to forget that individual observations have more variability than do the averages.

Moving average charts use the central limit theorem to make data approximately normal. There are two types of the moving average charts which are most commonly used: Exponentially weighted moving average charts (EWMA) and generally weighted moving average charts (GWMA).

2.3.1 *Exponentially Weighted Moving Average (EWMA) Control Charts*

The traditional EWMA control chart was introduced by Roberts in 1959 as below. The statistic that is calculated is:

$$EWMA_t = \lambda X_t + (1 - \lambda)EWMA_{t-1} \text{ for } t = 1, 2, \ldots, n \qquad (2.13)$$

where $EWMA_0$ is the mean of the historical data (target) and is equal to $\overline{\overline{X}}$, X_t refers to the observation at time t, n is the number of observations to be monitored, and $0 < \lambda \leq 1$ is a constant determining the depth of memory of the EWMA. The parameter λ determines the rate at which the older data enter into the calculation of the EWMA statistic where $\lambda = 1$ implies that only the most recent measurement from the observations influences the EWMA. In another words, a large value of λ that is closer to 1 gives more weight to recent data and a small value of λ that is closer to 0 gives more weight to the older data. The parameter λ is usually set between 0.2 and 0.3 although the choice is somewhat arbitrary (Montgomery 2001).

If X_t's are independent random variables with a known standard deviation of the population σ and a variance of σ^2/n, then the variance of the $EWMA_t$ becomes

$$\sigma^2_{EWMA_t} = \frac{\sigma^2}{n} \left(\frac{\lambda}{2 - \lambda} \right) \left[1 - (1 - \lambda)^{2t} \right] \qquad (2.14)$$

As t increases, $\sigma^2_{EWMA_t}$ reaches to a limiting value of

$$\sigma^2_{EWMA} = \frac{\sigma^2}{n} \left(\frac{\lambda}{2 - \lambda} \right) \qquad (2.15)$$

For a moderately large number of sample size, the control limits for the traditional EWMA control charts can be expressed as follows:

$$CL_{EWMA} = \overline{\overline{X}} \qquad (2.16)$$

$$UCL_{EWMA} = \overline{\overline{X}} + 3 \frac{\sigma}{\sqrt{n}} \sqrt{\frac{\lambda}{2 - \lambda}} \qquad (2.17)$$

$$LCL_{EWMA} = \overline{\overline{X}} - 3 \frac{\sigma}{\sqrt{n}} \sqrt{\frac{\lambda}{2 - \lambda}} \qquad (2.18)$$

If t is small, the control limits for the traditional EWMA control charts can be expressed as follows:

$$CL_{EWMA} = \overline{\overline{X}} \tag{2.19}$$

$$UCL_{EWMA} = \overline{\overline{X}} + 3\frac{\sigma}{\sqrt{n}}\sqrt{\frac{\lambda}{2-\lambda}\left[1-(1-\lambda)^{2t}\right]} \tag{2.20}$$

$$LCL_{EWMA} = \overline{\overline{X}} - 3\frac{\sigma}{\sqrt{n}}\sqrt{\frac{\lambda}{2-\lambda}\left[1-(1-\lambda)^{2t}\right]} \tag{2.21}$$

If σ is unknown and estimated from the samples, then $\overline{R}$ can be used for constructing traditional EWMA charts. In this case, the control limits are as follows:

$$CL_{EWMA} = \overline{\overline{X}} \tag{2.22}$$

$$UCL_{EWMA} = \overline{\overline{X}} + A_2\overline{R}\sqrt{\frac{\lambda}{2-\lambda}} \tag{2.23}$$

$$LCL_{EWMA} = \overline{\overline{X}} - A_2\overline{R}\sqrt{\frac{\lambda}{2-\lambda}} \tag{2.24}$$

where $\overline{R}$ is the mean of the ranges of the samples, and A_2 is a constant given in Appendix A.

2.3.1.1 A Numerical Example

Consider a process with the parameters of $EWMA_0 = 50.0$ and $s = 2.0539$ calculated from historical data. For the following 20 points observed, let us construct EWMA control charts.

51.9	47.1	53.1	49.4	50	47.6	50.2	50.1	51.4	50.6
49.8	47.5	9.9	50.9	47.6	51.5	52.8	52.3	54.9	50.1

With λ chosen to be 0.3 the parameter $\sqrt{\frac{\lambda}{2-\lambda}}$ is equal to 0.4201. CL, LCL and UCL for the EWMA chart can be calculated as follows.

$$CL_{EWMA} = \overline{\overline{X}} = 50.0$$

$$UCL_{EWMA} = \overline{\overline{X}} + 3\frac{\sigma}{\sqrt{n}}\sqrt{\frac{\lambda}{2-\lambda}} = 50.0 + 3(2.0539)(0.4201) = 52.5884$$

Table 2.7 EWMA statistics for 20 points

t	X	EWMA
0		50.00
1	51.9	50.57
2	47.1	49.53
3	53.1	50.60
4	49.4	50.24
5	50.0	50.17
6	47.6	49.40
7	50.2	49.64
8	50.1	49.78
9	51.4	50.26
10	50.6	50.36
11	49.8	50.20
12	47.5	49.39
13	49.9	49.54
14	50.9	49.95
15	47.6	49.24
16	51.5	49.92
17	52.8	50.78
18	52.3	51.24
19	54.9	52.34
20	50.1	51.67

$$LCL_{EWMA} = \overline{\overline{X}} + 3\frac{\sigma}{\sqrt{n}}\sqrt{\frac{\lambda}{2-\lambda}} = 50.0 - 3(2.0539)(0.4201) = 47.4115$$

EWMA statistics of the 20 points are calculated by using Eq. 2.13 and summarized in Table 2.7. Constructed EWMA chart is illustrated in Fig. 2.5.

2.3.2 Maximum Generally Weighted Moving Average (MaxGWMA) Control Charts

The EWMA chart is widely used to detect small shifts in process mean and it has successfully become a source of inspiration to the many researchers as in the reviews by Xie (1999), Han and Tsung (2004), Eyvazian et al. (2008), Li and Wang (2010), Zhang et al. (2010), Sheu et al. (2012). On the basis of maximum statistic values, Chen and Cheng developed a Maxtype chart which effectively controls both process mean and variability on a single chart (Chen and Cheng 1998); Xie further examined numerous EWMA-type control charts and resulted that the MaxEWMA chart is superior to others in detecting small shifts of the process mean and

Fig. 2.5 EWMA Chart for the given data

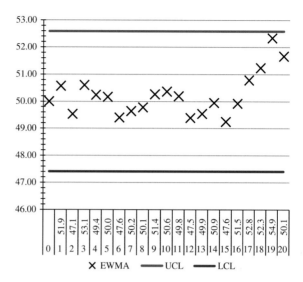

variability as well as in identifying the source and the direction of an out-of-control signal (Xie 1999). Sheu and Lin created the generally weighted moving average (GWMA) chart which can detect small shifts much quicker than the EWMA can (Shu et al. 2014). By combing the advantages of the MaxEWMA chart and GWMA chart, Sheu et al. proposed a new chart called the maximum generally weighted moving average (MaxGWMA) chart which was found to be more sensitive under abnormal variations of on-line manufacturing processes than the MaxEWMA chart (Sheu et al. 2012).

Let X be the key quality characteristic with a normal distribution $N(\mu_0, \sigma_0^2)$, where μ_0 is the process mean and σ_0 is the process standard deviation. If the new mean is $\mu_1 = \mu_0 \pm \delta\sigma_0$, then the process mean is said to have a shift of δ ($\delta \neq 0$) standard deviation. Similarly, if the new standard deviation is $\sigma_1 = (1 + \rho)\sigma_0$, then the process is said to have a shift of ρ standard deviation in variability. In real cases, μ_0 and σ_0 are usually unknown and can be estimated from the randomly collected sample data of which at least 20–25 in-control samples are recommended. Assume that m random subgroups and each subgroup containing n observation of x are collected. The sample average of the ith sample ($\bar{x}_i$) and the grand sample average $(\bar{\bar{x}})$ can be calculated by using formulas below.

$$\bar{x}_i = \frac{1}{n} \sum_{j=1}^{m} x_{ij} \text{ for } i = 1, 2, \ldots, m \tag{2.25}$$

$$\bar{\bar{x}} = \frac{1}{m} \sum_{i=1}^{n} x_{ij} \text{ for } i = 1, 2, \ldots, m \tag{2.26}$$

In the same way, the standard deviation of the ith sample (s_i) and the average of the m standard deviations ($\bar{s}$) can be calculated by using the following formulas.

$$s_i = \sqrt{\frac{\sum_{j=1}^{n}\left(x_{ij} - \bar{x}_i\right)^2}{n-1}} \tag{2.27}$$

$$\bar{s} = \frac{1}{m}\sum_{i=1}^{m} s_i \tag{2.28}$$

The unbiased estimators of the μ_0 and σ_0 are then given by

$$\mu_0 = E(\bar{x}) = \bar{\bar{x}} \tag{2.29}$$

$$\sigma_0 = E(\bar{s}) = \bar{s}/c_4 \tag{2.30}$$

where the value of the c_4 is a constant and can be found from Appendix A.

For the computation of the MaxGWMA statistic, two mutually independent statistics, M_i and S_i are defined as follows.

$$M_i = \frac{(\bar{x}_i - \mu_0)}{\sigma_0/\sqrt{n}} \tag{2.31}$$

$$S_i = \emptyset^{-1}\left\{F\left[\frac{(n-1)s_i^2}{\sigma_0^2}, n-1\right]\right\} \tag{2.32}$$

where $F(a, b)$ refers to the chi-square distribution of a with b degrees of freedom, and $\emptyset^{-1}$ is the inverse of the standard normal distribution.

Let A be an event of interest and t be the counting number of samples between two adjacent occurrences of A. $p_j = P(t > j)$ is the probability that A does not occur in the first j samples. The probability of p_j of the occurrence of A at the jth sample can be calculated by Shu et al. (2014)

$$p_j = P(t > j-1) = P_{j-1} - P_j \tag{2.33}$$

Remember that for $\forall j > 1$ and $j < i$, we have $P_j > P_i$.

GWMA statistics for the ith subgroup are given by

$$U_i = \sum_{j=1}^{i} p_j M_{i+1-j} \tag{2.34}$$

$$V_i = \sum_{j=1}^{i} p_j S_{i+1-j} \tag{2.35}$$

For the ease of computation, We chose $P_j = q^{j^\alpha}$, where q is called a design parameter that is a constant with the value in [0,1] and α is called an adjustment parameter determined by the practitioner (Sheu and Lin 2003). Obviously, the traditional EWMA chart is a special case of GWMA chart when $\alpha = 1$ and $q = 1 - \lambda$. Now, the probability p_j of the occurrence of A at the jth sample can be rewritten as

$$p_j = q^{(j-1)^\alpha} - q^{j^\alpha} \tag{2.36}$$

If the process is not shifting, with respect to the independence of the M_i and S_i, then GWMA statistics U_i and V_i are also mutually independent and follow the same standard normal distribution. Thus, their variances can be determined by

$$\sigma^2_{U_i} = \sigma^2_{V_i} = \eta_i = \sum_{j=1}^{i} p_j \tag{2.37}$$

The statistic (MG) used to construct the MaxGWMA chart is defined as

$$MG_i = \max(|U_i|, |V_i|) \tag{2.38}$$

A small value of MG_i indicates that the process mean and process variability are close to their respective targets, while a large value of MG_i indicates that the process mean and process variability are away from their respective targets.

Since MG_i is nonnegative, only upper control limit for the ith subgroup formulated below is used to monitor MG_i (Sheu et al. 2012).

$$UCL_i = E(MG_i) + L\sqrt{\sigma^2(MG_i)} \tag{2.39}$$

where L is a constant.

Based on desired in control ARL$_0$, sample size n, and optimal values of parameters q, α and L for an initial state of the MaxGWMA chart, the approximate value of UCL_i can be given as (Sheu et al. 2012)

$$UCL_i = (1.12838 + 0.60281L)\sqrt{\eta_i} \tag{2.40}$$

In the MaxGWMA chart, each of the MG_i values is compared with the UCL_i and the following judgement can be performed about whether the process is in control or out of control.

$$\text{Process Control for MaxGWMA} = \left\{ \begin{array}{lll} MG_i \leq UCL_i & ; & \text{in control} \\ MG_i > UCL_i & ; & \text{out of control} \end{array} \right\} \tag{2.41}$$

If there is a change in the process mean and/or process variability, Table 2.8 can be used to identify the situations (Sheu et al. 2012).

Table 2.8 Indications of the out of control points

Situation	Symbol	Indication
$MG_i = U_i$ and $|V_i| \leq UCL_i$	m^+	An increase in the process mean
$MG_i = -U_i$ and $|V_i| \leq UCL_i$	m_-	An decrease in the process mean
$MG_i = V_i$ and $|U_i| \leq UCL_i$	v^+	An increase in the process variability
$MG_i = -V_i$ and $|U_i| \leq UCL_i$	v_-	A decrease in the process variability
$U_i > UCL_i$ and $V_i > UCL_i$	$++$	An increase in both the process mean and the process variability
$-U_i > UCL_i$ and $-V_i > UCL_i$	$--$	A decrease in both the process mean and the process variability
$U_i > UCL_i$ and $-V_i > UCL_i$	$+-$	An increase in the process mean and a decrease in the process variability
$-U_i > UCL_i$ and $V_i > UCL_i$	$-+$	A decrease in the process mean and an increase in the process variability

2.4 Unnatural Patterns for Control Charts

The usual SPC control chart limit rules display at the 3-sigma level. In this case, a simple threshold test decides if a process is in or out of control. Once a process is brought under control using the simple 3-sigma level tests, quality professionals often want to increase the sensitivity of the control chart by detecting and correcting problems before the process excludes 3-sigma control limits. Based on the probability, more complex tests rely on more complicated decision-making criteria by examining the patterns of the points (sample characteristic) on the control chart and presenting a set of rules with respect to the very low probability of occurrence. These rules utilize historical data and look for a non-random (unnatural) pattern that can signify that the process is out of control, before reaching the normal ±3 sigma limits. In another words, a process may signal an out of control condition even its characteristic plots in control. The rules that characterize an out of control signal through the control chart limits are called "unnatural pattern rules" or "non-random pattern rules".

The most popular of unnatural (non-random) pattern rules are the Western Electric Rules, also known as the WECO Rules, or WE Runtime Rules. First implemented by the Western Electric Co. in the 1920s, these quality control guidelines were codified in the 1950s and form the basis for the other entire rule sets (Western Electric Company: Statistical Quality Control Handbook, Indianapolis, Indiana 1956). Different industries have developed their own variants based on the WECO Rules. Other sets of rules which are common enough to recognize an identifying name, i.e. named rules, are "Nelson Rules (1984)", "Juran Rules (2010)", "Duncan Rules (1986)", "Automotive

Fig. 2.6 Zones in a control chart

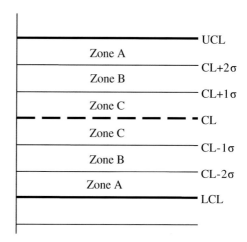

Industry Action Group (AIAG) Rules (Detroit 1995) ", "Gitlow Rules (1989) ", and "Westgard Rules (2014) ".

In general, when identifying these rules, the region between the usual ±3 sigma limits are divided into six region and the pattern is explained with respect to ±1, 2, and 3 sigma limits as shown in the Fig. 2.6.

Based on the zones illustrated by Fig. 2.6, some "Named Unnatural Pattern Rules" are explained in the following sections.

2.4.1 Western Electric Rules

In the Western Electric Rules, a process is accepted to signal an out of control if any of the following criteria are observed (Western Electric Company 1956):

1. One of the any point outside one of the 3-sigma control limits: If a point lies outside either of ±3 sigma limits, there is only a 0.27 % chance that this was caused by the normal process.
2. Two out of the three consecutive points outside of the 2-sigma control limits and on the same side of the center line: The probability that any point will fall outside the warning limit of 2-sigma is only 5 %. The chance that two out of three points in a row fall outside the warning limits is only about 1 %.
3. Four out of the five consecutive points outside of the 1-sigma control limits and on the same side of the center line: In normal processing, 68 % of points fall within 1-sigma of the mean. The probability that 4 of 5 points fall outside of one sigma is only about 3 %.
4. Eight consecutive points on the same side of the center line: The probability of getting eight points on the same side of the mean is only around 1 %.

Remember that these rules apply separately to both sides of the center line at a time. Therefore, in the WECO Rules there are eight actual alarm conditions. There are also additional WE Rules related with the trends of the points. These are often referred to as Western Electric Supplemental Rules.

5. Six points in a row increasing or decreasing: Sometimes this rule is changed to seven points rising or falling.
6. Fifteen points in a row within one sigma: In normal operation, 68 % of points will fall within one sigma of the mean. The probability that 15 points in a row will do so, is less than 1 %.
7. Fourteen points in a row alternating direction. The chances that the second point is always higher than (or always lower than) the preceding point, for all seven pairs is only about 1 %.
8. Eight points in a row outside one sigma. Since 68 % of points lie within one sigma of the mean, the probability that eight points in a row fall outside of the one-sigma line is less than 1 %.

2.4.2 Nelson Rules

The Nelson rules are almost identical to the combination of the WECO. The only difference is in Rule #4 where nine consecutive points on the same side of the center line is accepted as a signal (Nelson 1984).

2.4.3 Other Named Rules

In general, a given rule specifies two test conditions: Being a value of N points out of M consecutive points above and below of a specified sigma control limits. From this point of view, named rules mentioned in Sect. 2.4 are summarized and tabulated in Table 2.9.

2.5 Ranking Fuzzy Numbers and Direct Fuzzy Approach

Fuzzy numbers as they are used to represent uncertainties are an important issue in research in fuzzy set theory and their applications (Gülbay and Kahraman 2006). Because of the suitability for representing uncertain values, fuzzy numbers have been widely used in many applications. When quality characteristic and control limits are represented as fuzzy numbers, the main problem is to decide whether the quality characteristic lies within their respective fuzzy control limits or not in order to decide about the process: in-control or out of control. In such situations, a

Table 2.9 Summarization of some named rules in the form of N/M

Rule	Named rules						
	WECO	Nelson	Juran	Duncan	AIAG	Gitlow	Westgard
Outside ±3σ limits	1/1	1/1	1/1	1/1	1/1	1/1	1/1
Outside ±2σ limits	2/3	2/3	2/3	2/3		2/3	2/2 or 2/3
Outside ±1σ limits	4/5	4/5	4/5	4/5		4/5	4/4 or 3/4
On the same side of centerline	8/8	9/9	9/9		7/7	8/8	10/10
Increasing or decreasing in a row	6/6		6/6	7/7	7/7	8/8	7/7
Within ±1σ	15						
Outside ±1σ	8/8		8/8				
Outside ±2σ							1/1
Alternating	14/14						
Opposite sides of ±2σ							2/2

comparison of the fuzzy numbers is required. Various methods to manipulate fuzzy numbers have been developed to overcome the problem illustrated in Fig. 2.7 (Chen and Chen 2009; Chen and Sanguansat 2011; Deng et al. 2006; Wang and Lee 2008; Yager 1978; Zimmermann 1996).

The results of studies on ranking fuzzy numbers have been used in application areas especially where decision-making and data analysis have a vital importance. The ranking methods can be classified in three categories. The first category directly transforms each fuzzy number into a crisp real number and the second category compares a fuzzy number to all the other n − 1 fuzzy numbers to obtain its mapping into a positive real number. The third category differs substantially from the first two.

Fig. 2.7 Illustration of ranking two fuzzy numbers. **a** $\tilde{A} < \tilde{B}$. **b** Not clear

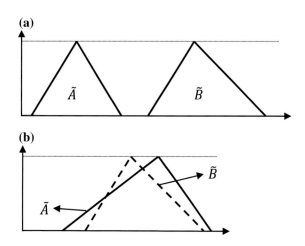

In this category, a method for pairwise ranking or preference for all pairs of fuzzy numbers is determined and then based on these pairwise orderings, a final order of the n fuzzy numbers is attempted (Shureshjani and Darehmiraki 2013). The significance of ranking fuzzy numbers for solving real world decision problems in a fuzzy environment has led to tremendous efforts being spent on the development of various ranking approaches (Bortolan and Degani 1985; Chen and Hwang 1992; Cheng 1998; Choobineh and Li 1993; Chu and Tsao 2002; Detyniecki and Yager 2001; Dias 1993; Dubois and Prade 1978, 1980; Fortemps and Roubens 1996; Jain 1976, 1978; Kim et al. 1998; Lee et al. 1994; Lee and Lee-Kwang 1999; Lee and Li 1998; Liu and Han 2005; Murakami 1983; Raj and Kumar 1999; Requena et al. 1994; Tran and Duckstein 2002; Wang et al. 2009; Zadeh 1965). To whom more interested to the ranking methods for fuzzy numbers, it is suggested to read (Brunelli and Mezeib 2013) for further knowledge.

For the fuzzy quality control chart studies we present a direct fuzzy comparison method to compare fuzzy numbers because the method enables the user to have a fuzzy decision about the comparison (Gülbay and Kahraman 2007).

Let $\tilde{X} = (X_a, X_b, X_c, X_d)$ be the fuzzy quality characteristic; $\widetilde{LCL} = (LCL_1, LCL_2, LCL_3, LCL_4)$ and $\widetilde{UCL} = (UCL_1, UCL_2, UCL_3, UCL_4,)$ be fuzzy lower control limit and fuzzy upper control limit, respectively, represented by trapezoidal fuzzy numbers. A decision about whether the process is in control can be made according to the percentage area of the sample which remains inside the $\widetilde{UCL}$ and/or $\widetilde{LCL}$. When the fuzzy sample is completely involved by the fuzzy control limits, the process is said to be "in-control". If a fuzzy sample is totally excluded by the fuzzy control limits, the process is said to be "out-of-control". Otherwise, a sample is partially included by the fuzzy control limits. In this case, if the percentage area which remains inside the fuzzy control limits (β_j) is equal or greater than a predefined acceptable percentage (β), then the process can be accepted as "rather in-control". Otherwise, it can be stated as "rather out of control". The possible decisions resulting from "Direct Fuzzy Approach (DFA) are illustrated in Fig. 2.8. The parameters to determine the sample's area outside the control limits for any a-level cut are LCL_1, LCL_2, UCL_3, UCL_4, a, b, c, d, and α. The shapes of the control limits and fuzzy samples are formed by the lines of $\overline{LCL_1 LCL_2}$, $UCL_3 UCL_4$, $\overline{ab}$, and $\overline{cd}$,. A flowchart to calculate area of the fuzzy sample outside the control limits is given in Fig. 2.9. The sample's area above the upper control limits, A_{out}^U, and sample area falling below the lower control limits, A_{out}^L, can be calculated according to the flowchart given in Fig. 88. The equations to compute A_{out}^U and A_{out}^L are given in Appendix B. Then, the total area outside the fuzzy control limits, A_{out}, is the sum of the areas below the fuzzy lower control limit and above the fuzzy upper control limit. The percentage sample area within the fuzzy control limits is calculated as

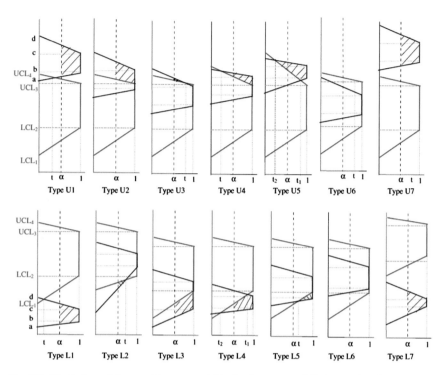

Fig. 2.8 Illustration of the possible areas outside the fuzzy control limits at α-level cut

$$\beta_j^\alpha = \frac{S_j^\alpha - A_{out}^\alpha}{S_j^\alpha} \tag{2.42}$$

where S_j^α is the sample's area at α-level cut. Remember that performing α-level cut is not a must but a preference if decided by a quality practitioner. Furthermore, the acceptable percentage (β) is set by the quality practitioner with respect to the tightness of the inspection.

2.6 Fuzzy Approaches for Control Charts for Variables

Many quality characteristics can be expressed in terms of a numerical measurement such as length, width, weight, temperature, volume etc. A process is either "in control" or "out of control" depending on numeric observation values. For many problems, control limits could not be so precise. Uncertainty comes from the measurement system including operators and gauges, and environmental conditions (Senturk and Erginel 2009). A research work incorporating uncertainty into decision analysis is basically done through the probability theory and/or the fuzzy set theory.

Fig. 2.9 Flowchart to compute the area of a fuzzy sample (a,b,c,d) falling outside the fuzzy control limits. (See Appendix B for the equations)

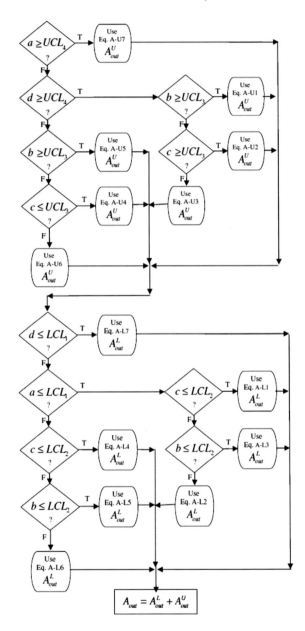

The former represents the stochastic nature of decision analysis while the latter captures the subjectivity of human behaviour. A rational approach toward decision-making should take human subjectivity into account, rather than employing only objective probability measures. The fuzzy set theory is a perfect means for modeling uncertainty (or imprecision) arising from mental phenomena which is

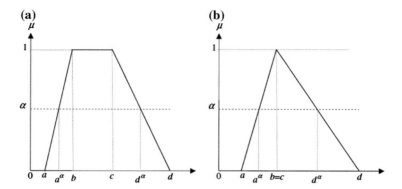

Fig. 2.10 Representation of a sample by trapezoidal and/or triangular fuzzy numbers: **a** Trapezoidal (a,b,c,d) and **b** triangular (a,b,b,d)

neither random nor stochastic. When human subjectivity plays an important role in defining the quality characteristics, the classical control charts may not be applicable since they require certain information. The judgment in classical process control results in a binary classification as "*in-control*" or "out-of-control" while fuzzy control charts may handle several intermediate decisions. Fuzzy control charts are inevitable to use when the statistical data in consideration are uncertain or vague; or available information about the process is incomplete or includes human subjectivity (Gülbay and Kahraman 2007). In the fuzzy case, each sample, or subgroup, is represented by a trapezoidal fuzzy number (a,b,c,d) or a triangular fuzzy number (a, b,d), or (a,c,d) with an α-cut (if necessary) as shown in Fig. 2.10.

X-R and X-s fuzzy control charts can be presented as given in Sects. 2.6.1 and 2.6.2 (Gülbay and Kahraman 2006).

2.6.1 Fuzzy $\overline{X}$ and R Control Charts

Let quality characteristic of a sample with a size of n be represented as fuzzy triangular numbers by $\overline{X}_i\left(X_{ija}, X_{ijb}, X_{ijc},\right) i = 1, 2, \ldots, m; j = 1, 2, \ldots, n$. Using the fuzzy arithmetic the mean of the each subgroup and grand average of the samples can be calculated by Equations below.

$$\tilde{\overline{X}}_i = \left(\frac{\sum_{j=1}^{n} X_{ija}}{n}, \frac{\sum_{j=1}^{n} X_{ijb}}{n}, \frac{\sum_{j=1}^{n} X_{ijc}}{n}\right) = (X_{ia}, X_{ib}, X_{ic})$$

$$i = 1, 2, \ldots, m; \quad j = 1, 2, \ldots, n$$

(2.43)

$$\tilde{\overline{\overline{X}}} = \left(\frac{\sum_{i=1}^{m} X_{ia}}{m}, \frac{\sum_{i=1}^{m} X_{ib}}{m}, \frac{\sum_{i=1}^{m} X_{ic}}{m}\right) = \left(\overline{\overline{X}}_a, \overline{\overline{X}}_b, \overline{\overline{X}}_c\right)$$

(2.44)

The fuzzy range of each subgroup can be represented by the equation below.

$$\tilde{R}_i = \tilde{X}_{ij,max} - \tilde{X}_{ij,min} \quad i = 1, 2, \ldots, m; \quad j = 1, 2, \ldots, n \qquad (2.45)$$

In crisp calculation, the maximum and minimum values of R can be easily determined. But it is not so easy to decide which fuzzy range observation is maximum and minimum. If represented fuzzy numbers are not intersecting, one can easily say that the fuzzy number with the most left support is smallest or minimum and the fuzzy number with the most right support is the greatest or maximum. In case where fuzzy observations have intersecting supports the problem about the ranking fuzzy numbers arises. Fuzzy numbers cannot be easily compared to each other. So, in decision analysis it is very difficult to distinguish the best possible course of action among alternatives defined by means of fuzzy numbers. Comparing and ranking fuzzy numbers in a given situation is complex and challenging (Yeh and Deng 2004; Sun and Wu 2006; Asady 2010). This is because fuzzy numbers usually represented by the possibility distribution (Zimmermann 2000; Dubois and Prade 1994) often overlap each other in many practical situations (Cheng 1998; Yeh and Deng 2004). It is difficult to clearly determine which fuzzy number is larger or smaller than another for a given situation, in particular when these two fuzzy numbers are similar (Kim and Park 1990; Deng 2007). Consequently, there are many fuzzy ranking methods, but an exhaustive review of ranking methods would be beyond the scope of this chapter. An attempt to list most of the ranking methods was made in Rao and Shankar (2011). DFA presented in Sect. 2.5 can also be used to find the greatest and smallest of the fuzzy numbers in any sample group.

Once the maximum and minimum fuzzy observation is decided, the fuzzy range can be determined by the following equations.

$$\tilde{R}_i = \tilde{X}_{ij,max} - \tilde{X}_{ij,min} = \tilde{X}_i \left(X_{ija}, X_{ijb}, X_{ijc}, \right)_{max} - \left(X_{ija}, X_{ijb}, X_{ijc}, \right)_{min} \qquad (2.46)$$

$$\tilde{R}_i = \left(\tilde{X}_{ija,max} - \tilde{X}_{ijc,min}, \tilde{X}_{ijb,max} - \tilde{X}_{ijb,min}, \tilde{X}_{ijc,max} - \tilde{X}_{ija,min} \right) = \left(R_{ia}, R_{ib}, R_{ic} \right) \qquad (2.47)$$

After calculating range of each subgroup, the fuzzy mean of the ranges can be defined as:

$$\tilde{\bar{R}} = \left(\frac{\sum_{i=1}^{m} R_{ia}}{m}, \frac{\sum_{i=1}^{m} R_{ib}}{m}, \frac{\sum_{i=1}^{m} R_{ic}}{m} \right) = \left(\bar{R}_a, \bar{R}_b, \bar{R}_c \right) \qquad (2.48)$$

Control limits for the fuzzy $\bar{\bar{X}}$ control charts, are then formulized as follows:

$$\widetilde{CL} = \tilde{\bar{\bar{X}}} = \left(\bar{\bar{X}}_a, \bar{\bar{X}}_b, \bar{\bar{X}}_c \right) = \left(CL_1, CL_2, CL_3 \right) \qquad (2.49)$$

Table 2.10 Summary of the fuzzy control limits for the $\overline{X} - R$ control chart

	$\overline{X}$ chart	R chart
Center Line $\widetilde{CL} = (CL_1, CL_2, CL_3)$	$\left(\overline{\overline{X}}_a, \overline{\overline{X}}_b, \overline{\overline{X}}_c\right)$	$\left(\overline{R}_a, \overline{R}_b, \overline{R}_c\right)$
Lower Control Limit $\widetilde{LCL} = (LCL_1, LCL_2, LCL_3)$	$\left(\overline{\overline{X}}_a - A_2\overline{R}_c, \overline{\overline{X}}_b - A_2\overline{R}_b, \overline{\overline{X}}_c - A_2\overline{R}_a\right)$	$\left(D_3\overline{R}_a, D_3\overline{R}_b, D_3\overline{R}_c\right)$
Upper Control Limit $\widetilde{UCL} = (UCL_1, UCL_2, UCL_3)$	$\left(\overline{\overline{X}}_a + A_2\overline{R}_a, \overline{\overline{X}}_b + A_2\overline{R}_b, \overline{\overline{X}}_c + A_2\overline{R}_c\right)$	$\left(D_4\overline{R}_a, D_4\overline{R}_b, D_4\overline{R}_c\right)$

$$\widetilde{UCL} = \overline{\overline{X}} + A_2\widetilde{\overline{R}} = \left(\overline{\overline{X}}_a, \overline{\overline{X}}_b, \overline{\overline{X}}_c\right) + A_2\left(\overline{R}_a, \overline{R}_b, \overline{R}_c\right)$$
$$= \left(\overline{\overline{X}}_a + A_2\overline{R}_a, \overline{\overline{X}}_b + A_2\overline{R}_b, \overline{\overline{X}}_c + A_2\overline{R}_c\right) \quad (2.50)$$
$$= (UCL_1, UCL_2, UCL_3)$$

$$\widetilde{LCL} = \overline{\overline{X}} - A_2\widetilde{\overline{R}} = \left(\overline{\overline{X}}_a, \overline{\overline{X}}_b, \overline{\overline{X}}_c\right) - A_2\left(\overline{R}_a, \overline{R}_b, \overline{R}_c\right)$$
$$= \left(\overline{\overline{X}}_a - A_2\overline{R}_c, \overline{\overline{X}}_b - A_2\overline{R}_b, \overline{\overline{X}}_c - A_2\overline{R}_a\right) \quad (2.51)$$
$$= (LCL_1, LCL_2, LCL_3)$$

Remember that the constants A_2 as well as D_3 and D_4 depend on the sample (number of observation in each sample) size and are tabulated for various sample sizes in Appendix A.

Fuzzy control limits for the R charts can be derived in the same way.

$$\widetilde{CL} = \widetilde{\overline{R}} = \left(\overline{R}_a, \overline{R}_b, \overline{R}_c\right) = (CL_1, CL_2, CL_3) \quad (2.52)$$

$$\widetilde{UCL} = D_4\widetilde{\overline{R}} = D_4\left(\overline{R}_a, \overline{R}_b, \overline{R}_c\right) = \left(D_4\overline{R}_a, D_4\overline{R}_b, D_4\overline{R}_c\right)$$
$$= (UCL_1, UCL_2, UCL_3) \quad (2.53)$$

$$\widetilde{LCL} = D_3\widetilde{\overline{R}} = \left(D_3\overline{R}_a, D_3\overline{R}_b, D_3\overline{R}_c\right) = (LCL_1, LCL_2, LCL_3) \quad (2.54)$$

Fuzzy control limits for the $\overline{X} - R$ control chart are summarized in Table 2.10.

2.6.2 Fuzzy $\overline{X}$ and S Control Charts

Determination of the control limits for paired $\overline{X}$ and s charts are based on the standard deviation as mentioned in Sect. 2.2.2.1. Hence, average standard deviation of the subgroups need to be firstly calculated.

Let quality characteristic of a sample with a size of n be represented as fuzzy triangular numbers by $\tilde{X}_i(X_{ija}, X_{ijb}, X_{ijc},)\, i = 1, 2, \ldots, m;\, j = 1, 2, \ldots, n$. Using the fuzzy arithmetics, the fuzzy standard deviation of the each subgroup and fuzzy average standard deviation of the samples can be derived by the Equations below.

$$\tilde{s}_i = \sqrt{\frac{\sum_{j=1}^{n}\left(\tilde{X}_{ij} - \tilde{\tilde{X}}_i\right)^2}{n-1}} = \sqrt{\frac{\sum_{j=1}^{n}\left[(X_{ija}, X_{ijb}, X_{ijc}) - (\overline{X}_{ia}, \overline{X}_{ib}, \overline{X}_{ic})\right]^2}{n-1}}$$
$$= (s_{ia}, s_{ib}, s_{ic}) \tag{2.55}$$

$$\tilde{\overline{s}} = \frac{\sum_{i=1}^{m}\tilde{s}_i}{m} = \left(\frac{\sum_{i=1}^{m}s_{ia}}{m}, \frac{\sum_{i=1}^{m}s_{ib}}{m}, \frac{\sum_{i=1}^{m}s_{ic}}{m}\right) = (\overline{s}_a, \overline{s}_b, \overline{s}_c) \tag{2.56}$$

The control limits of fuzzy $\overline{X}$ control chart based on standard deviation are obtained as follows:

$$\widetilde{CL} = \tilde{\overline{X}} = \left(\overline{\overline{X}}_a, \overline{\overline{X}}_b, \overline{\overline{X}}_c\right) = (CL_1,\ CL_2,\ CL_3) \tag{2.57}$$

$$\widetilde{UCL} = \tilde{\overline{X}} + A_3\tilde{\overline{s}} = \left(\overline{\overline{X}}_a, \overline{\overline{X}}_b, \overline{\overline{X}}_c\right) + A_3(\overline{s}_a, \overline{s}_b, \overline{s}_c)$$
$$= \left(\overline{\overline{X}}_a + A_3\overline{s}_a, \overline{\overline{X}}_b + A_3\overline{s}_b, \overline{\overline{X}}_c + A_3\overline{s}_c\right) \tag{2.58}$$
$$= (UCL_1,\ UCL_2,\ UCL_3)$$

$$\widetilde{LCL} = \tilde{\overline{X}} - A_3\tilde{\overline{s}} = \left(\overline{\overline{X}}_a, \overline{\overline{X}}_b, \overline{\overline{X}}_c\right) - A_3(\overline{s}_a, \overline{s}_b, s_c)$$
$$= \left(\overline{\overline{X}}_a - A_3\overline{s}_c, \overline{\overline{X}}_b - A_3\overline{s}_b, \overline{\overline{X}}_c - A_3\overline{s}_a\right) \tag{2.59}$$
$$= (LCL_1, LCL_2, LCL_3)$$

Similarly, the control limits of fuzzy s control chart are derived as follows:

$$\widetilde{CL} = \tilde{\overline{s}} = (\overline{s}_a, \overline{s}_b, \overline{s}_c) = (CL_1,\ CL_2,\ CL_3) \tag{2.62}$$

$$\widetilde{UCL} = B_4\tilde{\overline{s}} = B_4(\overline{s}_a, \overline{s}_b, \overline{s}_c) = (B_4\overline{s}_a, B_4\overline{s}_b, B_4\overline{s}_c) = (UCL_1, UCL_2, UCL_3)$$

$$\widetilde{LCL} = D_3\tilde{\overline{s}} = (B_3\overline{s}_a, B_3\overline{s}_b, B_3\overline{s}_c) = (LCL_1, LCL_2, LCL_3) \tag{2.61}$$

2.6.3   Fuzzy Exponentially Weighted Moving Average (FEWMA) Control Charts

Depending whether fuzzy process mean and fuzzy process standard deviation is known or not, FEWMA charts can be constructed as explained in Sects. 2.6.3.1 and 2.6.3.2

2.6.3.1   Fuzzy EWMA Control Charts When $\tilde{\sigma}$ Are Known

Let $\tilde{X}_i = (X_a, X_b, X_c)_i$ and $\tilde{\overline{\overline{X}}} = (\overline{X}_a, \overline{X}_b, \overline{X}_c)$ be the fuzzy observations for the ith sample and fuzzy grand averages of the t randomly collected sample data represented by triangular fuzzy numbers, respectively. Assume that fuzzy standard deviation $\tilde{\sigma}$ is known and represented by triangular fuzzy number as $\tilde{\sigma} = (\sigma_a, \sigma_b, \sigma_c)$

If the sample number t is moderately large, the parameter $\left[1 - (1 - \lambda)^{2t}\right]$ reaches to a limiting value of 1 and can be omitted from the formula. Hence, the control limits for the fuzzy EWMA control chart is given as follows:

$$\widetilde{CL}_{EWMA} = \tilde{\overline{\overline{X}}} = (\overline{X}_a, \overline{X}_b, \overline{X}_c) \tag{2.62}$$

$$\widetilde{UCL}_{EWMA} = \tilde{\overline{\overline{X}}} + \frac{3}{\sqrt{n}}\tilde{\sigma}\sqrt{\frac{\lambda}{2 - \lambda}} \tag{2.63}$$

$$\widetilde{LCL}_{EWMA} = \tilde{\overline{\overline{X}}} - \frac{3}{\sqrt{n}}\tilde{\sigma}\sqrt{\frac{\lambda}{2 - \lambda}} \tag{2.64}$$

Replacing the values of the $\tilde{\overline{\overline{X}}}$ and $\tilde{\sigma}$ to the equations above and performing simple fuzzy arithmetics, $\widetilde{UCL}_{EWMA}$ and $\widetilde{LCL}_{EWMA}$ for the moderately large number of samples can be rewritten as

$$\widetilde{UCL}_{EWMA} = \tilde{\overline{\overline{X}}} + \frac{3}{\sqrt{n}}\tilde{\sigma}\sqrt{\frac{\lambda}{2 - \lambda}} = (\overline{X}_a, \overline{X}_b, \overline{X}_c) + \frac{3}{\sqrt{n}}(\sigma_a, \sigma_b, \sigma_c)\sqrt{\frac{\lambda}{2 - \lambda}} \tag{2.65}$$

$$\widetilde{UCL}_{EWMA} = \left(\overline{X}_a + \frac{3\sigma_a}{\sqrt{n}}\sqrt{\frac{\lambda}{2 - \lambda}}, \overline{X}_b + \frac{3\sigma_b}{\sqrt{n}}\sqrt{\frac{\lambda}{2 - \lambda}}, \overline{X}_c + \frac{3\sigma_c}{\sqrt{n}}\sqrt{\frac{\lambda}{2 - \lambda}}\right) \tag{2.66}$$

$$\widetilde{\text{LCL}}_{\text{EWMA}} = \overline{\overline{\tilde{X}}} - \frac{3}{\sqrt{n}}\tilde{\sigma}\sqrt{\frac{\lambda}{2-\lambda}} = (\overline{X}_a, \overline{X}_b, \overline{X}_c) - \frac{3}{\sqrt{n}}(\sigma_a, \sigma_b, \sigma_c)\sqrt{\frac{\lambda}{2-\lambda}}$$

$$(2.67)$$

$$\widetilde{\text{LCL}}_{\text{EWMA}} = \left(\overline{X}_a + \frac{3\sigma_c}{\sqrt{n}}\sqrt{\frac{\lambda}{2-\lambda}}, \overline{X}_b + \frac{3\sigma_b}{\sqrt{n}}\sqrt{\frac{\lambda}{2-\lambda}}, \overline{X}_c + \frac{3\sigma_a}{\sqrt{n}}\sqrt{\frac{\lambda}{2-\lambda}}\right)$$

$$(2.68)$$

Similarly, if the sample number t is small, control limits for the fuzzy EWMA control chart can be given as follows:

$$\widetilde{\text{CL}}_{\text{EWMA}} = \overline{\overline{\tilde{X}}} = (\overline{X}_a, \overline{X}_b, \overline{X}_c) \tag{2.69}$$

$$\widetilde{\text{UCL}}_{\text{EWMA}} = \overline{\overline{\tilde{X}}} + \frac{3}{\sqrt{n}}\tilde{\sigma}\sqrt{\frac{\lambda}{2-\lambda}\left[1 - (1-\lambda)^{2t}\right]} \tag{2.70}$$

$$\widetilde{\text{LCL}}_{\text{EWMA}} = \overline{\overline{\tilde{X}}} - \frac{3}{\sqrt{n}}\tilde{\sigma}\sqrt{\frac{\lambda}{2-\lambda}\left[1 - (1-\lambda)^{2t}\right]} \tag{2.71}$$

By replacing the values of $\overline{\overline{\tilde{X}}}$ and $\tilde{\sigma}$, control limits for the fuzzy EWMA chart for small sample sizes can be given as

$$\widetilde{\text{UCL}}_{\text{EWMA}} = \left(\overline{X}_a + \frac{3\sigma_a}{\sqrt{n}}\sqrt{\frac{\lambda}{2-\lambda}\left[1 - (1-\lambda)^{2t}\right]}, \overline{X}_b\right.$$

$$+ \frac{3\sigma_b}{\sqrt{n}}\sqrt{\frac{\lambda}{2-\lambda}\left[1 - (1-\lambda)^{2t}\right]}, \overline{X}_c \qquad (2.72)$$

$$\left.+ \frac{3\sigma_c}{\sqrt{n}}\sqrt{\frac{\lambda}{2-\lambda}\left[1 - (1-\lambda)^{2t}\right]}\right)$$

$$\widetilde{\text{LCL}}_{\text{EWMA}} = \left(\overline{X}_a + \frac{3\sigma_c}{\sqrt{n}}\sqrt{\frac{\lambda}{2-\lambda}\left[1 - (1-\lambda)^{2t}\right]}, \overline{X}_b + \frac{3\sigma_b}{\sqrt{n}}\sqrt{\frac{\lambda}{2-\lambda}}, \overline{X}_c\right.$$

$$\left.+ \frac{3\sigma_a}{\sqrt{n}}\sqrt{\frac{\lambda}{2-\lambda}\left[1 - (1-\lambda)^{2t}\right]}\right) \qquad (2.73)$$

Readers who want to apply α-level cuts to the control limits can refer to (Şentürk et al. 2014; Gülbay et al. 2004).

2 Intelligent Process Control Using Control Charts—I …

2.6.3.2 Fuzzy EWMA Control Charts When $\tilde{\sigma}$ Are Unknown

Let $\tilde{R}_i = (R_a, R_b, R_c)_i$ and $\tilde{\bar{R}} = (\bar{R}_a, \bar{R}_b, \bar{R}_c)$ be the fuzzy range of the ith sample and fuzzy average range of the t samples for i = 1, 2, …, t. If fuzzy standard deviation, $\tilde{\sigma}$, is unknown, an unbiased estimator of the $\tilde{\sigma}$ can be determined from the ranges. Control limits for the fuzzy EWMA charts for the small sample sizes of t become as follows

$$\widetilde{CL}_{EWMA} = \tilde{\bar{\bar{X}}} = \left(\bar{X}_a, \bar{X}_b, \bar{X}_c\right) \tag{2.74}$$

$$\begin{aligned} \widetilde{UCL}_{EWMA} &= \tilde{\bar{\bar{X}}} + A_2\tilde{\bar{R}}\sqrt{\frac{\lambda}{2-\lambda}\left[1 - (1-\lambda)^{2t}\right]} \\ &= \left(\bar{X}_a, \bar{X}_b, \bar{X}_c\right) + A_2\left(\bar{R}_a, \bar{R}_b, \bar{R}_c\right)\sqrt{\frac{\lambda}{2-\lambda}\left[1 - (1-\lambda)^{2t}\right]} \end{aligned} \tag{2.75}$$

$$\begin{aligned} \widetilde{LCL}_{EWMA} &= \tilde{\bar{\bar{X}}} - A_2\tilde{\bar{R}}\sqrt{\frac{\lambda}{2-\lambda}\left[1 - (1-\lambda)^{2t}\right]} \\ &= \left(\bar{X}_a, \bar{X}_b, \bar{X}_c\right) - A_2\left(\bar{R}_a, \bar{R}_b, \bar{R}_c\right)\sqrt{\frac{\lambda}{2-\lambda}\left[1 - (1-\lambda)^{2t}\right]} \end{aligned} \tag{2.76}$$

Performing fuzzy arithmetic to the above equations, we obtain

$$\begin{aligned} \widetilde{UCL}_{EWMA} = \Bigg(&\bar{X}_a + A_2\bar{R}_a\sqrt{\frac{\lambda}{2-\lambda}\left[1 - (1-\lambda)^{2t}\right]}, \bar{X}_b \\ &+ A_2\bar{R}_b\sqrt{\frac{\lambda}{2-\lambda}\left[1 - (1-\lambda)^{2t}\right]}, \bar{X}_c \\ &+ A_2\bar{R}_c\sqrt{\frac{\lambda}{2-\lambda}\left[1 - (1-\lambda)^{2t}\right]} \Bigg) \end{aligned} \tag{2.77}$$

$$\begin{aligned} \widetilde{LCL}_{EWMA} = \Bigg(&\bar{X}_a - A_2\bar{R}_c\sqrt{\frac{\lambda}{2-\lambda}\left[1 - (1-\lambda)^{2t}\right]}, \bar{X}_b \\ &- A_2\bar{R}_b\sqrt{\frac{\lambda}{2-\lambda}\left[1 - (1-\lambda)^{2t}\right]}, \bar{X}_c \\ &- A_2\bar{R}_a\sqrt{\frac{\lambda}{2-\lambda}\left[1 - (1-\lambda)^{2t}\right]} \Bigg) \end{aligned} \tag{2.78}$$

For moderately large sample size of t, the parameter $\left[1 - (1-\lambda)^{2t}\right]$ tends to be 1 and can be ignored from the equations above.

2.6.4 Fuzzy Maximum Generally Weighted Moving Average (FMaxGWMA) Charts

Let $X_{ij}(i = 1, 2, \ldots, m; j = 1, 2, \ldots, n)$ be fuzzy observations. For any given $0 \leq \alpha \leq 1$, the corresponding real-values lower and upper bounds can be obtained as $\left(X_{ij}\right)_\alpha^L$ and $\left(X_{ij}\right)_\alpha^U$, respectively. A real-valued data for the lower and upper bounds of the $\tilde{\overline{X}}_i$ and $\tilde{s}_i$ can be written as (Shu et al. 2014)

$$\tilde{\overline{X}}_{i_\alpha}^U = \frac{1}{n}\sum_{j=1}^{n}\left(X_{ij}\right)_\alpha^U \quad \text{and} \quad \tilde{\overline{X}}_{i\alpha}^L = \frac{1}{n}\sum_{j=1}^{n}\left(X_{ij}\right)_\alpha^L \tag{2.79}$$

$$\tilde{s}_{i\alpha}^U = \sqrt{\frac{\sum_{j=1}^{n}\left(\left(X_{ij}\right)_\alpha^U - \tilde{\overline{X}}_{i\alpha}^U\right)^2}{n-1}} \quad \text{and} \quad \tilde{s}_{i\alpha}^L = \sqrt{\frac{\sum_{j=1}^{n}\left(\left(X_{ij}\right)_\alpha^L - \tilde{\overline{X}}_{i\alpha}^L\right)^2}{n-1}} \tag{2.80}$$

Then, we obtain unbiased estimators of the $\tilde{\sigma}_\alpha^U$ and $\tilde{\sigma}_\alpha^L$ as follows

$$\mu_\alpha^U = \tilde{\overline{\overline{X}}}_\alpha^U = \frac{1}{m}\sum_{i=1}^{m}\tilde{\overline{X}}_{i\alpha}^U \quad \text{and} \quad \mu_\alpha^L = \tilde{\overline{\overline{X}}}_\alpha^L = \frac{1}{m}\sum_{i=1}^{m}\tilde{\overline{X}}_{i\alpha}^L \tag{2.81}$$

$$\tilde{\overline{s}}_\alpha^U = \frac{1}{m}\sum_{i=1}^{m}\tilde{s}_{i\alpha}^U \quad \text{and} \quad \tilde{\overline{s}}_\alpha^L = \frac{1}{m}\sum_{i=1}^{m}\tilde{s}_{i\alpha}^L \tag{2.82}$$

$$\tilde{\sigma}_\alpha^U = \tilde{\overline{s}}_\alpha^U \big/ c_4 \quad \text{and} \quad \tilde{\sigma}_\alpha^L = \tilde{\overline{s}}_\alpha^L \big/ c_4 \tag{2.83}$$

Mutually independent statistics, M_i and S_i can also be rewritten in terms of the real-valued upper and lower bounds as

$$M_{i\alpha}^U = \frac{\tilde{\overline{X}}_{i\alpha}^U - \mu_\alpha^U}{\tilde{\sigma}_\alpha^U / \sqrt{n}} \quad \text{and} \quad M_{i\alpha}^L = \frac{\tilde{\overline{X}}_{i\alpha}^L - \mu_\alpha^L}{\tilde{\sigma}_\alpha^L / \sqrt{n}} \tag{2.84}$$

$$\tilde{S}_{i\alpha}^U = \emptyset^{-1}\left\{F\left[\frac{(n-1)\tilde{s}_{i\alpha}^U}{\left(\tilde{\sigma}_\alpha^U\right)^2}, n-1\right]\right\} \quad \text{and} \quad \tilde{S}_{i\alpha}^L = \emptyset^{-1}\left\{F\left[\frac{(n-1)\tilde{s}_{i\alpha}^L}{\left(\tilde{\sigma}_\alpha^L\right)^2}, n-1\right]\right\} \tag{2.85}$$

Finally, fuzzy GWMA statistics for the ith subgroup can be given by

$$\tilde{U}_{i\alpha}^U = \sum_{j=1}^{m}p_j\tilde{M}_{i+1-j_\alpha}{}^U \quad \text{and} \quad \tilde{U}_{i\alpha}^L = \sum_{j=1}^{m}p_j\tilde{M}_{i+1-j_\alpha}{}^L \tag{2.86}$$

$$\tilde{V}_{i_\alpha}^U = \sum_{j=1}^{m} p_j \tilde{S}_{i+1-j_\alpha}^U \quad \text{and} \quad \tilde{V}_{i_\alpha}^L = \sum_{j=1}^{m} p_j \tilde{S}_{i+1-j_\alpha}^L \tag{2.87}$$

Then, the fuzzy control limits of the F-MaxGWMA chart can be obtained from

$$\widetilde{UCL}_{i_\alpha}^U = \alpha UCL_i + (1 - \alpha)UCL_{i+1} \tag{2.88}$$

$$\widetilde{UCL}_{i_\alpha}^L = \alpha UCL_i + (1 - \alpha)UCL_{i-1} \tag{2.89}$$

where $UCL_0 = 0$.

The membership functions of $\widetilde{MG}_i$ required to be constructed for further identifying the manufacturing condition. Consider the closed interval C_α which is defined as:

$$C_\alpha = \left[\min\left\{ \max\left(\left|\tilde{U}_{i_\alpha}^U\right|, \left|\tilde{V}_{i_\alpha}^U\right| \right), \max\left(\left|\tilde{U}_{i_\alpha}^L\right|, \left|\tilde{V}_{i_\alpha}^L\right| \right) \right\}, \max\left\{ \max\left(\left|\tilde{U}_{i_\alpha}^U\right|, \left|\tilde{V}_{i_\alpha}^U\right| \right), \max\left(\left|\tilde{U}_{i_\alpha}^L\right|, \left|\tilde{V}_{i_\alpha}^L\right| \right) \right\} \right] \tag{2.90}$$

The membership functions of $\widetilde{MG}_i$ can be obtained by using the following expression

$$\xi_{\widetilde{MG}_i}(C) = \underbrace{\sup_{0 \le \alpha \le 1} \alpha} \, 1_{C_\alpha}(C) \tag{2.91}$$

Endpoints of the α-level closed interval $\widetilde{MG}_{i_\alpha} = \left[\widetilde{MG}_{i_\alpha}^L, \widetilde{MG}_{i_\alpha}^U \right]$ become

$$\widetilde{MG}_{i_\alpha}^L = \underbrace{\min_{\alpha \le \beta \le 1}} \min\left\{ \max\left(\left|\tilde{U}_{i_\beta}^U\right|, \left|\tilde{V}_{i_\beta}^U\right| \right), \max\left(\left|\tilde{U}_{i_\beta}^L\right|, \left|\tilde{V}_{i_\beta}^L\right| \right) \right\} \tag{2.92}$$

$$\widetilde{MG}_{i_\alpha}^U = \underbrace{\max_{\alpha \le \beta \le 1}} \max\left\{ \max\left(\left|\tilde{U}_{i_\beta}^U\right|, \left|\tilde{V}_{i_\beta}^U\right| \right), \max\left(\left|\tilde{U}_{i_\beta}^L\right|, \left|\tilde{V}_{i_\beta}^L\right| \right) \right\} \tag{2.93}$$

Now, to realize if the $\widetilde{MG}_i$ lie within their respective fuzzy control limits, comparisons of fuzzy numbers can be applied as mentioned in Sect. 2.5.

2.6.5 A Numerical Example for $\overline{X}$-R Control Chart

A company produces a material and wants to monitor its hardness measured by hardness testing equipment. Quality assistant takes a subgroup size of three, each also having three materials. For each material, the measured hardness values vary because of the material properties and gauge variability. To overcome the uncertainties caused by the non-uniform material properties, 3 readings for each sample are explained as a triangular fuzzy number as shown in Table 2.11. Company wants to construct x-R control charts for the uncertain hardness measurements using fuzzy control charts. Fuzzy mean and the fuzzy range for each sample are calculated by the equations given in Sect. 2.4.1, and shown in Table 2.11.

Fuzzy control limits are calculated according to the equations given in the previous sections. For n = 3, A_2 = 1.023, D_3 = 0, and D_4 = 2.574 are read from the coefficients table for variable control charts given in Appendix A.

Fuzzy control limits for $\overline{X}$ charts:

$$\widetilde{CL} = \tilde{\overline{X}} = (22.7, 24.7, 27.0)$$

$$\widetilde{UCL} = \left(\overline{\overline{X}}_a + A_2\overline{R}_a, \overline{\overline{X}}_b + A_2\overline{R}_b, \overline{\overline{X}}_c + A_2\overline{R}_c \right)$$
$$= (22.7 + 1.023(-1.5), 24.7 + 1.023(2.7), 27.0 + 1.023(6.9))$$

$$\widetilde{UCL} = (21.17, 27.46, 34.06)$$

$$\widetilde{LCL} = \left(\overline{\overline{X}}_a + A_2\overline{R}_c, \overline{\overline{X}}_b + A_2\overline{R}_b, \overline{\overline{X}}_c + A_2\overline{R}_a \right)$$
$$= (22.7 - 1.023(6.9), 24.7 + 1.023(2.7), 27.0 - 1.023(-1.5))$$

$$\widetilde{LCL} = (14.21, 21.94, 28.53)$$

Fuzzy control limits for R charts:

$$\widetilde{CL} = \tilde{\overline{R}} = (-1.5, 2.7, 6.9)$$

$$\widetilde{UCL} = \left(D_4\overline{R}_a, D_4\overline{R}_b, D_4\overline{R}_c \right) = (2.574(-1.5), 2.574(2.7), 2.574(6.9))$$
$$= (-3.86, 6.95, 17.76)$$

$$\widetilde{LCL} = \left(D_3\overline{R}_a, D_3\overline{R}_b, D_3\overline{R}_c \right) = (0, 0, 0)$$

Construction of fuzzy $\overline{X}$ control limits are shown in Fig. 2.11.

Assume that 21st sample's fuzzy average is $\overline{X}_{21} = (25.4, 28.5, 32.3)$ as shown in Fig. 2.12.

In order to decide whether $\tilde{X}_{21}$ plots an out of control or not we need to check if one of the conditions are met.

$$\widetilde{LCL} \leq \tilde{X}_{21} \leq \widetilde{UCL}$$

Table 2.11 Data for the testing hardness of a material

Sample number	Observations in each sample I Xa	I Xb	I Xc	II Xa	II Xb	II Xc	III Xa	III Xb	III Xc	Average Xa	Xb	Xc	Fuzzy mean I	II	III	Fuzzy range Max	Min	Xa	Xb	Xc
1	24.7	27.7	29.7	21.4	23.4	24.4	23.2	26.2	27.2	23.1	25.7	27.1	714.38	513.88	622.40	I	II	0.3	4.3	8.3
2	22.5	25.5	27.5	20.7	21.7	24.7	22.5	24.5	26.5	21.9	23.9	26.3	605.21	483.53	579.59	I	II	-2.2	3.8	6.8
3	23.3	26.3	29.3	22.4	25.4	28.4	21.2	24.2	27.2	22.3	25.3	28.3	657.81	613.07	555.26	I	III	-3.9	2.1	8.1
4	22.6	25.6	28.6	23.0	24.0	26.0	24.3	27.3	28.3	23.3	25.7	27.7	621.73	578.61	681.29	III	II	-1.7	3.3	5.3
5	23.2	24.2	27.2	22.4	23.4	26.4	21.0	22.0	25.0	22.2	23.2	26.2	599.84	559.61	496.17	I	III	-1.8	2.2	6.2
6	23.0	26.0	27.0	23.1	26.1	29.1	23.3	25.3	27.3	23.1	25.8	27.8	612.26	648.14	617.45	II	I	-3.8	0.2	6.2
7	24.5	26.5	27.5	23.9	24.9	27.9	24.8	26.8	29.8	24.4	26.1	28.4	666.18	634.09	706.86	III	II	-3.1	1.9	5.9
8	21.6	24.6	27.6	21.4	22.4	23.4	23.6	25.6	26.6	22.2	24.2	25.9	573.50	490.66	619.28	III	II	0.2	3.2	5.2
9	21.5	23.5	24.5	22.1	25.1	28.1	22.2	25.2	28.2	21.9	24.6	26.9	520.14	594.49	602.50	III	I	-2.3	1.7	6.7
10	24.3	25.3	27.3	20.8	21.8	23.8	24.0	26.0	27.0	23.0	24.4	26.0	642.39	475.21	639.93	I	II	0.5	3.5	6.5
11	25.3	28.3	31.3	21.9	23.9	26.9	23.6	25.6	27.6	23.6	26.0	28.6	763.41	564.10	632.05	I	II	-1.6	4.4	9.4
12	21.0	22.0	23.0	23.6	24.6	25.6	21.0	22.0	25.0	21.8	22.8	24.5	471.63	592.18	496.33	II	I	0.6	2.6	4.6
13	23.6	26.6	28.6	23.8	26.8	29.8	22.0	24.0	25.0	23.1	25.8	27.8	657.54	681.17	542.36	II	III	-1.2	2.6	7.8
14	23.8	26.8	29.8	21.7	22.7	23.7	23.4	24.4	27.4	22.9	24.6	26.9	680.73	503.43	607.53	I	II	0.1	4.1	8.1
15	23.9	24.9	25.9	21.6	24.6	25.6	24.2	25.2	28.2	23.2	24.9	26.6	608.38	546.77	648.90	III	II	-1.4	0.6	6.6
16	23.2	25.2	28.2	21.9	23.9	25.9	22.2	24.2	25.2	22.4	24.4	26.4	624.93	546.99	550.76	I	II	-2.7	1.3	6.3
17	24.8	26.8	29.8	21.6	24.6	25.6	22.4	24.4	27.4	22.9	25.3	27.6	708.47	548.11	584.45	I	II	-0.8	2.2	8.2
18	21.6	22.6	24.6	24.3	27.3	30.3	21.3	24.3	27.3	22.4	24.7	27.4	511.90	707.48	559.52	II	I	-0.3	4.7	8.7
19	20.2	22.2	25.2	22.2	23.2	26.2	22.6	23.6	25.6	21.7	23.0	25.7	484.02	549.68	559.43	III	I	-2.6	1.4	5.4
20	21.8	22.8	25.8	22.2	23.2	26.2	24.2	27.2	30.2	22.7	24.4	27.4	532.56	550.53	704.53	III	I	-1.6	4.4	8.4
Grand average X										22.7	24.7	27.0				Average range		-1.5	2.7	6.9

Fig. 2.11 Illustration of
Fuzzy $\overline{X}$ control limits

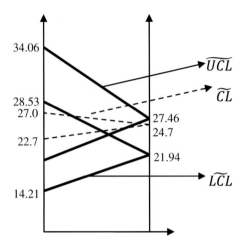

Fig. 2.12 Illustration of a
new fuzzy observation: $\overline{X}_{21}$

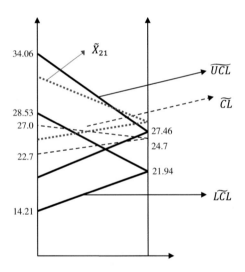

If we use ranking methods explained in Sect. 2.5, we obtain a crisp decision that
$\tilde{X}_{21}$ plots either out of control or in control condition. One of the best method is to
use direct fuzzy approach control presented in Gülbay and Kahraman (2007) which
allows quality professionals to decide and interpret the chart with the degree of
membership that a point shows out of control or in control. Furthermore, by
defining intermediate decisions between out of control and in control enables to the
usage of various actions to correct the process.

2.7 Fuzzy Unnatural Pattern Analyses for Control Charts for Variables

2.7.1 Probability of Fuzzy Events

The formula for calculating the probability of a fuzzy event A is a generalization of the probability theory: in the case which a sample space X is a continuum or discrete, the probability of a fuzzy event P(A) is given by Yen and Langari (1999):

$$P(\tilde{A}) = \begin{cases} \int \mu_A(x)P_X(x)dx & \text{if X is continuous,} \\ \sum_i \mu_A(x_i)P_X(x_i)dx & \text{if X is discrete.} \end{cases} \quad (2.94)$$

where P_X denotes a classical probability distribution function of X for continuous sample space and probability function for discrete sample space, and μ_A is a membership function of the event A.

The membership degree of a fuzzy sample that belongs to a region is directly related to its percentage area falling in that region, and therefore, it is continuous. For example, a fuzzy sample may be in zone B with a membership degree of 0.4 and in zone C with a membership degree of 0.6. While counting fuzzy samples in zone B, that sample is counted as 0.4.

2.7.2 Generation of Fuzzy Rules for Unnatural Patterns

The run rules are based on the premise that a specific run of data has a low probability of occurrence in a completely random stream of data. If a run occurs, then it is meant that something has changed in the process to produce such a nonrandom or unnatural pattern. Based on the expected percentages in each zone, sensitive run tests can be developed for analyzing the patterns of variation in the various zones. For the fuzzy control charts, based on the Western Electric rules (Western Electric Company 1956), the following fuzzy unnatural pattern rules can be defined. The probabilities of these fuzzy events are calculated using normal approach to binomial distribution (Gülbay and Kahraman 2006).

The probability of each fuzzy rule (event) below depends on the definition of the membership function which is subjectively defined so that the probability of each of the fuzzy rules is as close as possible to the corresponding classical rule for unnatural patterns. The idea behind this approach may justify the following rules (Gülbay and Kahraman 2006).

Rule 1: Any fuzzy data falling outside the three-sigma control limits with a ratio of more than predefined percentage (β) of sample area at desired α-level. The membership function for this rule can subjectively be defined as below:

$$\mu_1(x) = \begin{cases} 0 & ; & 0.85 \le x \le 1, \\ (x - 0.60)/0.25 & ; & 0.60 \le x \le 0.85, \\ (x - 0.10)/0.50 & ; & 0.10 \le x \le 0.60, \\ 1 & ; & 0 \le x \le 0.10, \end{cases} \qquad (2.95)$$

Rule 2: A total membership degree around 2 from three consecutive points in zone A or beyond. Probability of a sample being in zone A (0.0214) or beyond (0.00135) is 0.02275. Let the membership function for this rule be defined as follows:

$$\mu_2(x) = \begin{cases} 0 & ; & 0 \le x \le 0.59, \\ (x - 0.59)/1.41 & ; & 0.59 \le x \le 2 \\ 1 & ; & 2 \le x \le 3 \end{cases} \qquad (2.96)$$

The probability of the fuzzy event rule 2 is approximately 0.0015, which corresponds to the crisp case of this rule.

Rule 3: A total membership degree around 4 from five consecutive points in zone C or beyond:

$$\mu_3(x) = \begin{cases} 0 & ; & 0 \le x \le 2.42, \\ (x - 2.42)/1.58 & ; & 2.42 \le x \le 4, \\ 1 & ; & 4 \le x \le 5. \end{cases} \qquad (2.97)$$

The probability of the fuzzy event rule 3 is approximately 0.0027

Rule 4: A total membership degree around 8 from eight consecutive points on the same side of the centerline with the membership function below and its probability is 0.0039:

$$\mu_4(x) = \begin{cases} 0 & ; & 0 \le x \le 2.54, \\ (x - 2.54)/5.46 & ; & 2.54 \le x \le 8. \end{cases} \qquad (2.98)$$

Rule 5: A total membership degree around 7 from seven consecutive points on the same side of the center line. The fuzzy probability of this rule is 0.0079 when membership function is defined as below:

$$\mu_5(x) = \begin{cases} 0 & ; & 0 \le x \le 2.48, \\ (x - 2.48)/4.52 & ; & 2.48 \le x \le 7. \end{cases} \qquad (2.99)$$

Rule 6: At least a total membership degree around 10 from 11 consecutive points on the same side of the center line. The fuzzy probability of this rule is 0.0058 when the membership function is defined as below:

$$\mu_6(x) = \begin{cases} 0 & ; \quad 0 \le x \le 9.33, \\ (x - 9.33)/0.67 & ; \quad 9.33 \le x \le 10, \\ 1 & ; \quad 10 \le x \le 11. \end{cases} \qquad (2.100)$$

Rule 7: At least a total membership degree around 12 from 14 consecutive points on the same side of the center line. If the membership function is set as given below, then the fuzzy probability of the rule is equal to 0.0065.

$$\mu_7(x) = \begin{cases} 0 & ; \quad 0 \le x \le 11.33, \\ (x - 11.33)/0.67 & ; \quad 11.33 \le x \le 12, \\ 1 & ; \quad 12 \le x \le 14. \end{cases} \qquad (2.101)$$

Rule 8: At least a total membership degree around 14 from 17 consecutive points on the same side of the center line. The probability of this fuzzy event with the membership function below is 0.0062.

$$\mu_7(x) = \begin{cases} 0 & ; \quad 0 \le x \le 13.34, \\ (x - 13.34)/0.66 & ; \quad 13.34 \le x \le 14, \\ 1 & ; \quad 14 \le x \le 17. \end{cases} \qquad (2.102)$$

A framework for the application of the fuzzy unnatural pattern rules are as follows:

1. Determine $\pm 3\sigma$ fuzzy control limits.
2. Determine fuzzy regions of $\pm 1\sigma$ and $\pm 2\sigma$
3. For each sample, calculate the percentage of sample area that belongs to the regions of A, B, and C for both sides of the fuzzy center line.
4. For each fuzzy rule, check the last N points as defined in the rule and sum their percentage of sample area in the related region. Then, for that rule use its corresponding membership function to obtain the membership degree of the occurrence for the specified rule.
5. Repeat step 4 until all desired fuzzy rules are checked.

2.7.3 An Illustrative Example

Consider the case where a-three subgroup ($n = 3$) is taken for the construction of the fuzzy $\overline{X}$ control charts. For $n = 3$, the constant $A_2 = 1.023$. Grand average and average range for the 25 samples are calculated using the Eqs. 2.20–2.25 and given as

Table 2.12 Equations for calculation of fuzzy $\pm z\sigma$ control limits

z	Notation	Fuzzy $z\sigma$ limits
+3	$\widetilde{UCL}$	$\left(\overline{\overline{X}}_a + A_2\overline{R}_a, \overline{\overline{X}}_b + A_2\overline{R}_b, \overline{\overline{X}}_c + A_2\overline{R}_c\right)$
+2	$\widetilde{CL} + 2\sigma$	$\left(\overline{\overline{X}}_a + \left(\frac{2}{3}\right)A_2\overline{R}_a, \overline{\overline{X}}_b + \left(\frac{2}{3}\right)A_2\overline{R}_b, \overline{\overline{X}}_c + \left(\frac{2}{3}\right)A_2\overline{R}_c\right)$
+1	$\widetilde{CL} + 1\sigma$	$\left(\overline{\overline{X}}_a + \left(\frac{1}{3}\right)A_2\overline{R}_a, \overline{\overline{X}}_b + \left(\frac{1}{3}\right)A_2\overline{R}_b, \overline{\overline{X}}_c + \left(\frac{1}{3}\right)A_2\overline{R}_c\right)$
0	$\widetilde{CL}$	$\overline{\overline{X}} = \left(\overline{\overline{X}}_a, \overline{\overline{X}}_b, \overline{\overline{X}}_c\right)$
−1	$\widetilde{CL} - 1\sigma$	$\left(\overline{\overline{X}}_a - \left(\frac{1}{3}\right)A_2\overline{R}_c, \overline{\overline{X}}_b - \left(\frac{1}{3}\right)A_2\overline{R}_b, \overline{\overline{X}}_c - \left(\frac{1}{3}\right)A_2\overline{R}_a\right)$
−2	$\widetilde{CL} - 2\sigma$	$\left(\overline{\overline{X}}_a - \left(\frac{2}{3}\right)A_2\overline{R}_c, \overline{\overline{X}}_b - \left(\frac{2}{3}\right)A_2\overline{R}_b, \overline{\overline{X}}_c - \left(\frac{2}{3}\right)A_2\overline{R}_a\right)$
−3	$\widetilde{LCL}$	$\left(\overline{\overline{X}}_a - A_2\overline{R}_c, \overline{\overline{X}}_b - A_2\overline{R}_b, \overline{\overline{X}}_c - A_2\overline{R}_a\right)$

Table 2.13 Fuzzy $\pm z\sigma$ control limits and their regions (see Fig. 2.6)

z	Notation	Fuzzy $z\sigma$ limits	Region
+3	$\widetilde{UCL}$	(50.2, 70.3, 90.5)	A
+2	$\widetilde{CL} + 2\sigma$	(46.8, 65.2, 83.6)	B
+1	$\widetilde{CL} + 1\sigma$	(43.4, 60.1, 76.8)	C
0	$\widetilde{CL}$	(40.0, 55.0, 70.0)	C
−1	$\widetilde{CL} - 1\sigma$	(36.6, 79.9, 63.2)	B
−2	$\widetilde{CL} - 2\sigma$	(33.2, 44.8, 56.4)	A
−3	$\widetilde{LCL}$	(29.8, 39.7, 49.5)	

$$\overline{\overline{X}} = (40, 55, 70)$$

$$\overline{R} = (10, 15, 20)$$

Fuzzy $\pm z\sigma$ limits can be calculated using the equations given in Table 2.12.

Replacing the values of the $\overline{\overline{X}}, \overline{R}$, and the constant A_2 into the equations above, we obtain fuzzy regions as given in Table 2.13.

Let the next sample be $\tilde{X}_{26} = (48.0, 60.0, 70.0)$. Let's construct the $\overline{X}$ control chart and see at what membership degree $\tilde{X}_{26}$ belongs to the regions of A, B, and C for the both sides of the fuzzy center line.

In Fig. 2.14, fuzzy control limits and fuzzy regions (see Fig. 2.13) are simplified in order to show the region A above the centerline and $\tilde{X}_{26}$.

As can be seen from the Fig. 2.14, only a little part of the sample area of $\tilde{X}_{26}$ is out of the region A, namely most of its parts are in region A. The problem is to calculate the percentage area of $\tilde{X}_{26}$ which is inside the region A. The percentage

Fig. 2.13 Illustration of $\pm z\sigma$ control limits and $\tilde{X}_{26}$

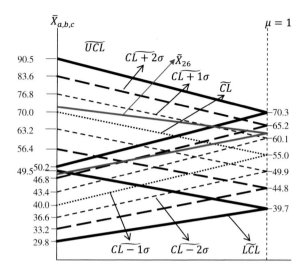

Fig. 2.14 Illustration of $+2\sigma$ and $+3\sigma$ control limits, and $\tilde{X}_{26}$

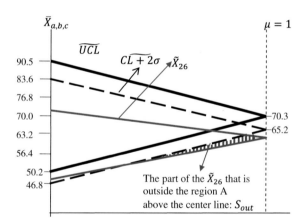

sample area within a specified region can be calculated using the formula given in Eq. 2.42.

$$\beta_j = \frac{S_j - S_{out}}{S_j} \tag{2.103}$$

where S_j is the sample area and S_{out} is the area of the sample outside the corresponding region. These calculations are a little hard, but by using simple software it can be easy to determine. Once control limits are specified a general formula can be derived for the area calculation and the percentage areas can be calculated using any spread sheets. The reader can refer to the Gülbay and Kahraman (2006) for the determination of the percentage areas. Suppose that β_{26}, β_{27}, and β_{28} are determined as 0.85, 0.50, 0,25. For these 3 consecutive samples the total degree of memberships is $0.85 + 0.50 + 0.25 = 1.60$ for being in region A above the center line. Fuzzy rule 2 can be checked now to decide at what membership degree that rule is performed. Remember that membership degree of the rule 2 was subjectively defined as:

$$\mu_2(x) = \begin{cases} 0 & ; \quad 0 \le x \le 0.59, \\ (x - 0.59)/1.41 & ; \quad 0.59 \le x \le 2 \\ 1 & ; \quad 2 \le x \le 3 \end{cases}$$

Then,

$$\mu_2(1.60) = \{\, (1.60 - 0.59)/1.41 = 0.716$$

The quality control professional can set a predefined value of μ to compare with the μ_i to accept or reject the occurrence of the rule and hence may justify a set of actions with respect to the calculated μ. The rest of the fuzzy rules are applied in the same way.

2.8 Conclusion

Control charts aim at detecting if any assignable cause exists in the considered process. If only random causes exist, no action is required. Otherwise, a corrective action is needed. We proposed fuzzy control charts to be used in case of incomplete and vague data for the process control. Fuzzy triangular fuzzy numbers have been preferred in the developed control charts because of their relative simplicity whereas the other types of fuzzy numbers can also be used. Trapezoidal fuzzy numbers or LR type fuzzy numbers can be replaced with triangular fuzzy numbers in these analyses. EWMA control charts are preferred when we need detecting small shifts. These charts have been also developed under fuzziness and a numerical example has been given. The new extensions of fuzzy sets such as type-2 fuzzy sets, Intuitionistic fuzzy sets, and hesitant fuzzy sets are the possible alternatives to extend our work. Each of these new extensions is also divided into a few types. For example, interval type Intuitionistic fuzzy sets for triangular Intuitionistic fuzzy sets are subalternatives for the new developments.

Appendix A

Table of coefficients for control charts for variables.

Observations in subgroup, n	A_2	A_3	c_4	B_3	B_4	B_5	B_6	d_2	d_3	D_1	D_2	D_3	D_4	E_2
2	1.880	2.659	0.798	0.000	3.267	0.000	2.606	1.128	0.853	0.000	3.686	0.000	3.267	2.660
3	1.023	1.954	0.886	0.000	2.568	0.000	2.276	1.693	0.888	0.000	4.358	0.000	2.574	1.772
4	0.729	1.628	0.921	0.000	2.266	0.000	2.088	2.059	0.880	0.000	4.698	0.000	2.282	1.457
5	0.577	1.427	0.940	0.000	2.089	0.000	1.964	2.326	0.864	0.000	4.918	0.000	2.114	1.290
6	0.483	1.287	0.952	0.030	1.970	0.029	1.874	2.534	0.848	0.000	5.078	0.000	2.004	1.184
7	0.419	1.182	0.959	0.118	1.882	0.113	1.806	2.704	0.833	0.204	5.204	0.076	1.924	1.109
8	0.373	1.099	0.965	0.185	1.815	0.179	1.751	2.847	0.820	0.388	5.306	0.136	1.864	1.054
9	0.337	1.032	0.969	0.239	1.761	0.232	1.707	2.970	0.808	0.547	5.393	0.184	1.816	1.010
10	0.308	0.975	0.973	0.284	1.716	0.276	1.669	3.078	0.797	0.687	5.469	0.223	1.777	0.975
11	0.285	0.927	0.975	0.321	1.679	0.313	1.637	3.173	0.787	0.811	5.535	0.256	1.744	0.945
12	0.266	0.886	0.978	0.354	1.646	0.346	1.610	3.258	0.778	0.922	5.594	0.283	1.717	0.921
13	0.249	0.850	0.979	0.382	1.618	0.374	1.585	3.336	0.770	1.025	5.647	0.307	1.693	0.899
14	0.235	0.817	0.981	0.406	1.594	0.399	1.563	3.407	0.763	1.118	5.696	0.328	1.672	0.881
15	0.223	0.789	0.982	0.428	1.572	0.421	1.544	3.472	0.756	1.203	5.741	0.347	1.653	0.864
16	0.212	0.763	0.984	0.448	1.552	0.440	1.526	3.532	0.750	1.282	5.782	0.363	1.637	0.849
17	0.203	0.739	0.985	0.466	1.534	0.458	1.511	3.588	0.744	1.356	5.820	0.378	1.622	0.836
18	0.194	0.718	0.985	0.482	1.518	0.475	1.496	3.640	0.739	1.424	5.856	0.391	1.608	0.824
19	0.187	0.698	0.986	0.497	1.503	0.490	1.483	3.689	0.734	1.487	5.891	0.403	1.597	0.813
20	0.180	0.680	0.987	0.510	1.490	0.504	1.470	3.735	0.729	1.549	5.921	0.415	1.585	0.803
21	0.173	0.663	0.988	0.523	1.477	0.516	1.459	3.778	0.724	1.605	5.951	0.425	1.575	0.794
22	0.167	0.647	0.988	0.534	1.466	0.528	1.448	3.819	0.720	1.659	5.979	0.434	1.566	0.786
23	0.162	0.633	0.989	0.545	1.455	0.539	1.438	3.858	0.716	1.710	6.006	0.443	1.557	0.778
24	0.157	0.619	0.989	0.555	1.445	0.549	1.429	3.895	0.712	1.759	6.031	0.451	1.548	0.770
25	0.153	0.606	0.990	0.565	1.435	0.559	1.420	3.931	0.708	1.806	6.056	0.459	1.541	0.763

Appendix B

The equations to compute sample area outside the control the limits.

$$
\begin{aligned}
A_{out}^{U} &= \frac{1}{2}\left[\left(d^{\alpha} - UCL_{4}^{\alpha}\right) + \left(d^{t} - UCL_{4}^{t}\right)\right]\left(\max(t - \alpha, 0)\right) \\
&+ \frac{1}{2}\left[\left(d^{z} - a^{z}\right) + (c - b)\right]\left(\min(1 - t, 1 - \alpha)\right)
\end{aligned}
\tag{2.104}
$$

where,

$$
t = \frac{UCL_{4} - a}{(b - a) + (c - b)} \text{ and } z = \max(t, \alpha)
$$

$$
A_{out}^{U} = \frac{1}{2}\left[\left(d^{\alpha} - UCL_{4}^{\alpha}\right) + (c - UCL_{3})\right](1 - \alpha)
\tag{2.105}
$$

$$
A_{out}^{U} = \frac{1}{2}\left(d^{\alpha} - UCL_{4}^{\alpha}\right)\left(\max(t - \alpha, 0)\right)
\tag{2.106}
$$

where

$$
t = \frac{UCL_{4} - d}{(UCL_{4} - UCL_{3}) - (d - c)}
$$

$$
A_{out}^{U} = \frac{1}{2}\left[(c - UCL_{3}) + \left(d^{z} - UCL_{4}^{z}\right)\right]\left(\min(1 - t, 1 - \alpha)\right)
\tag{2.107}
$$

where

$$
t = \frac{UCL_{4} - d}{(UCL_{4} - UCL_{3}) - (d - c)} \text{ and } z = \max(t, \alpha)
$$

$$
\begin{aligned}
A_{out}^{U} &= \frac{1}{2}\left[\left(d^{z_{2}} - UCL_{4}^{z_{2}}\right) + \left(d^{t_{1}} - UCL_{4}^{t_{1}}\right)\right]\left(\min(\max(t_{1} - \alpha, 0), t_{1} - t_{2})\right) \\
&+ \frac{1}{2}\left[\left(d^{z_{1}} - a^{z_{1}}\right) + (c - b)\right]\left(\min(1 - t_{1}, 1 - \alpha)\right)
\end{aligned}
$$

where

$$
\begin{aligned}
t_{1} &= \frac{UCL_{4} - a}{(b - a) + (UCL_{4} - UCL_{3})}, \\
t_{2} &= \frac{UCL_{4} - d}{(UCL_{4} - UCL_{3}) - (d - c)}, \\
z_{1} &= \max(\alpha, t_{1}), \text{ and } z_{2} = \max(\alpha, t_{2})
\end{aligned}
\tag{2.108}
$$

$$A_{out}^{U} = 0 \tag{2.109}$$

$$A_{out}^{U} = \frac{1}{2}\left[(d^{\alpha} - a^{\alpha}) + (c - b)\right](1 - \alpha) \tag{2.110}$$

$$A_{out}^{L} = \frac{1}{2}\left[\left(LCL_{1}^{\alpha} - a^{\alpha}\right) + \left(LCL_{1}^{t} - a^{t}\right)\right](\max(t - \alpha, 0)) \\ + \frac{1}{2}\left[(d^{z} - a^{z}) + (c - b)\right](\min(1 - t, 1 - \alpha)) \tag{2.111}$$

where

$$t = \frac{d - LCL_{1}}{(LCL_{2} - LCL_{1}) + (d - c)} \text{ and } z = \max(\alpha, t)$$

$$A_{out}^{L} = \frac{1}{2}\left[(d^{\alpha} - a^{\alpha}) + (c - b)\right](1 - \alpha) \tag{2.112}$$

$$A_{out}^{L} = \frac{1}{2}\left[\left(LCL_{1}^{\alpha} - a^{\alpha}\right) + (LCL_{2} - b)\right](1 - \alpha) \tag{2.113}$$

$$A_{out}^{L} = \frac{1}{2}\left[\left(LCL_{1}^{z_{2}} - a^{z_{2}}\right) + \left(LCL_{1}^{t_{1}} - a^{t_{1}}\right)\right](\min(\max(t_{1} - \alpha, 0), t_{1} - t_{2})) \\ + \frac{1}{2}\left[(d^{z_{1}} - a^{z_{1}}) + (c - b)\right](\min(1 - t, 1 - \alpha)) \tag{2.114}$$

where

$$t_{1} = \frac{d - LCL_{1}}{(LCL_{2} - LCL_{1}) + (d - c)},$$

$$t_{2} = \frac{a - LCL_{1}}{(LCL_{2} - LCL_{1}) - (b - a)}$$

$$z_{1} = \max(\alpha, t_{1}), \text{ and } z_{2} = \max(\alpha, t_{2})$$

$$A_{out}^{L} = \frac{1}{2}\left[\left(LCL_{1}^{z} - a^{z}\right) + (LCL_{2} - b)\right](\min(1 - t, 1 - \alpha)) \tag{2.115}$$

where

$$t = \frac{a - LCL_{1}}{(LCL_{2} - LCL_{1}) - (b - a)}, \text{ and } z = \max(\alpha, t)$$

$$A_{out}^{L} = 0 \tag{2.116}$$

$$A_{out}^L = \frac{1}{2}[(d^\alpha - a^\alpha) + (c - b)](1 - \alpha) \qquad (2.117)$$

References

Chen, G., Cheng, S.W.: Max-chart: combining X-bar and s-chart. Statistica Sinica **8**, 263–271 (1998)

Eyvazian, M., Naini, S.G.J., Vaghefi, A.: Monitoring process variability using exponentially weighted moving sample variance control charts. Int. J. Adv. Manuf. **39**, 261–270 (2008)

Han, D., Tsung, F.: A generalized EWMA control chart and its comparison with the optimal EWMA, CUSUM and GRL schemes. Ann Stat. **32**(1), 316–339 (2004)

Li, Z., Wang, Z.: An exponentially weighted moving average scheme with variable sampling intervals for monitoring linear profiles. Comput. Ind. Eng. **59**, 630–637 (2010)

Montgomery, D.C.: Introduction to Statistical Quality Control. Wiley, New York (2001)

Nelson, L.S.: The Shewhart control chart–tests for special causes. J. Qual. Technol. **16**(4), 238 (1984)

Sheu, S.H., Lin, T.C.: The generally weighted moving average control chart for detecting small shifts in the process mean. Qual. Eng. **16**, 209–231 (2003)

Sheu, S.H., Huang, C.J., Hsu, T.S.: Extended maximum generally weighted moving average control chart for monitoring process mean and variability. Comput. Ind. Eng. **62**, 216–225 (2012)

Shu, M.-H., Nguyen, T.-L., Hsu, B.-M.: Fuzzy MaxGWMA chart for identifying abnormal variations of on-line manufacturing processes with imprecise information. Expert Syst. Appl. **41**, 1342–1356 (2014)

Western Electric Company: Statistical Quality Control Handbook, Indianapolis, Indiana (1956)

Xie, H.: Contributions to qualimetry. Ph.D. thesis, Winnipeg, Canada: University of Manitoba (1999)

Zhang, J., Li, Z., Wang, Z.: A multivariate control chart for simultaneously monitoring process mean and variability. Comput. Stat. Data Anal. **54**, 2244–2252 (2010)

Juran's Quality Handbook: McGraw-Hill Professional; 6 edn. (2010). ISBN-10: 0071629734

Asady, B.: The revised method of ranking LR fuzzy numbers based on deviation degree. Expert Syst. Appl. **37**, 5056–5060 (2010)

Bortolan, G., Degani, R.: A review of some methods for ranking fuzzy numbers. Fuzzy Sets Syst. **15**, 1–19 (1985)

Brunelli, M., Mezeib, J.: How different are ranking methods for fuzzy numbers? A numerical study. Int. J. Approximate Reasoning **54**, 627–639 (2013)

Chen, S.-M., Chen, J.H.: Fuzzy risk analysis based on ranking generalized fuzzy numbers with different heights and different spreads. Expert Syst. Appl. **36**(3), 6833–6842 (2009)

Chen, S.-J., Hwang, C.-L.: Fuzzy Multiple Attribute Decision Making. Springer, New York (1992)

Chen, S.-M., Sanguansat, K.: Analyzing fuzzy risk based on a new fuzzy ranking method between generalized fuzzy numbers. Expert Syst. Appl. **38**(3), 2163–2171 (2011)

Cheng, C.H.: A new approach for ranking fuzzy numbers by distance method. Fuzzy Sets Syst. **95**, 307–317 (1998)

Choobineh, F., Li, H.: An index for ordering fuzzy numbers. Fuzzy Sets Syst. **54**, 287–294 (1993)

Chu, T.C., Tsao, C.T.: Ranking fuzzy numbers with an area between the centroid point and the original point. Comput. Math Appl. **43**, 111–117 (2002)

Deng, H.: A discriminative analysis of approaches to ranking fuzzy numbers in fuzzy decision making. In: Proceedings of the 4th IEEE International Conference on Fuzzy Systems and Knowledge Discovery, 26–29 Aug, Haikou, China (2007)

Deng, Y., Zhu, Z.F., Liu, Q.: Ranking fuzzy numbers with an area method using radius of gyration. Comput. Math Appl. **51**, 1127–1136 (2006)

Chrysler, Ford, GM: Measurement Systems Analysis Reference Manual. AIAG, Detroit, MI (1995)

Detyniecki, M., Yager, R.R.: Ranking fuzzy numbers using α-weighted valuations. Int. J. Uncertainty Fuzziness Knowl.-Based Syst. **8**(5), 573–592 (2001)

Dias, G.: Ranking alternatives using fuzzy numbers: a computational approach. Fuzzy Sets Syst. **56**, 247–252 (1993)

Dubois, D., Prade, H.: Operations on fuzzy numbers. Int. J. Syst. Sci. **9**, 613–626 (1978)

Dubois, D., Prade, H.: Fuzzy Sets and Systems: Theory and Applications. Academic Press, New York (1980)

Dubois, D., Prade, H.: Fuzzy sets—a convenient function for modeling vagueness and possibility. IEEE Trans. Fuzzy Syst. **2**, 16–21 (1994)

Duncan, A.J.: Quality Control and Industrial Statistics, 5th edn. Irwin, Homewood (1986)

Fortemps, P., Roubens, M.: Ranking and defuzzification methods based on area compensation. Fuzzy Sets Syst. **82**(3), 19–330 (1996)

Gitlow, H.S.: Tools and Methods for the Improvement of Quality (1989). ISBN-10: 0256056803

Gülbay, M., Kahraman, C.: Development of fuzzy process control charts and fuzzy unnatural pattern analyses. Comput. Stat. Data Anal. **51**, 434–451 (2006)

Gülbay, M., Kahraman, C.: An alternative approach to fuzzy control charts: direct fuzzy approach. Inf. Sci. **177**(6), 1463–1480 (2007)

Gülbay, M., Kahraman, C., Ruan, D.: α-cut fuzzy control charts for linguistic data. Int. J. Intell. Syst. **19**, 1173–1196 (2004)

Jain, R.: Decision-making in the presence of fuzzy variables. IEEE Trans. Syst. Man Cybern. **6**, 698–703 (1976)

Jain, R.: A procedure for multi-aspect decision making using fuzzy sets. Int. J. Syst. Sci. **8**, 1–7 (1978)

Kim, K., Park, K.S.: Ranking fuzzy numbers with index of optimism. Fuzzy Sets Syst. **35**, 143–150 (1990)

Kim, C.B., Seong, K.A., Lee-Kwang, H.: Design and implementation of fuzzy elevator group control system. IEEE Trans. Syst. Man Cybern. **28**, 277–287 (1998)

Lee, J.H., Lee-Kwang, H.: Distributed and cooperative fuzzy controller for traffic intersection group. IEEE Trans. Syst. Man Cybern. **29**, 263–271 (1999)

Lee, E.S., Li, R.J.: Comparison of fuzzy numbers based on the probability measure of fuzzy events. Comput. Math Appl. **15**, 887–896 (1998)

Lee, K.M., Cho, C.H., Lee-Kwang, H.: Ranking fuzzy values with satisfaction function. Fuzzy Sets Syst. **64**, 295–311 (1994)

Liu, X.-W., Han, S.-L.: Ranking fuzzy numbers with preference weighting function expectations. Comput. Math Appl. **49**, 1731–1753 (2005)

Murakami, S., Maeda, S., Imamura, S.: Fuzzy decision analysis on the development of centralized regional energy control system. In: Proceedings of the IFAC Symposium on Fuzzy Information, Knowledge Representation and Decision Analysis, pp. 363–368 (1983)

Raj, A., Kumar, D.N.: Ranking alternatives with fuzzy weights using maximizing set and minimizing set. Fuzzy Sets Syst. **105**, 365–375 (1999)

Rao, P.P.B., Shankar, N.R.: Ranking fuzzy numbers with a distance method using circumcenter of centroids and an index of modality. Adv. Fuzzy Syst. **2011**, Article ID 178308 (2011). doi:10. 1155/2011/178308

Requena, I., Delgado, M., Verdagay, J.I.: Automatic ranking of fuzzy numbers with the criterion of decision-maker learnt by an artificial neural network. Fuzzy Sets Syst. **64**, 1–19 (1994)

Senturk, S., Erginel, N.: Development of fuzzy and control charts using α-cuts. Inf. Sci. **179**(10), 1542–1551 (2009)

Şentürk, S., Erginel, N., Kaya, İ., Kahraman, C.: Fuzzy exponentially weighted moving average control chart for univariate data with a real case application. Appl. Soft Comput. **22**, 1–10 (2014)

Shureshjani, R.A., Darehmiraki, M.: A new parametric method for ranking fuzzy numbers. Indagationes Mathematicae **24**, 518–529 (2013)

Sun, H., Wu, J.: A new approach for ranking fuzzy numbers based on fuzzy simulation analysis method. Appl. Math. Comput. **174**, 755–767 (2006)

Tran, L., Duckstein, L.: Comparison of fuzzy numbers using a fuzzy distance measure. Fuzzy Sets Syst. **130**, 331–341 (2002)

Wang, Y.J., Lee, H.S.: The revised method of ranking fuzzy numbers with an area between the centroid and original points. Comput. Math Appl. **55**, 2033–2042 (2008)

Wang, Z.-X., Liu, Y.-J., Fan, Z.-P., Feng, B.: Ranking L-R fuzzy number based on deviation degree. Inf. Sci. **179**, 2070–2077 (2009)

Westgard, J.O., Westgard S.: Basic Quality Management Systems (2014). ISBN: 1-886958-28-9

Yager, R.R.: Ranking fuzzy subsets over the unit interval. In: Proceeding of the 17th IEEE International Conference on Decision and Control, San Diego, CA, pp. 1435–1437 (1978)

Yeh, C.H., Deng, H.: A practical approach to fuzzy utilities comparison in fuzzy multicriteria analysis. Int. J. Approximate Reasoning **35**(2), 179–194 (2004)

Yen, J., Langari, R.: Fuzzy Logic: Intelligence, Control, and Information. Prentice-Hall, Upper Saddle River (1999)

Zadeh, L.A.: Fuzzy sets. Inf. Control **8**, 338–353 (1965)

Zimmermann, H.-J.: Fuzzy Set Theory and its Applications, 3rd edn. Kluwer Academic Publishers, Boston (1996)

Zimmermann, H.-J.: An application-oriented view of modeling uncertainty. Eur. J. Oper. Res. **122**, 190–198 (2000)

Chapter 3
Intelligent Process Control Using Control Charts—II: Control Charts for Attributes

Seda Yanık, Cengiz Kahraman and Hafize Yılmaz

Abstract Control charts for attributes are used to detect nonrandom variation when the inspected quality characteristic cannot be represented numerically. Fuzzy attribute control charts allow flexibility in evaluating whether an item is conforming or nonconforming. Thus, it is preferred when there is ambiguity about the conformity of the item. In this chapter, crisp attribute control charts, fuzzy attribute control charts and some numerical examples are given.

Keywords Control charts for attributes · Fuzzy set theory · Fuzzy attribute control charts

3.1 Introduction

In manufacturing systems, variation is required to be caused only by chance. In statistics, it is called random variation and the process is said to be under control. However, commonly other causes may result in undesired variation of the process. In this case, process becomes out of control and the number of defects might increase substantially if the assignable causes are not identified and eliminated. Control charts are used as tools to detect assignable causes of variation. As a result, the process will be improved by the reduction of variation. Together with the decrease in the number of defects, productivity will be increased and the unnecessary process adjustments will be prevented.

S. Yanık (✉) · C. Kahraman
Department of Industrial Engineering, Istanbul Technical University,
34367 Macka Istanbul, Turkey
e-mail: sedayanik@itu.edu.tr

H. Yılmaz
Engineering Faculty Department of Industrial Engineering, Haliç University,
Istanbul, Turkey

© Springer International Publishing Switzerland 2016   71
C. Kahraman and S. Yanık (eds.), *Intelligent Decision Making in Quality Management*, Intelligent Systems Reference Library 97,
DOI 10.1007/978-3-319-24499-0_3

Attribute control charts are commonly used when the monitored quality characteristic may not be measured on a continuous scale. In such cases, nonconformities are monitored. Four types of attribute charts are used to define the nonconformities. p-charts deal with the ratio of the number of nonconforming items in a population to the total number of items in that population. As an alternative to calculating the fraction nonconforming, the np-charts track the number of nonconforming items that are directly observed. The p-chart and the np-chart are used to deal with nonconforming items while a c-chart tracks the total number of nonconformities in samples of constant size. However, when the sample size varies, u-chart is preferred.

Control charts are constructed using inspection data. Inspectors may sometimes have difficulty in identifying the attribute data as conforming or not. Due to such uncertainty of the attribute data, traditional control charts becomes insufficient. In this case, fuzzy control charts are pertinent control techniques used to capture this vagueness.

In this chapter, we aim to present an overview of the crisp and fuzzy attribute control charts. Based on the tightness of the inspection represented by a α-cut value, we then explore the α-cut fuzzy attribute control charts. The fuzzy attribute control charts is transformed to the crisp control charts via the fuzzy transformation techniques such as α-level fuzzy midrange, fuzzy median, fuzzy average and fuzzy mode. In this chapter, we review the fuzzy median transformation technique. We also present case studies from automobile supplier industry, garment production and paving tiles process.

The chapter is organized as follows: The traditional attribute control charts defined in Sect. 3.2. The fuzzy attribute control charts, α-cut fuzzy attribute control charts and α-level fuzzy median for α-cut fuzzy $\widetilde{np}$—control chart are presented in Sect. 3.3. Then, the conclusions are given in Sect. 3.4.

3.2 Control Charts for Attributes

A quality characteristic is an element defining the determined quality of a product or service (Mitra 2008). It can be physical characteristics as height, weight or time dependent characteristics as strength (Montgomery 2005). Attribute is a characteristic of an entity that cannot be measured quantitatively but can be observed with respect to its presence or absence (Braverman 1981). In quality control, the product with absent characteristic that does not meet certain prescribed specifications or standards is classified as a defective or a nonconforming item (Mitra 2008).

Attributes charts are generally not as informative as variables charts since the loss of information occurs during the classification process of items as conforming or nonconforming. However, they have a wide application area especially in service industries and in nonmanufacturing quality improvement efforts since the existence of numerous quality characteristics in these environments are not easily measured quantitatively (Montgomery 2005).

3.2.1 Control Chart for Fraction of Nonconformities: p-Chart

Ratio of the number of nonconforming items in a population to the total number of items in that population is described as the fraction nonconforming. The inspector may check items with one or several quality characteristics simultaneously and it is classified as nonconforming even if one of these characteristics of the item does not conform to standards.

The binomial distribution is the statistical basis underlying the control chart for fraction nonconforming. The assumption for the p-charts is that units are produced independently and there is a stable production process. Assume that p is the probability of a nonconforming unit with respect to the specifications. If a random sample of n units of product is selected, the probability of the number of units of product that are nonconforming is calculated using Eq. (3.1),

$$P\{D = x\} = \binom{n}{x} p^x (1 - p)^{n-x}, \quad x = 0, 1, \ldots, n \tag{3.1}$$

where D is the number of units of nonconforming products having a binomial distribution with parameters n and p. The mean of random variable D is np and variance of it is $np(1\text{-}np)$.

The ratio of the number units which are not conforming in the sample D to the sample size n, which is given in Eq. (3.2) is the sample fraction nonconforming;

$$\hat{p} = \frac{D}{n} \tag{3.2}$$

The mean and variance of $\hat{p}$ are calculated as using Eqs. (3.3) and (3.4);

$$\mu = p \tag{3.3}$$

and

$$\sigma_{\hat{p}}^2 = \frac{p(1 - p)}{n} \tag{3.4}$$

In some cases, the true fraction nonconforming p may be known or may be set as a specified standard value. In these cases, the control limits of the p-chart is formulated as Eqs. (3.5)–(3.7):

$$UCL = p + 3\sqrt{\frac{p(1-p)}{n}} \qquad (3.5)$$

$$Center\ Line(CL) = p \qquad (3.6)$$

$$LCL = p - 3\sqrt{\frac{p(1-p)}{n}} \qquad (3.7)$$

In process control, the sample fraction nonconforming $\hat{p}$ is monitored as samples of n units are taken. If there exist no observations of $\hat{p}$ beyond control limits, the process is in-control at the level of p. Otherwise, that the process fraction non-conforming level may be shifted to a new level.

The process fraction nonconforming p, may not be known in advance in some situations. In this case, it is estimated from the observed data. The estimation is conducted using n samples of each consisting of 20 or 25 observations. Assume D_i is the number of nonconforming units in sample i, the fraction nonconforming in the ith sample is computed using Eq. (3.8);

$$\hat{p}_i = \frac{D_i}{n} \quad i = 1, 2, \ldots, m \qquad (3.8)$$

Then, the average of the sample fractions nonconforming is calculated using Eq. (3.9)

$$\bar{p} = \frac{\sum_{i=1}^{m} D_i}{mn} = \frac{\sum_{i=1}^{m} \hat{p}_i}{m} \qquad (3.9)$$

Finally, $\bar{p}$ is used to estimate the fraction nonconforming p and the control limits of the fraction nonconforming control chart is calculated with Eqs. (3.10)–(3.12) (Montgomery 2005):

$$UCL = \bar{p} + 3\sqrt{\frac{\bar{p}(1-\bar{p})}{n}} \qquad (3.10)$$

$$Center\ Line(CL) = \bar{p} \qquad (3.11)$$

$$LCL = \bar{p} - 3\sqrt{\frac{\bar{p}(1-\bar{p})}{n}} \qquad (3.12)$$

Numerical example for p-charts
A company manufactures valves to use in automobiles. Ten samples of 10 valves each was taken from the production line and tested. The results are given in the following (Table 3.1):

The fraction defective is computed by dividing the number of defectives by the number in each sample, $n = 10$. The results are given in the following.

Table 3.1 Data for p-charts

Sample nr.	Nr. of defectives	Sample nr.	Nr. of defectives
1	4	6	1
2	1	7	1
3	0	8	2
4	0	9	3
5	2	10	0

Table 3.2 p-values of the samples

Sample no	1	2	3	4	5	6	7	8	9	10
p	0.4	0.1	0	0	0.2	0.1	0.1	0.2	0.3	0

From Table 3.2, the average proportion defective is calculated as $\bar{p} = 0.14$.
Using Eqs. (3.10)–(3.12), the control limits and center line are found as follows:

$$UCL = 0.14 + 3\sqrt{\frac{0.14 \times 0.86}{10}} = 0.4692$$

$$CL = 0.14$$

$$LCL = 0.14 - 3\sqrt{\frac{0.14 \times 0.86}{10}} = -0.1892 \rightarrow 0$$

Since all the data points in Table 3.2 are between the control limits, the obtained
trial control limits are now action control limits (Fig. 3.1).

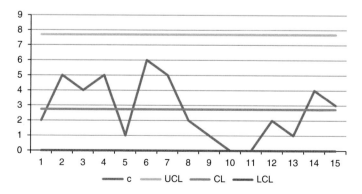

Fig. 3.1 Control chart for number of nonconformities

3.2.2 Control Chart for Number of Nonconformities: np-Chart

Another approach is directly using the number of nonconforming items for monitoring the process quality. Similar to the fraction nonconforming, the number of nonconforming items also is assumed to fit to a binomial distribution. This approach may be preferred rather than p-chart due to its ease of interpretation. The control limits of np-chart are computed using Eqs. (3.13)–(3.15).

$$UCL = np + 3\sqrt{np(1-p)} \tag{3.13}$$

$$UCL = np \tag{3.14}$$

$$UCL = np - 3\sqrt{np(1-p)} \tag{3.15}$$

In cases where sample sizes change, the control limits also change (Mitra 2008).

Numerical example for np-charts
The example in Sect. 3.2.1 will be handled for np-charts. The shortest way of obtaining an np-chart when you know the control limits of the p-chart is to multiply those limits by the sample size. Then we have

$$UCL = 0.4692 * 10 = 4.692$$

$$CL = 0.14 * 10 = 1.4$$

$$LCL = 0 * 10 = 0$$

Since all the data points in Table 3.1 are between the control limits, the obtained trial control limits are now action control limits.

3.2.3 Control Chart for Nonconformities: c-Chart

The p-chart and the np-chart are used to deal with nonconforming items while ac-chart tracks the total number of nonconformities in samples of constant size. However when the sample size varies, u-chart is preferred.

Assume that the defects or nonconformities occur in this inspection unit fit the Poisson distribution with the probability function as Eq. (3.16),

$$p(x) = \frac{e^{-\lambda}\lambda^x}{x!} \quad x = 0, 1, 2, \ldots \tag{3.16}$$

Where x is the number of nonconformities, λ is the parameter of the Poisson distribution and is greater than 0. Then the center line and control limits of control chart for number of nonconformities would be as Eqs. (3.17)–(3.19):

$$UCL = c + 3\sqrt{c} \tag{3.17}$$

$$Center\,Line(CL) = c \tag{3.18}$$

$$LCL = c - 3\sqrt{c} \tag{3.19}$$

Here it is assumed that a standard value for c is available. Also LCL is assumed 0 if the value of LCL is computed a negative value.

If there is not any given standard, c may be estimated as the observed average number of nonconformities in a preliminary sample of inspection units ($\bar{c}$). Then, the control chart parameters can be given as Eqs. (3.20)–(3.22),

$$UCL = \bar{c} + 3\sqrt{\bar{c}} \tag{3.20}$$

$$CenterLine(CL) = \bar{c} \tag{3.21}$$

$$LCL = \bar{c} - 3\sqrt{\bar{c}} \tag{3.22}$$

The control limits should be considered as trial control limits when no standard is given (Montgomery 2005).

Numerical example for c-charts

In a manufacturing paving-tile process, data regarding the number of nonconformities in 10 square meters of tiles is collected by the staff of quality control department. Type of defects such as scratch, burst, crack, stain and colour error are counted as nonconformity. Then, the total number of the nonconformities is noted as given Table 3.3.

From Table 3.3, the average proportion defective is calculated as $\bar{c} = 2,733$.

Table 3.3 Data for c-charts

Sample no	Nr. of nonconformities	Sample no	Nr. of nonconformities
1	2	9	1
2	5	10	0
3	4	11	0
4	5	12	2
5	1	13	1
6	6	14	4
7	5	15	3
8	2		

Using Eqs. (3.20)–(3.22), the control limits and center line are found as follows:

$$UCL = 2.733 + 3\sqrt{2.733} = 7.6932$$

$$Center\ Line(CL) = 2.733$$

$$LCL = 2.733 - 3\sqrt{2.733} = -2.2265$$

Since number of defects cannot be less than 0, LCL is accepted 0.

Since all the data points in Table 3.3 are between the control limits, the obtained trial control limits are now action control limits.

3.2.4 Control Charts for Nonconformities Per Unit: u-Charts

When the sample size is constant in an inspection, c-chart is employed to monitor and control the process. Operational constraints such as fluctuations in the availability of labour, machinery, and raw material may cause changes in the number of inspected items. When the sample size is not constant throughout the inspection, u-chart is preferred as an alternative to a c-chart.

A u-chart is employed to monitor the number of nonconformities per inspection unit. u_i is the fraction of the number of the nonconformities and sample size of the ith sample as in Eq. (3.23).

$$u_i = \frac{c_i}{n_i} \tag{3.23}$$

where c_i represents the number of nonconformities in the ith sample. And n_i is the sample size and it may vary throughout the inspection.

In a u-chart, the control limits change due to the different sizes of samples. However, the center line of a u-chart remains constant which allows meaningful comparisons to be made between subgroups of different sizes. The center line is constituted using the average number of nonconformities per inspection unit ($\bar{u}$) as in Eq. (3.24):

$$u = \frac{\sum_{i=1}^{m} c_i}{\sum_{i=1}^{m} n_i} = \frac{\sum_{i=1}^{m} u_i}{m} \tag{3.24}$$

where m is the number of samples.

Also, $\bar{u}$ is a Poisson random variable since it is a linear combination of independent and identically distributed Poisson random variables, c_i. Thus, the control limits are defined as in Eqs. (3.25)–(3.27);

Table 3.4 Data for u-charts

Sample Nr.	Sample size	Number of nonconformities	Sample Nr.	Sample size	Number of nonconformities
1	54	5	7	49	3
2	48	5	8	44	2
3	47	3	9	52	5
4	51	6	10	47	2
5	58	5	11	52	6
6	53	4	12	54	7

$$UCL = \bar{u} + 3\sqrt{\frac{\bar{u}}{n_i}} \tag{3.25}$$

$$Center\, Line(CL) = \bar{u} \tag{3.26}$$

$$LCL = \bar{u} - 3\sqrt{\frac{\bar{u}}{n_i}} \tag{3.27}$$

In order to construct a u-chart, the control limits have to be computed using the number of nonconformities. However, situations frequently arise where the inspector is not able to define the nonconformities precisely in an inspection. When the identification of the nonconformities becomes vague, representing the count of nonconformities as a single number will not be correct. When ill-defined structure of nonconformities is common in an inspection, using fuzzy numbers allows modelling the vagueness. Fuzzy set theory provides a framework and methods for incorporating vagueness in the u-charts.

Numerical example for u-charts

The production batches of a garment manufacturer in a specific day are monitored. The batches have different sample sizes and all the items of each batch are controlled with respect to its quality. Any type of defect is counted as nonconformity. Then, the total number of the nonconformities and the sample sizes are noted as given in Table 3.4.

The average fraction of the number of the nonconformities per sample size is computed by the total number of nonconformities and the total of the sample sizes of the batches as follows:

$$\bar{u} = \frac{\sum_{i=1}^{m} c_i}{\sum_{i=1}^{m} n_i} = \frac{609}{53} = 0.087$$

Then the center line and upper and lower control limited are calculated as in Table 3.5 and the control chart is depicted in Fig. 3.2.

Table 3.5 Parameters for u-chart

Sample Nr.	n_i	c_i	u_i	UCL	CL	LCL		
1	54	5	0.0926	0.2075	0.0870	−0.0334	→	0
2	48	5	0.1042	0.2148	0.0870	−0.0407	→	0
3	47	3	0.0638	0.2161	0.0870	−0.0421	→	0
4	51	6	0.1176	0.2110	0.0870	−0.0369	→	0
5	58	5	0.0862	0.2032	0.0870	−0.0292	→	0
6	53	4	0.0755	0.2086	0.0870	−0.0345	→	0
7	49	3	0.0612	0.2135	0.0870	−0.0394	→	0
8	44	2	0.0455	0.2204	0.0870	−0.0464	→	0
9	52	5	0.0962	0.2098	0.0870	−0.0357	→	0
10	47	2	0.0426	0.2161	0.0870	−0.0421	→	0
11	52	6	0.1154	0.2098	0.0870	−0.0357	→	0
12	54	7	0.1296	0.2075	0.0870	−0.0334	→	0

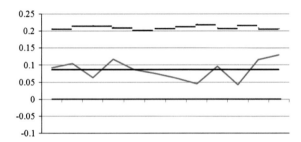

Fig. 3.2 Control chart for number of nonconformities per unit

3.3 Fuzzy Control Charts for Attributes

In this section, we develop the fuzzy $\tilde{p}$-control chart, $\tilde{u}$ control chart, $\tilde{c}$ control chart and $\widetilde{np}$-control chart.

3.3.1 Fuzzy Numbers and Fuzzy Control Charts

The attribute data might not be certain as for all types of processes. Situations frequently arise where nonconformities cannot be defined as a precise value. When the identification of the nonconformities becomes vague, representing it as linguistic values becomes more meaningful.

In such cases, it is necessary to adapt the fuzzy sets connected to linguistic values into scalars referred to as representative values to retain the standard format of control charts and to facilitate the plotting of observations on the chart. This adaptation may

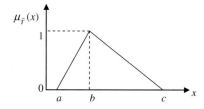

Fig. 3.3 A triangular fuzzy number $\tilde{T}$

be performed in various ways as long as the result is instinctively representative of the range of the base variable included in the fuzzy set. Fuzzy control charts are pertinent control techniques used to capture this vagueness.

Fuzzy set theory provides a framework and methods for incorporating vagueness. Hence, we have identified the nonconformities as triangular fuzzy numbers to construct fuzzy attribute control charts in this chapter.

The triangular fuzzy numbers defined in fuzzy set theory is as follows:

If R is a set of real numbers $F(R) = \{A|A: R \rightarrow [0, 1]$, when A is a continuous function$\}$, $F_T(R) = \{T_{a,b,c}|a, b, c \in R, a \leq b \leq c\}$, such,

$$\tilde{T}_{a,b,c}(x) = \begin{cases} (x-a)/(b-a), & \text{if} \quad a \leq x < b \\ (c-x)/(c-b), & \text{if} \quad b \leq x < c \\ 0, & \text{otherwise} \end{cases}$$

The membership function of the triangular fuzzy number $\tilde{T}$ is depicted in Fig. 3.3.

3.3.2 Fuzzy $\tilde{p}$-Control Chart

In a process using monitoring with p-control charts or np-control charts, due to the uncertainty of the attribute data, traditional control charts becomes insufficient. In this case, fuzzy control charts are pertinent control techniques used to capture this vagueness. Erginel (2013) construct fuzzy $\tilde{p}$ and $\widetilde{np}$ control charts using decision rules. In her paper, fully fuzzy control charts are introduced for $\tilde{p}$ control charts based on both constant sample size and variable sample size, and $\widetilde{np}$ control charts are subsequently introduced using decision rules for the process state conditions. Thus "rather in control" and "rather out of control" decisions can be considered for monitoring the process.

In the fuzzy case, triangular fuzzy number $(d_{a_j}, d_{b_j}, d_{c_j})$ and $(p_{a_j}, p_{b_j}, p_{c_j})$ is used to represent the number of nonconforming units and the fraction nonconforming, respectively. The fraction nonconforming is computed as Eq. (3.28). Also the fuzzy averages of the fraction nonconforming is denoted by (p_a, p_b, p_c) and calculated using Eq. (3.29), where $j = 1, 2, \ldots m$.

$$p_{a_j} = \frac{d_{a_j}}{n}, \quad p_{b_j} = \frac{d_{b_j}}{n}, \quad p_{c_j} = \frac{d_{c_j}}{n} \tag{3.28}$$

$$\bar{p}_a = \frac{\sum p_{a_j}}{m}, \quad \bar{p}_b = \frac{\sum p_{b_j}}{m}, \quad \bar{p}_c = \frac{\sum p_{c_j}}{m} \tag{3.29}$$

Fuzzy center line, fuzzy upper and fuzzy lower limits of fuzzy $\tilde{p}$-control chart are calculated as Eqs. (3.30)–(3.32) (Kahraman et al. 2010):

$$\widetilde{UCL}_p = \left(\bar{p}_a + 3\sqrt{\frac{\bar{p}_a(1 - \bar{p}_a)}{n}}, \quad \bar{p}_b + 3\sqrt{\frac{\bar{p}_b(1 - \bar{p}_b)}{n}}, \quad \bar{p}_c + 3\sqrt{\frac{\bar{p}_c(1 - \bar{p}_c)}{n}} \right) \tag{3.30}$$

$$\widetilde{CL}_p = (\bar{p}_a, \bar{p}_b, \bar{p}_c) \tag{3.31}$$

$$\widetilde{LCL}_p = (\bar{p}_a - 3\sqrt{\frac{\bar{p}_a(1 - \bar{p}_a)}{n}}, \quad \bar{p}_b - 3\sqrt{\frac{\bar{p}_b(1 - \bar{p}_b)}{n}}, \quad \bar{p}_c - 3\sqrt{\frac{\bar{p}_c(1 - \bar{p}_c)}{n}}) \tag{3.32}$$

Numerical example for fuzzy $\tilde{p}$-chart

A company manufactures valves to use in automobiles. Ten samples of 10 valves each was taken from the production line and tested. In this case the numbers of defectives are not certain since a product in a sample may be accepted as a defective or a nondefective. Defectiveness is a matter of degree in this case. The results are given in Table 3.6.

The average defective ratio is calculated using Eq. (3.29) and the control limits using Eqs. (3.30)–(3.32):

$$\widetilde{CL} = \tilde{\bar{p}} = \left(\frac{7}{100}, \frac{14}{100}, \frac{24}{100} \right) = (0.07, 0.14, 0.24)$$

Table 3.6 Data for $\tilde{u}$ chart

Sample nr.	Nr. of defectives	TFN	Sample nr.	Nr. of defectives	TFN
1	$\tilde{4}$	(3, 4, 5)	6	$\tilde{1}$	(0, 1, 2)
2	$\tilde{1}$	(0, 1, 2)	7	$\tilde{1}$	(0, 1, 2)
3	$\tilde{0}$	(0, 0, 1)	8	$\tilde{2}$	(1, 2, 3)
4	$\tilde{0}$	(0, 0, 1)	9	$\tilde{3}$	(2, 3, 4)
5	$\tilde{2}$	(1, 2, 3)	10	$\tilde{0}$	(0, 0, 1)

$$\widetilde{UCL}_p = (0.07 + 3\sqrt{\frac{0.07(1-0.07)}{10}}, 0.14 + 3\sqrt{\frac{0.14(1-0.14)}{10}}, 0.24$$
$$+ 3\sqrt{\frac{0.24(1-0.24)}{10}})$$

$$\widetilde{UCL}_p = (0.3121, 0.4692, 0.6452)$$

$$\widetilde{LCL}_p = (0.07 - 3\sqrt{\frac{0.07(1-0.07)}{10}}, 0.14 - 3\sqrt{\frac{0.14(1-0.14)}{10}}, 0.24$$
$$- 3\sqrt{\frac{0.24(1-0.24)}{10}})$$

$$\widetilde{LCL}_p \rightarrow (0,0,0)$$

If the sample size is not approximate to each other, it is required that the individual sample size for each of them should be used. The fuzzy fraction nonconforming for each sample and their fuzzy averages are calculated as Eqs. (3.33) and (3.34);

$$p_{a_j} = \frac{d_{a_j}}{n_j}, \quad p_{b_j} = \frac{d_{b_j}}{n_j}, \quad p_{c_j} = \frac{d_{c_j}}{n_j} \tag{3.33}$$

$$\bar{p}_a = \frac{\sum d_{a_j}}{\sum n_j}, \bar{p}_b = \frac{\sum d_{b_j}}{\sum n_j}, \bar{p}_c = \frac{\sum d_{c_j}}{\sum n_j} \tag{3.34}$$

where n_j is the jth sample size and $j = 1, 2, ..., m$.

The control limits are calculated in fuzzy $\widetilde{p}$-control chart for each n_j by using triangular membership functions and fuzzy averages of sample fraction nonconforming as Eqs. (3.35)–(3.37):

$$\widetilde{CL}_{p,j} = (\bar{p}_a, \bar{p}_b, \bar{p}_c) \tag{3.35}$$

$$\widetilde{UCL}_{p,j} = (\bar{p}_a + 3\sqrt{\frac{\bar{p}_a(1-\bar{p}_a)}{n_j}}, \quad \bar{p}_b + 3\sqrt{\frac{\bar{p}_b(1-\bar{p}_b)}{n_j}}, \quad \bar{p}_c + 3\sqrt{\frac{\bar{p}_c(1-\bar{p}_c)}{n_j}})$$
$$\tag{3.36}$$

$$\widetilde{LCL}_{p,j} = (\bar{p}_a - 3\sqrt{\frac{\bar{p}_a(1-\bar{p}_a)}{n_j}}, \quad \bar{p}_b - 3\sqrt{\frac{\bar{p}_b(1-\bar{p}_b)}{n_j}}, \quad \bar{p}_c - 3\sqrt{\frac{\bar{p}_c(1-\bar{p}_c)}{n_j}})$$
$$\tag{3.37}$$

3.3.2.1 α-Cut Fuzzy $\tilde{p}$-Control Charts

The mean of α-cut is defined with the elements having membership degrees greater than equal to α. The α-cuts of $\bar{p}_a^\alpha$ and $\bar{p}_c^\alpha$ are computed using Eq. (3.38):

$$\bar{p}_a^\alpha = \bar{p}_a + \propto (\bar{p}_b - \bar{p}_a) \quad \text{and} \quad \bar{p}_c^\alpha = \bar{p}_c - \propto (\bar{p}_c - \bar{p}_b) \tag{3.38}$$

The control limits of α-cut fuzzy $\tilde{p}$—are calculated with the Eqs. (3.39)–(3.41):

$$\widetilde{UCL}_p^\alpha = (\bar{p}_a^\alpha + 3\sqrt{\frac{\bar{p}_a^\alpha(1-\bar{p}_a^\alpha)}{n}}, \quad \bar{p}_b + 3\sqrt{\frac{\bar{p}_b(1-\bar{p}_b)}{n}}, \quad \bar{p}_c^\alpha + 3\sqrt{\frac{\bar{p}_c^\alpha(1-\bar{p}_c^\alpha)}{n}})$$

$$\tag{3.39}$$

$$\widetilde{CL}_p^\alpha = (\bar{p}_a^\alpha, \bar{p}_b, \bar{p}_c^\alpha) \tag{3.40}$$

$$\widetilde{LCL}_p^\alpha = (\bar{p}_a^\alpha - 3\sqrt{\frac{\bar{p}_a^\alpha(1-\bar{p}_a^\alpha)}{n}}, \quad \bar{p}_b - 3\sqrt{\frac{\bar{p}_b(1-\bar{p}_b)}{n}}, \quad \bar{p}_c^\alpha - 3\sqrt{\frac{\bar{p}_c^\alpha(1-\bar{p}_c^\alpha)}{n}})$$

$$\tag{3.41}$$

when sample sizes vary, α-cut control limits for fuzzy $\tilde{p}$-control chart are formulated by Eqs. (3.42)–(3.44):

$$\widetilde{UCL}_{p,j}^\alpha = (\bar{p}_a^\alpha + 3\sqrt{\frac{\bar{p}_a^\alpha(1-p_a^\alpha)}{n_j}}, \quad \bar{p}_b + 3\sqrt{\frac{\bar{p}_b(1-\bar{p}_b)}{n_j}}, \quad \bar{p}_c^\alpha + 3\sqrt{\frac{\bar{p}_c^\alpha(1-\bar{p}_c^\alpha)}{n_j}})$$

$$\tag{3.42}$$

$$\widetilde{CL}_{p,j}^\alpha = (\bar{p}_a^\alpha, \bar{p}_b, \bar{p}_c^\alpha) \tag{3.43}$$

$$\widetilde{LCL}_{p,j}^\alpha = (\bar{p}_a^\alpha - 3\sqrt{\frac{\bar{p}_a^\alpha(1-\bar{p}_a^\alpha)}{n_j}}, \bar{p}_b - 3\sqrt{\frac{\bar{p}_b(1-\bar{p}_b)}{n_j}}, \bar{p}_c^\alpha - 3\sqrt{\frac{\bar{p}_c^\alpha(1-\bar{p}_c^\alpha)}{n_j}})$$

$$\tag{3.44}$$

3.3.2.2 α-Level Fuzzy Median for α-Cut Fuzzy $\tilde{p}$-Control Chart

Fuzzy transformation techniques are used for obtaining crisp numbers from the fuzzy fraction nonconforming. α-level fuzzy midrange, fuzzy median, fuzzy average and fuzzy mode are the transformation techniques (Wang and Raz 1990). Fuzzy median transformation technique is used for transformation in this study.

Gulbay et al. (2004) stated that the fuzzy median (f_{med}) is expressed by the following Eq. (3.45):

$$\int_{a_\infty}^{f_{med}} \mu_F(x)dx = \int_{f_{med}}^{b_\infty} \mu_F(x)dx = \frac{1}{2}\int_{a_\infty}^{b_\infty} \mu_F(x)dx \tag{3.45}$$

where a and b are the end points in the base variable of the fuzzy set F such that $a < b$. For a sample j, α-level fuzzy median value ($S^\infty_{med-p,j}$) is calculated as Eq. (3.46):

$$S^\infty_{med-p,j} = \frac{1}{3}\left(p^\infty_{a,j} + p_{b,j} + p^\infty_{c,j}\right), j = 1,2,\ldots,m \tag{3.46}$$

The fuzzy control limits of α-level fuzzy median for α-cut fuzzy $\tilde{p}$-control chart are obtained as Eqs. (3.47)–(3.49) by using these formulations:

$$CL^\infty_{med-p} = \frac{1}{3}\left(p^\infty_a + p_b + p^\infty_c\right) \tag{3.47}$$

$$UCL^\infty_{med-p} = CL^\infty_{med-p} + 3\sqrt{\frac{CL^\infty_{med-p}(1 - CL^\infty_{med-p})}{n}} \tag{3.48}$$

$$LCL^\infty_{med-p} = CL^\infty_{med-p} - 3\sqrt{\frac{CL^\infty_{med-p}(1 - CL^\infty_{med-p})}{n}} \tag{3.49}$$

The condition of process control for each sample is given as:

$$Process\,control = \begin{cases} in\,control,\,for\,LCL^\infty_{med-p} \leq S^\infty_{med-p,j} \leq UCL^\infty_{med-p} \\ out\,of\,control,\,for\,otherwise \end{cases} \tag{3.50}$$

Upper, center and lower control limits of α-level fuzzy median for α-cut fuzzy $\tilde{p}$-control chart based on variable sample size are calculated by using fuzzy median transformation technique as Eqs. (3.51)–(3.53):

$$UCL^\infty_{med-p,j} = CL^\infty_{med-p} + 3\sqrt{\frac{CL^\infty_{med-p}(1 - CL^\infty_{med-p})}{n_j}} \tag{3.51}$$

$$CL^\infty_{med-p} = \frac{1}{3}\left(\bar{p}^\infty_a + \bar{p}_b + \bar{p}^\infty_c\right) \tag{3.52}$$

$$LCL^\infty_{med-p,j} = CL^\infty_{med-p} - 3\sqrt{\frac{CL^\infty_{med-p}(1 - CL^\infty_{med-p})}{n_j}} \tag{3.53}$$

α-level fuzzy median value for each sample is given as Eq. (3.54):

$$S_{med-p,j}^{\alpha} = \frac{1}{3}\left(p_{a,j}^{\alpha} + p_{b,j} + p_{c,j}^{\alpha}\right), j = 1, 2, \ldots, m \qquad (3.54)$$

The condition of process control for each sample is defined as Eq. (3.55):

$$Process\ control = \begin{cases} in\ control,\ for\ LCL_{med-p,j}^{\alpha} \leq S_{med-p,j}^{\alpha} \leq UCL_{med-p,j}^{\alpha} \\ out\ of\ control,\ for\ otherwise \end{cases} \qquad (3.55)$$

3.3.3 Fuzzy $\widetilde{np}$-Control Chart

In the fuzzy case, a triangular fuzzy number $(d_{a_j}, d_{b_j}, d_{c_j})$ represents the number of nonconforming units for each sample. A triangular fuzzy number $(n\bar{p}_a, n\bar{p}_b, n\bar{p}_c)$ denotes the average sample number of nonconforming units as Eq. (3.56):

$$n\bar{p}_a = \frac{\sum_{j=1}^{m} d_{a_j}}{m}, \quad n\bar{p}_b = \frac{\sum_{j=1}^{m} d_{b_j}}{m}, \quad n\bar{p}_c = \frac{\sum_{j=1}^{m} d_{c_j}}{m} \qquad (3.56)$$

The limits of fuzzy $\widetilde{np}$-control chart are obtained with the following Eqs. (3.57)–(3.59);

$$\widetilde{UCL}_{np} = \left(n\bar{p}_a + 3\sqrt{n\bar{p}_a(1 - n\bar{p}_a)}, n\bar{p}_b + 3\sqrt{n\bar{p}_b(1 - n\bar{p}_b)}, n\bar{p}_c + 3\sqrt{n\bar{p}_c(1 - n\bar{p}_c)}\right) \qquad (3.57)$$

$$\widetilde{CL}_{np} = \left(n\bar{p}_a, n\bar{p}_b, n\bar{p}_c\right) \qquad (3.58)$$

$$\widetilde{LCL}_{np} = \left(n\bar{p}_a - 3\sqrt{n\bar{p}_a(1 - n\bar{p}_a)}, n\bar{p}_b - 3\sqrt{n\bar{p}_b(1 - n\bar{p}_b)}, n\bar{p}_c - 3\sqrt{n\bar{p}_c(1 - n\bar{p}_c)}\right) \qquad (3.59)$$

Numerical example for fuzzy $\widetilde{np}$ chart
Consider the example given in Sect. 3.3.2. In this case, the obtained control limits for fuzzy $\tilde{p}$ chart in Sect. 3.3.2 can be easily converted to fuzzy $\widetilde{np}$ chart control limits by multiplying these limits by the sample size:

$$\widetilde{CL}_{np} = n\tilde{p} = 10 \times (0.07, 0.14, 0.24) = (0.7, 1.4, 2.4)$$

$$\widetilde{UCL}_{np} = n \times \widetilde{UCL}_p = 10 \times (0.3121, 0.4692, 0.6452) = (3.121, 4.692, 6.452)$$

$$\widetilde{LCL}_{np} = n \times \widetilde{LCL}_p \rightarrow (0, 0, 0)$$

3.3.3.1   α-Cut Fuzzy $\widetilde{np}$ Control Chart

The limits of α-cut fuzzy $\widetilde{np}$-control chart are calculated as Eqs. (3.60)–(3.62):

$$\widetilde{UCL}_{np}^{\alpha} = \left(n\bar{p}_a^{\alpha} + 3\sqrt{n\bar{p}_a^{\alpha} + \left(1 - n\bar{p}_a^{\alpha}\right)}, n\bar{p}_b + 3\sqrt{n\bar{p}_b(1 - n\bar{p}_b)}, n\bar{p}_c^{\alpha} + 3\sqrt{n\bar{p}_c^{\alpha}\left(1 - n\bar{p}_c^{\alpha}\right)} \right)$$

$$(3.60)$$

$$\widetilde{CL}_{np} = \left(n\bar{p}_a^{\alpha}, n\bar{p}_b, n\bar{p}_c \right) \tag{3.61}$$

$$\widetilde{LCL}_{np} = \left(n\bar{p}_a^{\alpha} - 3\sqrt{n\bar{p}_a^{\alpha}\left(1 - n\bar{p}_a^{\alpha}\right)}, n\bar{p}_b - 3\sqrt{n\bar{p}_b(1 - n\bar{p}_b)}, n\bar{p}_c^{\alpha} \right.$$
$$\left. - 3\sqrt{n\bar{p}_c^{\alpha}\left(1 - n\bar{p}_c^{\alpha}\right)} \right) \tag{3.62}$$

Numerical example for α-Cut fuzzy $\widetilde{np}$-control chart
A company manufactures valves to use in automobiles. Ten samples of 10 valves each was taken from the production line and tested. In this case the numbers of defectives are not certain since a product in a sample may be accepted as a defective or a non-defective. Defectiveness is a matter of degree in this case. The results are given in Table 3.7:

First we will calculate the average number of defectives. This is also equal to the center line.

$$\widetilde{CL}_{np}^{\alpha} = \widetilde{np} = \left(\frac{7 + 7\alpha}{10}, \frac{24 - 10\alpha}{10} \right) = (0.7 + 0.7\alpha; 2.4 - \alpha)$$

$$\tilde{p} = \left(\frac{7 + 7\alpha}{100}, \frac{24 - 10\alpha}{100} \right) = (0.07 + 0.07\alpha; 0.24 - 0.1\alpha)$$

$$\tilde{q} = \left(1 - \frac{7 + 7\alpha}{100}, 1 - \frac{24 - 10\alpha}{100} \right) = (0.93 - 0.07\alpha; 0.76 + 0.1\alpha)$$

$$\widetilde{UCL}_{np}^{\alpha} = \left((0.7 + 0.7\alpha) + 3\sqrt{10(0.07 + 0.07\alpha)(0.93 - 0.07\alpha)}, \right.$$
$$\left. (2.4 - \alpha) + 3\sqrt{10(0.24 - 0.1\alpha)(0.76 + 0.1\alpha)} \right)$$

$$\widetilde{LCL}_{np}^{\alpha} = \left((0.7 + 0.7\alpha) - 3\sqrt{10(0.7 + 0.7\alpha)(0.93 - 0.07\alpha)}, \right.$$
$$\left. (2.4 - \alpha) - 3\sqrt{10(0.24 - 0.1\alpha)(0.76 + 0.1\alpha)} \right)$$

Table 3.7 Data and α-cut values of $\widetilde{np}$ control chart

Sample nr.	Nr. of defectives	α-Cut	Sample nr.	Nr. of defectives	α-Cut
1	$\tilde{4}$	(3 + α; 5-α)	6	$\tilde{1}$	(α; 2-α)
2	$\tilde{0}$	(α; 2-α)	7	$\tilde{1}$	(α; 2-α)
3	$\tilde{0}$	(0; 1-α)	8	$\tilde{2}$	(1 + α; 3-α)
4	$\tilde{0}$	(0; 1-α)	9	$\tilde{3}$	(2 + α; 4-α)
5	$\tilde{2}$	(1 + α; 3-α)	10	$\tilde{0}$	(0; 1-α)

These α-Cut control limits give the same fuzzy numbers of fuzzy control limits for $\alpha = 0$, $\alpha = 1$ and $\alpha = 0$, respectively:

$$\widetilde{CL}_{np}^{\propto} = (0.7, 1.4, 2.4)$$

$$\widetilde{UCL}_{np}^{\propto} = (0.7 + 3\sqrt{10 \times 0.07 \times 0.93}; 1.4 + 3\sqrt{10 \times 0.14 \times 0.86}; 2.4$$
$$+ 3.\sqrt{10 \times 0.24 \times 0.76}) = (3.121; 4.692; 6.452)$$
$$\widetilde{LCL}_{np}^{\propto} = (0.7 - 3\sqrt{10 \times 0.07 \times 0.93}; 1.4 - 3\sqrt{10 \times 0.14 \times 0.86}; 2.4$$
$$- 3.\sqrt{10 \times 0.24 \times 0.76}) \rightarrow (0; 0; 0)$$

3.3.3.2 α-Level Fuzzy Median for α-Cut Fuzzy $\widetilde{np}$-Control Chart Based on Constant Sample Size

The control limits of α-cut fuzzy $\widetilde{np}$-control chart with α-level fuzzy median is given by Eqs. (3.63)–(3.65):

$$UCL_{med-np}^{\propto} = CL_{med-np}^{\propto} + 3\sqrt{CL_{med-np}^{\propto} \frac{(1 - CL_{med-np}^{\propto})}{n}} \qquad (3.63)$$

$$CL_{med-np}^{\propto} = \frac{1}{3}\left(n\overline{p}_a^{\propto} + p_b + n\overline{p}_c^{\propto}\right) \qquad (3.64)$$

$$LCL_{med-np}^{\propto} = CL_{med-np}^{\propto} - 3\sqrt{CL_{med-np}^{\propto} \frac{(1 - CL_{med-np}^{\propto})}{n}} \qquad (3.65)$$

α-level fuzzy median value $S_{med-np,j}^{\propto}$ for jth sample is calculated as Eq. (3.66):

$$S_{med-np,j}^{\propto} = \frac{1}{3}\left(np_{a,j}^{\propto} + np_{b,j} + np_{c,j}^{\propto}\right) \qquad (3.66)$$

The condition of process control for each sample is given as Eq. (3.67):

$$Process\,control = \begin{cases} incontrol, for\,LCL_{med-np,j}^{\propto} \le S_{med-np,j}^{\propto} \le UCL_{med-np,j}^{\propto} \\ out\,of\,control, for\,otherwise \end{cases} \qquad (3.67)$$

3.3.4 Fuzzy $\widetilde{c}$ Control Chart

Trapezoidal fuzzy number (a, b, c, d) or a triangular fuzzy number (a, b, b, d) can be used for representing each sample or subgroup in the fuzzy case. Also it is obvious that a trapezoidal fuzzy number transforms triangular when $b = c$. A triangular fuzzy number is also represented as a trapezoidal fuzzy number like (a, b, b, d) or (a, c, c, d) for the simplicity of representation and calculation. $(\bar{a}, \bar{b}, \bar{c}, \bar{d})$ is the representation of center line $(\widetilde{CL})$ that is the mean of fuzzy samples and $\bar{a}, \bar{b}, \bar{c}, \bar{d}$ are the arithmetic means of the values a, b, c, and d, respectively. Center line can be written in the fuzzy case as Eq. (3.68).

$$\widetilde{CL} = \left(\frac{\sum_{j=1}^{n} a_j}{n}, \frac{\sum_{j=1}^{n} b_j}{n}, \frac{\sum_{j=1}^{n} c_j}{n}, \frac{\sum_{j=1}^{n} d_j}{n} \right) = (\bar{a}, \bar{b}, \bar{c}, \bar{d}) \tag{3.68}$$

$\widetilde{CL}$ can be represented by a fuzzy number and $[\bar{b}, \bar{c}]$ is the the closed interval of its fuzzy mode(Gülbay and Kahraman 2006, 2007). $\widetilde{LCL}$ $\widetilde{CL}$ and $\widetilde{UCL}$ are expressed in formulas as Eqs. (3.69)–(3.71):

$$\widetilde{CL} = (\bar{a}, \bar{b}, \bar{c}, \bar{d}) = (CL_1, CL_2, CL_3, CL_4) \tag{3.69}$$

$$\widetilde{LCL} = \widetilde{CL} - 3\sqrt{\widetilde{CL}} = \left(\bar{a} - 3\sqrt{\bar{a}}, \bar{b} - 3\sqrt{\bar{b}}, \bar{c} - 3\sqrt{\bar{c}}, \bar{d} - 3\sqrt{\bar{d}} \right)$$
$$= (LCL_1, LCL_2, LCL_3, LCL_4) \tag{3.70}$$

$$\widetilde{UCL} = \widetilde{CL} + 3\sqrt{\widetilde{CL}} = \left(\bar{a} + 3\sqrt{\bar{a}}, \bar{b} + 3\sqrt{\bar{b}}, \bar{c} + 3\sqrt{\bar{c}}, \bar{d} + 3\sqrt{\bar{d}} \right)$$
$$= (UCL_1, UCL_2, UCL_3, UCL_4) \tag{3.71}$$

Also fuzzy control limits can be rewritten using α-cut representations as Eqs. (3.72)–(3.74).

$$\widetilde{CL^{\alpha}} = \left(\bar{a}^{\alpha}, \bar{b}, \bar{c}, \bar{d}^{\alpha} \right) = (CL_1^{\alpha}, CL_2, CL_3, CL_4^{\alpha}) \tag{3.72}$$

$$\widetilde{LCL^{\alpha}} = \widetilde{CL^{\alpha}} - 3\sqrt{\widetilde{CL^{\alpha}}} = \left(\bar{a}^{\alpha} - 3\sqrt{\bar{a}^{\alpha}}, \bar{b} - 3\sqrt{\bar{b}}, \bar{c} - 3\sqrt{\bar{c}}, \bar{d}^{\alpha} - 3\sqrt{\bar{d}^{\alpha}} \right)$$
$$= (LCL_1^{\alpha}, LCL_2, LCL_3, LCL_4^{\alpha}) \tag{3.73}$$

$$\widetilde{UCL^{\alpha}} = \widetilde{CL^{\alpha}} + 3\sqrt{\widetilde{CL^{\alpha}}} = \left(\bar{a}^{\alpha} + 3\sqrt{\bar{a}^{\alpha}}, \bar{b} + 3\sqrt{\bar{b}}, \bar{c} + 3\sqrt{\bar{c}}, \bar{d}^{\alpha} + 3\sqrt{\bar{d}^{\alpha}} \right)$$
$$= (UCL_1^{\alpha} UCL_2, UCL_3, UCL_4^{\alpha}) \tag{3.74}$$

While a^{α} and d^{α} are calculated as Eqs. (3.75)–(3.76):

Fig. 3.4 Representation of fuzzy control limits (Gülbay and Kahraman 2006a, b)

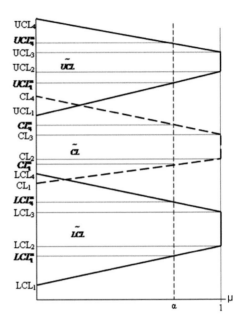

$$a^{\propto} = a + \propto (b - a) \tag{3.75}$$

$$d^{\propto} = d - \propto (d - c) \tag{3.76}$$

Figure 3.4 illustrates results of these equations. The value of $\propto$-cut can be defined by the quality manager and described as the tightness of the inspection. The higher $\propto$ value means the tighter inspection. Also if $\propto$-cut value is chosen as 1, control chart turns into the classical case (Gulbay and Kahraman 2007).

It is necessary to adapt the fuzzy sets connected to linguistic values into scalars referred to as representative values to retain the standard format of control charts and to facilitate the plotting of observations on the chart. This adaptation may be performed in various ways as long as the result is instinctively representative of the range of the base variable included in the fuzzy set. Four ways are fuzzy mode, α-level fuzzy midrange, fuzzy median, and fuzzy average. In addition, these methods are similar with regard to the measures of central tendency used in descriptive statistics. It should be indicated that there is not any theoretical basis for selection of these methods. Ease of computation or preference of the user should be considered (Wang and Raz 1990). Conversion of fuzzy sets into crisp values causes loss of information in linguistic data. Keeping fuzzy sets as themselves and comparing fuzzy samples with the fuzzy control limits are preferred to retain information of linguistic data. Hence, a method that is based on the area measurement called direct fuzzy approach (DFA) is suggested for the fuzzy control charts. α- level fuzzy control limits $\widetilde{UCL}^{\propto}$, $\widetilde{CL}^{\propto}$ and $\widetilde{LCL}^{\propto}$ can be formulated with fuzzy arithmetic as follows.

Numerical example for fuzzy $\widetilde{c}$ chart

In a manufacturing paving tile process, data regarding the number of nonconformities per 10 square meters of tiles is collected by quality control department. Types of defects such as scratch, burst, crack, stain, colour error are counted as nonconformities. In this case the numbers of defects are not certain since it depends on the controller's decision, thus it involves vagueness for a controller. Then, the total number of the nonconformities is given in Table 3.8.

Thus, the number of nonconformities can be expressed as fuzzy numbers and the center line, the upper and lower limits are computed as fuzzy numbers as given in Table 3.8. The control chart with fuzzy limits is illustrated in Fig. 3.5.

$$\widetilde{CL}_c = (\bar{a}, \bar{b}, \bar{c}) = \left(\frac{28}{15}, \frac{41}{15}, \frac{56}{15}\right) = (1.8667, 2.7333, 3.7333)$$

$$\widetilde{LCL}_c = \left(1.8667 - 3\sqrt{1.8667}, 2.7333 - 3\sqrt{2.7333}, 3.7333 - 3\sqrt{3.773}\right)$$

$$\widetilde{LCL}_c \rightarrow (0,0,0)$$

$$\widetilde{UCL}_c = \left(1.8667 + 3\sqrt{1.8667}, 2.7333 + 3\sqrt{2.7333}, 3.7333 + 3\sqrt{3.7333}\right)$$

$$\widetilde{UCL}_c = (5.9655, 7.6932, 9.5299)$$

Also, fuzzy control limits can be revised with α-cut representations as in Table 3.9. Here α equals to 0.7. The α-cut of the control limits' membership functions are shown in Fig. 3.6.

$$\widetilde{CL}^\alpha = (\bar{a}^\alpha, \bar{b}, \bar{d}^\alpha) = \left(\frac{37.1}{15}, \frac{41}{15}, \frac{45.5}{15}\right) = (2.4733, 2.7333, 3.0333)$$

Table 3.8 Data for $\widetilde{c}$ control chart

Sample nr.	Nr. of nonconformities	TFN	Sample nr.	Nr. of nonconformities	TFN
1	$\widetilde{2}$	(1, 2, 3)	9	$\widetilde{1}$	(0, 1, 2)
2	$\widetilde{5}$	(4, 5, 6)	10	$\widetilde{0}$	(0, 0, 1)
3	$\widetilde{4}$	(3, 4, 5)	11	$\widetilde{0}$	(0, 0, 1)
4	$\widetilde{5}$	(4, 5, 6)	12	$\widetilde{2}$	(1, 2, 3)
5	$\widetilde{1}$	(0, 1, 2)	13	$\widetilde{1}$	(0, 1, 2)
6	$\widetilde{6}$	(5, 6, 7)	14	$\widetilde{4}$	(3, 4, 5)
7	$\widetilde{5}$	(4, 5, 6)	15	$\widetilde{3}$	(2, 3, 4)
8	$\widetilde{2}$	(1, 2, 3)			

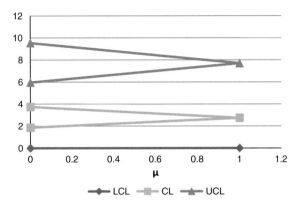

Fig. 3.5 Fuzzy membership functions of the control limits

Table 3.9 α-cut of $\tilde{c}$ control chart

Sample nr.	c	c_a	c_b	c_c	a^α	c^α
1	2	1	2	3	1.7	2.3
2	5	4	5	6	4.7	5.3
3	4	3	4	5	3.7	4.3
4	5	4	5	6	4.7	5.3
5	1	0	1	2	0.7	1.3
6	6	5	6	7	5.7	6.3
7	5	4	5	6	4.7	5.3
8	2	1	2	3	1.7	2.3
9	1	0	1	2	0.7	1.3
10	0	0	0	1	0	0.3
11	0	0	0	1	0	0.3
12	2	1	2	3	1.7	2.3
13	1	0	1	2	0.7	1.3
14	4	3	4	5	3.7	4.3
15	3	2	3	4	2.7	3.3

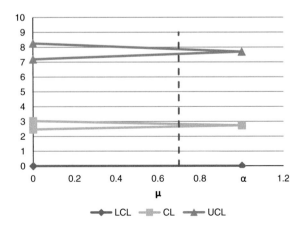

Fig. 3.6 α-cut of fuzzy control limits

$$\widetilde{LCL}^\propto = \left(2.4733 - 3\sqrt{2.4733}, 2.7333 - 3\sqrt{2.7333}, 3.0333 - 3\sqrt{3.0333}\right)$$

$$\widetilde{LCL}^\propto \rightarrow (0,0,0)$$

$$\widetilde{UCL}^\propto = \left(2.4733 + 3\sqrt{2.4733}, 2.7333 - 3\sqrt{2.7333}, 3.0333 + 3\sqrt{3.0333}\right)$$

$$\widetilde{UCL}^\propto = (7.1914, 7.6932, 8.2583)$$

The percentage area of the sample which remains inside the $\widetilde{UCL}$ and/or $\widetilde{LCL}$ defined asfuzzy numbers can be used to determine whether the process is in control or not. The process is said to be "in-control" when the fuzzy sample is totally involved by the fuzzy control limits while the process is said to be "out of control" when a fuzzy sample is completely excluded by the fuzzy control limits. When a sample is partially included by the fuzzy control limits the percentage area which remains inside the fuzzy control limits (β_j) should be checked. If it is equal or smaller than a predefined acceptable percentage (β), then the process can be stated as "rather out of control", otherwise it can be accepted as "rather in-control". In Fig. 3.7, possible decisions resulting from DFA are given. LCL_1, LCL_2, UCL_3, UCL_4, a, b, c, d, and α are the parameters for determination of the sample area outside the control limits for α-level fuzzy cut. The lines of $\overline{LCL_1 LCL_2}$, $\overline{UCL_3 UCL_4}$, $\overline{ab}$, and $\overline{cd}$ forms the shape of the control limits and fuzzy sample. A_{out}^U and A_{out}^L, that defines sample area above the upper control limits and sample area falling below the lower control limits respectively, are calculated. Then, A_{out} is the sum of the areas above fuzzy upper control limit and below fuzzy lower control limit and equals to the total sample area outside the fuzzy control limits. Percentage sample area within the control limits is computed as Eq. (3.77).

Fig. 3.7 Illustration of all possible sample areas outside the fuzzy control limits at α-level cut (Gülbay and Kahraman 2006a, b)

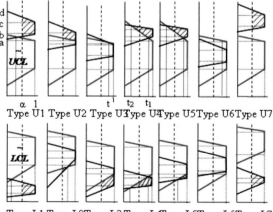

Type U1 Type U2 Type U3 Type U4 Type U5 Type U6 Type U7

Type L1 Type L2 Type L3 Type L4 Type L5 Type L6 Type L7

$$\beta_j^\alpha = \frac{S_j^\alpha - A_{out,j}^U}{S_j^\alpha} \tag{3.77}$$

where S_j^α is the sample area at α-level cut. The interested reader can refer to Gülbay and Kahraman (2006a, b) for the derivation of the formula and detailed information. The possibility of obtaining linguistic decisions like "rather in control" or "rather out of control" is provided using DFA methodology. Further intermediate levels of process control decisions can be obtained by using β as follows.

$$Process\ Control = \begin{cases} in\ control,\ 0,85 \le \beta_j \le 1 \\ rather\ in\ control\ 0,60 \le \beta_j < 0,85 \\ rather\ out\ of\ control,\ 0,10 \le \beta_j < 0,60 \\ out\ of\ control,\ 0,\ 10 \le \beta_j < 0,60 \end{cases} \tag{3.78}$$

3.3.5 Fuzzy $\tilde{u}$ Control Chart

Various types of fuzzy numbers are used in fuzzy set theory to express the vagueness. In this study, we will represent the imprecise number of nonconformities by triangular fuzzy numbers $(u_{a_i}, u_{b_i}, u_{c_i})$.

In order to construct a fuzzy u-chart, the fuzzy control limits are computed using fuzzy numbers. The average of the number of fuzzy nonconformities is calculated as Eq. (3.79) (Senturk et al. 2011):

$$\bar{u}_a = \frac{\sum u_{a_j}}{m}, \bar{u}_b = \frac{\sum u_{b_j}}{m}, \bar{u}_c = \frac{\sum u_{c_j}}{m} \tag{3.79}$$

The fuzzy $\tilde{\alpha}$-control chart limits are given as Eqs. (3.80)–(3.82):

$$\widetilde{UCL}_u = \left(\bar{u}_a + 3\sqrt{\frac{\bar{u}_a}{n_j}}, \bar{u}_b + 3\sqrt{\frac{\bar{u}_b}{n_j}}, \bar{u}_c + 3\sqrt{\frac{\bar{u}_c}{n_j}}\right) \tag{3.80}$$

$$\widetilde{CL}_u = (\bar{u}_a, \bar{u}_b, \bar{u}_c) \tag{3.81}$$

$$\widetilde{LCL}_u = \left(\bar{u}_a - 3\sqrt{\frac{\bar{u}_a}{n_j}}, \bar{u}_b - 3\sqrt{\frac{\bar{u}_b}{n_j}}, \bar{u}_c - 3\sqrt{\frac{\bar{u}_c}{n_j}}\right) \tag{3.82}$$

Numerical example for fuzzy $\tilde{u}$ chart
In the garment quality control example given in Sect. 3.2.4, the defects faced may not be specified as defects precisely. The below defect types may involve vagueness for a controller.

Some of these defects of the garments are listed below:

- Wrong gradation of sizes
- Uneven sizes
- Difference in fabric colours/shading variation
- Pilling of the material
- Seams not lined up
- Twisted, roped seams
- Irregular or uneven top stitching
- Uneven parts
- Faulty zippers (e.g. wavy, tape not matching colour specs)
- Irregular hemming
- Loose buttons
- Improper button holes
- Inappropriate trimming
- Spots, soil, stains, dust.

The data related to the number of nonconformities in a batch can be expressed as fuzzy numbers and given in Table 3.10. The center line, the upper and lower limits are computed as fuzzy numbers as given in Table 3.11. The control chart with fuzzy limits is illustrated in Fig. 3.8.

$$\bar{u}_a = \frac{\sum u_{a_j}}{m} = \frac{0,90}{12} = 0.07$$

$$\bar{u}_b = \frac{\sum u_{b_j}}{m} = \frac{1,11}{12} = 0.09$$

$$\bar{u}_c = \frac{\sum u_{c_j}}{m} = \frac{1,42}{12} = 0.12$$

Table 3.10 Data for $\tilde{c}$ control chart

Sample nr.	n	c_a	c_b	c_c	u_a	u_b	u_c	Sample nr.	n	c_a	c_b	c_c	u_a	u_b	u_c
1	54	4	5	6	0.07	0.09	0.11	6	49	3	3	5	0.06	0.06	0.10
2	48	4	5	5	0.08	0.10	0.10	7	44	2	3	4	0.05	0.07	0.09
3	47	3	3	4	0.06	0.06	0.09	8	52	5	5	8	0.10	0.10	0.15
4	51	6	6	7	0.12	0.12	0.14	9	47	2	4	5	0.04	0.09	0.11
5	58	5	6	8	0.09	0.10	0.14	10	52	4	6	6	0.08	0.12	0.12
6	53	3	4	6	0.06	0.08	0.11	11	54	5	7	9	0.09	0.13	0.17

Table 3.11 Fuzzy control limits for $\tilde{c}$ control chart

UCL$_a$	UCL$_b$	UCL$_c$	CL$_a$	CL$_b$	CL$_c$	LCL$_a$			LCL$_b$			LCL$_c$		
0.19	0.22	0.26	0.07	0.09	0.12	−0.04	→	0	−0.03	→	0	−0.02	→	0
0.19	0.22	0.27	0.07	0.09	0.12	−0.04	→	0	−0.04	→	0	−0.03	→	0
0.19	0.23	0.27	0.07	0.09	0.12	−0.04	→	0	−0.04	→	0	−0.03	→	0
0.19	0.22	0.26	0.07	0.09	0.12	−0.04	→	0	−0.04	→	0	−0.03	→	0
0.18	0.21	0.25	0.07	0.09	0.12	−0.03	→	0	−0.03	→	0	−0.02	→	0
0.19	0.22	0.26	0.07	0.09	0.12	−0.04	→	0	−0.03	→	0	−0.02	→	0
0.19	0.22	0.27	0.07	0.09	0.12	−0.04	→	0	−0.04	→	0	−0.03	→	0
0.20	0.23	0.27	0.07	0.09	0.12	−0.05	→	0	−0.04	→	0	−0.04	→	0
0.19	0.22	0.26	0.07	0.09	0.12	−0.04	→	0	−0.03	→	0	−0.02	→	0
0.19	0.23	0.27	0.07	0.09	0.12	−0.04	→	0	−0.04	→	0	−0.03	→	0
0.19	0.22	0.26	0.07	0.09	0.12	−0.04	→	0	−0.03	→	0	−0.02	→	0
0.19	0.22	0.26	0.07	0.09	0.12	−0.04	→	0	−0.03	→	0	−0.02	→	0

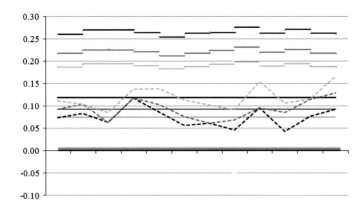

Fig. 3.8 Fuzzy control chart for number of nonconformities per unit

3.3.5.1 α-Cut Fuzzy $\tilde{u}$ Control Chart

When α-cut is adapted to the fuzzy sets, the values of u_a^α and u_c^α are determinedas Eq. (3.83):

$$\bar{u}_a^\alpha = \bar{u}_a + \propto (\bar{u}_b - \bar{u}_a) \text{ and } \bar{u}_c^\alpha = \bar{u}_c - \propto (\bar{u}_c - \bar{u}_b) \qquad (3.83)$$

α-cut fuzzy $\tilde{u}$-control chart is obtained by Eqs. (3.84)-(3.86)

$$\widetilde{UCL}_u^\alpha = \bar{u}_a^\alpha + 3\sqrt{\frac{\bar{u}_a^\alpha}{n_j}}, \bar{u}_b + 3\sqrt{\frac{\bar{u}_b}{n_j}}, \bar{u}_c^\alpha + 3\sqrt{\frac{\bar{u}_c^\alpha}{n_j}}) \qquad (3.84)$$

$$\widetilde{CL}_u = (u_a^\propto, \overline{u}_b, u_c^\propto) \tag{3.85}$$

$$\widetilde{LCL}_u = (\overline{u}_a^\propto - 3\sqrt{\frac{\overline{u}_a^\propto}{n_j}}, \overline{u}_b - 3\sqrt{\frac{\overline{u}_b}{n_j}}, \overline{u}_c^\propto - 3\sqrt{\frac{\overline{u}_c^\propto}{n_j}}) \tag{3.86}$$

Numerical example for α-cut fuzzy $\widetilde{u}$-control chart
The α-cut for α = 0.7 of the center line, upper and lower limits of the example in Sect. 3.3.5 are calculated as in Table 3.12:

$$\overline{u}_a^{0.7} = \overline{u}_a + 0.7(\overline{u}_b - \overline{u}_a) = 0.075 + 0.7(0.093 - 0.075) = 0.09$$

$$\overline{u}_c^{0.7} = \overline{u}_c - 0.7(\overline{u}_c - \overline{u}_b) = 0.119 - 0.7(0.119 - 0.093) = 0.10$$

3.3.5.2 α-Level Fuzzy Median for α-Cut Fuzzy $\widetilde{u}$ Control Chart

α-cut fuzzy $\widetilde{u}$-control chart is transformed to crisp numbers via the fuzzy transformation techniques. α-level fuzzy midrange, fuzzy median, fuzzy average and fuzzy mode (Wang and Raz 1990) are the transformation techniques. For a sample j, α-level fuzzy median value ($S_{med-u,j}^\propto$) is calculated as Eq. (3.87):

$$S_{med-u,j}^\propto = \frac{1}{3}(u_{a,j}^\propto + u_{b,j}^\propto + u_{c,j}^\propto) \tag{3.87}$$

By using these formulations, the fuzzy center line, fuzzy upper and fuzzy lower limits of α-level fuzzy median for α-cut fuzzy $\widetilde{u}$-control chart is obtained by Eqs. (3.88)–(3.90):

$$UCL_{med-u}^\propto = CL_{med-u}^\propto + 3\sqrt{\frac{CL_{med-u}^\propto}{n_j}} \tag{3.88}$$

$$CL_{med-u}^\propto = \frac{1}{3}(\overline{u}_a^\propto + \overline{u}_b^\propto + \overline{u}_c^\propto) \tag{3.89}$$

$$UCL_{med-u}^\propto = CL_{med-u}^\propto - 3\sqrt{\frac{CL_{med-u}^\propto}{n_j}} \tag{3.90}$$

The condition of process control for each sample is defined as: Eq. (3.91):

$$Process\ control = \begin{cases} in\ control,\ for\ LCL_{med-u}^\propto \leq S_{med-u,j}^\propto \leq UCL_{med-u}^\propto \\ out\ of\ control,\ for\ otherwise \end{cases} \tag{3.91}$$

Table 3.12 Data and fuzzy control limits for α-cut of $\tilde{u}$ control chart

n	c_a	c_b	c_c	u_a	u_b	u_c	UCL^{η}_a	UCL^{η}_b	UCL^{η}_c	CL^{η}_a	CL^{η}_b	CL^{η}_c	LCL^{η}_a	LCL^{η}_b	LCL^{η}_c
54	4	5	6	0.07	0.09	0.11	0.21	0.22	0.23	0.09	0.09	0.10	−0.03 → 0	−0.03 → 0	−0.03 → 0
48	4	5	5	0.08	0.10	0.10	0.22	0.22	0.24	0.09	0.09	0.10	−0.04 → 0	−0.04 → 0	−0.04 → 0
47	3	3	4	0.06	0.06	0.09	0.22	0.23	0.24	0.09	0.09	0.10	−0.04 → 0	−0.04 → 0	−0.04 → 0
51	6	6	7	0.12	0.12	0.14	0.21	0.22	0.23	0.09	0.09	0.10	−0.04 → 0	−0.04 → 0	−0.03 → 0
58	5	6	8	0.09	0.10	0.14	0.20	0.21	0.23	0.09	0.09	0.10	−0.03 → 0	−0.03 → 0	−0.02 → 0
53	3	4	6	0.06	0.08	0.11	0.21	0.22	0.23	0.09	0.09	0.10	−0.03 → 0	−0.03 → 0	−0.03 → 0
49	3	3	5	0.06	0.06	0.10	0.21	0.22	0.24	0.09	0.09	0.10	−0.04 → 0	−0.04 → 0	−0.04 → 0
44	2	3	4	0.05	0.07	0.09	0.22	0.23	0.24	0.09	0.09	0.10	−0.05 → 0	−0.04 → 0	−0.04 → 0
52	5	5	8	0.10	0.10	0.15	0.21	0.22	0.23	0.09	0.09	0.10	−0.04 → 0	−0.03 → 0	−0.03 → 0
47	2	4	5	0.04	0.09	0.11	0.22	0.23	0.24	0.09	0.09	0.10	−0.04 → 0	−0.04 → 0	−0.04 → 0
52	4	6	6	0.08	0.12	0.12	0.21	0.22	0.23	0.09	0.09	0.10	−0.04 → 0	−0.03 → 0	−0.03 → 0
54	5	7	9	0.09	0.13	0.17	0.21	0.22	0.23	0.09	0.09	0.10	−0.03 → 0	−0.03 → 0	−0.03 → 0

Table 3.13 Data and fuzzy control limits for α-level fuzzy median for α-cut of $\tilde{u}$ control chart

n	c_a	c_b	c_c	u_a	u_b	u_c	$u^{\prime 7}{}_a$	$u^{\prime 7}{}_b$	$u^{\prime 7}{}_c$	$S^{\prime 7}$ med-u	$UCL^{\prime 7}$ med	$CL^{\prime 7}$ med	$LCL^{\prime 7}$ med
54	4	5	6	0.07	0.09	0.11	0.09	0.09	0.10	0.09	0.22	0.09	$-0.03 \rightarrow 0$
48	4	5	5	0.08	0.10	0.10	0.10	0.10	0.10	0.10	0.23	0.09	$-0.04 \rightarrow 0$
47	3	3	4	0.06	0.06	0.09	0.06	0.06	0.07	0.07	0.23	0.09	$-0.04 \rightarrow 0$
51	6	6	7	0.12	0.12	0.14	0.12	0.12	0.12	0.12	0.22	0.09	$-0.03 \rightarrow 0$
58	5	6	8	0.09	0.10	0.14	0.10	0.10	0.11	0.11	0.21	0.09	$-0.03 \rightarrow 0$
53	3	4	6	0.06	0.08	0.11	0.07	0.08	0.09	0.08	0.22	0.09	$-0.03 \rightarrow 0$
49	3	3	5	0.06	0.06	0.10	0.06	0.06	0.07	0.07	0.22	0.09	$-0.04 \rightarrow 0$
44	2	3	4	0.05	0.07	0.09	0.06	0.07	0.08	0.07	0.23	0.09	$-0.04 \rightarrow 0$
52	5	5	8	0.10	0.10	0.15	0.10	0.10	0.11	0.10	0.22	0.09	$-0.03 \rightarrow 0$
47	2	4	5	0.04	0.09	0.11	0.07	0.09	0.09	0.08	0.23	0.09	$-0.04 \rightarrow 0$
52	4	6	6	0.08	0.12	0.12	0.10	0.12	0.12	0.11	0.22	0.09	$-0.03 \rightarrow 0$
54	5	7	9	0.09	0.13	0.17	0.12	0.13	0.14	0.13	0.22	0.09	$-0.03 \rightarrow 0$
						Avr.	0.09	0.09	0.10				

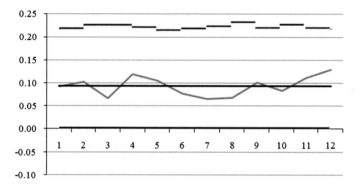

Fig. 3.9 α-level fuzzy median for α-cut fuzzy $\tilde{u}$ control chart

Numerical example for α-level fuzzy median for α-cut fuzzy $\tilde{u}$ control chart
For each sample j, α-level fuzzy median value ($S^{\infty}_{med-u,j}$) and the corresponding center line, upper and lower limits for $\alpha = 0.7$ is calculated as in Table 3.13 and the control chart is depicted in Fig. 3.9.

3.4   Conclusion

In this study, we presented a review of control charts for attributes, specifically fraction nonconforming (p-chart), number of nonconforming (np-chart), nonconformities (c-chart) and nonconformities per unit (u-charts). In real life, attribute data may be uncertain because nonconformities can not be defined as a precise value.

To deal with this, fuzzy control charts are proposed in the literature. We present a review of fuzzy control charts, α-cut of the charts and the α-level fuzzy median for α-cut fuzzy charts for each attribute chart. Then we apply these methods numerical examples. For further research, other fuzzy number types such as type II fuzzy sets can be employed at the fuzzification phase. Besides, different transformation techniques proposed in the literature can be used at the defuzzification phase of fuzzy attribute control charts.

References

Braverman, J.D.: Fundamentals of Statistical Quality Control. Reston Pub. Co., Virginia, Reston, Va. (1981)

Erginel, N.: Fuzzy rule-based p and np control charts. J. Intell. Fuzzy Syst. (2013) forthcoming

Gülbay, M., Kahraman, C., Ruan, D.: α-cut fuzzy control charts for linguistic data. Int. J. Intell. Syst. **19**, 1173–1195 (2004)

Gülbay, M., Kahraman, C.: An alternative approach to fuzzy control charts: Direct fuzzy approach. Inf. Sci. **177**, 1463–1480 (2007)

Gülbay, M., Kahraman, C.: Development of fuzzy process control charts and fuzzy unnatural pattern analyses. Comput. Stat. Data Anal. **51**, 434–451 (2006a)

Gülbay, M., Kahraman, C.: Design of fuzzy process control charts for linguistic and imprecise data. In: Kahraman, C. (ed.) Fuzzy Applications in Industrial Engineering, pp. 59–86. Springer, Berlin (2006b)

Kahraman, C., Gülbay, M., Erginel, N., Şentürk, S.: Fuzzy statistical process control techniques in production systems. In: Kahraman, C., Yavuz, M. (eds). Production Engineering and Management Under Fuzziness Studies in Fuzziness And Soft Computing. Springer, Berlin, pp. 431–456 (2010)

Mitra, A.: Fundamentals of Quality Control and Improvement, 3rd edn. Wiley, New Jersey (2008)

Montgomery, D.C.: Introduction to Statistical Quality Control, 5th edn. Wiley, USA (2005)

Şentürk, S., Erginel, N., Kaya, İ., Kahraman, C.: Design of fuzzy α control charts. J. Multiple-Valued Logic Soft Comput. **17**, 459–473 (2011)

Wang, J.H., Raz, T.: On the construction of control charts using linguistic variables. Intell. J. Prod. Res. **28**, 477–487 (1990)

Chapter 4
Special Control Charts Using Intelligent Techniques: EWMA Control Charts

Bulut Aslan, Yeliz Ekinci and Ayhan Özgür Toy

Abstract In this chapter we consider the economical design of EWMA zone control charts for set of machines operating under JPS (Jidoka Production System). We provide an extensive literature review of intelligent systems in quality control deductively to fit our purposes. It starts with an overview of quality control charts; then, reviews charts designed for special purposes such as EWMA, CUSUM and zone control charts. Finally, as particularly related to this study, reviews of economical design and intelligent applications of EWMA are provided. We discuss and review Jidoka Production System and motivation of operating such a system. We suggest an intelligent control and repair system such that in a production system, machines are individually controlled and repaired when an out-of-control signal is triggered in the zone with the tight control limits, however a system-wide shut down and repair is conducted when the out-of-control signal is from beyond the inner (tight) control limits which is considered as an opportunity for repair and calibration of all machines. We illustrate and investigate the behaviour of control parameters, namely sample size, sampling interval and control limits, via a numerical study of a three-machine system through simulation. We also provide insights for implementation of several metaheuristics for the system setting discussed in this chapter.

Keywords Control charts · EWMA · CUSUM · Intelligent · Jidoka production system

B. Aslan (✉) · Y. Ekinci · A.Ö. Toy
Department of Industrial Engineering, Istanbul Bilgi University,
34060 Eyüp/Istanbul, Turkey
e-mail: bulutaslan@gmail.com

Y. Ekinci
e-mail: yeliz.ekinci@bilgi.edu.tr

A.Ö. Toy
e-mail: ozgur.toy@bilgi.edu.tr

© Springer International Publishing Switzerland 2016
C. Kahraman and S. Yanık (eds.), *Intelligent Decision Making in Quality Management*, Intelligent Systems Reference Library 97,
DOI 10.1007/978-3-319-24499-0_4

101

4.1 Introduction

This study suggests an intelligent system approach to the quality control problem of a production environment. This intelligent system provides controlling particular decisions automatically, which are once made manually by operators at the job level. An Exponentially Weighted Moving Average (EWMA) quality control chart, integrated with zone control policy, is intelligently controlled at the system level to minimize the rate of total cost. This intelligent system will be useful for the companies which use Jidoka Production System and implement statistical process control tools. The statistical process control (SPC) is a widely used method in quality control. SPC has been introduced by Walter Shewhart (Shewart, 1924) in 1920s and it is defined as a technique of monitoring, controlling, and improving a process through statistical analysis. The advantages of SPC are listed by Parkash et al. (2013) among which there are *improving process performance by reducing product variability*, *minimizing rework and loss of sales*, and *elimination of unnecessary quality checks and higher quality product by reducing variability and defects*. De Vries and Reneau (2010) state that there are two aspects of the SPC approach: (i) to assist the continuous improvement of performance by further reduction of unexplained variability; (ii) to detect any variability in the system as quickly as possible.

Benneyan et al. (2003) define the basic principles of SPC as: (i) individual measurement from any process exhibits variation (ii) variability due to common cause can be predicted and represented by statistical models such as Gaussian, binomial, or Poisson distribution, (iii) variability due to assignable cause displays deviation in some observable way from the aforementioned random distribution models, (iv) statistical limits can be established to test the data in order to provide evidence of any change. We refer the reader to Montgomery (1991) for a detailed description and explanation of SPC.

There are seven basic tools for quality improvement that are used for statistical process control. These are; check sheet, defect concentration diagram, histogram, Pareto chart, scatter diagram/chart, cause and effect or fishbone diagram, control chart. Quality control charts are the most widely used technique and are of concern in this study.

Quality control charts started to be used in the manufacturing industries in the 1920s, but later implemented in lots of other application areas such as health care management, epidemiology, animal production systems, service, financial and agriculture systems, (see De Vries and Reneau 2010; Woodall 2006; Thor et al. 2007; Quesenberry 1997; Montgomery 2009; Solodky et al. 1998 and Diaz and Neuhauser 2005 for various applications of SPC). We will dwell into control charts in subsequent sections.

In some processes it may be required to implement more sensitive control charts, i.e., to detect smaller or moderate sized shifts in the process mean, for which Exponentially Weighted Moving Average (EWMA) control charts and Cumulative Sum (CUSUM) control charts have been developed. A large body of literature has been built on EWMA control charts following the pioneering work by Roberts (1959).

We will present some of recent works on EWMA control charts in the subsequent sections. We also refer the reader to Page (1954) for CUSUM control charts.

Recognition of any systematic or non-random patterns in observations is of interest in some cases. In order to pinpoint these patterns, establishment of control limits at different expanding levels in multiples of process standard deviation is necessary. Zone control charts are the choice when the concern is to detect systematic patterns. The zone control charts can be traced back to The *Statistical Quality Control Handbook* (Western Electric 1956).

Implementation of control charts in production systems generally assumes that machines are individually operated and controlled. Therefore, control charts are designed for each machine considered in isolation. However, this assumption does not always hold. Recent trends in manufacturing view machines operating in coordination as a single system, and hence, suggest a central control mechanism. *Jidoka* (translated as autonomation) is such a defect detection system, which automatically or manually stops the production operation whenever an abnormal or defective condition arises. In the concept of jidoka when a team member encounters a problem in his or her workstation, he/she is responsible for indicating the problem by pulling an *andon cord*, which can stop the line. Hence, when any machine issues an alarm, a system-wide shut down is triggered and the production is ceased until the inspection/restoration of the triggering machine. Here, we assume that holding WIP (Work-In-Process) inventory between the machines, that keeps upstream and downstream of the line working during restoration of an intermediate machine, is not feasible or undesirable. Such a production system is denoted as *Jidoka Production System* (JPS) by Berk and Toy (2009).

Designing a control chart means to determine the operating parameters, (i.e., when to sample, how much to sample and the reference values (control limits) for an inference about the sample) of a control chart. There are economical, statistical and economic-statistical procedures of determining these operating parameters (refer to Montgomery 1980a, b for explanation of these procedures). Our focus in this chapter is on the economical design of control charts, specifically economically designing EWMA zone control charts for set of machines operating under JPS.

We suggest a control and repair policy such that in a production system, machines are individually controlled and repaired when an out-of-control signal is triggered in the zone with the tight control limits, however a system-wide shut down and repair is conducted when the out-of-control signal is from beyond the inner (tight) control limits which is considered as an opportunity for repair and calibration of all machines. We illustrate and investigate the behaviour of control parameters, namely sample size, sampling interval, control limits and EWMA smoothing parameter.

In the sequel, we will provide literature review on intelligent systems in quality control. Then, we introduce Jidoka Production System as an intelligent QC approach with our assumptions and the model prior to presenting our methodology and numerical study. Finally we present our conclusions.

4.2 Intelligent Systems in Quality Control

Intelligent systems (IS) is a broad term, covering a range of computing techniques that have emerged from research into artificial intelligence (Hopgood 2012). It is about generating representations, procedures and strategies to handle tasks that were once thought only do-able by humans (Schalkoff 2009). The tools of particular interest are roughly divided among *knowledge-based systems, computational intelligence* and *hybrid* systems (Schalkoff 2009; Negnevitsky 2011; Hopgood 2012). Knowledge-based systems include expert and rule-based systems, object-oriented and frame-based systems, and intelligent agents. Computational intelligence includes neural networks, genetic algorithms and other optimization algorithms (i.e., metaheuristics). Techniques for handling uncertainty, such as fuzzy logic, fit into both categories. Knowledge-based systems, computational intelligence, and their hybrids are collectively referred to here as IS. When knowledge is not explicitly stated but represented by numbers, computational intelligence takes place to improve any system accuracy; where numerical techniques such as genetic algorithms and neural networks.

This study brings in an IS approach to the quality control problem of a production environment. Particular decisions, which are once made manually by operators at the job level, are now automatically controlled and taken into consideration at the system-level. An EWMA chart, integrated with zone control policy, is intelligently controlled at the system level to minimize the rate of total cost.

The following literature review is organized in a deductive way to clearly underline this study's contribution. It starts with a generic introduction to quality control charts; then, an overview of QC charts designed for special purposes such as EWMA, CUSUM and zone control charts is provided. Subsequently, as particularly related to this study, studies on economical design and finally intelligent applications of EWMA are reviewed. Each subsection is chronologically organized on its own.

4.2.1 Quality Control Charts

Quality control charts, developed by Shewhart (1924), is a statistical tool applied to detect a shift on the process. In this tool observations are plotted over time to see whether a process is running as it should be. In this context, a process is said to be "in-control" if the probability distribution representing the quality characteristic is constant over time. If there is some change in this distribution the process is said to be "out-of-control". Control charts help to distinguish between common causes and assignable causes and quickly detect occurrences of unplanned assignable causes so corrective action may be undertaken and the assignable causes are removed.

Quality control charts are commonly classified into two types: (i) if a control chart is used to track a quality characteristic which can be measured and expressed as a number on some continuous scale of measurement, it is usually called a *variable control chart*. Conveniently, the quality characteristic is described with a measure of central tendency (e.g., mean) and a measure of dispersion (e.g., range or standard deviation). The $\bar{X}$ chart is the most widely used chart for controlling the former, whereas the charts based on the latter, such as R-chart or s-chart, are used to control process variability; (ii) In cases where quality characteristics are not measured on a continuous or quantitative scale, control charts are constructed based on quality conformances and called *attribute control charts*. Each unit is categorized as either conforming or nonconforming on the basis of possession of certain attributes or the number of nonconformities (defects) appearing on a unit of product.

De Vries and Reneau (2010) state that good design of control charts depends on grouping of observations, *the distribution of the process observations*, *the size of the process shift of interest*, and *the costs of Type I and Type II errors*. There are two types of variable control charts that are widely used, namely, $\bar{X}$ chart and R-chart. The $\bar{X}$ chart is used to monitor the process mean and the R-chart is used to monitor variability. For both of them, samples are taken over time and values of a statistic are plotted. The first quality control chart developed by Shewhart is of type $\bar{X}$ and is slow in detecting the smaller mean shifts. The "variable sampling intervals (VSI) $\bar{X}$ charts" and "variable parameters (VP) $\bar{X}$ charts" detect small shifts in the process mean faster than the standard $\bar{X}$ chart (see Costa 1999; Reynolds et al. 1988; Lee et al. 2012 for discussion). CUSUM and EWMA charts are also good alternatives when the aim is detecting small shifts. However Lee et al. (2012) argue that their control procedures are not as easy to set up as Shewhart control chart.

A control chart consists of two parts: (i) a series of measurements plotted in time order, and (ii) the control chart "template" which consists of three horizontal lines called the center line (typically, the mean), the upper control limit (UCL), and the lower control limit (LCL) The values of the UCL and LCL are usually calculated from the inherent variation in the data. The lower control limit and the upper control limit determine the bounds of the in-control region, i.e. as long as the measurement of the sample taken falls in between these two lines the process is assumed to be in the in-control status. Only random causes exist if the sample data points fall between the two control limits. If the process shifts to the out-of-control status then we expect that most of the observations are outside the control limits. Moreover, when in-control, the data points plotted on the chart should be distributed without a pattern between the control limits. Sometimes the data points are located only on one side of the center line and close to each other. This also may be an evidence for a systematic variation, hence, out-of-control state. It is assumed that, at the start of a production run after the last restoration, the production process is in the in control status, producing items of acceptable quality. After a period of time in production, the process may shift to the out-of-control status.

The placing of the control limits on a control chart depends on the cost of false alarms (i.e., Type I errors) and the cost after a shift occurs but not detected (i.e.,

Type II errors). The optimal placing of control limits and the optimal frequency of collecting observations has triggered much research (Montgomery 2009). If the limits are set too narrow there is a high probability of a "Type I error"—mistakenly inferring assignable cause variation exists when, in fact, a predictable extreme value is being observed which is expected periodically from common cause variation. On the other hand, if the limits are set too wide there is a high probability of a "Type II error".

The control limits are usually set at ±3 standard errors of the plotted statistic from a center line at its historical average value. Conventionally, "control limits at 3-standard errors from the mean" is robust for observations from most kinds of distributions but could result in poor performance when unplanned changes are costly and need to be detected quickly. The formula for the calculation of the standard error is usually based on a distributional assumption. The control limits for the Shewhart charts are relatively easy to calculate. Finding the control limits for the CUSUM and EWMA given a desired rate of false alarms requires special software or tables. Hawkins and Olwell (1998) present algorithms for control limits on CUSUM charts.

Process engineers prefer to evaluate the effectiveness of a control procedure in terms of cost-based performance rather than risk-based performance measures such as the Type I and Type II errors. A typical example of the cost- based approach for a control chart is the economic design of the control chart (Park 2013).

4.2.2   EWMA and CUSUM Quality Control Charts

Alluded to above, EWMA and CUSUM control charts are implemented to detect small and moderate-sized shifts in the process. In EWMA, each point represents the weighted average of current and certain number of previous observation values, giving weight on the observations based on the recency. Shewart-type control charts are not very efficient in detecting small shifts since they only consider the final observation and do not consider accumulated information of the multiple observations; EWMA and CUSUM control charts are the methods to overcome this difficulty.

Neubauer (1997) discusses the properties of EWMA control charts and comparison with other quality control procedures. He specifically states, "*The EWMA chart offers a flexible instrument for visualizing imprecision and inaccuracy*". The properties of the EWMA chart with the constant control limits have also been studied by Robinson and Ho (1978), Waldmann (1986), Lucas and Saccucci (1990) and Gan (1991). The four-step procedure for implementing the EWMA chart established by Crowder (1989) has been discussed in Neubauer (1997).

The smoothing operation in EWMA control charts is achieved through a "smoothing parameter, λ" which guarantees giving less and less weight to observations as they are further removed in time. In brief, after multiplication by a factor λ, the current measurement is added to the sum of all former measurements, which is weighted with $(1-\lambda)$. Thus, at each time epoch $t(t = 1,2,\ldots)$, the test statistic Z_t can be obtained by Eq. 4.1:

$$Z_t = [\lambda \bar{X}_t + (1 - \lambda)Z_{t-1}] \tag{4.1}$$

where $\bar{X}_t$ is the mean of current sample and $\lambda \in [0, 1]$. The computed Z_t values are displayed on a control chart over time. Since the statistic used is the sample mean in this particular example, this control chart is called the EWMA-$\bar{x}$ chart. Note that setting the smoothing parameter to unity ($\lambda = 1$) in EWMA control chart yields a Shewhart-type control chart.

In EWMA control charts test statistic computed above is plotted on a chart and compared with the control limits. A common approach to determining the control limits (see e.g. Montgomery 1991, p. 300) first requires the calculation of the variance of the test statistic, Z_t, which is given in Eq. 4.2:

$$\sigma_t^2 = [\sigma^2/n][\lambda/(2 - \lambda)][1 - (1 - \lambda^{2t}) \tag{4.2}$$

where we assume that the individual observations (sample mean) are independent random variables with variance σ^2.

The variance calculation above enables us to derive the following control limits for the EWMA control chart (Eqs. 4.3 and 4.4):

$$UCL = \mu_0 + L\sigma\sqrt{\left(\frac{\lambda}{2 - \lambda}\right)[1 - (1 - \lambda)^{2t}]} \tag{4.3}$$

$$LCL = \mu_0 - L\sigma\sqrt{\left(\frac{\lambda}{2 - \lambda}\right)[1 - (1 - \lambda)^{2t}]} \tag{4.4}$$

where L is the coefficient which defines "the width of the control limits". As the sample number, t, gets larger control limits converges to, so called, the steady-state EWMA control limits, which are given by Eqs. 4.5 and 4.6:

$$UCL = \mu_0 + L\sigma\sqrt{\left(\frac{\lambda}{2 - \lambda}\right)} \tag{4.5}$$

$$LCL = \mu_0 - L\sigma\sqrt{\left(\frac{\lambda}{2 - \lambda}\right)} \tag{4.6}$$

It is important to note, however, that the expression for the variance of Z_t for $t > 1$ is derived by ignoring the truncation effects of the control limits employed at the previous $t-1$ samples. Thus, using a constant multiple of the standard deviation to give the control limits is done somewhat arbitrarily.

Vargas et al. (2004) provides insights for choosing smoothing parameter, λ, and for the coefficient of the width of the control limit interval, L. They state that, in practice, values between 0.05 and 0.25 for the smoothing parameter work well, with the popular choices being 0.05; 0.10 and 0.20; and the usual three-sigma limits ($L = 3$) work reasonably well particularly with the larger values of λ. Montgomery (1996) suggests that when λ is small, ($\lambda \leq 0.1$), there is an advantage in reducing the width of limits, using a value of L between 2.6 and 2.8.

We next provide a brief review of CUSUM control charts. The CUSUM and EWMA control charts differ from each other in incorporation of the smoothing parameter in EWMA control charts, which allows the adjustment of shift sensitivity. The CUSUM control chart was initially proposed by Page (1954) and has been widely studied and implemented then on. CUSUM chart attributes equal weight to all observations independent of their recency.

Like EWMA control charts CUSUM control charts also incorporate past observations and are therefore sensitive to detecting small shifts in the process. The CUSUM charts are available for distributions such as the Normal, Binomial, Poisson, and Weibull.

In CUSUM charts, each plotted point represents the algebraic sum of the previous observations and the most recent deviation from the target (Parkash et al. 2013). Assuming that samples of size $n \geq 1$ are collected, $\bar{x}_j$ is the average of the jth sample and μ_0 is the value wanted for the process average, the CUSUM control chart is formed by the formula resulting quantity along with sample i (Vargas et al. 2004) (Eq. 4.7):

$$C_i = \sum_{j=1}^{i} (\bar{x}_j - \mu_0) \tag{4.7}$$

where C_i is the cumulative sum including the ith sample, since they combine information from several samples. If the process keeps in control at the target value μ_0, the cumulative sums describe a random way. On the other hand, if the average changes to any value above $\mu_1 > \mu_0$, then an ascendant tendency will develop at the cumulative sum C_i. Reciprocally, if the average changes to some value below $\mu_1 < \mu_0$, the cumulative sum C_i will have a negative direction. Considering this, if at the demarcated points a tendency up or down appears, it must be considered as an evidence of process average change, and a search for the assignable causes must be done (Vargas et al. 2004). The main advantage of CUSUM charts is that it is very effective for small shifts and samples of size $n = 1$. The main disadvantage is that they are relatively slow to respond to large shifts and special patterns are hard to see and analyze.

Montgomery (2013) states that the general consensus is that the practical performances of the CUSUM and EWMA are quite similar and neither of them has a clear advantage over the other. Thus, users only need to implement one or the other to monitor their process. The CUSUM chart has a well-known optimality property: if a shift occurs in steady state, the CUSUM to which it is tuned has a faster average response than does any other chart (Hawkins and Olwell 1998; Hawkins and Wu 2014). Vargas et al. (2004) present a comparative study of the performance of the CUSUM and EWMA control charts. Their objective is to verify when CUSUM and EWMA control charts do the best control region, in order to detect small changes in the process mean. One of the results they come up with after several simulations is that the CUSUM control chart practically does not sign points out of control for the levels of variation between ±1.0 standard deviation; for these variation levels the EWMA control chart is more efficient. In a recent study, Hawkins and Wu (2014) conclude that, though the CUSUM outperforms the EWMA, if the actual shift is smaller than that used in the design, the EWMA may respond faster. Recently, synthetic EWMA (SynEWMA) and synthetic CUSUM (SynCUSUM) control charts have been proposed based on simple random sampling (SRS) by integrating the EWMA and CUSUM control charts with the conforming run length control chart, respectively. Haq et al. (2014) state that these synthetic control charts provide overall superior detection over a range of mean shift sizes.

Recently, Abbas et al. (2012) introduced the design structure of a mixed EWMA-CUSUM (MEC) control chart for improved monitoring of the process parameters. In their study, the EWMA statistic is used as the input for the CUSUM structure. Zaman et al. (2014) propose a reverse version of this mixing, that is, a mixed CUSUM-EWMA (MCE) control chart. In this new setup, the CUSUM statistic will serve the input for the EWMA structure. MCE control chart is used to monitor the location of a process. The performance of the proposed mixed CUSUM-EWMA control chart is measured through the average run length, extra quadratic loss, relative average run length, and a performance comparison index study. The analysis has revealed that the proposed MCE control chart is very sensitive for the detection of small and moderate shifts and offers a quite efficient structure as compared with existing counterparts. The relative performance of the proposed chart as compared with the other charts varies depending on the amounts of shifts.

4.2.3   Zone Control Charts

It may be of interest to recognize any systematic or non-random patterns in observations. Establishing control limits at different expanding levels in multiples of process standard deviation facilitates to pinpoint any patterns. Zone control charts are the choice when the concern is to detect systematic patterns. The zone control charts can be traced back to The *Statistical Quality Control Handbook* (Western Electric 1956) in which it has been suggested that the process is concluded to be out-of-control if either (1) one point plots outside the three-sigma control limits, (2) two out of three consecutive points plot beyond the two-sigma warning limits, (3) four out of five consecutive points plot at a distance of one sigma or beyond from the center line, or (4) eight consecutive points plot on one side of the center line.

Jaehn (1987, 1989) suggests a zone chart with eight zones, four on each side of the center line. Scores are assigned to each zone, and the procedure signals an out-of-control status when the total score exceeds a threshold value. Likewise, Flaig (2004) proposed a zone control chart that partitions the normal process distribution into regions and assigns a score to each region. A normal process distribution is partitioned as follows (Table 4.1):

An illustration of the above rules is depicted in Fig. 4.1 (excerption from Flaig 2004). In the figure, the chart starts with current score of zero and a cumulative score of zero. As additional observations are recorded, the cumulative score is adjusted using the following rules (Flaig 2004).

1. if observation i and observation $i-1$ are on different sides of the center line, then the cumulative score at i is set to the zone score at i
2. if observation i and observation $i-1$ are on the same side of the center line, then add the current score at i to the cumulative score at $i-1$.
3. If the cumulative score reaches 8, then an out of control signal is generated.

Performance of zone control charts, in the presence of a constant process mean, has been studied by Davis et al. (1990, 1994). The latter paper showed that in the presence of a constant process mean, a zone control chart with appropriate parameters has superior performance compared to the corresponding Shewhart chart with some combination of supplementary runs rules (Davis and Krehbiel 2002). Davis et al. (1994) proposed a general model for the zone control chart. In this model, any zone control chart is based on a cumulative score, which begins at zero. The cumulative score for the zone control chart with score vector S = (A1, A2, A3, M) is incremented based on how many standard errors (s.e.) a sample mean is away from the target mean, with zone scores assigned as follows (Table 4.2):

Table 4.1 Zone control chart scores

Zone	Score
Centerline	0
Between center and one sigma	1
Between one and two sigma	2
Between two and three sigma	4
Beyond three sigma	8

Table 4.2 Zone control chart scores in the study of Davis et al. (1994)

Zone	Score
Within one s.e. of target	A_1
Between one and two s.e. from target	A_2
Between two and three s.e.from target	A_3
Beyond three s.e. from target	M

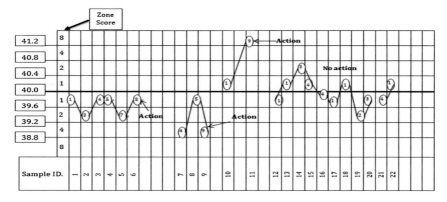

Fig. 4.1 Zone chart (Flaig 2004)

The cumulative score after any given sample consists of the score for the most recent sample mean added to the previous cumulative score. The only exception is when two successive sample means fall on opposite sides of the target; in this case, the cumulative score is reset to the score for only the most recent sample mean. Thus, if the process is in control, the sample means will fall on either side with a probability of 1/2 and the cumulative score will reset fairly frequently instead of continuing to get larger. If the cumulative score reaches or exceeds M, the chart generates an out-of-control signal (Davis and Krehbiel 2002; Davis et al. 1994).

In another paper, Davis and Krehbiel (2002) studied the average run length performances of Shewhart charts with supplementary runs rules and zone control charts when the process mean changes linearly over time. Shewhart charts with all possible combinations of the typical runs rules are compared to zone control charts with identical false alarm rates. The zone control charts generally outperform the Shewhart charts in detecting a process mean that is changing linearly over time.

4.2.4   Economical Design of Quality Control Charts

Economic design of control charts has been extensively studied since the pioneering work of Duncan (1956). Duncan (1956) studied on the economic design of $\bar{X}$ charts used to maintain current control of process, developed a cost model and solved for the design parameters (sample size, sampling interval, and control limit coefficient). Since then, there has been an increased interest in the economic design of the control charts in the late 1980s and early 1990s in accordance with developments in lean management of production systems (e.g., Montgomery 1980a, b; Vance 1983; Woodall 1986; Pignatiello and Tsai 1988; Niaki et al. 2010, 2013). A detailed explanation of construction of economically designed control charts is provided by Montgomery (1980a, b) and Ho and Case (1994b).

In economic design, parameters are determined such that the total cost associated with the implementation of quality control policy is minimized. In a regular $\bar{X}$ control chart, these parameters consist of the sample size, the time interval between two consecutive sampling, the coefficient (multiple of standard deviation) that specifies the Upper Control Limit, and the Lower Control Limit.

The economic models are generally formulated using the total cost per unit time function. Overall production time is divided into stochastically identical cycles. Each cycle starts with the production in the in-control status. When, at some sampling instance, control chart indicates an out-of-control status (denoted as an alarm) a search for the assignable cause is conducted and if discovered the process is stored to the in-control status. The time between these two time points is called a cycle. Hence the expected cost within this cycle is computed and divided by the expected duration of the cycle. Minimization of this cost rate yields the design parameters of the control chart.

The study of Lorenzen and Vance (1986) proposed a general cost model that applied to all control charts, regardless of the statistic used. Later, several studies (Montgomery et al. 1995; Reynolds et al. 1988; Costa 1993;Torng et al. 2009a, b; Nenes 2011; Lee et al. 2012) focused on the design of various control charts and minimize the costs of process control (Niaki et al. 2011).

The weaknesses of the economic design of control charts approach have been highlighted by Woodall et al. (1986) and Woodall (1987). Despite the problems, the economic design of control charts is appealing to process engineers since the effectiveness of the control chart procedure is explained in terms of cost (Park 2013).

There have been studies in the literature, which consider economical design of control charts other than $\bar{X}$ charts. Next we will dwell into economic design of EWMA control charts by providing reviews of some milestone and recent papers. The very first work on the economical design of EWMA control charts is by Ho and Case (1994b). Tolley and English (2001) studied the economic design of EWMA control chart and EWMA-$\bar{X}$ chart, and provide a comparison of the two. One of the studies in the last decade belongs to Park et al. (2004), who looked into the economic design of an adaptive EWMA chart. In this study, the user changes the sampling interval and/or sample size dynamically based on the current chart statistic. Chou et al. (2008) presented economic design of Variable Sampling Intervals EWMA (VSI EWMA) control charts with sampling at fixed times. The two more recent studies dealing with joint economic design of EWMA charts for the process mean and dispersion have been performed by Serel and Moskowitz (2008) and Serel (2009). In the study of Serel (2009) the case where the assignable cause changes only the process mean or dispersion is explored. The economic design of the single control chart used for monitoring the process parameter (mean or variance) influenced by the assignable cause is the main concern in this study. It suggests that using a different type of quality loss function (linear versus quadratic) leads to a significant change in sampling interval while affecting the sample size and control limits very little. It is also observed that the overall costs are insensitive to the choice of Shewhart or EWMA charts. Various authors have and studied meta-heuristic applications in economical design of EWMA control chart, we discuss these works in later sections.

To the best of our knowledge, the only work that considers economic design of zone control charts is by Ho and Case (1994a). In their study, based on factors such as performance, simplicity, efficiency, ease of use, and ease of understanding, they recommend the joint Zone Control Chart.

4.2.5 Intelligent Applications of Ewma QC Chart

In this study and in most of the studies on economical design of quality control charts, classical optimization methods are used. Typically, the optimization problem of EWMA charts contains both, continuous (e.g., smoothing parameter) and discrete (e.g., sample size) decision variables. This produces a discontinuous non-convex solution space exists; standard non-linear programming techniques may prove to be ineffective (He et al. 2002; Aparisi and Garcia-Diaz 2007). Hence, here comes the computational intelligence into place which is employed in order to solve that optimization problem. Among several intelligent applications of EWMA quality control charts, metaheuristic applications have outnumbered the rest in the quality control area. Besides, we review a few but worthwhile efforts deploying evolutionary methods and neural networks in EWMA quality control chart design.

In the recent years metaheuristic approaches have been widely used to investigate the economic design of EWMA control charts. For instance, a study by Niaki et al. (2011) models and solves the economic and the economic-statistical design problems of Multivariate EWMA (MEWMA) control charts by a Particle Swarm Optimization (PSO) approach. Yet another economical design was conducted by Chou et al. (2008) who focused on variable sampling intervals of EWMA charts with sampling at fixed times using genetic algorithms. In fact, Genetic Algorithms are one of the most widely used heuristics. Aparisi and García-Díaz (2004) employed an GA optimization of Average Run Length (ARL) of EWMA and multivariate EWMA charts with respect to *smoothing parameter* and *control limits,* given that *sample size* n is 1. The authors further improved this setting where detection of small shifts is not necessary, while shift detecting is still important (Aparisi and García-Díaz 2007).

Sample size is added as another integer decision variable; and a third zone is defined where it is insignificant whether the process shift is detected or not.

Seeking a different heuristic approach, Niaki and Ershadi (2012) applied an ant-colony optimization of the economic-statistical design model of the MEWMA control chart in which the main parameters of the employed ant colony algorithm are tuned by means of a response surface methodology approach. Similarly, Zhou and Zhu (2008) used Grid Search minimization of the (hourly) cost of the integrated model (SPC and Preventive Maintenance) with respect to *sample size, sampling interval, control limits,* and *inspection interval.* To improve that Charongrattanasakul and Pongpullponsak (2011) contributed to the usual zone control policy, by adding another zone, called *warning zone*, and increased the set of four decision variables of Zhou and Zhu (2008) to six variables. In other words, the (hourly) cost of the integrated model (SPC and Preventive Maintenance) was minimized using GA with respect to the decision variables of *sample size, sampling interval, warning control limits, the number of subintervals between two consecutive sampling times,* and *the number of samples taken before Planned Maintenance.* This case revealed an interesting point such that the addition of the warning zone caused an overall increase of the total costs, due to increased ability of defective product detection which resulted to the increase of repairing and maintenance of machines; thus the hourly cost got higher. Other successful applications of Genetic Algorithms (GA) in the economic designs of control charts can be found in Saghaei et al. (2013), Celano and Fichera (1999), Chou and Chen (2006), Vommi and Seetala (2007), Torng et al. (2009a, b), and Lin et al. (2012). Besides these heuristics widely used in several areas of Operations Management, the real AI applications of EWMA charts achieved their potential in the semiconductor industry.

Since early 90s, EWMA has been popular in the semiconductor industry to maintain process targets over extended periods for improved product quality and decreased machine downtime (Spanos 1992; Su and Hsu 2004b). However, the several process factors (alternating in time) are thought to determine the 'best' value for an EWMA smoothing parameter. Smith and Boning (1997a, 1997b) proposed a self-tuning EWMA controller which dynamically updates its smoothing parameter by estimating the disturbance state and using the Artificial Neural Network (ANN) function mapping to provide updates to the controller parameters. Similarly, Su and Hsu (2004a) applied ANN for online tuning of EWMA smoothing parameter. The underlying approach indicated that the network learns very quickly when taking autocorrelation function and sample partial autocorrelation function patterns as the input features. Fan and Wang (2008) further contributed to this setting by incorporating a multivariate double EWMA component (i.e., EWMA having two smoothing parameters, coined for semiconductor industry by Butler and Stefani 1994). All these studies were conducted and results obtained in a simulation environment.

The next section introduces the concept of *jidoka* and its use within the model of this study as an intelligent quality control procedure.

4.3 Jidoka Production System as an Intelligent QC Approach

The proposed intelligent system provides controlling particular decisions automatically, which are once made manually by operators at the job level in a Jidoka Production System. The entrance of Japanese goods in western markets was a subject of discussion in American business in the 1970s (Schonberger 2007) and since then, academia has used many terms to explain this phenomenon, including Toyota Production System (TPS) also known as Lean Manufacturing (Monden 1983; Hoss and ten Caten 2013). Toyota Production System proposes factory designers to combine inspections with operations (Kim and Gershwin 2005). By this way, the production quality is increased and much more benefits are gained. Hoss and ten Caten (2013)suggest implementation of functions such as just in time (JIT) and *jidoka* to achieve high quality, low cost, and low lead time.

TPS is frequently modelled as a house with two pillars (see Fig. 4.2). The top of the house consists "highest quality, lowest cost, shortest lead time", whereas one of the two pillars represents just-in-time (JIT), and the other pillar the concept of jidoka. Jidoka (translated as autonomation) is a defect detection system, which automatically or manually stops the production operation whenever an abnormal or defective condition arises. The manufacturing system introduced will not stand without both of the pillars. Yet many researchers and practitioners focus on the mechanisms of implementation—one piece flow, pull production, tact time, standard work, kanban—without linking those mechanisms back to the pillars that hold up the entire system. While the majority of the studies in the literature have focused on the problems of the first pillar (JIT), two research articles by Kim and Gershwin (2005) and Berk and Toy (2009) are two notable exceptions for the other pillar (Hoss and ten Caten 2013). We can state that a lot of failed implementations can be traced back to not building this second pillar. In the concept of *jidoka* when a team member encounters a problem in his or her workstation, he/she is responsible for correcting the problem by pulling an andon cord, which can stop the line. The objective of jidoka can be summed up as: Ensuring quality 100 % of the time, preventing equipment breakdowns, and working efficiently.

TPS advocates think that mechanical and human *jidoka* prevent the waste that would result from producing a series of defective items. Hence, *jidoka* can be

Fig. 4.2 The house of toyota production system (Ohno 1988)

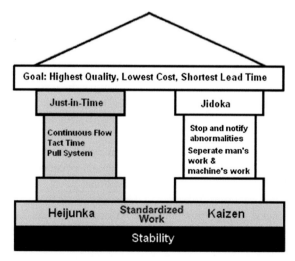

defined as a means to improve quality and increase productivity at the same time (Shingo 1989; Toyota Motor Corporation 1996; Kim and Gershwin 2005). Berk and Toy (2009)consider design of control charts in the presence of machine stoppages that are exogenously imposed (as under *jidoka* practices). Here, each stoppage creates an opportunity for inspection/repair at reduced cost. They first model a single machine facing opportunities arriving according to a Poisson process, develop the expressions for its operating characteristics and construct the optimization problem for economic design of a control chart. Afterwards, they consider a multiple machine setting where alarms about the quality status of the machines cause system-wide stoppages as it is the case under *jidoka* practices. Their findings indicate that ignoring exogenous inspection/repair opportunities and employing the classical QC chart parameters may result in significant cost increases.

4.3.1 Model Description

In the model described herein, we consider a multiple machine environment. In this environment, machines are operated and controlled individually. Each machine is controlled through a separate control chart. EWMA procedure is employed with multiple control limits; hence there are zones in the control charts. We use 2-zone control chart. We specify tight control limits (UCL_t and LCL_t) and loose control limits (UCL_l and LCL_l). The zone between the tight control limits is the inner zone and the zone between tight and loose control limits identifies the outer zone. Inner zone is where the process is considered to be in-control. A different inspection and repair policy is defined for outer zone and outside the control limits. A test statistic, Z_t, (exponentially weighted moving-average values of the sample means) which lies in the outer zone and outside the control limits is denoted as an *alarm* for the system.

Production system considered herein is assumed to be part of *Jidoka* Production System. Whenever a machine raises an alarm, all the system stops, i.e. the production is ceased. In the policy we suggest, these overall system stoppages due to individual machine alarms create opportunities for other machines to be inspected and repaired. This opportunity is taken only if the test statistic, Z_t lies beyond the loose control limits. That is, all the machines in the system are inspected and repaired, yielding in-control machines at the next system re-start instance. The time and cost of inspection/repair for each machine at a system stoppage are computed as follows: The alarm-raising machine(s) incur the downtime cost related to its own inspection/repair time. Among the remainder, the machine(s) with the longest inspection/repair time will have an available duration (as an opportunity for preventive maintenance) equal to the inspection/repair time of the alarm-raising machine(s). There will be no additional delay if these durations are equal to each other; otherwise, there will be some extra delay and machine experiencing the longest inspection/repair time will incur the additional downtime cost. However, when the system is stopped due to a Z_t value in the outer zone, only the machine(s) raising alarm is (are) inspected and repaired. As shutdowns are not utilized for inspecting other machines, the downtime cost is charged only to the self-stopping machine in this case.

In the model description below we retain the notation used in Berk and Toy (2009). Let M be the set of machines working in coordination, and let m denote the number of machines. The set of machines may comprise non-identical machines. Hence, when the machines have different reliability, restoration times, sampling interval etc., inspection opportunities may be beneficial, i.e., resulting in cost reduction, for some machines, it may be not beneficial, i.e., increasing the cost for the others.

Every machine $i(\in M)$, is subject to control with an EWMA control chart. A sample of size $y^{(i)}$ is taken with $h^{(i)}$ time intervals, test statistic Z_t is computed. The EWMA values of the quality specification of the sample weighted by λ, is plotted on a control chart. If the EWMA value of machine i falls outside the loose control limits (k_2) defined by Eq. 4.8.

$$\mu_0^{(i)} \pm k_2^{(i)} \frac{\sigma^{(i)}}{\sqrt{y^{(i)}}} \sqrt{\frac{\lambda}{2-\lambda}} \qquad (4.8)$$

all machines are stopped. If it falls between the loose and tight control limits (k_1) for machine i, the latter defined by

$$\mu_0^{(i)} \pm k_1^{(i)} \frac{\sigma^{(i)}}{\sqrt{y^{(i)}}} \sqrt{\frac{\lambda}{2-\lambda}} \qquad (4.9)$$

only that particular machine is stopped. A sample results in a false alarm with probability α (Type I error) and a true alarm with probability $(1-\beta)$ (complement of Type II error). Clearly, α and β are related to k_1, k_2, λ, y and for a normal variate Z. If the process faces an exogenous shutdown triggered by another alarm-raising machine, its operator uses this stoppage as an opportunity to carry out an inspection of the process although no signals have been received from the control chart to initiate one. On inspection, if the process is found to be in the out-of-control status, the opportunity is said to be a true opportunity, which is followed by a complete restoration of the process to the in-control status; otherwise, the opportunity is a false opportunity, which requires no adjustment. Type I and Type II errors for multiple machine with two zone control limits (α_1, α_2 and β_1, β_2) are as follows (Eqs. 4.10, 4.11, 4.12, 4.13):

$$\alpha_1 = 2\Phi\left[-k_1\sqrt{\frac{\lambda}{2-\lambda}}\right] \qquad (4.10)$$

$$\alpha_2 = 2\Phi\left[-k_2\sqrt{\frac{\lambda}{2-\lambda}}\right] \qquad (4.11)$$

$$\beta_1 = \Phi\left[k_1\sqrt{\frac{\lambda}{2-\lambda}} - \delta\sqrt{y^{(i)}}\right] - \Phi\left[-k_1\sqrt{\frac{\lambda}{2-\lambda}} - \delta\sqrt{y^{(i)}}\right] \qquad (4.12)$$

$$\beta_2 = \Phi\left[k_2\sqrt{\frac{\lambda}{2-\lambda}} - \delta\sqrt{y^{(i)}}\right] - \Phi\left[-k_2\sqrt{\frac{\lambda}{2-\lambda}} - \delta\sqrt{y^{(i)}}\right] \qquad (4.13)$$

Next we define the costs associated with the quality control of the systems. We consider three categories of quality costs: (i) sampling cost, (ii) the cost of operating in the out-of-control state, and (iii) the inspection and repair cost.

Sampling Cost: The sampling cost has two components, fixed and variable. Fixed component is denoted by u and is incurred at each sampling instance whereas variable component, y, is associated with the sample size and incurred at each sampling instance, as well. Hence, the total sampling cost is given by $u + by$.

Cost of Operating in out-of-control state: once the process shifts to the out-of-control state, any product processed in a stage (machine) is considered to be defective hence requires rework. The cost associated with these sub-standard products is denoted by a per time unit.

Inspection and repair cost: When a machined is stopped by an alarm, an inspection is conducted immediately and if shift is detected a repair operation is conducted. Since the system stops until inspection and repair operations completed we assume that there is a lost profit cost associated with idle time. The idle time is related with the system status at the stoppage instance; i.e., if the process has shifted it requires a time for both inspection and repair, however if the process has not shifted the idle time is only as much as the inspection time. The elapsed time until the process shift is distributed exponentially with mean θ. This cost component is computed as πL_s, where π is the profit (lost) per unit of time and L_s denotes the downtime of the process. Depending on the status of the machine at the stoppage instance repair cost varies, as well. Such that shifted process costs more than the non-shifted process, in terms of labour hours and spare parts.

Our objective is to economically and jointly design the control charts of all machines resulting a minimum system-wide cost rate. Control parameters of this system are same as the regular control charts. Specifically, we decide on the sample size (y), sampling frequency (h), control limits (k_1, k_2) and smoothing parameter (λ) which minimizes the expected cost per unit time.

4.4 Numerical Study

Three-machine setting with two zone control policy is simulated using the parameter sets below which results in 54 experimental instances. In each combination, an exhaustive search over the variable set (within the provided intervals) has been performed in order to find the 'best (minimum)' overall system cost rate.

The results are compared with the case where autonomation (or JPS) is not applied. Namely, the same experimental instances are simulated for the three independent machine system, i.e., for each machine a separate control chart is maintained, hence only self-stoppages are allowed.

The parameter ranges for our experimental set: variable cost (per sample) cost of sampling, $b = 0.1$; fixed cost of sampling $u = 5$; true and false alarm repair costs, $R_T = R_F = 0$; true and false alarm idle times, $L_T = L_F = L$, with $L \in \{0.05, 0.10, 0.15\}$; profit per unit time, $\pi \in \{500, 1500\}$; process mean shift rate, $\theta \in \{0.01, 0.05, 0.1\}$; and the per time cost of operating in out-of-control state, $a \in \{50, 250, 500\}$.

The domains for our decision variables, in which we performed an exhaustive search, are as follows: [1, 10] for the sample size (y), [0.01, 10] for sampling frequency (h), [1, 2] and [k_1, 3] for the control limits (k_1, k_2), and [0, 0.5] for

Table 4.3 Comparative results with improvement percentage for $\pi = 500$

Case #	L	θ	a	No JPS policy	System cost Under JPS policy	Improvement (%)
1	0.05	0.01	50	10.98	8.58	22
2			250	24.01	18.45	23
3			500	31.76	25.25	20
4		0.05	50	23.72	21.44	10
5			250	55.68	44.59	20
6			500	77.11	59.89	22
7		0.1	50	34.38	29.34	15
8			250	71.77	59.86	17
9			500	107.2	100.12	7
10	0.1	0.01	50	14.36	10.98	24
11			250	32.2	25.38	21
12			500	44.62	31.7	29
13		0.05	50	32.65	25.43	22
14			250	72.24	58.1	20
15			500	103.4	74.33	28
16		0.1	50	44.46	37.03	17
17			250	96.78	77.19	20
18			500	131.3	106.34	19
19	0.15	0.01	50	17.86	14.28	20
20			250	36.26	28.03	23
21			500	52.54	37.77	28
22		0.05	50	37.3	33.54	10
23			250	85.05	67.31	21
24			500	119.05	91.13	23
25		0.1	50	51.2	47.07	8
26			250	117.05	106.56	9
27			500	152.72	116.2	24

smoothing parameter (λ). Considering the time-related variables and parameters, the cases were simulated for a reasonably long final simulation run time and replicated 4.5 times on average. On the completion of each replication (50,000 time units), the domains of all decision variables were shrunk to narrower domains and another replication was performed. An improvement in total quality cost was generally observed after each replication, yet whenever the cost got worse the simulation was run for 150,000 time units using the parameter set resulting the lowest cost. Therefore, the number of replications depended on the quality cost of the penultimate replication.

Tables 4.3 and 4.4 presents the resulting costs of 54 different cases, which are the combinations of different parameters explained above. The costs under JPS policy and classical policy are given in the tables together with the percentage

Table 4.4 Comparative results with improvement percentage for $\pi = 1000$

Case #	L	θ	a	No	System cost	Improvement
				JPS policy	Under JPS policy	(%)
28	0.05	0.01	50	17.16	13.54	21
29			250	38.97	29.4	25
30			500	55.03	42.03	24
31		0.05	50	37.63	33.26	12
32			250	87.45	71.43	18
33			500	126.63	102.17	19
34		0.1	50	54.96	45.72	17
35			250	122.83	102.7	16
36			500	166.72	142.92	14
37	0.1	0.01	50	25.6	21.99	14
38			250	54.66	43.9	20
39			500	76.06	56.73	25
40		0.05	50	50.23	46.87	7
41			250	121.21	95.68	21
42			500	170.32	127.29	25
43		0.1	50	65.93	62.76	5
44			250	167.55	135.66	19
45			500	235.49	188.66	20
46	0.15	0.01	50	30.94	28.9	7
47			250	66.65	51.57	23
48			500	91.31	67.88	26
49		0.05	50	59.35	54.61	8
50			250	148.03	130.6	12
51			500	207.68	175.27	16
52		0.1	50	76.29	80.94	−6
53			250	198.38	169.32	15
54			500	276.1	225.02	19

of improvement. It can be clearly observed that, only one case does not show improvement under the JPS policy. The reason for this specific case is that, the computation time was significantly long for this parameter set and an adequate exhaustive search was not performed. The minimum improvement is 5 %, average improvement is 18 % and maximum is 29 % for all 54 cases. Hence we can conclude that JPS policy is significantly superior than the classical policy. The minimum improvement (5 %) is obtained in the case where $L = 0.10$, $\pi = 1500$, $\theta = 0.10$, and $a = 50$. The maximum improvement (29 %) is obtained in the case where $L = 0.10$, $\pi = 500$, $\theta = 0.01$, and $a = 500$.

We observe that system costs increase when a increases while other parameters are kept constant. Another result that can be derived from Table 4.4 is that improvement in the cost decreases when π increases from 500 to 1500 except for the cases where $L = 0.05$. Table 4.3 shows the results where $\pi = 500$ and Table 4.4 shows the results where $\pi = 1000$. The average improvement in Table 4.3 is 19 % while it is 16 % in Table 4.4. It is also seen from the results that, the improvement percentage decreases when θ increases. For instance, the average improvement percentage for the first three combinations, where $\theta = 0.01$ is approximately 22 % while it is 17 % for the next three cases where $\theta = 0.05$, and 13 % for the cases where $\theta = 0.1$. Similar pattern can be seen for the other case groups. However, we do not observe any significant pattern in the improvement percentages with respect to L.

4.5   Conclusion

In this article, we give an extensive literature review on quality control charts and we specifically consider the intelligent systems of economical design of EWMA zone control charts in the presence of machine stoppages that are exogenously imposed. Each stoppage creates an opportunity for inspection/repair at reduced cost. We consider a multiple machine setting where alarms about the quality status of the machines cause system-wide stoppages as it is the case under *Jidoka* practices. In a numerical study for three machine setting to investigate the cost advantages of employing the models herein versus the classical model where JPS is not used. Our findings indicate that ignoring exogenous inspection/repair opportunities and employing the classical QC chart parameters may result in significant cost increases. The average improvement percentage in cost under classical policy and JPS policy is 18 %, which shows that the performance of our model is really high. Hence we can conclude that JPS policy is significantly successful than the classical policy.

There are a number of extensions to our basic model. Herein, we consider only the design of $\bar{X}$ control charts in our numerical study, but our model can be applied to other variable and attribute-control charts. Similarly, different design criteria (semieconomic and statistical) can be considered, as well. For the future research, further applications of metaheuristics are also encouraged for the economic design of quality control charts.

References

Abbas, N., Riaz, M., Does, R.J.M.M.: Mixed exponentially weighted moving average–cumulative sum charts for process monitoring. Qual. Reliab. Eng. Int. **29**(3), 345–356 (2012)

Aparisi, F., García-Díaz, J.C.: Optimization of univariate and multivariate exponentially weighted moving-average control charts using genetic algorithms. Comput. Oper. Res. **31**(9), 1437–1454 (2004)

Aparisi, F., García-Díaz, J.C.: Design and optimization of EWMA control charts for in-control, indifference, and out-of-control regions. Comput. Oper. Res. **34**(7), 2096–2108 (2007)

Benneyan, J.C., Lloyd R.C., Plsek, P.E.: Statistical process control as a tool for research and healthcare improvement. Qual. Saf. Health Care **12**, 458–464 (2003)

Berk, E., Toy, A.O.: Quality control chart design under jidoka. Naval Res. Logist. **56**, 465–477 (2009)

Butler, S.W., Stefani, J.A.: Supervisory run-to-run control of polysilicon gate etch using in situ ellipsometry. IEEE Trans. Semicond. Manuf. **7**(2), 193–201 (1994)

Celano, G., Fichera, S.: Multiobjective economic design of an X control chart. Comput. Ind. Eng. **37**, 129–132 (1999)

Charongrattanasakul, P., Pongpullponsak, A.: Minimizing the cost of integrated systems approach to process control and maintenance model by EWMA control chart using genetic algorithm. Expert Syst. Appl. **38**(5), 5178–5186 (2011)

Chou, C.-Y., Chen, C.-H.: Economic design of variable sampling intervals T2 control charts using genetic algorithms. Expert Syst. Appl. **30**, 233–242 (2006)

Chou, C.Y., Cheng, J.C., Lai, W.T.: Economic design of variable sampling intervals EWMA charts with sampling at fixed times using genetic algorithms. Expert Syst. Appl. **34**, 419–426 (2008)

Costa, A.F.B.: Joint economic design of $\bar{X}$ and R control charts for process subject to two independent assignable causes. IIE Trans. **25**, 27–33 (1993)

Costa, A.F.B.: Joint X chart with variable parameters. J. Qual. Technol. **31**, 408–416 (1999)

Crowder, S.V.: Design of exponentially weighted moving average scheme. J. Qual. Technol. **21**, 155–162 (1989)

Davis, R.B., Krehbiel, T.C.: Shewhart and zone control chart performance under linear trend. Commun. Stat. —Simul. Comput. **31**(1), 91–96 (2002)

Davis, R.B., Jin, C., Guo, Y.: Improving the performance of the zone control chart. Commun. Stat. —Theory Methods **23**, 3557–3565 (1994)

Davis, R.B., Homer, A., Woodall, W.H. Performance of the zone control chart. Commun. Stat.—Theory Methods **19**, 1581–1587 (1990)

De Vries, A., Reneau, J.K.: Application of statistical process control charts to monitor changes in animal production systems. J. Anim. Sci. **88**, E11–E24 (2010)

Diaz, M., Neuhauser, D.: Pasteur and parachutes: when statistical process control is better than a randomized controlled trial. Qual. Saf. Health Care **14**, 140–143 (2005)

Duncan, A.J.: The economic design of $\bar{X}$ charts used to maintain current control of process. J. Am. Stat. Assoc. **51**, 228–242 (1956)

Fan, S.-K.S., Wang, C.-Y.: On-line tuning system of multivariate dEWMA control based on a neural network approach. Int. J. Prod. Res. **46**(13), 3459–3484 (2008)

Flaig, J.J.: Zone control charts. http://www.e-AT-USA.com (2004)

Gan, F.F.: Computing the percentage points of the run length distribution of an exponentially weighted moving average control chart. J. Qual. Technol. **23**, 359–362 (1991)

Haq, A., Brown, J., Moltchanova, E.: New synthetic EWMA and synthetic CUSUM control charts for monitoring the process mean. Qual. Reliab. Eng. Int. (2014). doi:10.1002/qre.1747

Hawkins, D.M., Olwell, D.H.: Cumulative Sum Charts and Charting for Quality Improvement. Springer, New York (1998)

Hawkins, D.M., Wu, Q.: The CUSUM and the EWMA head-to-head. Qual. Eng. **26**, 215–222 (2014)

He, H., Grigoryan, A., Sigh, M.: Design of double- and triple-sampling X-bar control charts using genetic algorithms. Int. J. Prod. Res. **40**(6), 1387–1404 (2002)

Ho, C., Case, K.E.: An economic design of the zone control chart for jointly monitoring process centering and variation. Comput. Ind. Eng. **26**(2), 213–221 (1994a)

Ho, C., Case, K.E.: Economic design of control charts: a literature review for 1981–1991. J. Qual. Technol. **26**, 39–53 (1994b)

Hopgood, A.A. Intelligent Systems for Engineers and Scientists, CRC Press, Boca Raton (2012)

Hoss, M., ten Caten, C.S.: Lean schools of thought. Int. J. Prod. Res. **51**(11), 3270–3282 (2013)

Jaehn, A.H.: Zone control charts—SPC made easy. Quality **26**, 51–53 (1987)

Jaehn, A.H.: Zone control charts find new applications. ASQC Qual. Congr. Trans. 890–895 (1989)

Kim, J., Gershwin, S.B.: Integrated quality and quantity modeling of a production line. OR Spectrum **27**, 287–314 (2005)

Lee, P.-H., Torng, C.-C., Liao, L.-F.: An economic design of combined double sampling and variable sampling interval X control chart. Int. J. Prod. Econ. **138**, 102–106 (2012)

Lin, S.-N., Chou, C.-Y., Wang, S.-L., Liu, H.-R.: Economic design of autoregressive moving average control chart using genetic algorithms. Expert Syst. Appl. **39**, 1793–1798 (2012)

Lorenzen, T.J., Vance, L.C.: The economic design of control charts: a unified approach. Technometrics **28**, 3–10 (1986)

Lucas, J.M., Saccucci, M.S.: Exponentially weighted moving average control schemes: properties and enhancements (with discussion). Technometrics **32**, 1–29 (1990)

Monden, Y.: Toyota Production System: Practical Approach to production Management. Management Institute Engineering and Management Press, Norcross (1983)

Montgomery D.C.: The economic design of control charts: a review and literature survey. J. Qual. Technol. **1**, 24–32 (1980a)

Montgomery, D.C.: The economic design of control charts: a review and literature survey. J. Qual. Technol. **12**, 75–81 (1980b)

Montgomery, D.C.: Introduction to Statistical Quality Control. Wiley, New York (1996)

Montgomery, D.C.: Introduction to Statistical Quality Control, 2nd edn. Wiley, New York (1991)

Montgomery, D.C., Torng, J.C.-C., Cocgran, J.K., Lawrence, F.P.: Statistically constrained economic design of the EWMA control chart. J. Qual. Technol. **27**(3), 250–256 (1995)

Montgomery D.C.: Introduction to Statistical Quality Control, 6th edn. Wiley, New York (2009)

Montgomery, D.C.: Introduction to Statistical Quality Control, 7th edn. Wiley, New York (2013)

Nenes, G.: A new approach for the economic design of fully adaptive control charts. Int. J. Prod. Econ. **131**(2), 631–642 (2011)

Neubauer, A.S.: The EWMA control chart: properties and comparison with other quality-control procedures by computer simulation. Clin. Chem. **43**(4), 594–601 (1997)

Negnevitsky, M. Artificial Intelligence: a Guide to Intelligent Systems, Addison-Wesley, Boston (2011)

Niaki, S.T.A., Ershadi, M.J., Malaki, M.: Economic and economic-statistical designs of MEWMA control charts, a hybrid taguchi loss, markov chain and genetic algorithm approach. Int. J. Adv. Manuf. Technol. **48**, 283–296 (2010)

Niaki, S.T.A., Gazaneh, F.M. Toosheghanian, M. Economic design of variable sampling interval X̄ control charts for monitoring correlated non normal samples. Commun. Stat.—Theory Methods **42**(18), 3339–3358 (2013)

Niaki, S.T.A., Malaki, M., Ershadi, M.J.: A particle swarm optimization approach on economic and economic-statistical designs of MEWMA control charts. ScientiaIranica E **18**(6), 1529–1536 (2011)

Niaki, S.T.A., Ershadi, M.J.: A hybrid ant colony Markov chain, and experimental design approach for statistically constrained economic design of MEWMA control charts. Expert Syst. with Appl. **39**(3), 3265–3275 (2012)

Ohno, T.: Toyota Production System, Productivity Press, Boca Raton (1988)

Page, E.S.: Continuous inspection schemes. Biometrika **41**, 100–115 (1954)

Park, C.: Economic design of charts when signals may be misclassified and the bounded reset chart. IIE Trans. **45**(4), 436–448 (2013)

Park, C., Lee, J., Kim, Y.: Economic design of a variable sampling rate EWMA chart. IIE Trans. **36**, 387–399 (2004)

Parkash, V., Kumar, D., Rajoria, R.: Statistical Process Control. IJRET: Int. J. Res. Eng. Technol. **2**(8), 70–72 (2013)

Pignatiello, J.J., Tsai, A.: Optimal economic design of control chart when cost model parameters are not known. IIE Trans. **20**, 103–110 (1988)

Quesenberry, C.P.: SPC Methods for Quality Improvement. Wiley, New York (1997)

Reynolds Jr, M.R., Amin, R.W., Arnold, J.C., Nachlas, J.A.: X charts with variable sampling intervals. Technometrics **30**, 181–192 (1988)

Roberts, S.W.: Control chart tests based on geometric moving averages. Technometrics **1**, 239–250 (1959)

Robinson, P.B., Ho, T.Y.: Average run lengths of geometric moving average charts by numerical methods. Technometric **20**, 85–93 (1978)

Saghaei, A., Ghomi, S.M.T.F., Jaberi, S.: Economic design of exponentially weighted moving average control chart based on measurement error using genetic algorithm. Reliab. Eng. Int. Qual. doi:10.1002/qre.1538

Schonberger, R.J.: Japanese production management: an evolution—with mixed success. J. Oper. Manag. **25**(2), 403–419 (2007)

Serel, A.S.: Economic design of EWMA control charts based on loss function. Math. Comput. Model. **49**, 745–759 (2009)

Serel, D.A., Moskowitz, H.: Joint economic design of EWMA control charts for mean and variance. Eur. J. Oper. Res. **184**, 157–168 (2008)

Schalkoff, R.J.: Intelligent Systems: Principles. Jones & Bartlett Publishers, Paradigms and Pragmatics (2009)

Shewhart, W.A.: Some Applications of Statistical Methods. ASQ Publications (1924)

Shingo, S.: A Study of the Toyota Production System from an Industrial Engineering Viewpoint. Productivity Press, Portland (1989)

Smith, Taber H., Boning, Duane S.: Artificial neural network exponentially weighted moving average controller for semiconductor processes. J. Vac. Sci. Technol. A **15**, 1377–1384 (1997a)

Smith, T.H., Boning, D.S.: A self-tuning EWMA controller utilizing artificial neural network function approximation techniques. Compon. Packag. Manuf. Technol. Part C, IEEE Trans. **20** (2), 121–132 (1997b)

Solodky, C., Chen, H., Jones, P.K., Katcher, W., Neuhauser, D.: Patients as partners in clinical research: a proposal for applying quality improvement methods to patient care. Med. Care 36 (Suppl.):AS13 AS20 (1998)

Spanos, C.J.: Statistical process control in semiconductor manufacturing. Proc. IEEE 819–830 (1992)

Su, C.-T., Hsu, C.-C.: On-line tuning of a single EWMA controller based on the neural technique. Int. J. Prod. Res. **42**(11), 2163–2178 (2004a)

Su, C.-T., Hsu, C.-C.: A time-varying weights tuning method of the double EWMA controller. Omega **32**(6), 473–480 (2004b)

Thor, J., Lundberg, J., Ask, J., Olsson, J., Carli, C., PukkHarenstam, K., Brommels, M.: Application of statistical process control in healthcare and improvement: a systematic review. Qual. Saf. Health Care **16**(387), 399 (2007)

Tolley, G.O., English, J.R.: Economic designs of constrained EWMA and combined EWMA-X-bar control schemes. IIE Trans. **33**, 429–436 (2001)

Torng, C.-C., Lee, P.-H., Liao, N.-Y.: An economic-statistical design of double sampling X̄ control chart. Int. J. Prod. Econ. **120**(2), 495–500 (2009a)

Torng, C.-C., Lee, P.-H., Liao, H.-S., Liao, N.-Y.: An economic design of double sampling X̄ charts for correlated data using genetic algorithms. Expert Syst. Appl. **36**(10), 12621–12626 (2009b)

Toyota Motor Corporation (1996). The Toyota production system

Vance, L.C.: A bibliography of statistical quality control chart techniques, 1970–1980. J. Qual. Technol. **15**, 59–62 (1983)

Vargas, V.C.C., Lopes, L.F.D., Souza, A.M.: Comparative study of the performance of the CuSum and EWMA control charts. Comput. Ind. Eng. **46**(4), 707–724 (2004)

Vommi, V.B., Seetala, S.N.: A simple approach for robust economic design of control charts. Comput. Oper. Res. **34**(7), 2001–2009 (2007)

Waldman, K.H.: Bounds for the distribution of the run length of geometric moving average charts. J. Roy. Statist. Soc. Ser. B (Appl. Statist.), **35**, 151–158 (1986)

Electric, Western: Statistical Quality Control Handbook. AT& T, Indianapolis (1956)

Woodall, W.H.: Weakness of the economic design of control charts. Technometrics **28**, 408–409 (1986)

Woodall, W.H.: Conflicts between Deming's philosophy and the economic design of control charts. In: Lenz, H.J., Wetherill, G.B., Wilrich, P.T. (eds.) Frontiers in Statistical Quality Control 3, pp. 155–168. Physica-Verlag, Heidelberg (1987)

Woodall, W.H.: The use of control charts in health-care and public-health surveillance. J. Qual. Technol. **38**, 89–104 (2006)

Woodall, W.H., Lorenzen, T.J., Vance, L.C.: Weaknesses of the economic design of control charts. Technometrics **28**, 408–410 (1986)

Zaman, B., Riaz, M., abbas, N., Does, R.J.M.M.: Mixed cumulative sum–exponentially weighted moving average control charts: an efficient way of monitoring process location. Qual. Reliab. Eng. Int. doi:10.1002/qre.1678

Zhou, W.-H., Zhu, G.-L.: Economic design of integrated model of control chart and maintenance. Math. Comput. Model. **47**, 1389–1395 (2008)

Chapter 5
Trends on Process Capability Indices in Fuzzy Environment

Abbas Parchami and B. Sadeghpour-Gildeh

Abstract After the fuzzy set theory was introduced and developed, many studies have been realized to combine quality control methods and fuzzy set theory. This chapter is including the categorization of most essential works on fuzzy process capability indices in the following four main categories:

(1) Lee et al.'s method and its extensions: This class deals with the method of modeling and estimating the membership function of process capability indices where all data and specifications are fuzzy numbers;

(2) Parchami et al.'s method and its extensions: This class deals with the problem of obtaining fuzzy process capability indices based on fuzzy specification limits and crisp data by extension principle approach;

(3) Kaya and Kahraman's method and its extensions: This class deals with the problem of estimating the classical process capability indices by a triangular shaped fuzzy number when both specifications and data are crisp;

(4) Yongting's method and its extensions: This class deals with introducing process capability indices based on fuzzy quality where the data and parameters are crisp.

After presenting the basic idea of the main works, all related studies briefly reviewed in each class. Some numerical examples are presented to show the applicability of the proposed methods.

Keywords Confidence interval · Statistical quality control · Fuzzy data · Fuzzy specification limits · Fuzzy event · Fuzzy quality

A. Parchami (✉)
Department of Statistics, Faculty of Mathematics and Computer Science,
Shahid Bahonar University of Kerman, 6169-133 Kerman, Iran
e-mail: parchami@uk.ac.ir

B. Sadeghpour-Gildeh
Department of Statistics, Faculty of Mathematical Science,
Ferdowsi University of Mashhad, Mashhad, Iran

© Springer International Publishing Switzerland 2016
C. Kahraman and S. Yanık (eds.), *Intelligent Decision Making in Quality Management*, Intelligent Systems Reference Library 97,
DOI 10.1007/978-3-319-24499-0_5

5.1   Preliminaries: Process Capability Indices

A common way to measure performance of a manufacturing process is using process capability indices based on a random sample which taken from the production line. In fact, process capability ratio or process capability index (PCI) is a shorthand numerical comparison, which measured the capability and effectiveness of the quality characteristic with respect to the specification limits. In other words, PCI is a statistical measure to calculate the ability of a process to produce output within specification limits. Several PCIs introduced in the literature such as C_p, C_{pk}, C_{pm}, C_{pmk} and so on (Kotz 1993). When univariate measurements concerned, we will denote the corresponding random variable (quality characteristic) by X. Expected value and standard deviation of X will be denoted by μ and σ, respectively. The commonly recognized PCIs are:

$$C_p = \frac{USL - LSL}{6\sigma},$$

(5.1)

where USL and LSL are respectively the upper and lower specification limits. This C_p is used when $\mu = M$ with $M = (U+L)/2$.

$$C_{pk} = \frac{USL - LSL - 2|\mu - M|}{6\sigma} = \frac{\min\{USL - \mu, \mu - LSL\}}{3\sigma},$$

(5.2)

and

$$C_{pm} = \frac{USL - LSL}{6\sqrt{\sigma^2 + (\mu - T)^2}} = \frac{USL - LSL}{6\sqrt{E[(X - T)^2]}},$$

(5.3)

where T is the target value and $E[.]$ denotes the expected value. There is also the hybrid index

$$C_{pmk} = \frac{USL - LSL - 2|\mu - M|}{6\sqrt{\sigma^2 + (\mu - T)^2}} = \frac{USL - LSL - 2|\mu - M|}{6\sqrt{E[(X - T)^2]}}.$$

(5.4)

Usually, $T = M$. If $T \neq M$ the situation sometimes described as "asymmetric tolerances", see (Boyles 1994). Introduction of C_p ascribed to Juran (1974); that of C_{pk} to Kane (1986); that of C_{pm} for the most part to Hsiang and Taguchi (1985), and C_{pmk} to Pearn et al. (1992). Substituting the sample mean and standard deviation provides a point estimate for any of PCIs. For more details on conventional and classical PCIs see (Kotz and Johnson 2002), (Montgomery 2005) and (Kotz and Lovelace 1998).

Although PCIs are effective tools for quality assurance and they have been proposed to provide numerical measures on process capability in a precise environment, but we may confront imprecise concepts in a manufacturing process. If we introduce vagueness into some crisp assumptions (such as data, quality set, specification limits and target value), then we face quite new and interesting processes, where the ordinary capability indices are not appropriate for measuring the capability of these processes. However, classical PCIs extension to fuzzy environment is concomitant with some computational difficulties. Classical statistical quality control is based on crisp data, random variables, control charts, decision rules, capability indices, and so on. As there are many different situations in which the above assumptions are rather unrealistic, there have been some attempts to analyze these situations with the fuzzy set theory. In the present chapter, we try to briefly overview some works on the applications of fuzzy set theory and fuzzy logic for extending process capability indices in quality control researches. For simplify, we classify this overview in four categories that are presented in four Sects. 5.2, 5.3, 5.4 and 5.5. It must be mentioned that the assumptions of theses four categories are different and so they are not comparable.

At the end of this section, we clarify our notation for triangular fuzzy numbers which used through paper. As an especial case of fuzzy numbers, triangular fuzzy number defined by membership function

$$T(x) = \begin{cases} \frac{x-a}{b-a} & \textit{if} \quad a \leq x < b \\ \frac{c-x}{c-b} & \textit{if} \quad b \leq x < c \\ 0 & \textit{elsewhere} \end{cases} \tag{5.5}$$

and it symbolically noted by $T(a,b,c)$. The real number b called the core value and the positive real numbers $b - a$ and $c - b$ called left and right spreads of triangular fuzzy number, respectively. $F_T(R)$ denotes the set of all triangular fuzzy numbers where R is the set of all real numbers.

Four basic methods on generalization of process capability indices for fuzzy environment and their extensions are reviewed and discussed in this paper.

The rest of the paper is organized as follows. An introduction and a brief review of Lee et al.'s (1999) method, Parchami et al.'s (2005) method, Kaya and Kahraman's (2009) method and Yongting's (1996) method are presented in Sects. 5.2, 5.3, 5.4 and 5.5, respectively. The last section concludes the paper and gives suggestions for further research.

5.2   Lee et al.'s Method and Its Extensions

Lee et al. (1999) generalized the capability index C_p by extension principles based on fuzzy specifications and fuzzy data. Under a similar conditions, Lee (2001) follows his approach to generalize capability index C_{pk}. Based on triangular fuzzy observations $\tilde{x}_j = T(o_j, p_j, q_j) \in F_T(R), j = 1, \ldots, n$, and considering triangular fuzzy target value $\tilde{t} = T(w, y, z) \in F_T(R)$ and also triangular fuzzy specification limits $\widetilde{LSL} = T(l,m,n) \in F_T(R)$ and $\widetilde{USL} = T(o,p,q) \in F_T(R)$, Lee proposed the following approximation for the membership function of C_{pk} index

$$U_{C_{pk}}(I) \simeq \begin{cases} \frac{-B_1}{2A_1} + \left[\left(\frac{B_1}{2A_1}\right)^2 - \frac{C_1-I}{A_1}\right]^{1/2} & if \quad C_1 \leq I \leq C_3 \\ \frac{B_2}{2A_2} - \left[\left(\frac{B_2}{2A_2}\right)^2 - \frac{C_2-I}{A_2}\right]^{1/2} & if \quad C_3 \leq I \leq C_2 \\ 0 & elsewhere, \end{cases} \quad (5.6)$$

in which

$$A_1 = (b-a)(e-d), \quad A_2 = (c-b)(f-e),$$
$$B_1 = a(e-d) + d(b-a), \quad B_2 = c(f-e) + f(c-b),$$
$$C_1 = ad, \quad C_2 = cf, \quad C_3 = be,$$

$$a = 1 - \left(\frac{\sum_{j=1}^n o_j}{n} - z\right)\left(\frac{2}{q-l}\right),$$

$$b = 1 - \left(\frac{\sum_{j=1}^n p_j}{n} - y\right)\left(\frac{2}{p-m}\right),$$

$$c = 1 - \left(\frac{\sum_{j=1}^n q_j}{n} - w\right)\left(\frac{2}{o-n}\right),$$

$$d = (o-n)\left(\frac{1}{6C_2}\right),$$

$$e = (p-m)\left(\frac{1}{6C_3}\right),$$

$$f = (q-l)\left(\frac{1}{6C_1}\right).$$

After computing the membership function of fuzzy PCI, he fuzzified the proposed fuzzy PCI for making final decision in the examined manufacturing process. The major advantage of the proposed method is using extension principle approach. Complex calculations, low speed of process and presenting non-exact approximates for capability indices are weakness points of Lee's method which cause increasing the progress of the proposed method.

A similar approach to solve this problem based on extension principle presented by Shu and Wu (2009) by fuzzy data. In their approach, which is easier and fasten than Lee's method, the α-cuts of fuzzy index C_{pk} was calculated based on the α-cuts of fuzzy data for $0 \leq \alpha \leq 1$. Meanwhile, they investigated on the capability of the LCD monitors assembly line using their generalized indices. In this regard, the capability test on the generalized capability index C_p with fuzzy data have been investigated by Tsai and Chen (2006).

5.3 Parchami et al.'s Method and Its Extensions

A process with fuzzy specification limits, which Parchami et al. (2005) called a fuzzy process for short, is one which approximately satisfies the normal distribution condition and its specification limits are fuzzy. They extend the classical PCIs (1-4) by extension principle for fuzzy processes as follows

$$\tilde{C}_p = T\left(\frac{a_u - c_l}{6\sigma}, \frac{b_u - b_l}{6\sigma}, \frac{c_u - a_l}{6\sigma}\right), \tag{5.7}$$

$$\tilde{C}_{pk} = T\left(\frac{a_u - c_l - 2|\mu - m|}{6\sigma}, \frac{b_u - b_l - 2|\mu - m|}{6\sigma}, \frac{c_u - a_l - 2|\mu - m|}{6\sigma}\right) \tag{5.8}$$

$$\tilde{C}_{pm} = T\left(\frac{a_u - c_l}{6\sqrt{\sigma^2 + (\mu - t)^2}}, \frac{b_u - b_l}{6\sqrt{\sigma^2 + (\mu - t)^2}}, \frac{c_u - a_l}{6\sqrt{\sigma^2 + (\mu - t)^2}}\right) \tag{5.9}$$

and

$$\tilde{C}_{pmk} = T\left(\frac{a_u - c_l - 2|\mu - m|}{6\sqrt{\sigma^2 + (\mu - t)^2}}, \frac{b_u - b_l - 2|\mu - m|}{6\sqrt{\sigma^2 + (\mu - t)^2}}, \frac{c_u - a_l - 2|\mu - m|}{6\sqrt{\sigma^2 + (\mu - t)^2}}\right), \tag{5.10}$$

where t is target value, $m = (b_u + b_l)/2$, $a_u \geq c_l$ and the fuzzy numbers $U(a_u, b_u, c_u) = T(a_u, b_u, c_u) \in F_T(R)$ and $L(a_l, b_l, c_l) = T(a_l, b_l, c_l) \in F_T(R)$ are the upper and lower engineering specification limits, respectively. It is obvious that the proposed fuzzy indices $\tilde{C}_p, \tilde{C}_{pk}, \tilde{C}_{pm}, \tilde{C}_{pmk}$ are exactly triangular fuzzy numbers and they are applied when the data are crisp and specification limits are two triangular fuzzy numbers.

Example 1 For a special product suppose that the lower and upper specification limits are considered to be "approximately 4" and "approximately 8", which are characterized by $L(2, 4, 6) = T(2, 4, 6) \in F_T(R)$ and $U(7, 8, 9) = T(7, 8, 9) \in F_T(R)$; respectively (see the left graph of Fig. 5.1). Assume that the process mean and the process standard deviation are 6 and $\frac{2}{3}$, respectively. Also, let the target value be equal to 7. By Eq. (5.7) one can easily calculate the fuzzy index $\tilde{C}_p = T\left(\frac{1}{4}, 1, \frac{7}{4}\right)$. Therefore, $\tilde{C}_p$ is "approximately one", as shown in the right graph of Fig. 5.1. Also by Eq. (5.9), one can similarly calculate the fuzzy index $\tilde{C}_{pm} = T\left(\frac{1}{2\sqrt{13}}, \frac{2}{\sqrt{13}}, \frac{7}{2\sqrt{13}}\right)$. Moreover, considering Eqs. (5.7–5.10), we can expect that $\tilde{C}_{pk} = \tilde{C}_p = T(0.25, 1, 1.75)$ and $\tilde{C}_{pmk} = \tilde{C}_{pm} = T(0.139, 0.555, 0.971)$ which are drawn in the right graph of Fig. 5.1, since $m = \mu$.

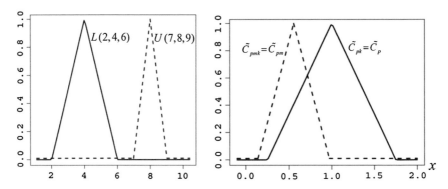

Fig. 5.1 The membership functions of fuzzy specification limits (*left graph*) and the membership functions of fuzzy process capability indices based on Parchami et al.'s (2005) method (*right graph*) in Example 1

In recent years, some papers have been concentrated on different statistical fields of fuzzy process which we briefly review them. Moeti et al. (2006) introduced the fuzzy process capability indices based on *LR* specification limits. Ramezani et al. (2011) and Parchami et al. (2006), (2011) constructed several fuzzy confidence regions for fuzzy PCIs $\tilde{C}_p$ and $\tilde{C}_{pm}$, respectively. Testing the capability of fuzzy processes are investigated by Parchami and Mashinchi (2009) where specification limits are triangular fuzzy numbers. Extending other classical and conventional PCIs are followed by Kaya and Kahraman (2010) based on this method. Also, after extending this method by Kaya and Kahraman (2008) for trapezoidal fuzzy specification limits, they applied their extended PCIs to compare several educational and teaching processes (also see Mashinchi et al. 2005).

5.4 Kaya and Kahraman's Method and Its Extensions

An another prevalent method for PCIs estimation is constructed on the basis of Buckley's estimation approach. Buckley (2004), (2006) propose a general estimation approach to estimate any unknown parameter by a triangular shaped fuzzy number whose α-cuts are equal to the $100(1 - \alpha)\%$ confidence intervals of the parameter. Recently, several authors used Buckley's estimation approach to PCIs estimation by a triangular shaped fuzzy number when both specifications and data are crisp. Parchami and Mashinchi (2007) estimated classical PCIs C_p, C_{pk} and C_{pm} by Buckley's approach and they proposed a method for the comparison of the estimated PCIs. For instance in their approach, the α-cut of the fuzzy estimation for C_p is equivalent to

$$\left[\hat{C}_p \sqrt{\frac{\chi^2_{n-1,\alpha/2}}{n-1}}, \ \hat{C}_p \sqrt{\frac{\chi^2_{n-1,1-\alpha/2}}{n-1}} \right], \quad 0 < \alpha < 1, \tag{5.11}$$

in which $\hat{C}_p = \frac{USL-LSL}{6s}$ is the point estimation of C_p and $\chi^2_{n,\alpha}$ is the α-quantile of Chi-square distribution with n degrees of freedom. So, the proposed estimations for PCIs contain both point and interval estimates and so provide more information for the practitioner. Kahraman and Kaya (2009) introduced fuzzy PCIs for quality control of irrigation water. Wu (2009) proposed an approach for testing process performance C_{pk} based on Buckley's estimator with crisp data and crisp specification limits. Also, after introducing Buckley's fuzzy estimation for capability index, Wu and Liao (2009) investigated on testing process yield assuming fuzzy critical value and fuzzy p-value. It must be clarified that both data and specification limits have considered crisp in two recent works and the presented concepts are also illustrated in a case study on the light emitting diodes manufacturing process. In this regard, Kaya and Kahraman (2009) introduced fuzzy robust capability indices and they evaluated the air pollution's Istanbul by their fuzzy PCIs. For instance, the α-cut of the presented fuzzy estimation in Eq. (5.11) modified in their method as follows

$$\left[\hat{C}_p \sqrt{\frac{\chi^2_{n-1,\alpha/2}}{n-1}} + \left(\hat{C}_p - \hat{C}_p \sqrt{\frac{\chi^2_{n-1,0.5}}{n-1}} \right), \right.$$
$$\left. \hat{C}_p \sqrt{\frac{\chi^2_{n-1,1-\alpha/2}}{n-1}} + \left(\hat{C}_p - \hat{C}_p \sqrt{\frac{\chi^2_{n-1,0.5}}{n-1}} \right) \right], \quad 0 < \alpha \leq 1. \tag{5.12}$$

Example 2 Suppose that the lower and upper specification limits for a product are $L = 4$ and $U = 8$, respectively. By assuming $\mu = 6$, we take a random sample $X_1, X_2, \cdots, X_{41}$ from $N(6, \sigma^2)$ to estimate index C_p and assume that the estimated process standard deviation is $2/3$. According to Eq. (5.11), the α-cut of Parchami and Mashinchi's (2007) fuzzy estimation for C_p is equivalent to

$$\left[\sqrt{\frac{\chi^2_{40,\alpha/2}}{40}}, \ \sqrt{\frac{\chi^2_{40,1-\alpha/2}}{40}} \right], \quad 0 < \alpha < 1, \tag{5.13}$$

in which $\hat{C}_p = \frac{8-4}{6 \times \frac{2}{3}} = 1$ is computed. The graph of the membership function of Parchami and Mashinchi's (2007) fuzzy estimation is drawn in Fig. 5.2 by line. We would never expect the classical precise point estimate $\hat{C}_p = 1$ to be exactly equal to the parameter value, so we often compute a $(1-\alpha)100\%$ confidence interval for C_p. The fuzzy estimate obtained by this approach contains more information than a point

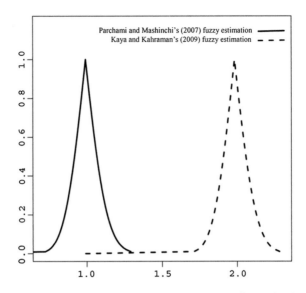

Fig. 5.2 The membership functions of fuzzy estimate for capability index by Parchami and Mashinchi (2007) and Kaya and Kahraman's (2009) methods in Example 2

or interval estimate, in the sense that the fuzzy estimate contains the point estimate and all $(1 - \alpha)100\%$ confidence intervals all at once for $0 < \alpha < 1$, which can be useful from practical point of view. From Parchami and Mashinchi's (2007) fuzzy estimate, one can conclude that the classical crisp estimate $\hat{C}_p = 1$ belongs to their fuzzy estimate with grade of membership one. It is obvious that their fuzzy set contains more elements other than "1" with corresponding grades of membership. For example, one can says that $\hat{C}_p = 0.946$ belongs to their fuzzy estimate with grade of membership 0.68. Meanwhile according to Eq. (5.12), the α-cut of the modified Kaya and Kahraman's (2009) fuzzy estimation can be calculated as follows

$$\left[\sqrt{\frac{\chi^2_{40,\alpha/2}}{40}} + \left(1 - \sqrt{\frac{\chi^2_{40,0.5}}{40}}\right), \sqrt{\frac{\chi^2_{40,1-\alpha/2}}{40}} + \left(1 - \sqrt{\frac{\chi^2_{40,0.5}}{40}}\right) \right], \quad 0 < \alpha \leq 1,$$

(5.14)

where $\chi^2_{40,0.5} = 39.34$ is the median of Chi-square distribution with 40 degrees of freedom. The graph of the membership function of Kaya and Kahraman's (2009) fuzzy estimation is drawn in Fig. 5.2 by dash line.

As another work on this topic, Hsu and Shu (2008) studied on fuzzy estimation of capability index C_{pm} to assess manufacturing process capability with imprecise data. Kaya and Kahraman (2011b) estimated classical capability indices via triangular shaped fuzzy numbers by replacing Buckley's fuzzy estimations of process mean and process standard deviation. Analyzing fuzzy PCIs followed by Kaya and Kahraman (2011a) based on fuzzy measurements and also they drawn fuzzy control charts for fuzzy measurements. Moradi and Sadeghpour-Gildeh (2013) worked on fuzzy one-sided process capability plots for the family of one-sided specification limits.

5.5 Yongting's Method and Its Extensions

Yongting (1996), for the first time, defines fuzzy quality by substituting the indicator function $I_{\{x\mid x\in[LSL,USL]\}}$ with the membership function of the fuzzy set $\tilde{Q}$, where the membership function $\tilde{Q}(x)$ represents the degree of conformity of the measured quality characteristic with standard quality (or briefly, the degree of quality). Note that by using fuzzy quality idea, the range of quality characteristic function will be changed from $\{0,1\}$ into $[0,1]$, see Fig. 5.3.

Also, Yongting (1996) introduced the capability index

$$C_{\tilde{p}(Y)} = \begin{cases} \int\limits_{-\infty}^{+\infty} \tilde{Q}(x)f(x)dx & \text{continuous random variable} \\ \sum\limits_{i=1}^{N} \tilde{Q}(x_i)P(x_i) & \text{discrete random variable} \end{cases} \qquad (5.15)$$

based on fuzzy quality for precise data in which f and P are p.d.f. and p.m.f. of the quality characteristic, respectively. Sadeghpour-Gildeh (2003) compared capability indices C_p, C_{pk} and $C_{\tilde{p}(Y)}$ with respect to the measurement error occurrence.

Example 3 Suppose that a random sample is taken from an assembly line of a special product under the normality assumption. The mean and standard deviation of the observed data are $\bar{x} = 0.7$ and $s = 0.15$, respectively. First, we consider a non-symmetric triangular fuzzy quality with the following membership function for product (see Fig. 5.4)

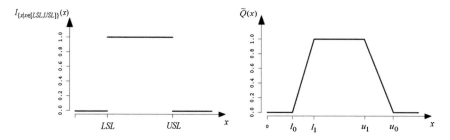

Fig. 5.3 Characterized classical quality with the indicator function of non-defective products (*left figure*), and characterized fuzzy quality with the fuzzy set of non-defective products (*right figure*)

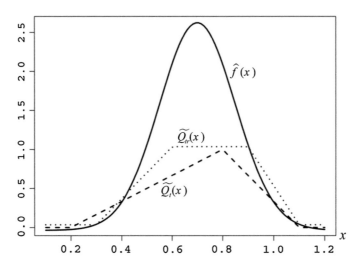

Fig. 5.4 The membership functions of triangular and trapezoidal fuzzy qualities and the estimated probability density function of the quality characteristic in Example 3

$$\widetilde{Q}_t(x) = \begin{cases} \frac{x-0.2}{0.6} & if \quad 0.2 \le x < 0.8 \\ \frac{1.1-x}{0.3} & if \quad 0.8 \le x < 1.1 \\ 0 & elsewhere. \end{cases}$$

In this situation, one can estimate Yongting's capability index by Eq. (5.15) as follow

$$\widehat{C_{\widetilde{p}(Y)}} = \int_{-\infty}^{+\infty} \widetilde{Q}_t(x)\widehat{f}(x)dx$$

$$= \frac{1}{\sqrt{2\pi}s} \int_{-\infty}^{+\infty} \widetilde{Q}_t(x) \exp\left(-\frac{(x-\bar{x})^2}{2s^2}\right) dx$$

$$= \frac{1}{0.15\sqrt{2\pi}} \left[\int_{0.2}^{0.8} \frac{x-0.2}{0.6} \exp\left(-\frac{(x-0.7)^2}{2 \times 0.15^2}\right) dx \right.$$

$$\left. + \int_{0.8}^{1.1} \frac{1.1-x}{0.3} \exp\left(-\frac{(x-0.7)^2}{2 \times 0.15^2}\right) dx \right]$$

$$= 0.543 + 0.178 = 0.721.$$

Now, let us to consider a trapezoidal fuzzy quality with the following membership function for this product

$$\widetilde{Q}_{tr}(x) = \begin{cases} \frac{x-0.3}{0.3} & if \quad 0.3 \leq x < 0.6 \\ 1 & if \quad 0.6 \leq x < 0.9 \\ \frac{1.1-x}{0.2} & if \quad 0.9 \leq x < 1.1 \\ 0 & elsewhere. \end{cases}$$

Similarly, one can estimate Yongting's capability index as follow

$$\widehat{C_{\tilde{p}}}(Y) = \frac{1}{\sqrt{2\pi}s} \int\limits_{-\infty}^{+\infty} \widetilde{Q}_{tr}(x) \exp\left(-\frac{(x-\bar{x})^2}{2s^2}\right) dx$$

$$= \frac{1}{0.15\sqrt{2\pi}} \left[\int\limits_{0.3}^{0.6} \frac{x-0.3}{0.3} \exp\left(-\frac{(x-0.7)^2}{2 \times 0.15^2}\right) dx \right.$$

$$\left. + \int\limits_{0.6}^{0.9} \exp\left(-\frac{(x-0.7)^2}{2 \times 0.15^2}\right) dx + \int\limits_{0.9}^{1.1} \frac{1.1-x}{0.2} \exp\left(-\frac{(x-0.7)^2}{2 \times 0.15^2}\right) dx \right]$$

$$= 0.178 + 0.656 + 0.060 = 0.894.$$

Therefore, the process is more capable under considering the trapezoidal fuzzy quality $\widetilde{Q}_{tr}$ with respect to considering the triangular fuzzy quality $\widetilde{Q}_t$.

Amirzadeh et al. (2009) constructed a new control chart based on Yongting's fuzzy quality, and meanwhile they shown that the developed control chart has a better response to variations in both the mean and the variance of the process. Parchami and Mashinchi (2010) proved that Yongting's introduced PCI is an extension for the probability of the product is qualified. Therefore, his capability index is not a suitable extension for C_p index, since C_p is not a probability and is not always in [0,1]. Then, Parchami and Mashinchi (2010) presented the revised version of Yongting's fuzzy quality on the basis of two fuzzy specification limits $\widetilde{LSL}$ and $\widetilde{USL}$ which are able to characterize two non-precise concepts of "approximately bigger than" and "approximately smaller than" in a fuzzy process, respectively. An instance for fuzzy quality is depicted in Fig. 5.6 in which the fuzzy quality is characterized by two membership functions of fuzzy specification limits $\widetilde{LSL}$ and $\widetilde{USL}$. Figure 5.5 is shown as an instance of the classical quality by characterizing two indicator functions $I_{\{x \mid x \geq LSL\}}$ and $I_{\{x \mid x \leq USL\}}$. Note that equation $\widetilde{Q} = \widetilde{USL} \cap \widetilde{LSL}$, or equivalently $\widetilde{Q}(x) = \min\{\widetilde{USL}(x), \widetilde{LSL}(x)\}$, presents the governed relation between membership functions of fuzzy specification limits in Fig. 5.6 and the membership function of Yongting's fuzzy quality in the right graph

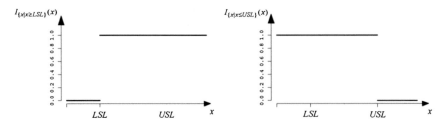

Fig. 5.5 Characterized classical quality with two indicator functions of "bigger than *USL*" (*left figure*) and "smaller than *USL*" (*right figure*)

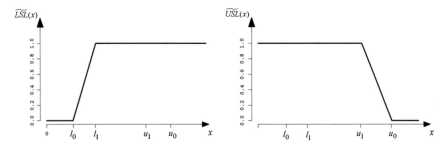

Fig. 5.6 Characterized fuzzy quality with two membership functions of "approximately bigger than" (*left figure*) and "approximately smaller than" (*right figure*)

of Fig. 5.3. Similarly, $I_{\{x|\, x \in [LSL, USL]\}}(x) = \min\{I_{\{x|\, x \le USL\}}(x), I_{\{x|\, x \ge LSL\}}(x)\}$ presents the relation between depicted indicator functions in Fig. 5.5 and depicted indicator function of classical quality $[LSL, USL]$ in the left graph of Fig. 5.3.

Motivations and merits of using fuzzy quality by considering fuzzy specification limits $\widetilde{LSL}$ and $\widetilde{USL}$, instead of applying classical quality are discussed in Parchami et al. (2014a). Parchami and Mashinchi (2010) introduced an extended version for traditional PCIs to present an alternative approach to measure the capability based on two new revised fuzzy specification limits. Their extended PCIs are used to give a numerical measure about whether a production method is capable of producing items within the fuzzy specification limits $\widetilde{LSL}$ and $\widetilde{USL}$. This new idea, provides a new methodology for measuring the fuzzy quality and also constructing confidence intervals for various PCIs, for example see Parchami and Mashinchi (2011). As an extended version of Yongting's index, Sadeghpour-Gildeh and Moradi (2012) proposed a general multivariate PCI based on fuzzy tolerance region which has not some of restriction of conventional PCIs. Another generalized version of the classical PCIs (1-4) is introduced by Parchami et al. (2014b) to measure the capability of a fuzzy-valued process in producing products based on fuzzy quality.

5.6   Conclusions and Future Research Directions

Traditional process capability indices are based on crispness of data, parameters, lower and upper specification limits, target value, and so on. As there are many different situations in which the above assumptions are rather unjustified and unrealistic. This chapter is including the classification of most essential researches on process capability indices extension for applying in fuzzy environment. After presenting the basic idea of the main works, all related studies briefly overviewed in each category. Also, some numerical examples are investigated to show how the proposed methods can be implemented in real-world cases. Some potential subjects for further research are presented in follow: (1) extending other non-extended capability indices in each category; (2) study on alternative approaches for point and interval estimation of the extended capability indices for each category; (3) construct statistical testing fuzzy quality based on Bayes, Minimax, Neyman–Pearson, Likelihood ratio, sequential and p-value approaches; (4) applying the extended capability indices and the presented methods in real-world cases.

Acknowledgments   The authors would like to thank the referees and Professor C. Kahraman for their constructive suggestions and comments.

References

Amirzadeh, V., Mashinchi, M., Parchami, A.: Construction of p-charts using degree of nonconformity, Information Sciences 179. Issues **1–2**, 150–160 (2009)

Boyles, R.A.: Process capability with asymmetric tolerances. Commun. Stat. Simul. Comput. **23**, 613–643 (1994)

Buckley, J.J.: Fuzzy Statistics, Studies in Fuzziness and Soft Computing. Springer, New York (2004)

Buckley, J.J.: Fuzzy Probability and Statistics. Springer-Verlag, Berlin, Heidelberg (2006)

Hsiang, T.C., Taguchi, G.: Tutorial on Quality Control and Assurance-The Taguchi Methods, p. 188. Nevada, Joint Meetings of the American Statistical Association, Las Vegas (1985)

Hsu, B.M., Shu, M.H.: Fuzzy inference to assess manufacturing process capability with imprecise data. Eur. J. Oper. Res. **186**(2), 652–670 (2008)

Juran, J.M.: Juran's Quality Control Handbook, 3rd edn. McGraw-Hill, New York (1974)

Kahraman, C., Kaya, İ.: Fuzzy Process Capability Indices for Quality Control of Irrigation Water. Stoch. Env. Res. Risk Assess. **23**, 451–462 (2009)

Kane, V.E.: Process capability indices. J. Qual. Technol. **18**, 41–52 (1986)

Kaya, İ., Kahraman, C.: Fuzzy process capability analyses: an application to teaching processes. J. Intell. Fuzzy Syst. **19**, 259–272 (2008)

Kaya, İ., Kahraman, C.: Fuzzy robust process capability indices for risk assessment of air pollution. Stoch. Env. Res. Risk Assess. **23**, 529–541 (2009)

Kaya, İ., Kahraman, C.: Development of fuzzy process accuracy index for decision making problems. Inf. Sci. **180**, 861–872 (2010)

Kaya, İ., Kahraman, C.: Process capability analyses based on fuzzy measurements and fuzzy control charts. Expert Syst. Appl. **38**, 3172–3184 (2011a)

Kaya, İ., Kahraman, C.: Process capability analyses with fuzzy parameters. Expert Syst. Appl. **38**(9), 11918–11927 (2011b)

Kotz, S.: Process Capability Indices. Chapman and Hall, New York (1993)
Kotz, S., Johnson, N.: Process capability indices- a review, 1992-2000. J. Qual. Technol. **34**, 2–19 (2002)
Kotz, S., Lovelace, C.R.: Process Capability Indices in Theory and Practice. Oxford University Press Inc., New York (1998)
Lee, H.T.: C_{pk} index estimation using fuzzy numbers. Eur. J. Oper. Res. **129**, 683–688 (2001)
Lee, Y.H., Wei, C.C., Chang, C.L.: Fuzzy design of process tolerances to maximize process capability. Int. J. Adv. Manuf. Technol. **15**, 655–659 (1999)
Mashinchi, M., Parchami, A., Maleki, H.R.: Application of fuzzy capability indices in educational comparison. In: Proceedings of 3rd International Qualitative Research Convention, Sofitel Palm Ressort, Senai, Johor, Malaysia, 21–23 August 2005, Paper No. 42 on a CD-Rom (2005)
Moeti, M.T., Parchami, A., Mashinchi, M.: A note on fuzzy process capability indices. Scientia Iranica **13**(4), 379–385 (2006)
Montgomery, D.C.: Introduction to Statistical Quality Control, 5th edn. Wiley, New York (2005)
Moradi, V., Sadeghpour Gildeh, B.: Fuzzy process capability plots for families of one-sided specification limits. Adv. Manuf. Technol. **64**, 357–367 (2013)
Parchami, A., Mashinchi, M.: Fuzzy estimation for process capability indices. Inf. Sci. **177**(6), 1452–1462 (2007)
Parchami, A., Mashinchi, M.: Testing the capability of fuzzy processes. Qual. Technol. Quant. Manage. **6**(2), 125–136 (2009)
Parchami, A., Mashinchi, M.: A new generation of process capability indices. J. Appl. Stat. **37**(1), 77–89 (2010)
Parchami, A., Mashinchi, M.: Interval estimation of an extended capability index with application in educational systems. Turk. J. Fuzzy Syst. **2**(2), 64–76 (2011)
Parchami, A., Mashinchi, M., Maleki, H.R.: Fuzzy confidence intervals for fuzzy process capability index. J. Intell. Fuzzy Syst. **17**(3), 287–295 (2006)
Parchami, A., Mashinchi, M., Mashinchi, M. H.: Approximate Confidence Interval for Generalized Taguchi Process Capability Index. In: IEEE International Conference on Fuzzy Systems, July 27–30, Taipei, Taiwan, 2968–2971 (2011)
Parchami, A., Mashinchi, M., Yavari, A.R., Maleki, H.R.: Process capability indices as fuzzy numbers. Austrian J. Stat. **34**(4), 391–402 (2005)
Parchami, A., Sadeghpour-Gildeh, B., Mashinchi, M.: A Note on Quality in Manufacturing Processes: Why Fuzzy Quality?, Submitted, 2014 (2014a)
Parchami, A., Sadeghpour-Gildeh, B., Nourbakhsh, M., Mashinchi, M.: A new generation of process capability indices based on fuzzy measurements. J. Appl. Stat. **41**(5), 1122–1136 (2014)
Pearn, W.L., Kotz, S., Johnson, N.L.: Distribution and inferential properties of capability indices. J. Qual. Technol. **24**, 41–52 (1992)
Ramezani, Z., Parchami, A., Mashinchi, M.: Fuzzy confidence regions for the Taguchi capability index. Int. J. Syst. Sci. **42**(6), 977–987 (2011)
Sadeghpour Gildeh, B. (2003). Comparison of C_p, C_{pk} and C_p-tilde process capability indices in the case of measurement error occurrence. IFSA World Congress, Istanbul, Turkey, 563–567
Sadeghpour Gildeh, B., Moradi, V.: Fuzzy tolerance region and process capability analysis. Adv. Intell. Soft Comput. **147**, 183–193 (2012)
Shu, M.H., Wu, H.C.: Quality-based supplier selection and evaluation using fuzzy data. Comput. Ind. Eng. **57**, 1072–1079 (2009)
Tsai, C.C., Chen, C.C.: Making decision to evaluate process capability index C_p with fuzzy numbers. Int. J. Adv. Manuf. Technol. **30**, 334–339 (2006)
Wu, C.W.: Decision-making in testing process performance with fuzzy data. Eur. J. Oper. Res. **193**, 499–509 (2009)
Wu, C.W., Liao, M.Y.: Estimating and testing process yield with imprecise data. Expert Syst. Appl. **36**, 11006–11012 (2009)
Yongting, C.: Fuzzy quality and analysis on fuzzy probability. Fuzzy Sets Syst. **83**, 283–290 (1996)

Chapter 6
An Integrated Framework to Analyze the Performance of Process Industrial Systems Using a Fuzzy and Evolutionary Algorithm

Harish Garg

Abstract In the design of critical combinations and complex integrations of large engineering systems, their reliability, availability and maintainability (RAM) analysis of the inherent processes in the system and their related equipments are needed to be determined. Although there have been tremendous advances in the art and science of system evaluation, yet it is very difficult to assess these parameters with a very high accuracy or precision. Basically, this inaccuracy in assessment stems mainly from the inaccuracy of data, lack of exactness of the models and even from the limitations of the current methods themselves and hence management decisions are based on experience. Thus the objective of this chapter is to present a methodology for increasing the performance as well as productivity of the system by utilizing these uncertain data. For this an optimization problem is formulated by considering RAM parameters as an objective function. The conflicting nature between the objectives is resolved by defining their nonlinear fuzzy goals and then aggregate by using a product aggregator operator. The failure rates and repair times of all constituent components are obtained by solving the reformulated fuzzy optimization problem with evolutionary algorithms. In order to increase the performance of the system, the obtained data are used for analyzing their behavior pattern in terms of membership and non-membership functions using intuitionistic fuzzy set theory and weakest t-norm based arithmetic operations. A composite measure of RAM parameters named as the RAM-Index has been formulated for measuring the performance of the system and hence finding the critical component of the system based on its performance. Finally the computed results of the proposed approach have been compared with the existing approaches for supremacy the approach. The suggested framework has been illustrated with the help of a case.

H. Garg (✉)
School of Mathematics, Thapar University Patiala, Patiala 147004, Punjab, India
e-mail: harishg58iitr@gmail.com
URL: http://sites.google.com/site/harishg58iitr/

© Springer International Publishing Switzerland 2016
C. Kahraman and S. Yanık (eds.), *Intelligent Decision Making in Quality Management*, Intelligent Systems Reference Library 97, DOI 10.1007/978-3-319-24499-0_6

141

6.1 Introduction

Today's competitive business environment requires manufacturers to design, develop, test, manufacture, and deploy high-reliability products in less time at lower cost. For achieving this, billions of dollars are being spent annually worldwide to develop reliable and efficient products. With the advance in technology, a designer always wants to manufacture the equipment and systems of greater capital cost, complexity and capacity which results in increasing the reliability of the system. Also at the same time the unfortunate penalty of low availability and high maintenance cost need to be improved for their survival. Thus, for this reason, there is a growing interest in the investigations of the reliability, availability and maintainability (RAM) principles in various industrial systems during last decades which affects on the system performance directly. A brief literature review regarding the reliability/availability evaluation using evolutionary as well as fuzzy methodology is given below.

6.1.1 Reliability/Availability Analysis Using Evolutionary Algorithm

With the advances in technology and need of the modern society, the job of the system analyst and plant personnel becomes so challenging in order to maintain the profile of the system so that it becomes operating continuously for a longer time. This is happening so because failure is an inevitable phenomenon for all industrial systems. Therefore, it is difficult, if not impossible, to construct their mathematical and statistical model so as to reduce the number of likelihood failures. Thus there is a need of developing a suitable methodology for analyzing the performance of the complex systems so that necessary action should be initiated for enhancing the performance as well as achieving the goal of higher targets. For this, generally, system performance can be improved either by incremental improvements of component reliability/availability or by provision of redundant components in parallel; both methods result in an increase in system cost. Traditionally analytical and Monte-Carlo simulation techniques have been used for analyzing the system reliability. While analytical techniques are potentially faster, they tend to get difficult as system size and complexity increases. Monte Carlo methods, on the other hand, afford tremendous modeling flexibility, and can be used for systems with large size and complexity. However, Monte Carlo methods tend to be extremely time consuming, particularly for reliable systems. Therefore, optimization methods are necessary to obtain allowable costs at the same time as high availability levels. Extensive reliability design techniques have been introduced by the researchers during the past two decades for solving the optimization problem on the specific applications. Comprehensive overviews of these models have been addressed in Gen and Yun (2006), Kuo et al. (2001).

As demonstrated in the literature, the aforementioned optimization techniques are successfully applied to various reliability optimization problems and show a significant difference in getting an optimal or near optimal solution. However, the previously-developed algorithms, as stochastic optimization techniques, heuristic algorithm have some weakness such as the lower robustness, premature convergence of the solution, not using a prior knowledge, not exploiting local search information, difficultly in dealing with large scale reliability problems. Also, the heuristic techniques require derivatives for all non-linear constraint functions that are not derived easily because of the high computational complexity. To overcome this difficulty metaheuristics/evolutionary algorithms have been selected such as Genetic Algorithm (GA) (Goldberg 1989; Holland 1975), Particle Swarm optimization (PSO) (Eberhart and Kennedy 1995; Kennedy and Eberhart 1995), Artificial bee colony (ABC) (Karaboga 2005; Karaboga and Akay 2009; Karaboga and Basturk 2007) etc., and have proved itself to be able to approach the optimal solution against these problems.

In that direction, Bris et al. (2003) attempted to optimize the maintenance policy, for each component of the system, minimizing the cost function, with respect to the availability constraints using genetic constraints. Barabady and Kumar (2005a, b) had presented a methodology for improving the availability of a repairable system by using the concept of important measures. The empirical data of two crushing plants at the Jajarm bauxite mine of Iran are used as a case study for reliability and availability analysis. Zavala et al. (2005) proposed a PSO-based algorithm, to solve a bi-objective redundant reliability problem with the aim of maximizing the system reliability, and minimizing redundant components' cost for three types of systems as series, parallel, and k-out-of-N systems. Juang et al. (2008) proposed a genetic algorithm based optimization model to optimize the availability of a series parallel system where the objective is to determine the most economical policy of component's MTBF and MTTR. Liberopoulos and Tsarouhas (2002) presented a case study of chipitas food processing system, based on the simplified assumption that the failure and repair times of the workstations of the lines have exponential distributions. Kumar et al. (2007) developed an optimization model for optimizing the reliability, maintainability and supportability under performance based logistics using goal programming. Their model simultaneously considered multiple system engineering metrics during the design stage of the product development. Khan et al. (2008) presented a two step risk-based methodology to estimate optimal inspection and maintenance intervals which maximize the system's availability. Sharma and Kumar (2008) presented the application of RAM analysis in a process industry by using a Markovian approach as a tool to model the system behavior. Rajpal et al. (2006) explored the application of artificial neural networks to model the behavior of a complex, repairable system. A composite measure of RAM parameters called as the RAM—Index has been proposed for measuring the system performance by simultaneously considers all the three key indices which influence the system performance directly. Their index was static in nature while Garg et al. (2012, 2013) introduced RAM-Index which was time dependent and used historical uncertain data for its evolution. Yeh et al. (2011) presented an approximate model for predicting the network reliability by combining the ABC algorithm and Monte Carlo simulation. Yeh and Hsieh (2011) and Hsieh and Yeh (2012) presented a penalty guided artificial bee colony algorithm to solve system reliability redundancy allocation problems with a mix of components. Garg and Sharma (2012) had discussed the two-phase approach for analyzing the reliability and maintainability analysis of the industrial system. The crankcase unit of the two wheeler manufacturing industry has been taken as an illustrative example and gave a recommendation to the system analyst. Garg and Sharma (2013) have investigated the multi-objective reliability-redundancy allocation problem by using PSO and GA while Garg et al. (2012, 2014) have solved the reliability optimization problem with ABC algorithm and compared their performance with other evolutionary algorithm.

6.1.2 *Reliability Analysis Using a Fuzzy Algorithm*

Engineering systems are usually complex, involve a lot of detail, and operate in unpredictable environments which leads to the job of system analysts has become more challenging, as they have to study, characterize, measure and analyze the uncertain systems' behavior, using various techniques, which require the component failure and repair pattern. Further, age, adverse operating conditions and the vagaries of the system, affect each unit of the system differently. Thus, one comes across the problem of uncertainty in reliability assessment. To this effect, fuzzy-theoretic approach (Zadeh 1965) has been used to handle the subjective information or uncertainties during the evaluation of the reliability of a system than the probabilistic approach. After their successful applications, a lot of progress has been made in both theory and application and hence several researches were conducted on the extensions of the notion of fuzzy sets. Among these extensions the one that have drawn the attention of many researches during the last decades is the theory of intuitionistic fuzzy sets (IFS) introduced by Attanassov (1986, 1989). The concepts of IFS can be viewed as an appropriate/alternative approach to define a fuzzy set in the case where available information is not sufficient for the definition of an imprecise concept by means of a conventional fuzzy set. IFS add an extra degree to the usual fuzzy sets in order to model hesitation and uncertainty about the membership degree of belonging. In fuzzy sets, the degree of acceptance is considered only but IFS is characterized by a membership function and a non-member function so that the sum of both values is less than or equal to one. Gau and Buehrer (1993) extended the idea of fuzzy sets by vague sets. Bustince and Burillo (1996) showed that the notion of vague sets coincides with that of IFSs. Therefore, it is expected that IFSs could be used to simulate any activities and processes requiring human expertise and knowledge, which are inevitably imprecise or not totally reliable. As far as reliability field is concerned, IFSs has been proven to be highly useful to deal with uncertainty and vagueness, and a lot of work has been done to develop and enrich the IFS theory given in Chang et al. (2006), Chen (2003), Garg and Rani

(2013), Garg et al. (2014), Kumar et al. (2006) Kumar and Yadav (2012) Taheri and Zarei (2011) and their corresponding references.

All the above researchers have used only reliability index during their analysis. But it is quite common that other reliability parameters such as failure rate, repair time, mean time between failures (MBTF) etc. are simultaneously affect the system behavior and hence on its performance. This idea is highlighted by Knezevic and Odoom (2001) and Garg (2013) in the fuzzy and intuitionistic fuzzy set theory respectively. In their approaches, system are modeled with the help of Petri nets and uncertainties which are present in the data are handled with the help of triangular fuzzy numbers and hence various reliability parameters of interest are computed in the form of membership and nonmembership functions. But it has been analyzed from their study that their approach is limited to a small size structural system. Thus when their approaches are applied to a complex structural system then the computed reliability indices contains a wide range of uncertainties in the form of support (spread) (Garg et al. 2013; Garg and Sharma 2012). This is due to the use of various fuzzy arithmetic operations involved in the analysis. Thus these approaches are no longer suitable for constructing the membership functions of IFS and hence do not give the accurate trend of the system as the uncertainty level increases. Therefore, there is a need of suitable methodology that can be used for computing the membership function of the reliability index up to a desired degree of accuracy. For this, by taking the advantages of evolutionary algorithms, the formulated reliability optimization problem has been solved with the Cuckoo search algorithm and compares their results with other algorithms. Since most of the collected data are imprecise and vague, so increase the relevance of the study, the obtained desired parameters are represented in the form of fuzzy numbers by taking different level of uncertainties. Based on these numbers, an analysis has been conducted for finding the most critical component of the system so that proper maintenance actions should be implemented for increasing the performance of the system.

Thus in the nutshell, the motive of this chapter is to devise a method to chalk out the performance measures of any repairable system by utilizing limited, vague and imprecise data. For this, the methodology has been proposed which is an amalgam of two techniques, EAs and intuitionistic fuzzy set theory, which can be described in stepwise as, (i) develop an optimization model by considering reliability, availability and maintainability of the system. The conflict naturalists between the objectives are resolved with the help of defining their fuzzy goals by using a nonlinear (sigmoidal) functions) (ii) obtain optimal MTBF and MTTR for the main component of the system using EAs and optimize the reliability parameters, and (iii) use their optimal parameters for computing various performance measures such as failure rate, repair time, ENOF etc. by using intuitionistic fuzzy set theory and weakest t-norm based arithmetic operations. Sensitivity analysis has been conducted on system MTBF for various combinations of reliability parameters. Finally, a composite measure of RAM parameters called RAM-Index has been used for finding the critical components of the system based on their variations of failure rate and repair time individually as well as simultaneously on its index. Results obtained from proposed technique are compared with the existing fuzzy and intuitionistic fuzzy set theory result. Plant personnel may use the results and can give guidelines to improve the system's performance by adopting suitable maintenance strategies. An example of the washing unit in a paper mill is taken into account to demonstrate the proposed approach.

6.2   Overview of IFS and EAs

A brief overview about the intuitionistic fuzzy set theory (IFS) and evolutionary algorithm (EA) have been given here.

6.2.1   Intuitionistic Fuzzy Set Theory

Intuitionistic fuzzy set (IFS) is one of the widely used and successful extension of the concept of fuzzy set. In order to model the hesitation and uncertainty about the degree of membership, Atanassov in (1986) add an extra degree, called as degree of non-membership, to the notion of the fuzzy set. Mathematically, if we define X be a universe of discourse then

$$\tilde{A} = \{ <x, \mu_{\tilde{A}}(x), v_{\tilde{A}}(x) > | x \in X \} \tag{6.1}$$

where $\mu_{\tilde{A}}, v_{\tilde{A}} : X \to [0, 1]$ be the degree of membership and nonmembership of the element x in the fuzzy set $\tilde{A}$, respectively such that $\mu_{\tilde{A}}(x) + v_{\tilde{A}}(x) \leq 1$. The function $\pi_{\tilde{A}}(x) = 1 - \mu_{\tilde{A}}(x) - v_{\tilde{A}}(x)$ is called the degree of hesitation or uncertainty level of the element x in the set $\tilde{A}$. Especially, if $\pi_{\tilde{A}}(x) = 0$ for all $x \in X$, then the IFS is reduced to a fuzzy set.

(α, β)–cut of the IFS set is defined as

$$A_{(\alpha,\beta)} = \{ x \in X \,|\, \mu_{\tilde{A}}(x) \geq \alpha \quad and \quad v_{\tilde{A}}(x) \leq \beta \} \tag{6.2}$$

In other words, $A_{(\alpha,\beta)} = A_\alpha \cap A^\beta$ where $A_\alpha = \{ x \in X \,|\, \mu_{\tilde{A}}(x) \geq \alpha \}$ and $A^\beta = \{ x \in X | v_{\tilde{A}}(x) \leq \beta \}$

Definition: Convex Intuitionistic fuzzy set An IFS $\tilde{A}$ in universe X is convex if and only if membership functions of $\mu_{\tilde{A}}(x)$ and $v_{\tilde{A}}(x)$ of $\tilde{A}$ are fuzzy—convex and fuzzy—concave respectively i.e.,

$$\mu_{\tilde{A}}(\lambda x_1 + (1 - \lambda)x_2) \geq \min(\mu_{\tilde{A}}(x_1), \mu_{\tilde{A}}(x_2)) \forall x_1, x_2 \in U, 0 \leq \lambda \leq 1 \tag{6.3}$$

and

$$v_{\tilde{A}}(\lambda x_1 + (1 - \lambda)x_2) \leq \max(v_{\tilde{A}}(x_1), v_{\tilde{A}}(x_2))\forall x_1, x_2 \in U, 0 \leq \lambda \leq 1 \quad (6.4)$$

Definition: Normal Intuitionistic fuzzy set Let $\tilde{A}$ be an IFS with universe $\mathbb{R}$, then $\tilde{A}$ is said to be normalized if there exist at least two points $x_1, x_2 \in \mathbb{R}$ such that $\mu_{\tilde{A}}(x_1) = 1$ and $v_{\tilde{A}}(x_2) = 1$ otherwise it is said to subnormal IFS.

Definition: Intuitionistic fuzzy number (IFN) An IFN $\tilde{A}$ is a normal, convex membership function on the real line $\mathbb{R}$ with bounded support i.e. $\{x \in X | v_{\tilde{A}}(x) < 1\}$ is bounded and $\mu_{\tilde{A}}$ is upper semi-continuous and $v_{\tilde{A}}$ is lower semi-continuous. Let $\tilde{A}$ be IFS denoted by $\tilde{A} = \,<[(a, b, c); \mu, v]>$, where $a, b, c \in \mathbb{R}$ then the set $\tilde{A}$ is said to be triangular intuitionistic fuzzy number (TIFN) if its membership function is given by

$$\mu_{\tilde{A}}(x) = \begin{cases} \mu\left(\frac{x-a}{b-a}\right); & a \leq x \leq b \\ \mu; & x = b \\ \mu\left(\frac{c-x}{c-b}\right); & b \leq x \leq c; \\ 0 & \text{otherwise} \end{cases}$$

$$1 - v_{\tilde{A}}(x) = \begin{cases} (1 - v)\left(\frac{x-a}{b-a}\right); & a \leq x \leq b \\ 1 - v; & x = b \\ (1 - v)\left(\frac{c-x}{c-b}\right); & b \leq x \leq c \\ 0; & \text{otherwise} \end{cases}$$

where the parameter b gives the modal values of A and a, c are the lower and upper bounds of available area for the evaluation data. A triangular vague set defined by the triplet (a, b, c) with α-cuts, given in Fig. 6.1 is defined below for membership and non-membership functions respectively.

Fig. 6.1 Representation of α-cut of the IFS set

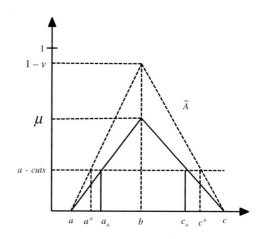

$$A_\alpha = [a_\alpha, c_\alpha] \quad \text{and} \quad A^\alpha = [a^\alpha, c^\alpha] \tag{6.5}$$

Here a_α, a^α are the increasing functions, c_α, c^α are decreasing functions of cut set given as follows.

$$a_\alpha = a + \frac{\alpha}{\mu_i}(b-a); \quad a^\alpha = a + \frac{\alpha}{1-v_i}(b-a)$$

$$c_\alpha = c - \frac{\alpha}{\mu_i}(c-b); \quad c^\alpha = c - \frac{\alpha}{1-v_i}(c-b)$$

Definition: T-norm and weakest t-norm A triangular norm (t-norm) T is a binary operation on $[0,1]$, i.e. a function $T : [0,1]^2 \to [0,1]$ such that (i) T is associative, (ii) T is commutative, (iii) T is nondecreasing, and (iv) T has 1 as a neutral element such that $T(x,1) = x$ for each $x \in [0,1]$.

A t-norm is called the weakest t—norm iff

$$T(x,y) = \begin{cases} 0; & \max(x,y) < 1 \\ \min(x,y); & \text{otherwise} \end{cases} \tag{6.6}$$

The basic arithmetic operations i.e. addition, subtraction, multiplication and division of IFNs depends upon the arithmetic of the interval of confidence. The four main basic arithmetic operations for the n triangular IFSs using T_ω—based approximate intuitionistic fuzzy arithmetic operations and with α—cuts arithmetic operations on triangular fuzzy numbers (TFNs), with $\mu = \min(\mu_i)$ and $v = \max(v_i)$, are defined as follow.

1. Addition of T_w ($\oplus$)

$$\tilde{A}_1 \oplus^\alpha_{T_w} \cdots \oplus^\alpha_{T_w} \tilde{A}_n = \begin{cases} [\sum_{i=1}^n a_{i1}^{(\alpha)}, \sum_{i=1}^n a_{i3}^{(\alpha)}] & \text{if} \quad \tilde{A}_i \in \text{TFNs} \\ \left[\sum_{i=1}^n a_{i2} - \max_{1 \le i \le n}\left((a_{i2} - a_{i1}^{(\alpha)})\right), \right. \\ \left. \sum_{i=1}^n a_{i2} + \max_{1 \le i \le n}\left((a_{i3}^{(\alpha)} - a_{i2})\right)\right] & \text{otherwise} \end{cases} \tag{6.7}$$

2. Subtraction of T_w ($\ominus$):

$$\tilde{A}_1 \ominus^\alpha_{T_w} \cdots \ominus^\alpha_{T_w} \tilde{A}_n = \begin{cases} [a_{11}^{(\alpha)} - \sum_{i=2}^n a_{i3}^{(\alpha)}, a_{13}^{(\alpha)} - \sum_{i=2}^n a_{i1}^{(\alpha)}] & \text{if} \quad \tilde{A}_i \in \text{TFNs} \\ \left[a_{12} - \sum_{i=2}^n a_{i2} - \max_{1 \le i \le n}\left((a_{i2} - a_{i1}^{(\alpha)})\right), \right. \\ \left. a_{12} - \sum_{i=2}^n a_{i2} + \max_{1 \le i \le n}\left((a_{i3}^{(\alpha)} - a_{i2})\right)\right] & \text{otherwise} \end{cases} \tag{6.8}$$

3. Multiplication of T_w ($\otimes$): Here, multiplication of the approximate fuzzy operations are shown for $\tilde{A}_i \in \mathbb{R}^+$

$$
\tilde{A}_1 \otimes_{T_w}^{\alpha} \cdots \otimes_{T_w}^{\alpha} \tilde{A}_n =
\begin{cases}
\left[\prod_{i=1}^{n} a_{i1}^{(\alpha)}, \prod_{i=1}^{n} a_{i3}^{(\alpha)} \right] & \text{if } \tilde{A}_i \in \text{TFNs} \\
\left[\prod_{i=1}^{n} a_{i2} - \max_{1 \leq i \leq n} \left((a_{i2} - a_{i1}^{(\alpha)}) \prod_{\substack{j=1 \\ j \neq i}}^{n} a_{j2} \right), \right. & \\
\left. \prod_{i=1}^{n} a_{i2} + \max_{1 \leq i \leq n} \left((a_{i3}^{(\alpha)} - a_{i2}) \prod_{\substack{j=1 \\ j \neq i}}^{n} a_{j2} \right) \right] & \text{otherwise}
\end{cases}
$$

(6.9)

4. Division of T_w ($\o$): Here, division of the approximate fuzzy operations are shown for $\tilde{A}_i \in \mathbb{R}^+$

$$
\tilde{A}_1 \o_{T_w}^{\alpha} \cdots \o_{T_w}^{\alpha} \tilde{A}_n = \tilde{A}_1 \otimes_{T_w}^{\alpha} \frac{1}{\tilde{A}_2} \cdots \otimes_{T_w}^{\alpha} \frac{1}{\tilde{A}_n}; \quad \text{if } 0 \notin \tilde{A}_i, i \geq 2 \qquad (6.10)
$$

6.2.2 Evolutionary Algorithms: GA, PSO, ABC, CS

6.2.2.1 Genetic Algorithm

Genetic Algorithms (GAs) (Goldberg 1989; Holland 1975) are adaptive heuristic search algorithms introduced in the evolutionary themes of natural selection. The fundamental concept of the GA design is to model processes in a natural system that is required for evolution, specifically those that follow the principles posed by Charles Darwin to find the survival of the fittest. GAs constitutes an intelligent development of a random search within a defined search space to solve a problem. GAs was first pioneered by John Holland in the 1960s, and has been widely studied, experimented, and applied in numerous engineering disciplines. GAs was introduced as a computational analogy of adaptive systems. They are modeled loosely on the principles of the evolution through natural selection, employing a population of individuals that undergo selection in the presence of variability-inducing operators such as mutation and recombination (crossover). A fitness function is used to evaluate individuals, and reproductive success varies with fitness. The pseudo code of the GA algorithm is described in Algorithm 1:

Algorithm 1 Pseudo code of Genetic algorithm (GA)

1: Objective function: $f(\mathbf{x})$
2: Define Fitness F (eg. F $\propto$ f(x) for maximization)
3: Initialize population
4: Initial probabilities of crossover (p_c) and mutation (p_m)
5: **repeat**
6:     Generate new solution by crossover and mutation
7:     if p_c >rand, Crossover; end if
8:     if p_m >rand, Mutate; end if
9:     Accept the new solution if its fitness increases.
10:    Select the current best for the next generation.
11: **until** requirements are met

6.2.2.2 Particle Swarm Optimization Algorithm

In 1995, Eberhart and Kennedy (1995), Kennedy and Eberhart (1955) developed PSO, a population based on stochastic optimization strategy, inspired by social behavior of a flock of birds, schools of fish, a swarm of bees and even sometimes social behavior of human. Though PSO is similar to Genetic Algorithms (GA) in terms of population initialization with random solutions and searching for global optima in successive generations, PSO does not undergo crossover and mutation, whereas the particles move through the problem space following the current optimum particles. The underlying concept is that, for every time instant, the velocity of each particle also known as the potential solution, changes between its personnel best (pbest) and global best (gbest) locations. Mathematically, swarm of particles is initialized randomly over the search space and move through D—dimensional space to search new solutions. Let x_k^i and v_k^i respectively be the position and velocity of ith particle in the search space at kth iteration then the position of this particle at $(k+1)$th iteration is updated through the equation,

$$x_{k+1}^i = x_k^i + v_{k+1}^i \tag{6.11}$$

where v_{k+1}^i is the updated velocity vector of ith particle at $(k+1)$th iteration and are adjusted according to their swarm own experience and that of its neighbors and are given as follow.

$$v_{k+1}^i = \underbrace{w \cdot v_k^i}_{\text{inertia}} + \underbrace{c_1 \cdot r_1 \cdot (p_k^i - x_k^i)}_{\text{personal influence}} + \underbrace{c_2 \cdot r_2 \cdot (p_k^g - x_k^i)}_{\text{social influence}} \tag{6.12}$$

where v_k^i is the velocity vector at kth iteration, r_1 and r_2 represent random numbers between 0 and 1; p_k^i represents the best ever position of ith particle, and p_k^g corresponds to the global best position in the swarm up to kth iteration.

The essential steps of the particle swarm optimization can be summarized as the pseudo code given in Algorithm 2.

Algorithm 2 Pseudo code of Particle swarm optimization (PSO)

1: Objective function: $f(\mathbf{x})$, $\mathbf{x} = (x_1, x_2, \ldots, x_D)$;
2: Initialize particle position and velocity for each particle and set $k = 1$.
3: Initialize the particle's best known position to its initial position i.e. $p_k^i = x_k^i$.
4: **repeat**
5: Update the best known position (p_k^i) of each particle and swarm's best known position (p_k^g).
6: Calculate particle velocity according to the velocity equation (12).
7: Update particle position according to the position equation (11).
8: **until** requirements are met.

6.2.2.3 Artificial Bee Colony Algorithm

The artificial bee colony (ABC) optimization algorithm was first developed by Karaboga in 2005. Since then Karaboga and Basturk and their coauthors (2005), Karaboga and Akay (2009) have systematically studied the performance of the ABC algorithm and its extension on unconstrained optimization problems. In ABC algorithm, the bees in a colony are divided into three groups: employed bees (forager bees), onlooker bees (observer bees) and scouts. For each food source, there is only one employed bee. That is to say, the number of employed bees is equal to a number of food sources. The employed bee of a discarded food site is forced to become a scout for searching new food source randomly. The whole process of the algorithm may also be explained through the Algorithm 2.2.3. In this, the first stage is the initialization stage in which food source positions are randomly selected by the bees and their nectar amounts (i.e. fitness function, f) is determined. Then, these bees come into the hive and share the nectar information of the sources with the bees waiting for the dance area with a probability $p_h = f_h / \sum_{h=1}^{N} f_h$ where N is the number of food sources and $f_h = f(x_h)$ is the amount of nectar evaluated by its employed bee. After a solution is generated, that solution is improved by using a local search process called greedy selection process carried out by an onlooker and employed bees and is given by

$$Z_{hj} = x_{hj} + \phi(x_{hj} - x_{kj}) \tag{6.13}$$

where $k \in \{1, 2, \ldots, N\}$ and $j \in \{1, 2, \ldots, D\}$ are randomly chosen index such that k is different from h and ϕ is a random number between $[-1, 1]$ and Z_h is the solution in the neighborhood of x_h. Here, except for the selected parameter j, all other parametric value of Z_h are same as that of x_h. If a particular food source solution does not improve for a predetermined iteration number then it becomes a scout and hence discovers a new food source with the randomly generated food source within its domain. So this randomly generated food source is equally assigned to this scout and changing its status from scout to employ and hence other iteration/cycle of the algorithm begins until the termination condition, maximum cycle number (MCN) or relative error, is not satisfied.

Algorithm 3 Pseudo code of Artificial Bee Colony (ABC) optimization

1: Objective function: $f(\mathbf{x})$, $\mathbf{x} = (x_1, x_2, \ldots, x_D)$;
2: Initialization Phase
3: **repeat**
4:      Employed Bee Phase
5:      Onlooker Bee Phase
6:      Scout Bee Phase
7:      Memorize the best position achieved so far.
8: **until** requirements are met.

6.2.2.4 Cuckoo Search Algorithm

CS is a meta-heuristic search algorithm which has been proposed recently by Yang and Deb (2009) getting inspired from the reproduction strategy of cuckoos. At the most basic level, cuckoos lay their eggs in the nests of other host birds, which may be of different species. The host bird may discover that the eggs are not its own so it either destroys the eggs or abandons the nest all together. This has resulted in the evolution of cuckoo eggs which mimic the eggs of local host birds. CS is based on three idealized rules:

(i)   Each cuckoo lays one egg at a time, and dumps it in a randomly chosen nest.
(ii)   The best nests with high quality of eggs (solutions) will carry over to the next generations.
(iii)   The number of available host nests is fixed, and a host can discover an alien egg with a probability $p_a \in [0, 1]$. In this case, the host bird can either throw the egg away or abandon the nest so as to build a completely new nest in a new location.

To make the things even more simple, the last assumption can be approximated by the fraction of p_a of n nests that are replaced by new nests with new random solutions. The fitness function of the solution is defined in a similar way as in other evolutionary techniques. In this technique, egg presented in the nest will represent the solution while the cuckoo egg represent the new solution. The aim is to use the new and potentially better solutions (cuckoos) to replace worse solutions that are in the nests. Based on these three rules, the basic steps of the cuckoo search is described in Algorithm 4.

Algorithm 4 Pseudo code of Cuckoo Search (CS)

1: Objective function: $f(\mathbf{x})$, $\mathbf{x} = (x_1, x_2, \ldots, x_D)$;
2: Generate an initial population of n host nests x_i ; $i = 1, 2, \ldots, N$;
3: While (t < MaxGeneration) or (stop criterion)
4:      Get a cuckoo randomly (say, i)
5:      Generate a new solution by performing Lévy flights;
6:      Evaluate its fitness f_i
7:      Choose a nest among n (say, j) randomly;
8:      if $(f_i > f_j)$
9:          Replace j by new solution
10:     end if
11:     A fraction(p_a) of the worse nests are abandoned and new ones are built;
12:     Keep the best solutions/nests;
13:     Rank the solutions/nests and find the current best;
14:     Pass the current best solutions to the next generation;
15: end while

The new solution $x_i^{(t+1)}$ of the cuckoo search is generated, from its current location x_i^t and probability of transition, with the following equation

$$x_i^{(t+1)} = x_i^{(t)} + \alpha \oplus \text{Lévy}(\lambda) \tag{6.14}$$

where α, $(\alpha > 0)$ represents a step size and we can use $\alpha = O(L/10)$ where L is the characteristic scale of the problem of interest. This step size should be related to the problem specification and t is the current iteration number. The product $\oplus$ represents entry-wise multiplications as similar to other evolutionary algorithms like PSO but random walk via Lévy flight is much more efficient in exploring the search space as its step length is much longer in the long run.

The Lévy flight essentially provides a random walk whose random step length drawn from a Lévy distribution

$$\text{Lévy} \sim u = t^{-\lambda}, (1 < \lambda \leq 3) \tag{6.15}$$

which has an infinite variance with an infinite mean. Here the steps essentially form a random walk process with a power-law step length distribution with a heavy tail.

6.3 Methodology

The present methodology is divided into two folds for analyzing the behavior of an industrial system. In the first fold, optimal design parameters for system performance has been computed by formulating and solving reliability optimization model with EAs. On the other hand, second fold deals with the determination of the various reliability parameters by using the obtained optimal desired parameters—MTBF and MTTR in terms of membership and non-membership functions of IFS.

The following tools are adopted for this purpose, which may give better results (closer to real conditions)

- The reliability optimization model has been constructed for optimal design of systems parameters i.e. MTBF and MTTR by considering reliability, availability and maintainability functions as an objective.
- Sigmoidal membership functions has been used for handling the conflictness between the objectives.
- CS is used for finding the optimal (or near to) values as it always give a global solution as compared to other EAs.
- For increasing the efficiency of the methodology, the weakest t-norm based arithmetic operations has been used for computing the various reliability parameters in terms of membership functions.
- Sensitivity and performance analysis of the components of the system has been addressed for ranking the components as per preferential order for increasing the productivity of the system

The strategy followed through this approach is shown by the flow chart in Fig. 6.2 and both the phases are described as below under the following assumptions.

(i) component failure and repair rates are statistically independent, constant and obey exponential distribution.
(ii) after repair, the repaired component is considered as good as new.
(iii) separate maintenance facility is available for each component
(iv) standby units are of the same nature and capacity as the active unit.
(v) system structure is precisely known.

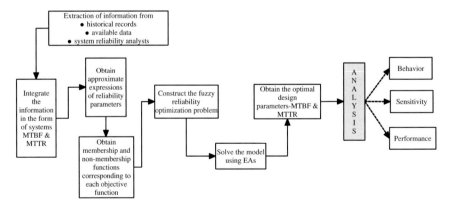

Fig. 6.2 Flow chart of the methodology

6.3.1 Obtaining the Optimal Values of Design Parameters

The main motive of this fold is to compute the design parameters—MTBF and MTTR—of each component of the system so that the design efficiency will be maximized. System reliability, maintainability and availability have assumed great significance in recent years due to a competitive environment and overall operating and production costs. Performance of equipment depends on the reliability and availability of the equipment used, operating environment, maintenance efficiency, operation process and technical expertise of operators, etc. When the reliability and availability of systems are low, efforts are needed to improve them by reducing the failure rate or increasing the repair rate for each component or subsystem. Thus, reliability, availability and maintainability are the important key features for keeping the production and productivity of the system high. The given industrial system is divided into its constituent components and based on the reliability block diagram (RBD), the expressions for the availability, failure rate and repair rates are obtained from Birolini (2007). The basic parameters for series and parallel system are shown in Table 6.1. In this table, λ_i and μ_i represent respectively the failure and repair rates for the ith component of system while λ_s and μ_s represent the same for system's. Av_s and Av_i represent the system and ith component availability. Based on the expressions in Table 6.1, the approximate reliability (R_s), availability (Av_s) and maintainability (M_s) expression for the system can be written as:

$$R_s = \exp(-\lambda_s t) \qquad (6.16)$$

$$Av_s = f(\text{MTBF}_1, \ldots \text{MTBF}_n, \text{MTTR}_1, \ldots \text{MTTR}_n) \qquad (6.17)$$

$$M_s = 1 - \exp(-\mu_s t) \qquad (6.18)$$

The conflict between the objectives $(f_t$'s$)$ are resolved by defining their fuzzy goals corresponding to $f_t(x) \leq m_t$ and $f_t(x) \geq M_t$ where m_t and M_t are the lower and upper bound of the objective functions respectively. For defining of this, we make use of the standard logarithm sigmoid function $\psi(a) = \frac{1}{1+e^{-a}}$ and arbitrarily take the

Table 6.1 Basic parameters of availability for series-parallel systems

Type of system	Expression
Series configuration	$Av_s = Av_1 \cdot Av_2 \cdots Av_n \approx 1 - \left(\frac{\lambda_1}{\mu_1} + \frac{\lambda_2}{\mu_2} + \cdots + \frac{\lambda_n}{\mu_n}\right)$
	$\lambda_s \approx \lambda_1 + \lambda_2 + \cdots + \lambda_n; \ \mu_s \approx \frac{\lambda_1 + \lambda_2 + \cdots + \lambda_n}{\frac{\lambda_1}{\mu_1} + \frac{\lambda_2}{\mu_2} + \cdots + \frac{\lambda_n}{\mu_n}}$
Parallel configuration	$Av_s \approx 1 - \frac{\lambda_1 \cdot \lambda_2 \cdots \lambda_n}{\mu_1 \cdot \mu_2 \cdots \mu_n}$
	$\lambda_s \approx \frac{\lambda_1 \cdot \lambda_2 \cdots \lambda_n (\mu_1 + \mu_2 \cdots \mu_n)}{\mu_1 \cdot \mu_2 \cdots \mu_n}; \ \mu_s \approx \mu_1 + \mu_2 + \cdots + \mu_n$

domain of this function as $[-5, 5]$. The corresponding membership functions are given as (Garg and Sharma 2013)

$$\mu_{f_t}(x) = \begin{cases} 1, & f_t(x) \leq m_t \\ \frac{\psi(5)-\psi(\{f_t(x)-\frac{M_t+m_t}{2}\}\delta_t)}{\psi(5)-\psi(-5)}, & m_t \leq f_t(x) \leq M_t \\ 0, & f_t(x) \geq M_t \end{cases} \tag{6.19}$$

and

$$\mu_{f_t}(x) = \begin{cases} 1, & f_t(x) \geq M_t \\ \frac{\psi(\{f_t(x)-\frac{M_t+m_t}{2}\}\delta_t)-\psi(-5)}{\psi(5)-\psi(-5)}, & m_t \leq f_t(x) \leq M_t \\ 0, & f_t(x) \leq m_t \end{cases} \tag{6.20}$$

where $\delta_t = \frac{10}{M_t-m_t}$. The membership function μ_{f_t} are on the same scale and are discontinuous at the points m_t, f_t, M_t. Here $(M_t + m_t)/2$ is the crossover point of the sigmoidal membership functions.

Using the achieved objective functions of the system, the optimization model is formulated as

$$\begin{aligned} Maximize\ \mu_D &= \mu_{R_s} \times \mu_{A_s} \times \mu_{M_s} \\ subject\ to\ \text{LbMTBF}_i &\leq \text{MTBF}_i \leq \text{UbMTBF}_i \\ \text{LbMTTR}_i &\leq \text{MTTR}_i \leq \text{UbMTTR}_i \\ i &= 1, 2 \ldots n \quad \text{All variables} \geq 0 \end{aligned} \tag{6.21}$$

where $\text{LbMTBF}_i, \text{UbMTBF}_i, \text{LbMTTR}_i, \text{UbMTTR}_i$ are respectively the lower and upper bound of MTBF and MTTR for ith component of the system. The optimization model (6.21) thus obtained is solved by the evolution strategies techniques, namely as GA, PSO, ABC and CS.

6.3.2 Analyzing the Behavior of the System

In this fold, the optimal values of design parameters, obtained in previous folds/phase are used to calculate the various reliability parameters using weakest t —norm based arithmetic operations on vague lambda-tau methodology, so as to increase the efficiency of the methodology. The procedural steps of the methodology can be described as follows:

Step 1 The technique start with the information extraction phase in which data related to the failure rate and repair time of the main component of the system are collected or extracted from various resources. In the present study, the data related to failure rate and repair time, are obtained using phase Sect. 6.3.1 of the proposed technique

Step 2 To handle the uncertainties or vagueness in the data, the obtained data are converted into intuitionistic triangular vague numbers with some spread as suggested by the DMs on both sides of the data. For instance, the failure rate and repair time for the ith component of the system are converted into ITFNs with $\pm 15\ \%$ spreads are depicted in Fig. 6.3 where $\widetilde{\lambda}_{ij}$ and $\widetilde{\tau}_{ij}$ are the vague failure rate and repair time, of component i, with $j = 1, 2, 3$, being lower, middle (crisp) and upper limit of a triangular membership function, respectively. As soon as the input data are represented in the form intuitionistic fuzzy numbers then their corresponding values for their top event of the system are calculated using the extension principle coupled with $\alpha-$ cuts and interval weakest t- norm based arithmetic operations on conventional AND/OR expression, as listed in Table 6.2. The weakest t-norm based interval expression for the triangular vague number, for the failure rate $\widetilde{\lambda}$ and repair time $\widetilde{\tau}$, for AND/OR-transitions are as follow

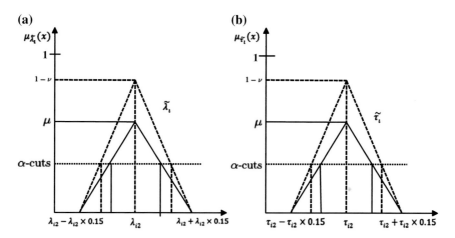

Fig. 6.3 Input intuitionistic triangular fuzzy number. **a** Membership Functions of $\widetilde{\lambda}_i$ **b** Membership Functions of $\widetilde{\tau}_i$

Table 6.2 Basic expressions of lambda tau methodology

Gate	λ_{AND}	τ_{AND}	λ_{OR}	τ_{OR}
Expression	$\prod_{j=1}^{n} \lambda_j [\sum_{i=1}^{n} \prod_{j=1}^{n} \tau_j]$ $\quad i \neq j$	$\dfrac{\prod_{i=1}^{n} \tau_i}{\sum_{j=1}^{n} [\prod_{i=1}^{n} \tau_i]}$ $\quad i \neq j$	$\sum_{i=1}^{n} \lambda_i$	$\dfrac{\sum_{i=1}^{n} \lambda_i \tau_i}{\sum_{i=1}^{n} \lambda_i}$

For truth membership functions:

Expressions for AND-Transitions

$$\lambda^{(\alpha)} = \left[\prod_{i=1}^{n}\{(\lambda_{i2} - \max_{1 \leq i \leq n}\left((\lambda_{i2} - \lambda_{i1}^{(\alpha)})\right)\} \cdot \sum_{j=1}^{n}\left[\prod_{\substack{i=1 \\ i \neq j}}^{n}\{(\tau_{i2} - \max_{1 \leq i \leq n}\left((\tau_{i2} - \tau_{i1}^{(\alpha)})\right)\} \right], \right.$$

$$\left. \prod_{i=1}^{n}\{\lambda_{i2} + \max_{1 \leq i \leq n}\left((\lambda_{i3}^{(\alpha)} - \lambda_{i2})\right)\} \cdot \sum_{j=1}^{n}\left[\prod_{\substack{i=1 \\ i \neq j}}^{n}\{(\tau_{i2} + \max_{1 \leq i \leq n}((\tau_{i3}^{(\alpha)} - \tau_{i2})) \right] \right]$$

$$\tau^{(\alpha)} = \left[\frac{\prod_{i=1}^{n}\{\tau_{i2} - \max_{1 \leq i \leq n}\left((\tau_{i2} - \tau_{i1}^{(\alpha)})\right)}{\sum_{j=1}^{n}[\prod_{\substack{i=1 \\ i \neq j}}^{n}\{\tau_{i2} + \max_{1 \leq i \leq n}\left((\tau_{i3}^{(\alpha)} - \tau_{i2})\right)\}]}, \frac{\prod_{i=1}^{n}\{\tau_{i2} + \max_{1 \leq i \leq n}\left((\tau_{i3}^{(\alpha)} - \tau_{i2})\right)\}}{\sum_{j=1}^{n}[\prod_{\substack{i=1 \\ i \neq j}}^{n}\{(\tau_{i2} - \max_{1 \leq i \leq n}\left((\tau_{i2} - \tau_{i1}^{(\alpha)})\right)\}]} \right]$$

Expressions for OR-Transitions

$$\lambda^{(\alpha)} = \left[\sum_{i=1}^{n}\{\lambda_{i2} - \max_{1 \leq i \leq n}\left((\lambda_{i2} - \lambda_{i1}^{(\alpha)})\right)\}, \sum_{i=1}^{n}\{\lambda_{i2} + \max_{1 \leq i \leq n}\left((\lambda_{i3}^{(\alpha)} - \lambda_{i2})\right)\} \right]$$

$$\tau^{(\alpha)} = \left[\frac{\sum_{i=1}^{n}[\{\lambda_{i2} - \max_{1 \leq i \leq n}\left((\lambda_{i2} - \lambda_{i1}^{(\alpha)})\right)\} \cdot \{\tau_{i2} - \max_{1 \leq i \leq n}\left((\tau_{i2} - \tau_{i1}^{(\alpha)})\right)\}]}{\sum_{i=1}^{n}\{\lambda_{i2} + \max_{1 \leq i \leq n}\left((\lambda_{i3}^{(\alpha)} - \lambda_{i2})\right)\}}, \right.$$

$$\left. \frac{\sum_{i=1}^{n}[\{\lambda_{i2} + \max_{1 \leq i \leq n}\left((\lambda_{i3}^{(\alpha)} - \lambda_{i2})\right)\} \cdot \{\tau_{i2} + \max_{1 \leq i \leq n}\left((\tau_{i3}^{(\alpha)} - \tau_{i2})\right)\}]}{\sum_{i=1}^{n}\{\lambda_{i2} - \max_{1 \leq i \leq n}\left((\lambda_{i2} - \lambda_{i1}^{(\alpha)})\right)\}} \right]$$

For false membership functions (i.e. non-membership functions):
Expressions for AND-Transitions

$$\lambda^{(\beta)} = \left[\prod_{i=1}^{n} \{(\lambda_{i2} - \max_{1 \le i \le n}((\lambda_{i2} - \lambda_{i1}^{(\beta)}))\} \cdot \sum_{j=1}^{n} [\prod_{\substack{i=1 \\ i \ne j}}^{n} \{(\tau_{i2} - \max_{1 \le i \le n}((\tau_{i2} - \tau_{i1}^{(\beta)}))\} , \right.$$

$$\left. \prod_{i=1}^{n} \{\lambda_{i2} + \max_{1 \le i \le n}((\lambda_{i3}^{(\beta)} - \lambda_{i2}))\} \cdot \sum_{j=1}^{n} [\prod_{\substack{i=1 \\ i \ne j}}^{n} \{(\tau_{i2} + \max_{1 \le i \le n}((\tau_{i3}^{(\beta)} - \tau_{i2}))] \right]$$

$$\tau^{(\beta)} = \left[\frac{\prod_{i=1}^{n} \{\tau_{i2} - \max_{1 \le i \le n}((\tau_{i2} - \tau_{i1}^{(\beta)}))}{\sum_{j=1}^{n} [\prod_{\substack{i=1 \\ i \ne j}}^{n} \{\tau_{i2} + \max_{1 \le i \le n}((\tau_{i3}^{(\beta)} - \tau_{i2}))\}]} , \frac{\prod_{i=1}^{n} \{\tau_{i2} + \max_{1 \le i \le n}((\tau_{i3}^{(\beta)} - \tau_{i2}))\}}{\sum_{j=1}^{n} [\prod_{\substack{i=1 \\ i \ne j}}^{n} \{(\tau_{i2} - \max_{1 \le i \le n}((\tau_{i2} - \tau_{i1}^{(\beta)}))\}]} \right]$$

Expressions for OR-Transitions

$$\lambda^{(\beta)} = \left[\sum_{i=1}^{n} \{\lambda_{i2} - \max_{1 \le i \le n}((\lambda_{i2} - \lambda_{i1}^{(\beta)}))\}, \sum_{i=1}^{n} \{\lambda_{i2} + \max_{1 \le i \le n}((\lambda_{i3}^{(\beta)} - \lambda_{i2}))\} \right]$$

$$\tau^{(\beta)} = \left[\frac{\sum_{i=1}^{n} [\{\lambda_{i2} - \max_{1 \le i \le n}((\lambda_{i2} - \lambda_{i1}^{(\beta)}))\} \cdot \{\tau_{i2} - \max_{1 \le i \le n}((\tau_{i2} - \tau_{i1}^{(\beta)}))\}]}{\sum_{i=1}^{n} \{\lambda_{i2} + \max_{1 \le i \le n}((\lambda_{i3}^{(\beta)} - \lambda_{i2}))\}} , \right.$$

$$\left. \frac{\sum_{i=1}^{n} [\{\lambda_{i2} + \max_{1 \le i \le n}((\lambda_{i3}^{(\beta)} - \lambda_{i2}))\} \cdot \{\tau_{i2} + \max_{1 \le i \le n}((\tau_{i3}^{(\beta)} - \tau_{i2}))\}]}{\sum_{i=1}^{n} \{\lambda_{i2} - \max_{1 \le i \le n}((\lambda_{i2} - \lambda_{i1}^{(\beta)}))\}} \right]$$

Step 3   In order to analyze the system behavior quantitatively, various reliability parameters such as system failure rate, repair time, MTBF, reliability etc. are analyzed in terms of membership and non-membership functions at various membership grades with an increment of 0.1 confidence level

Step 4   In order to obtain a crisp result from fuzzy output, defuzzification is carried out. In the literature various techniques for defuzzification such as centroid, bisector, middle of the max, weighted average exists. The criterion's for their selection are disambiguated (result in unique value), plausibility (lie approximately in the middle of the area) and computational simplicity (Ross 2004). In the present study, the centroid method is used for defuzzification as it gives mean value of the parameters

6.4 An Illustrative Example

To demonstrate the application of the proposed methodology, a case from a paper mill, situated in the northern part of India is taken which produces approximately 200 tons of paper per day. The paper mills are large capital oriented engineering systems, comprising of various subsystems namely, feeding, pulping, washing, screening, bleaching and paper formulation system, arranged in a predefined configuration (Garg 2013; Garg and Sharma 2012). The present analysis is based on the study of one of the important unit i.e. washing unit whose brief description is as follows.

6.4.1 System Description

The Washing of prepared pulp is done in three to four stages, shown in systematic diagram in Fig. 6.4, to get it free from blackness and to prepare the fine fibers of the pulp. The system consists of four main subsystems, defined as:

- **Filter (A)**: It consists of single unit which is used to drain black liquor from the cooked pulp.
- **Cleaners (B)**: In this subsystem three units of cleaners are arranged in parallel configuration. Each unit may be used to clean the pulp by centrifugal action. Failure of anyone will reduce the efficiency of the system as well as quality of paper.
- **Screeners (C)**: Herein two units of screeners are arranged in series. These are used to remove oversized, uncooked and odd shaped fibers from pulp through straining action. Failure of any one will cause the complete failure of the system.
- **Deckers (D)**: Two units of deckers are arranged in parallel configuration. The function of deckers is to reduce the blackness of pulp. Complete failure of decker occurs when both the components will fail.

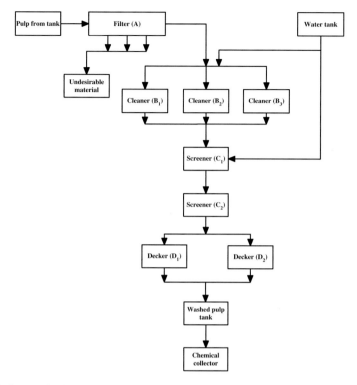

Fig. 6.4 Systematic diagram of the washing system

6.4.2 *Formulation of Optimization Model*

Let MTBF$_i$ and MTTR$_i$ be the mean time between failures and mean time to repair of the ith component of the system then the approximate expressions of system parameters in the form of reliability, availability and maintainability are expressed as below

$$R_s = \exp(-\lambda_s t)$$

$$A_s = 1 - \left[\frac{MTTR_1}{MTBF_1} + \left(\frac{MTTR_2}{MTBF_2}\right)^3 + 2 \cdot \frac{MTTR_3}{MTBF_3} + \left(\frac{MTTR_4}{MTBF_4}\right)^2\right]$$

$$M_s = 1 - \exp(-t/\tau_s)$$

where λ_s and τ_s are given as

$$\lambda_s = \lambda_1 + \lambda_2 \lambda_3 \lambda_4 (\tau_2 \tau_3 + \tau_3 \tau_4 + \tau_4 \tau_2) + \lambda_5 + \lambda_6 + \lambda_7 \lambda_8 (\tau_7 + \tau_8)$$

$$\tau_s = \frac{\lambda_1 \tau_1 + \lambda_2 \lambda_3 \lambda_4 \tau_2 \tau_3 \tau_4 + \lambda_5 \tau_5 + \lambda_6 \tau_6 + \lambda_7 \lambda_8 \tau_7 \tau_8}{\lambda_s}$$

Table 6.3 Variance range of MTBF and MTTR of components

Component	MTBF in hrs		MTTR in hrs	
	Lb	Ub	Lb	Ub
Filter	2995	3150	2	4
Cleaners	1850	1950	2	5
Screeners	1880	1920	2	4
Deckers	1860	1910	2	5

As the information collected related to systems' parameter—MTBF and MTTR, are mostly imprecise in nature because these data are collected from various historical records, logbooks etc. which represents the past behavior of the system but unable to represent the future behavior. Thus for handling this issue and to resolve the conflictness between the objective, the membership functions corresponding to objectives are defined by using log-sigmoidal membership functions as given in Eq. (6.20) and hence an optimization model (6.21) is formulated for the considered system. Variance range of the main components' of the system in the form of MTBF and MTTR are summarized in Table 6.3.

6.4.2.1 Parametric Setting

In all algorithms, the values of the common parameters used in each algorithm such as population size and total evaluation number are chosen to be the same. Population size and the maximum evaluation number are taken as $20 \times D$ and 500 respectively for the function, where D is the dimension of the problem. The method has been implemented in Matlab (MathWorks) and the program has been run on a T6400 @ 2 GHz Intel Core (TM) 2 Duo processor with 2 GB of Random Access Memory (RAM). In order to eliminate stochastic discrepancy, 30 independent runs has been made that involves 30 different initial trial solutions. The termination criterion has been set either limited to a maximum number of generations or to the order of relative error equal to 10^{-6}, whichever is achieved first. The other specific parameters of algorithms are given below:

GA Settings: In our experiments, we employed a real coded standard GA having an evaluation, fitness scaling, crossover, mutation units. Single point crossover operation with the rate of 0.85 was employed. Mutation operation restores genetic diversity lost during the application of reproduction and crossover. Mutation rate in our experiments was 0.02.

PSO Settings: Cognitive and social components (c_1 and c_2 in (6.12)) are constants that can be used to change the weighting between personal and population experience, respectively. In our experiments cognitive and the social components were both set to 1.5. Inertia weight (w), which determines how the previous velocity of the particle influences the velocity in the next iteration, was defined as the linear decreases from initial weight $w_{max} = 0.9$ to final weight $w_{min} = 0.4$ with the relation $w = w_{max} - (w_{max} - w_{min})(iter/iter_{max})$. Here $iter_{max}$ represents the maximum generation number and 'iter' is used a generation number as recommended in Clerc and Kennedy (2002), Shi and Eberhart (1998).

ABC Settings: Except common parameters (population number and maximum evaluation number), the basic ABC used in this study employs only one control parameter, which is called *limit*. A food source will not be exploited anymore and is assumed to be abandoned when *limit* is exceeded for the source. This means that the solution of which "trial number" exceeds the limit value cannot be improved anymore. The *limit* value is defined by using the dimension of the problem and the colony size as (Karaboga and Akay 2009) *limit* $= SN \times D$, where SN is the number of food sources or employed bees.

CS Settings: Except common parameters, CS employ only one control parameter called probability (p_a) of a host for discovering an alien egg. Here p_a is set to be randomly 0.25 (Yang and Deb 2009).

6.4.2.2 Computational Results

By using these settings, the optimal design parameters for the system performance optimization are obtained and their corresponding results are tabulated in Table 6.4. The estimation of optimal design parameters will generally help the maintenance engineers to understand the behavioral dynamics of the system. However, by using these optimal designs—MTBF and MTTR—results, the plant personnel may change their initial goals so as to reduce the operational and maintenance cost by adopting suitable maintenance strategies from their design results. This methodology will assist the plant managers to carry out design modification, if any, required to achieve minimum failures, and to help in maintenance (repair and replacement actions) decision making.

The statistical simulation results after 30 independent results in terms of values of the mean, best, worst, standard deviation (S.D) and median of the objective functions are obtained by CS algorithm and compared with respect to other algorithms are summarized in Table 6.5. It has also been observed from the table that the S.D. by proposed one are pretty low, and it further implies that the approach seems reliable to solve the reliability optimization problems.

In order to analyze whether the results as obtained in the above tables are statistically significantly with each other or not, we performed t—test on pair of algorithms. For this firstly equality of variances will be tested, since the t—test assumes equality of variances, by using an F—test on the pair of algorithms. For this, two tailed F—test has been performed with significant level of $\alpha = 0.05$ for checking equality of variances of the two results based on their variances values after 30 independent runs. The two-tailed version tests against the alternative that the variances are not equal. Under the null hypothesis, no difference in population variances, the calculated values of F-statistics are 1.227887, 1.189696 and 1.697688 respectively for GA, PSO and ABC when pair with CS. As the critical

Table 6.4 Optimal design parameters for the System

Method → Components ↓	GA		PSO		ABC		CS	
	MTBF	MTTR	MTBF	MTTR	MTBF	MTTR	MTBF	MTTR
Digester	3103.629984	2.153718	3121.631510	2.055150	3104.871834	2.010746	3116.751764	2.026474
Knotter	1910.838263	3.539449	1857.405859	2.363101	1880.633596	2.027245	1938.555352	4.245118
Decker	1899.390049	2.345644	1896.305893	2.003680	1880.054203	2.000236	1906.594880	2.003558
Opener	1895.796417	3.835511	1866.889014	2.123994	1869.688537	2.891751	1865.039620	2.407312
Obj. function	0.9965286		0.9970178		0.9970131		0.9970375	

Table 6.5 Statistics analysis for the optimization problem

Methods	Mean	Best	Worst	Median	SD $(\times 10^{-5})$
GA	0.9963972	0.9965286	0.9961616	0.9963215	3.5109
PSO	0.9969629	0.9970177	0.9969189	0.9969829	3.4017
ABC	0.9969330	0.9970130	0.9968258	0.9969427	4.8542
CS	0.9969831	0.9970375	0.9968965	0.9969861	2.8593

values for testing null hypothesis against the alternative hypothesis at level of significance $\alpha = 0.05$ are given by

$$F > F_{29,29}(\alpha/2) = F_{29,29}(0.025) = 0.475964$$
$$and \quad F < F_{29,29}(1 - \alpha/2) = F_{29,29}(0.975) = 2.100995$$

Since, the calculated value of F-statistics (= 1.227887, 1.189696 and 1.697688) lies between 0.475964 and 2.100995, it is not significant and hence null hypothesis of equality of population variances may be accepted at level of significance $\alpha = 0.05$. Now a single-tail t-test has been performed with the null hypothesis that their mean difference is zero at 5 % significance level in the case of CS results with other results. The results computed are tabulated in Table 6.6 and it indicates that the value of their t-stat is much greater than the t-critical values. Also the p-value obtained during the test is less than the significance level. Thus it is highly significant and null hypothesis i.e. mean of the two algorithms is identical is rejected. Hence the two types of means differ significantly. Further, since mean of the performance function value of the system with CS is greater than others, we conclude that CS is definitely better than others results and this difference is statistically significant.

Table 6.6 t-test for Statistical analysis

	GA	PSO	ABC	CS
Mean	0.99639724	0.9969629	0.9969330	0.9969831
SD	3.5109×10^{-4}	3.4017×10^{-5}	4.8542×10^{-5}	2.8593×10^{-5}
Variance $(\times 10^{-8})$	0.123264	0.115715	0.235632	0.081756
Observation	30	30	30	30
Pooled variance $(\times 10^{-8})$	0.1025101	0.0987357	0.1586942	
Hypothesized mean difference	0	0	0	
Degree of freedom	58	58	58	
t—stat	70.868982	2.4897725	4.8708273	
$P(T \leq t)$ one tail	0	0.007835	4.475094×10^{-6}	
T-critical one tail	1.6772241	1.6772241	1.6772241	

6.4.3 Behavior Analysis

The behavior of the system has been analyzed by using the above computed design parameters in the vague set [0.6, 0.8] i.e. degree of acceptance $\mu = 0.6$ and degree of rejection is $v = 1 - 0.8 = 0.2$ so that efficiency of the vague lambda-tau methodology may increase. In this, the computed failure rate and repair time of each of the components are represented in the form of vague triangular numbers with ± 15 % spread and hence various reliability parameters of the system are computed in the form of membership and non-membership functions with the left and right spreads. These behavior plots are shown graphically in Fig. 6.5 along with the existing methodologies results.

(i) The results computed by the traditional or crisp methodology are independent of the uncertainty level α. Hence their results will be suitable only for a system with precise data.

(ii) The results computed by FLT methodology (Knezevic and Odoom 2001) are not that much practical as it contains a wide range of uncertainties during the analysis. Also domain of confidence level is taken to be one and there is a $0°$ of hesitation between the membership functions.

(iii) The above shortcomings during the analysis has been taken into account by Garg (2013) in their analysis and hence proposed a new technique named as Vague Lambda-Tau methodology (VLTM). In their approach, the domain of confidence level is taken to be ≤ 0.8 instead of one and the intuitionistic fuzzy set theory has been used for representing the uncertainties in the data in the form of membership and non-membership functions. In their approach, interval level uncertainty has been considered with $0.2°$ of hesitation between the membership functions. However, their results gave more maintenance strategy for the decision maker for increasing the performance of the system as it gives an interval value of a reliability parameter for a particular level of significance (α) as compared to point value. Since in their analysis fuzzy arithmetic operations have been used for computing the system's reliability parameters and hence the level of uncertainties has not been reduced so much.

(iv) The proposed approach provides an improvement over the above shortcoming by considering $0.2°$ of hesitation between the degree of membership and non-membership functions. In the proposed approach the domain of confidence level is clearly $\alpha \leq 0.8$. The graphical results show that if the uncertainty in input data is described by means of triangular fuzzy numbers, then the possibility distribution of failure rate and repair time is a distorted triangle because after applying the fuzzy operations, the linear sides of triangle changes to parabolic one. These results obtained by weakest t-norm based arithmetic operations on vague set theory are more suitable than the other existing methods. To sustain the analysis for different spreads say ± 15, ± 25 and ± 50 % and to import the results to the system analysts it is necessary that the obtained fuzzy output is converted into crisp value so that decision maker/system analyst may implement these results into the system. For this

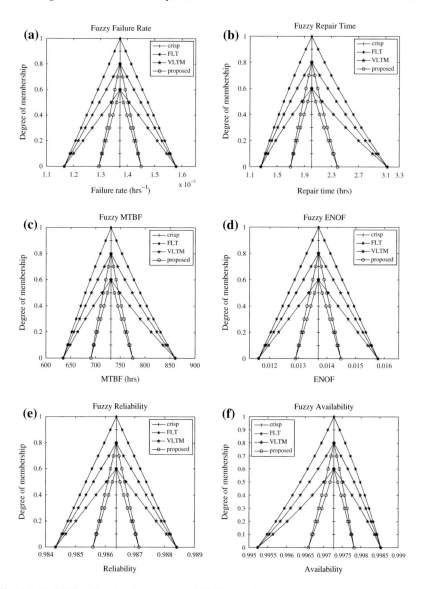

Fig. 6.5 Reliability plots for the system at ±15 % spreads

defuzzification has been done by using the center of gravity method and their corresponding values at different level of uncertainties along with their crisp values are tabulated in Table 6.7. It has been concluded from the table that crisp values do not change with the change of spread while defuzzified values change with change of spreads.

Table 6.7 Defuzzified values of the reliability parameters

Spread	Technique		Failure rate ($\times 10^{-3}$)	Repair time	MTBF	ENOF	Reliability	Availability
	Crisp		1.3725926	0.0081063	30.58174	0.0136952	0.9863683	0.9972699
	Defuzzified values for reliability indices							
±15 %	I		1.3725926	2.1010431	739.06560	0.0136948	0.9863688	0.9970756
	II	A:	1.3725929	2.1016184	739.11816	0.0136948	0.98636890	0.9970745
		B:	1.3725927	2.1012252	739.08223	0.0136948	0.9863688	0.9970753
	III	A:	1.3725448	2.0256469	731.81321	0.0136950	0.9863684	0.9972364
		B:	1.3725444	2.0255711	731.80823	0.0136950	0.9863684	0.9972366
±25 %	I		1.3726777	2.2761652	754.99016	0.0136941	0.9863699	0.9967136
	II	A:	1.3726785	2.2779755	755.15528	0.0136941	0.9863699	0.9967099
		B:	1.3726780	2.2767388	755.04247	0.0136941	0.9863699	0.9967124
	III	A:	1.3726328	2.0592971	734.02191	0.0136947	0.9863687	0.9971736
		B:	1.3726324	2.0590829	734.00783	0.0136947	0.9863687	0.9971740
±50 %	I		1.3730771	3.3173456	848.32954	0.0136898	0.9863747	0.9945536
	II	A:	1.3730803	3.3314687	849.59655	0.0136898	0.98637480	0.9945225
		B:	1.3730781	3.3218515	848.73335	0.0136898	0.9863747	0.9945437
	III	A:	1.3729649	2.2224341	744.68031	0.0136931	0.9863700	0.9968725
		B:	1.3729614	2.2215040	744.61915	0.0136931	0.9863700	0.9968742

I: FLT (Knezevic and Odoom 2001), II: VLTM (Garg 2013), III: proposed approach
A: membership function, B: nonmembership function

6.4.3.1 Sensitivity Analysis

To analyze the impact of the reliability parameters on system MTBF, an analysis has been done in which various combinations of reliability, availability and failure rate parameters has been taken. Throughout the combinations, ranges of repair time and ENOF are fixed and have been varied respectively in the range computed from their membership functions at cut level $\alpha = 0$. For instance, the first three combinations of the reliability parameters states that when reliability and availability of the system has been fixed to 0.9855 and 0.9964 respectively and failure rate are changed from 0.0008 to 0.0013 and further to 0.0018 then the predicted range of the system MTBF has been reduced to 56.7185, 56.7614 and 56.8040 % from Garg (2013) approach when proposed approach has been applied. A similar effect is observed for other combinations too and their ranges are tabulated in Table 6.8. The major advantage of this analysis is that based on their results the system analyst may preserve the particular index and hence seen the effect of taking wrong combinations of the reliability parameters on its MTBF. Also it shows that how the slightest change of failure rate will effect on system MTBF and hence on its performance.

6.4.3.2 Performance Analysis Using RAM-Index

As the time passes then the reliability of the system would gradually decrease if no preventive maintenance action has been taken within a regular interval of time. Thus it is necessary for the system analyst to perform a necessary maintenance action in order to increase the performance of the system. But it is difficult, if not impossible, to find the component from the system on which more attention should be given for saving the money, time and manpower so that the efficiency of the system may increase. For such analysis, a composite measure of the system reliability, availability and maintainability parameter named as the RAM—Index has been used for finding the critical component, as per preferential order, of the system.

The mathematical expression of the RAM-Index is defined as

$$RAM(t) = w_1 \times R_s(t) + w_2 \times A_s(t) + w_3 \times M_s(t) \tag{6.22}$$

where $w_i \in (0, 1), i = 1, 2, 3$ are the weights corresponding to reliability, availability and maintainability respectively such that $\sum_{i=1}^{3} w_i = 1$. Here $w_1 = 0.36$, $w_2 = 30$ and $w_3 = 0.34$ have been used during the analysis. The major advantage of using this index is that by varying the components failure and repair rate parameters, the impact onto the system's performance by the change in its behavior can be analyzed effectively to make the future course of action. Since RAM parameters are represented in the form of membership functions and hence consequently RAM-Index will come as a fuzzy membership function. In order to analyze the system performance, firstly the effect of uncertainties on RAM-Index

Table 6.8 Change in MTBF for various combination of reliability parameters

Methods	Range of MTBF	(0.9855, 0.0008, 0.9964)	(0.9855, 0.0013, 0.9964)	(0.9855, 0.0018, 0.9964)	(0.9863, 0.0008, 0.9972)	(0.9863, 0.0013, 0.9972)	(0.9863, 0.0018, 0.9972)	(0.9869, 0.0008, 0.9977)	(0.9869, 0.0013, 0.9977)	(0.9869, 0.0018, 0.9977)
I	Min:	886.06863	545.35771	393.93063	836.80659	515.02378	372.00920	799.89237	492.29557	355.58589
	Max:	1574.22222	969.83379	701.21671	1486.29795	915.48618	661.79206	1420.47933	874.83218	632.32234
II	Min:	1118.23302	688.28734	497.20038	1056.04747	649.98732	469.51615	1009.45175	621.29304	448.77806
	Max:	1416.07603	871.82499	629.93564	1337.23570	823.22040	594.76916	1278.17571	786.82113	568.44131

I: VLTM approach II: proposed approach

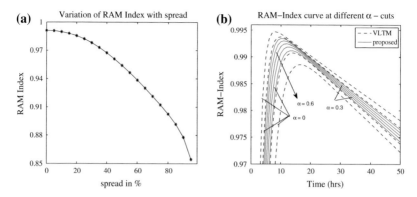

Fig. 6.6 Variation of the RAM-Index plots

has been investigated by varying their spread from 0 to 100 % and their corresponding variation of their index has been plotted in Fig. 6.6a which indicates that RAM-Index decreases with the increase in the uncertainty level. It means to achieve higher performance of the systems, involved uncertainties should be minimized. On the other hand, at different α—cut (0, 0.3, 0.6) the long-run period behavior of the RAM-Index for the system has been shown in Fig. 6.6b which shows that

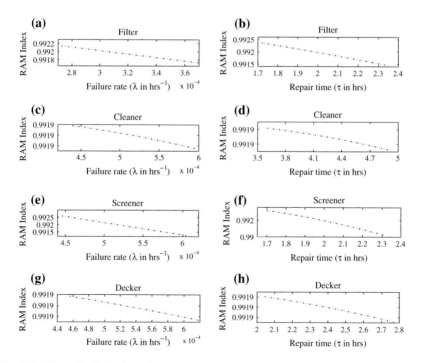

Fig. 6.7 Effect of individual varying components parameters on RAM-Index

RAM-Index of the system increases within the time interval from t = 0 to 13 h and attain its maximum value at t = 13 h in the interval 0.9918217–0.9929697 and after that system performance reduces exponentially. Thus it is found that for increasing the performance of the system, a necessary action should be taken after time t = 13 h.

As the performance of the system is directly depends upon its components and hence the effect on its index has been investigated by varying the failure rate and repair time of each component separately at t = 10 h and simultaneously fixing the other component parameter in Fig. 6.7. In this figure, each plot contains two sub-plots against variations in failure rate and repair time of the each component while their corresponding maximum and minimum values are summarized in Table 6.8. On the other hand, the effect of the simultaneous variations of failure rate and repair time of each component is shown in Fig. 6.8. It may be observed from the Fig. 6.8b that the variation in the failure rate and repair time of the cleaner components shows the significant impact on the performance of the system i.e. an increase in their failure rate from $(0.4394331$ to $0.5945271) \times 10^{-3}$ h^{-1} and repair time from 3.60835 to 4.88188 h reduce the system index by 2.6484 %. On the other hand, the variation in the failure rate and repair time of the filter components shows the insignificant impact on the performance of the system. Similar effect on system RAM-Index by the variation of the other component failure rate and repair times is

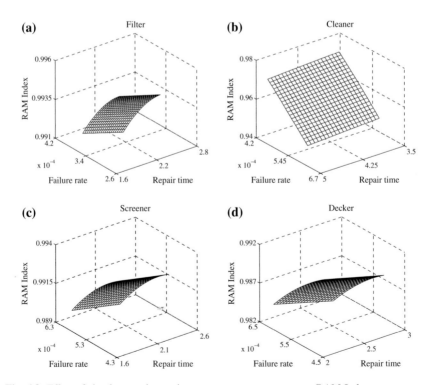

Fig. 6.8 Effect of simultaneously varying components parameters on RAM-Index

Table 6.9 Effect of variations of system's components' failure and repair times on its RAM-Index for washing system

Component	Range of failure rate $\lambda \times 10^{-3} (h^{-1})$	RAM-Index	Range of repair time $\tau(h)$	RAM-Index
Filter	0.2728972–0.3692139	Min: 0.99173248	1.7225029–2.33044510	Min: 0.99147413
		Max: 0.99213980		Max: 0.99235452
Cleaner	0.4394331–0.5945271	Min: 0.99193595	3.6083503–4.8818857	Min: 0.99193595
		Max: 0.99193597		Max: 0.99193598
Screener	0.4462899–0.6038041	Min: 0.99128629	1.7030243–2.3040917	Min: 0.99028515
		Max: 0.99258587		Max: 0.99313700
Decker	0.4563434–0.6174058	Min: 0.99193573	2.0462152–2.7684088	Min: 0.99193456
		Max: 0.99193616		Max: 0.99193703

Table 6.10 Effect of simultaneously variations of system's components' failure and repair times on its RAM-Index for washing system

Component	Range of failure rate $\lambda \times 10^{-3} (h^{-1})$	Range of repair time $\tau(h)$	RAM-Index
Filter	0.2728972–0.3692139	1.7225029–2.33044510	Min: 0.99103063
			Max: 0.99561977
Cleaner	0.4394331–0.5945271	3.6083503–4.8818857	Min: 0.94760573
			Max: 0.97270171
Screener	0.4462899–0.6038041	1.7030243–2.3040917	Min: 0.98920784
			Max: 0.99357250
Decker	0.4563434–0.6174058	2.0462152–2.7684088	Min: 0.98273166
			Max: 0.99170908

analyzed from the Fig. 6.8. The magnitude of the effect of variation in failure rate and repair times of various subsystems of the system on its performance is summarized in Table 6.10. On the basis of results tabulated, it can be analyzed that for improving the performance of the system, more attention should be given to the components as per the preferential order; cleaner, decker, screener and filter.

6.5 Conclusion

This chapter deals with the evaluation of the various reliability parameters of the industrial systems by using uncertain, vague and imprecise data. For this, a structural framework has been developed by the author, based on CS and vague set theory, to model, analyze and predict the system behavior by utilizing quantified, limited and uncertain data. The washing system of the paper industry has been taken as an illustrative example to demonstrate the approach. For this, optimal design parameters—MTBF and MTTR—of the system have been obtained firstly using reliability, availability and maintainability as an objective. The conflicting nature between the objectives is resolved by defining their nonlinear fuzzy goals and then aggregate by using a product aggregator operator. The stability of these optimal parameters is justified by means of pooled t-test statistics. These optimal design parameters will generally help the maintenance engineers to understand the behavioral dynamics of the system and to reallocate the resources. The information system stored the system designs and parameters in the knowledge base and can be retrieved by significant features, which facilitates the designer and increases design efficiency.

Due to complexity in the system configuration, the data obtained from historical records, is imprecise and inaccurate. Keeping this point in view, efficiency for analyzing the behavior of the system is increased by using computed design parameters in terms of membership and non-membership functions using weakest t-norm based arithmetic operations on vague set theory. The development of intuitionistic fuzzy numbers from the available data and using vague possibility theory can greatly increase the relevance of reliability study. The computed results are compared with the existing methodology results and have been observed that the proposed technique has compressed range of uncertainties during the analysis as compared to others and consequently the proposed approach is more flexible for the decision maker to make a more sound and effective decision in a lesser time. The crisp and defuzzified values of various reliability parameters are summarized in a tabular form. Sensitivity as well as performance analysis of the system performance index has been investigated which help the plant personnel to rank the system components. Based on their analysis, the components of the system which has excessive failure rates, long repair times or high degree of uncertainty associated with these values are identified and reported in preferential order as cleaner, decker, screener and a filter.

6.6 Future Research Direction

The present work can be done equally well to evaluate the system behavior in other process industries such as thermal power plant, sugar plant etc. as the considered methodology can overcome various kinds of problem in the area of quality, reliability and maintainability, which strongly needs the management attention. Also we can extend the present work for time varying component failure rate instead of constant rate i.e. from exponential distribution to Weibull or Normal distribution functions. The work can also be extended to devise suitable methodology for [(i)]

(i)   Conducting cost analysis.
(ii) Developing inventory and spare parts maintenance management system.
(iii) Redundancy allocation problem.
(iv) suitable maintenance strategies after understanding the behavior dynamics associated with functioning of the system.

Also, the general idea presented here could also be applicable to many other systems like complex, circular, series-parallel, k-out-of-n systems and so on. The investigations on these different systems will be carried out in our future work.

References

Attanassov, K.T.: Intuitionistic fuzzy sets. Fuzzy Sets Syst. **20**, 87–96 (1986)

Attanassov, K.T.: More on intuitionistic fuzzy sets. Fuzzy Sets Syst. **33**(1), 37–46 (1989)

Barabady, J., Kumar, U.: Maintenance schedule by using reliability analysis: a case study at jajarm bauxite mine of iran. In: 20th World Mining Congress and EXPO2005, pp. 79–86. Tehran, Iran (2005a)

Barabady, J., Kumar, U.: Reliability and maintainability analysis of crushing plants in jajarm bauxite mine of iran. In: Proceedings of the Annual Reliability and Maintainability Symposium, pp. 109–115 (2005b)

Birolini, A.: Reliability Engineering: Theory and Practice, 5th edn. Springer, New York (2007)

Bris, R., Chatelet, E., Yalaoui, F.: New method to minimize the preventive maintenance cost of series-parallel systems. Reliab. Eng. Syst. Saf. **82**, 247–255 (2003)

Bustince, H., Burillo, P.: Vague sets are intuitionistic fuzzy sets. Fuzzy Sets Syst. **79**(3), 403–405 (1996)

Chang, J.R., Chang, K.H., Liao, S.H., Cheng, C.H.: The reliability of general vague fault tree analysis on weapon systems fault diagnosis. Soft. Comput. **10**, 531–542 (2006)

Chen, S.M.: Analyzing fuzzy system reliability using vague set theory. Int. J. Appl. Sci. Eng. **1**(1), 82–88 (2003)

Clerc, M., Kennedy, J.F.: The particle swarm: explosion, stability, and convergence in a multi-dimensional complex space. IEEE Trans. Evol. Comput. **6**(1), 58–73 (2002)

Dinesh-Kumar, U., Marquez, J.E.R., Nowicki, D., Verma, D.: A goal programming model for optimizing reliability, maintainability and supportability under performance based logistics. Int. J. Reliab., Qual. Saf. **14**(3), 251–261 (2007)

Eberhart, R., Kennedy, J.: A new optimizer using particle swarm theory. In: Proceedings of the Sixth International Symposium on Micro Machine and Human Science, pp. 39–43 (1995)

Garg, H.: Reliability analysis of repairable systems using Petri nets and Vague Lambda-Tau methodology. ISA Trans. **52**(1), 6–18 (2013)

Garg, H., Rani, M.: An approach for reliability analysis of industrial systems using PSO and IFS technique. ISA Trans. **52**(6), 701–710 (2013)

Garg, H., Rani, M., Sharma, S.P.: Fuzzy RAM analysis of the screening unit in a paper industry by utilizing uncertain data. Int. J. Qual., Stat. Reliab., Article ID: 203,842, 14 p (2012)

Garg, H., Rani, M., Sharma, S.P.: Predicting uncertain behavior of press unit in a paper industry using artificial bee colony and fuzzy Lambda-Tau methodology. Appl. Soft Comput. **13**(4), 1869–1881 (2013)

Garg, H., Rani, M., Sharma, S.P.: An approach for analyzing the reliability of industrial systems using soft computing based technique. Expert Syst. Appl. **41**(2), 489–501 (2014a)

Garg, H., Rani, M., Sharma, S.P., Vishwakarma, Y.: Intuitionistic fuzzy optimization technique for solving multi-objective reliability optimization problems in interval environment. Expert Syst. Appl. **41**, 3157–3167 (2014b)

Garg, H., Sharma, S.P.: Stochastic behavior analysis of industrial systems utilizing uncertain data. ISA Trans. **51**(6), 752–762 (2012)

Garg, H., Sharma, S.P.: A two-phase approach for reliability and maintainability analysis of an industrial system. Int. J. Reliab., Qual. Saf. Eng. (IJRQSE) **19**(3) (2012). doi:10.1142/S0218539312500131

Garg, H., Sharma, S.P.: Multi-objective reliability-redundancy allocation problem using particle swarm optimization. Comput. Ind. Eng. **64**(1), 247–255 (2013)

Garg, H., Sharma, S.P., Rani, M.: Cost minimization of washing unit in a paper mill using artificial bee colony technique. Int. J. Syst. Assur. Eng. Manag. **3**(4), 371–381 (2012)

Gau, W.L., Buehrer, D.J.: Vague sets. IEEE Trans. Syst., Man, Cybern. **23**, 610–613 (1993)

Gen, M., Yun, Y.S.: Soft computing approach for reliability optimization: state-of- the-art survey. Reliab. Eng. Syst. Saf. **91**(9), 1008–1026 (2006)

Goldberg, D.E.: Genetic Algorithm in Search, Optimization and Machine Learning. Addison-Wesley, MA (1989)

Holland, J.H.: Adaptation in Natural and Artificial Systems. The University of Michigan Press, Ann Arbor (1975)

Hsieh, T.J., Yeh, W.C.: Penalty guided bees search for redundancy allocation problems with a mix of components in series–parallel systems. Comput. Oper. Res. **39**(11), 2688–2704 (2012)

Juang, Y.S., Lin, S.S., Kao, H.P.: A knowledge management system for series-parallel availability optimization and design. Expert Syst. Appl. **34**, 181–193 (2008)

Karaboga, D.: An idea based on honey bee swarm for numerical optimization. Tech. rep., TR06, Erciyes University, Engineering Faculty, Computer Engineering Department (2005)

Karaboga, D., Akay, B.: A comparative study of artificial bee colony algorithm. Appl. Math. Comput. **214**(1), 108–132 (2009)

Karaboga, D., Basturk, B.: A powerful and efficient algorithm for numerical function optimization: artificial bee colony (ABC) algorithm. J. Global Optim. **39**, 459–471 (2007)

Kennedy, J., Eberhart, R.C.: Particle swarm optimization. In: IEEE International Conference on Neural Networks, vol. 4, pp. 1942–1948. Piscataway, Seoul (1995)

Khan, F.I., Haddara, M., Krishnasamy, L.: A new methodology for risk-based availability analysis. IEEE Trans. Reliab. **57**(1), 103–112 (2008)

Knezevic, J., Odoom, E.R.: Reliability modeling of repairable systems using Petri nets and Fuzzy Lambda-Tau Methodology. Reliab. Eng. Syst. Saf. **73**(1), 1–17 (2001)

Kumar, A., Yadav, S.P., Kumar, S.: Fuzzy reliability of a marine power plant using interval valued vague sets. Int. J. Appl. Sci. Eng. **4**(1), 71–82 (2006)

Kumar, M., Yadav, S.P.: A novel approach for analyzing fuzzy system reliability using different types of intuitionistic fuzzy failure rates of components. ISA Trans. **51**(2), 288–297 (2012)

Kuo, W., Prasad, V.R., Tillman, F.A., Hwang, C.: Optimal Reliability Design—Fundamentals and Applications. Cambridge University Press, Cambridge (2001)

Liberopoulos, G., Tsarouhas, P.: Systems analysis speeds up chipita's food processing line. Interfaces **32**(3), 62–76 (2002)

Rajpal, P.S., Shishodia, K.S., Sekhon, G.S.: An artificial neural network for modeling reliability, availability and maintainability of a repairable system. Reliab. Eng. Syst. Saf. **91**(7), 809–819 (2006)

Ross, T.J.: Fuzzy Logic with Engineering Applications, 2nd edn. Wiley, New York (2004)

Sharma, R.K., Kumar, S.: Performance modeling in critical engineering systems using RAM analysis. Reliab. Eng. Syst. Saf. **93**(6), 913–919 (2008)

Shi, Y., Eberhart, R.C.: Parameter selection in particle swarm optimization. evolutionary programming VII. In: EP 98, pp. 591–600. Springer, New York (1998)

Taheri, S., Zarei, R.: Bayesian system reliability assessment under the vague environment. Appl. Soft Comput. **11**(2), 1614–1622 (2011)

Yang, X.S., Deb, S.: Cuckoo search via lévy flights. In: Proceedings of World Congress on Nature & Biologically Inspired Computing (NaBIC 2009), pp. 210–214. IEEE Publications, USA (2009)

Yeh, W.C., Hsieh, T.J.: Solving reliability redundancy allocation problems using an artificial bee colony algorithm. Comput. Oper. Res. **38**(11), 1465–1473 (2011)

Yeh, W.C., Su, J.C.P., Hsieh, T.J., Chih, M., Liu, S.L.: Approximate reliability function based on wavelet latin hypercube sampling and bee recurrent neural network. IEEE Trans. Reliab. **60**(2), 404–414 (2011)

Zadeh, L.A.: Fuzzy sets. Inf. Control **8**, 338–353 (1965)

Zavala, A., Diharce, E., Aguirre, A.: Particle evolutionary swarm for design reliability optimization. In: Lecture Notes in Computer Science 3410, Presented at the Third International Conference on Evolutionary Multi-Criterion Optimization, pp. 856–869, Guanajuato (2005)

Chapter 7
A Fuzzy Design of Single and Double Acceptance Sampling Plans

Cengiz Kahraman, Ebru Turanoglu Bekar and Ozlem Senvar

Abstract In this chapter, we briefly introduce the topic of acceptance sampling. We also examine acceptance sampling plans with intelligent techniques for solving complex quality problems. Among intelligent techniques, we focus on the application of the fuzzy set theory in the acceptance sampling. Moreover, we propose multi-objective mathematical models for fuzzy single and fuzzy double acceptance sampling plans with illustrative examples. The study illustrates how an acceptance sampling plan should be designed under fuzzy environment.

Keywords Acceptance sampling · Double sampling · Single sampling · Fuzzy sets

7.1 Introduction

In manufacturing industries, sampling inspection is a common practice for quality assurance and cost reduction. Acceptance sampling is a practical and economical alternative to costly 100 % inspection. Acceptance sampling offers an efficient way to assess the quality of an entire lot of product and to decide whether to accept or reject it. The basic decisions in sampling inspection are how many manufactured items to be sampled from each lot and how many identified defective items in the sample to accept or reject each lot (Wang and Chankong 1991). The application of

C. Kahraman
Department of Industrial Engineering, Istanbul Technical University, Istanbul, Turkey

E.T. Bekar (✉)
Department of Industrial Engineering, Izmir University, Uckuyular, Izmir, Turkey
e-mail: ebru.turanoglu@izmir.edu.tr

O. Senvar
Université de Technologie de Troyes, Institut Charles Delaunay, Laboratoire d'Optimisation des Systèmes Industriels, Troyes, France

© Springer International Publishing Switzerland 2016
C. Kahraman and S. Yanık (eds.), *Intelligent Decision Making in Quality Management*, Intelligent Systems Reference Library 97,
DOI 10.1007/978-3-319-24499-0_7

acceptance sampling minimizes product destruction during inspection and testing as well as increases the inspection quantity and effectiveness.

Practically, acceptance sampling is a form of testing that involves taking random samples of lots or batches of finished products and measuring them against predetermined standards. Acceptance sampling pertains to incoming batches of raw materials or purchased parts and to outgoing batches of finished goods.

Acceptance sampling is useful when one or more of the following conditions is available: a large number of items must be processed in a small amount of time; the cost of passing defective items is low; destructive testing is required; or the inspectors may experience boredom or fatigue in inspecting large numbers of items.

Acceptance sampling plans are useful tools for quality control practices, which involve quality contracting on product orders between the vendor and the buyer. Those sampling plans provide the vendor and the buyer rules for lot sentencing while meeting their preset requirements on product quality. Nowadays, sampling plans are the primary tools for quality and performance management in industry. Sampling plans are used to decide either to accept or reject a received batch of items. In any acceptance sampling plan, there are two possible error, which are producer's risk and consumer's risk. Producer's risk is the rejection of a good lot. Consumer's risk is the acception of a bad lot.

Acceptance sampling plans provide the vendor and buyer the decision rules for product acceptance to meet the present product quality requirement. In practice, proper design of an acceptance sampling planning is based on the true quality level required by customers. However, it is sometimes not possible to determine this quality level with certain values. Especially in production, it is not easy to determine the parameters of acceptance sampling such as proportion of defective items, sample size, acceptable defective items.

Classical acceptance sampling plans have been studied by many researchers. In different acceptance sampling plans the proportion of defective items, is considered as a crisp value. The proportions of defective items are estimated or provided by experiment. According to Fountoulaki et al. (2008), approaches employing machine learning techniques in acceptance sampling are limited and mainly focused on the design of acceptance sampling plans. Sampath (2009) emphasized that in the manufacturing processes, quantities such as the proportion of defective items in a production lot may not be precisely known and usually the practitioners have to compromise with some imprecise or approximate values. Prior knowledge of such quantities is required to evaluate the quality of a produced lot.

The vagueness present from personal judgment, experiment or estimation can be treated formally with the help of fuzzy set theory. Among other intelligent techniques, fuzzy set theory is known as a powerful mathematical tool for modeling uncertainity in classical attribute quality characteristics (Jamkhaneh et al. 2009).

There are many other investigations and many other publications related to acceptance sampling plans. In this chapter, we briefly introduce the topic of acceptance sampling. Also, we examine acceptance sampling plans with intelligent techniques for solving important as well as fairly complex problems related to acceptance sampling. A lot or batch of items can be inspected in several ways including the use of single, double, multiple, sequential sampling. Among other intelligent techniques, we focus on the application of fuzzy set theory in the acceptance sampling. We propose mathematical models for fuzzy single and fuzzy double acceptance sampling plans with illustrative examples.

The rest of this chapter is organized as follows. In Sect. 7.2, acceptance sampling basic concepts, terminology and plans are given. In Sect. 7.3, intelligent techniques in acceptance sampling are briefly reviewed. Design of fuzzy acceptance sampling plans and their illustrative examples are provided in Sect. 7.4. In Sect. 7.5 proposed fuzzy multi-objective mathematical models are explained with illustratives examples. Finally conclusion, discussions as well as recommendations for further studies are provided in the last section.

7.2   Acceptance Sampling Basic Concepts and Terminology

Acceptance sampling inspection is part of statistical practice concerned with sampled items to produce some quality information about the inspected products, especially to check whether products have met predetermined quality specifications (Schilling 1982). The complexity of the sampling inspection process gives rise to challenges for the definition of data quality elements, determination of sample item, sample size and acceptance number, and a combination of quality levels required by the producer and the consumer. Here are the top 10 reasons why acceptance sampling is still necessary:

- Tests are destructive, necessitating sampling.
- Process not in control, necessitating sampling to evaluate product.
- 100 % sampling is inefficient, 0 % is risky.
- Special causes may occur after process inspection.
- Need for assurance while instituting process control.
- Rational subgroups for process control may not reflect outgoing quality.
- Deliberate submission of defective material.
- Process control may be impractical because of cost, or lack of sophistication of personnel.
- 100 % inspection does not promote process/product improvement.
- Customer mandates sampling plan.

The principle of acceptance sampling to control quality is the fact that it is not checked all units (N), but only selected part (n). Acceptance sampling plan is a specific plan that clearly states the rules for sampling and the associated criteria for acceptance or otherwise. Acceptance sampling plans can be applied for inspection of end items, components, raw materials, operations, materials in process, supplies in storage, maintenance operations, data or records and administrative procedures. There are two essential issues in acceptance sampling inspection theory. The first is the determination of the acceptance sampling plan, which is characterized by sample size and acceptance number. The main goal of designing an optimal sampling plan is to obtain a high accuracy of product inspection and to reduce the inspection cost (Von Mises 1957). The second is to determine the method to select samples from the lot, which refers to the sampling method. Commonly used sampling methods include simple random sampling, system sampling, stratified sampling, and cluster sampling (Cochran 1977; Degroot 1986; Wang et al. 2010).

Acceptance-sampling plans classify to different ways. One major classification is by attributes and variables. Acceptance-sampling plans by attributes are single sampling plan, double sampling plan, multiple-sampling plan, and sequential sampling plan (Schilling 1982).

Single sampling is undoubtedly the most used of any sampling procedure. The simplest form of such a plan is single sampling by attributes which relates to dichotomous situations, i.e., those in which inspection results can be classified into only two classes of outcomes. This includes go, no-go gauging procedures as well as other classifications. Applicable to all sampling situations, the attributes single sampling plan has become the benchmark against which other sampling plans are judged. It is employed in inspection by counting the number of defects found in sample (Poisson distribution) or evaluating the proportion defective from processes or large lots (binomial distribution) or from individual lots (hypergeometric distribution). It involves taking a random sample size n from a lot size N. The number of defectives (or defects) d found is compared to an acceptance number c. If the number found is less than or equal to c, the lot is accepted. If the number found is greater than c, the lot is rejected.

Often a lot of items are so good or so bad that we can reach a conclusion about its quality by taking a smaller sample than would have been used in a single sampling plan. In double sampling if the results of the first sample are not definitive in leading to acceptance or rejection, a second sample is taken which then leads to a decision on the disposition of the lot. In brief, if the number of defects in this first sample (d_1) is less than or equal to some lower limit (c_1), the lot can be accepted. If the number of defects first and second sample (d_2) exceeds an upper limit (c_2), the whole lot can be rejected. But if the number of defects in the n1 sample is between c_1 and c_2, a second sample is drawn. The cumulative results determine whether to accept or reject the lot. The concept is called double sampling.

Multiple sampling involves the inspection of specific lots on the basis of k successive samples as needed to make a decision, where k varies from 1 to K (i.e. a whole number). It is an extension of double sampling, with smaller samples used sequentially until a clear decision can be made. In multiple sampling by attributes, more than two samples can be taken in order to reach a decision to accept or reject the lot. The main advantage of multiple sampling plans is a reduction in sample size for the same protection.

Single, double, and multiple plans assess one or more successive samples to determine lot acceptability. Sequential sampling involves making a decision as to disposition of the lot or resample successively as each item of the sample is taken and it may be regarded as multiple-sampling plan with sample size one and no upper limit on the number of samples to be taken. It is often applied where sample size critical so that a minimum sample must be taken. Under sequential sampling, samples are taken, one at time, until a decision is made on the lot or process sampled. After each item is taken a decision is made to (7.1) accept, (7.2) reject, or (7.3) continue sampling. Samples are taken until an acceptance or rejection decision is made. Thus, the procedure is open ended, the sample size not being determined until the lot is accepted or rejected. Selection of the best sampling approach (single, double, multiple or sequential) depends on the types of products being inspected and their expected quality level. A very low-quality batch of goods, for example, can be identified quickly and more cheaply with sequential sampling. This means that the inspection, which may be costly and/or destructive, can end sooner. On the other hand, in many cases a single sampling plan is easier and simpler for workers to conduct even though the number sampled may be greater than under other plans.

7.3   Operating Characteristic Curves

The operating characteristic (OC) curve plots the probability of acceptance against possible values of proportion defective. OC curve describes how well an acceptance plan discriminates between good and bad lots. A curve pertains to a specific plan, that is, a combination of n and c. It is intended to show the probability that the plan will accept lots of various quality levels. The curves for different sampling plans are shown in Fig. 7.1. The OC curve sketches the performance of a plan for various possible proportions defective. It is plotted using appropriate probability functions for the sampling situation involved. The curve shows the ability of a sampling plan to discriminate between high quality and low quality lots. With acceptance sampling, two parties are usually involved: the producer of the product and the

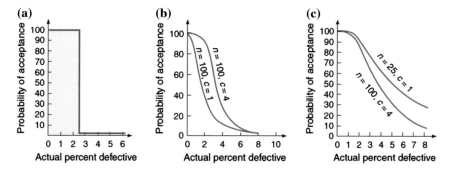

Fig. 7.1 a Perfect discrimination for inspection plan. **b** OC curves for two different acceptable levels of defects (c = 1, c = 4) for the same sample size (n = 100). **c** OC curves for two different sample sizes (n = 25, n = 100) but same acceptance percentages (4 %). Larger sample size shows better

consumer of the product. When specifying a sampling plan, each party wants to avoid costly mistakes in accepting or rejecting a lot.

The producer wants to avoid the mistake of having a good lot rejected (producer's risk) because he or she usually must replace the rejected lot. Conversely, the customer or consumer wants to avoid the mistake of accepting a bad lot because defects found in a lot that has already been accepted are usually the responsibility of the customer (consumer's risk). The producer's risk α is the probability of not accepting a lot of acceptable quality level (AQL) quality and the consumer's risk β is the probability of accepting a lot of limiting quality level (LQL) quality. The term acceptable quality level (AQL) is commonly used as the 95 % point of probability of acceptance, although most definitions do not tie the term to a specific point on the OC curve and simply associate it with a "high" probability of acceptance. The term is used here as it was used by the Columbia Statistical Research Group in preparing the (Von Mises 1957) input to the JAN-STD-105 standard. LTPD refers to the 10 % probability point of the OC curve and is generally associated with percent defective. The advent of plans controlling other parameters of the distribution led to the term limiting quality level (LQL), usually preceded by the percentage point controlled. Thus, "10 % limiting quality" is the LTPD (Schilling 1982).

In most sampling plans, when a lot is rejected, the entire lot is inspected and all of the defective items are replaced. Use of this replacement technique improves the average outgoing quality in terms of percent defective.

The average outgoing quality (AOQ) can be explained as the expected quality of outgoing product following the use of an acceptance sampling plan for a given value of the incoming quality. For the lots accepted by the sampling plan, no screening will be done and the outgoing quality will be the same as that of the incoming quality p. For those lots screened, the outgoing quality will be zero, meaning that they contain no nonconforming items. Since the probability of accepting a lot is P_a, the outgoing lots will contain a proportion of pP_a defectives. If the nonconforming units found in the sample of size n are replaced by good ones, the average outgoing quality (AOQ) will be (Kahraman and Kaya 2010):

$$AOQ = \frac{N-n}{N}pP_a \qquad (7.1)$$

For large N,

$$AOQ \cong pP_a \qquad (7.2)$$

The maximum value of AOQ over all possible values of fraction defective, which might be submitted, is called the AOQ limit (AOQL). It represents the maximum long-term average fraction defective that the consumer can see under operation of the rectification plan. It is sometimes necessary to determine the average amount of inspection per lot in the application of such rectification schemes, including 100 % inspection of rejected lots. This average, called the average total inspection (ATI), is made up of the sample size n on every lot plus the remaining (N-n) units on the rejected lots, so that the ATI for single sampling is calculated as following Eqs. (7.3 and 7.4).

$$ATI = n + (1 - P_a)(N - n) \tag{7.3}$$

$$ATI = P_a n + (1 - P_a)N \tag{7.4}$$

The ATI for the double sampling plan can be calculated from the following Eqs. (7.5–7.7). In Eq. (7.5), the average sample number (ASN) is the mean number of items inspected per lot. The concept of ASN is very useful in determining the average number of samples that will be inspected in using more advanced sampling plans. For a single sampling plan, one takes only a single sample of size n and hence the ASN is simply the sample size n. In single sampling, the size of the sample inspected from the lot is always constant, whereas in double sampling, in double sampling plans, for example, the second sample is taken only if results from the first sample are not sufficiently definitive to lead to acceptance or rejection outright. In such a situation the inspection may be concluded after either one or two samples are taken and so the concept of ASN is necessary to evaluate the average magnitude of inspection in the long run.

$$ATI = ASN + (N - n_1) P(d_1 > c_2) + (N - n_1 - n_2) P(d_1 + d_2 > c_2) \tag{7.5}$$

where

$$P(d_1 > c_2) = 1 - P(d_1 \le c_2) \tag{7.6}$$

$$P(d_1 + d_2 > c_2) = 1 - P_a - P(d_1 > c_2) \tag{7.7}$$

A general formula for the average sample number in double sampling is

$$ASN = n_1 P_1 + (n_1 + n_2)(1 - P_1) = n_1 + n_2(1 - P_1) \tag{7.8}$$

where P_1 is the probability of making a lot dispositioning decision on the first sample. This is calculated as following equation:

$$P_1 = P\{\text{lot is accepted on the first sample}\} + P\{\text{lot is rejected on the first sample}\} \tag{7.9}$$

Acceptance sampling is useful for screening incoming lots. When the defective parts are replaced with good parts, acceptance sampling helps to increase the quality of the lots by reducing the outgoing percent defective. Sampling plans and OC curves facilitate acceptance sampling and provide the manager with tools to evaluate the quality of a production run or shipment.

7.4   Literature Review on Acceptance Sampling

In recent years, there are some studies concentrated on acceptance sampling in the literature. Figure 7.2 shows the publication frequencies of acceptance sampling according to years between 2004 (including 2004 and earlier) and 2013.

Some of these publications are journal articles, books/e-books, and so on. Figure 7.3 shows the distribution of these publications according to publication categories. According to this figure, most of the studies on acceptance sampling are published in journals with a rate of 69 %. For example, Baklizi (2003) developed acceptance sampling plans assuming that the life test is truncated at a pre-assigned time. The minimum sample size necessary to ensure the specified average life was obtained and the operating characteristic values of the sampling plans and producer's risk were presented. Kuo (2006) developed an optimal adaptive control policy for joint machine maintenance and product quality control. It included the interactions between the machine maintenance and the product sampling in the search for the best machine maintenance and quality control strategy to find the optimal value function and identify the optimal policy more efficiently in the value iteration algorithm of the dynamic programming.

Borget et al. (2006) applied a single sampling plan by attributes with an acceptance quality level of 2.2 % was evaluated. A prognostic study using a logistic regression model was performed for some drugs to identify risk factors associated with the non-conformity rate of preparations to determine if it was necessary to assay all therapeutic batches produced, or to calculate an individual control rate for each cytotoxic drug, according to various parameters (like number of batches or drug stability). The sampling plan allowed a reduction of almost 8000 analyses with respect to the number of batches analysed for 6 drugs. Pearn and Wub (2007) proposed an effective sampling plan based on process capability index, C_{pk}, to deal with product acceptance determination for low fraction non-conforming products

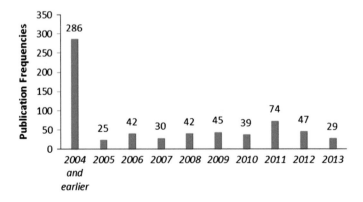

Fig. 7.2 Publication frequencies of acceptance sampling

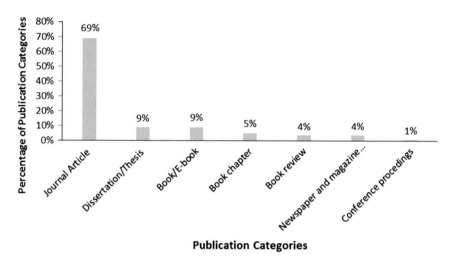

Publication Categories

Fig. 7.3 Percentages of publication categories of acceptance sampling

based on the exact sampling distribution rather than approximation. Practitioners could use this proposed method to determine the number of required inspection units and the critical acceptance value, and make reliable decisions in product acceptance. Aslam (2008) evolved a reliability acceptance plan assuming that the lifetime of a product follows the generalized Rayleigh distribution with known value of the shape parameter. Tsai et al. (2009) explained ordinary and approximate acceptance sampling procedures under progressive censoring with intermittent inspections for exponential lifetimes. Jozani and Mirkamali (2010) demonstrated the use of maxima nomination sampling (MNS) technique in design and evaluation of single AQL, LTPD, and EQL acceptance sampling plans for attributes. They exploited the effect of sample size and acceptance number on the performance of their proposed MNS plans using operating characteristic (OC) curve. Aslam et al. (2010) developed the double sampling and group sampling plans and determined the design parameters satisfying both the producer's and consumer's risks simultaneously for the specified reliability levels in terms of the mean ratio to the specified life. Nezhad et al. (2011) introduced a novel acceptance-sampling plan is proposed to decide whether to accept or reject a receiving batch of items. In this plan, the items in the receiving batch are inspected until a nonconforming item is found. When the sum of two consecutive values of the number of conforming items between two successive nonconforming items falls underneath of a lower control threshold, the batch is rejected. If this number falls above an upper control threshold, the batch is accepted, and if it falls within the upper and the lower thresholds then the process of inspecting items continues. Fernández and Pérez-González (2012) presented for determining optimal failure-censored reliability sampling plans for log-location–scale lifetime models. The optimization procedure to decide the acceptability of a product is usually sufficiently accurate for the most widely used parametric lifetime models, such as the Weibull and log-normal distributions, and fairly robust to small deviations in the prior knowledge. Hsieh and Lu (2013) developed a risk-embedded model via conditional value-at-risk that allows a decision maker to choose an acceptance sampling plan with minimal expected excess cost in accordance with his or her attitude towards risk to gain insights into the role of a decision maker's risk aversion in the determination of Bayesian acceptance sampling plans.

7.5 Intelligent Techniques in Acceptance Sampling

In sampling inspection, the fundamental decisions are how many manufactured items to be sampled from each lot and how many identified defective items in the sample to accept or reject each lot. The problem of determining an optimal sampling plan is NP-complete (2008). The reason is the combinatorial nature of alternative solutions on the sample sizes and acceptance criteria possessing the combinatorial nature. From this standpoint, in recent years, there are a number of researches that merge acceptance sampling with intelligent techniques. In this section, we briefly examine the researches regarding acceptance sampling plans with intelligent techniques for solving important as well as fairly complex problems related to acceptance sampling.

Wang and Chankong (1991) proposed a neurally-inspired approach to generating acceptance sampling inspection plans. They formulated a Bayesian cost model of multi-stage-multi-attribute sampling inspections for quality assurance in serial production systems. The proposed model can accommodate various dispositions of rejected lot such as scraping and screening. Besides, the model can reflect the relationships between stages and among attributes. To determine the sampling plans based on the formulated model, they developed a neurally-inspired stochastic algorithm, which simulates the state transition of a primal-dual stochastic neural network to generate the sampling plans. The simulated primal network is responsible for generation of new states whereas the dual network is for recording the generated solutions. Starting with an arbitrary feasible solution, this algorithm is able to converge to a near optimal or an optimal sampling plan with a sequence of monotonically improved solutions.

Tabled sampling schemes such as MIL-STD-105D offer limited flexibility to quality control engineers in designing sampling plans to meet specific needs. Vasudevan et al. (2012) attempted to find a closed form solution for the design of a single sampling plan for attributes to determine the accepted quality level (AQL) indexed single sampling plan using an artificial neural network (ANN). They used the data from tabled sampling schemes and determined the sample size and the acceptance number by training ANNs, namely with feed forward neural networks with sigmoid neural function by a back propagation algorithm for normal, tightened, and reduced inspections. From these trained ANNs, they obtained the relevant weight and bias values and the closed form solutions to determine the sampling plans using these values. They provided the examples for using these closed form solutions to determine sampling plans for normal, tightened, and reduced inspections. The proposed method does not involve table look-ups or complex calculations. Sampling plan can be determined by using this method, for any required acceptable quality level and lot size. They provided suggestions to duplicate this idea for applying to other standard sampling table schemes process.

Cheng and Chen (2007) suggested a Genetic Algorithm (GA) mechanism to reach a closed form solution for the design of a double sampling plan. In order to design the double sampling plan, the operating characteristic curve has to satisfy some specific criteria. As the parameters of the sampling plan have to be integers, the solution has to be optimal in each case. The GA mechanism is responsible for providing the optimal solution in contrast to the trial-and-error method that has been used so far. This approach seeks for the minimum sample number, even when the initial criteria are not satisfied. Its disadvantage is the relatively large number of the proposed solutions, from which the quality engineer has to decide the optimal one by changing the criteria that were predetermined at the beginning of the process.

Designing double sampling plan requires identification of sample sizes and acceptance numbers. Sampath and Deepa (2012) designed a genetic algorithm for the selection of optimal acceptance numbers and sample sizes for the specified producer's risk and consumer's risk. Implementation of the algorithm has been illustrated numerically for different choices of quantities involved in a double sampling plan.

Fountoulaki et al. (2008) proposed methodology for Acceptance Sampling by Variables, dealing with the assurance of products quality, using machine learning techniques to address the complexity and remedy the drawbacks of existing approaches. Their methodology exploited ANNs to aid decision making about the acceptance or rejection of an inspected sample. For any type of inspection, ANNs are trained by data from corresponding tables of a standard's sampling plan schemes. Once trained, ANNs can give closed-form solutions for any acceptance quality level and sample size, thus leading to an automation of the reading of the sampling plan tables, without any need of compromise with the values of the specific standard chosen each time. Their methodology provides enough flexibility to quality control engineers during the inspection of their samples, allowing the consideration of specific needs, while it also reduces the time and the cost required for these inspections.

In acceptance sampling plans, the decisions on either accepting or rejecting a specific batch is still a challenging problem. In order to provide a desired level of protection for customers as well as manufacturers, Fallahnezhad and Niaki (2012) proposed a new acceptance sampling design to accept or reject a batch based on Bayesian modeling to update the distribution function of the percentage of non-conforming items. They utilized the backwards induction methodology of the decision tree approach to determine the required sample size. They carried out a sensitivity analysis on the parameters of the proposed methodology showing the optimal solution is affected by initial values of the parameters. Furthermore, they determined an optimal (n, c) design when there is a limited time and budget available and they specified the maximum sample size in advance.

In many practical cases it is difficult to classify inspected items as conforming or nonconforming. This problem rather frequently can be faced when quality data come from users who express their assessments in an informal way, using such expressions like "almost good", "quite good", "not so bad", and etc.

Ohta and Ichihashi (1988), Kanagawa and Ohta (1990), Tamaki et al. (1991), and Grzegorzewski (1998), Grzegorzewski et al. (2001) discussed single sampling by attributes with relaxed requirements.

Ohta and Ichihashi (1988) presented a procedure for designing a single sampling plan using fuzzy membership functions for both the producer's risk and consumer's risk, with the aim of finding a reasonable solution for the trade-off between the sampling size and the producer's and consumer's risks. This design methodology is deficient in the sense that it does not explicitly takes into account of minimizing the sample size n. The desire for smaller sample size is imposed by choosing triangular membership functions for the risks. However, this choice does not make sense for the part of the membership functions where the risks are higher than their nominal values.

Kanagawa and Ohta (1990) selected trapezoidal membership functions for risks, and taking into account a membership function for the grade of satisfaction for the sample size. They stated that the membership function must be a monotonically decreasing function of the sample size n, however, no method for constructing this function is proposed.

Sampling plan by attributes for vague data were considered by Hryniewicz. Hryniewicz (1992) attempted to cope with the statistical analysis of such quality data.

If the quality characteristic monitored is a variable, acceptable quality level and rejectable quality level (AQL and RQL) are identified for evaluating the acceptance or rejection of an inspection lot. Otherwise, when the quality characteristic is an attribute, the number of defectives is compared to a specific limit of number of allowed defectives for the decisions of accept/reject. In the former case, acceptable quality level and rejectable quality level may not be specified as a crisp value because rigid values of AQL and RQL may not necessarily give a sampl plan. Besides, these values are commonly not very precise but rather descriptive. Thus, the crisp values of AQL and RQL may be relaxed as fuzzy values.Much more practical procedure, namely the fuzzy version of an acceptance sampling plan by variables, has been proposed by Grzegorzewski (2002). Grzegorzewski et al. (2001), Grzegorzewski (2002) considered sampling plan by variables with fuzzy requirements. General results from the theory of fuzzy statistical tests have been used for the construction of fuzzy sampling plans when the quality characteristic of interest is described by a fuzzy normal distribution.

Hryniewicz (2003) has shown why in the case of imprecise input information optimal inspection intervals are usually determined using additional preference measures than strict optimization techniques.

Krätschmer (2005) proposed a mathematically sound basis for the sampling inspection by attributes in fuzzy environment. According to Hryniewicz (2008), no new practical SQC procedures have been proposed using that general model.

Jamkhaneh et al. (2009) proposed a method for designing acceptance single sampling plans with fuzzy quality characteristic with using fuzzy Poisson distribution. They presented the acceptance single sampling plan when the fraction of nonconforming items is a fuzzy number and modeled by fuzzy Poisson distribution. Their plans are well defined since if the fraction of defective items is crisp they reduce to classical plans. They showed that the operating characteristic curve of the plan is like a band having a high and low bounds whose width depends on the ambiguity proportion parameter in the lot when that sample size and acceptance numbers is fixed. They showed that the plan operating characteristic bands are convex with zero acceptance number.Then, they compared the operating characteristic bands of using of binomial with the operating characteristic bands of using of Poisson distribution.

Sampath (2009) considered the properties of single sampling plan under situations involving both impreciseness and randomness using the Theory of Chance. For fuzzy random environment, the process of drawing an operating characteristic curve and the issue of identifying optimal sampling plans are also addressed in the study called hybrid single sampling plan.

In a single stage sampling plan, the decision to accept or reject a lot is made based on inspecting a random sample of certain size from the lot. Conventional designs may result in needlessly large sample size. The sample size n can be reduced by relaxing the conditions on the producer's and consumer's risks. Ajorlou and Ajorlou (2009) proposed a method for constructing the membership function of the grade of satisfaction for the sample size n based on the shape of the sampling cost function. They found a reasonable solution to the trade-off between relaxing the conditions on the actual risks and the sample size n. For three general sampling cost functions, they derived the membership function of the grade of satisfaction for the sample size.

Kahraman and Kaya (2010) handled two main distributions of acceptance sampling plans which are binomial and Poisson distributions with fuzzy parameters and they derived their acceptance probability functions. Then fuzzy acceptance sampling plans were developed based on these distributions.

Jamkhaneh et al. (2011) discussed the single acceptance sampling plan, when the proportion of nonconforming products is a fuzzy number and also they showed that the operating characteristic (OC) curve of the plan is a band having high and low bounds and that for fixed sample size and acceptance number, the width of the band depends on the ambiguity proportion parameter in the lot. Consequently they explained when the acceptance number equals zero, this band is convex and the convexity increases with n.

Turanoglu et al. (2012) analyzed when main parameters of acceptance sampling plan were assumed triangular and trapezoidal fuzzy numbers and also operating characteristic curve (OC), AOQ, average sample number (ASN), and ATI were obtained for single and double sampling plans under fuzzy environment.

In the latter case, when the fraction of the defective items is needed to be used due to the nature of the quakity characteristic, the non-conforming items may not be specified exactly. Thu, the fraction of the non-conforming items, the fraction of the non-conforming items is generally not known exactly in practical cases. The general approach is to replace the value with a crisp estimate value. Due to the uncertainty of the estimation or the experimentation procedure for the estimation, there exists a vagueness of the value of the fraction of defective items. In order to model the vagueness, fuzzy set theory has been used in the literature. The number of defective items in a sample has a binomial distribution. When we use a fuzzy approach in order to model the uncertainty, the binomial distribution is defined with a fuzzy parameter $\tilde{p}$. If the number of defective items in a sample is small, the common approach is to use fuzzy Poisson distribution to approximate the fuzzy binomial (Turanoglu 2012).

Acceptance sampling applications are classified into two based on the nature of the quality characteristics inspected. If the items can only be identified as disjoint categories such as good and bad, acceptance sampling by attributes are applied. In cases where quality characteristics can be continuously measured such as weight, strength, we apply acceptance sampling by variables. The fuzzy approaches for both of these types of acceptance sampling have been studied in the literature. Sampling by attributes with relaxed requirements were discussed by Ohta and Ichihashi (1988), Kanagawa and Ohta (1990), Tamaki et al. (1991), and Grzegorzewski (1998), Grzegorzewski et al. (2001), Hryniewicz (1992). Grzegorzewski et al. (2001), Grzegorzewski (2002), Jamkhaneh and Gildeh (2010) considered sampling plan by variables with fuzzy requirements.

Another classification of acceptance sampling applications is based on the number of samples taken until a decision is made related to the lot. A sequential sampling consists of a sequence of samples from the lot and the number of samples to be taken is identified based on the results of the sampling process. A random sample if drawn from the lot and the actual quality level of the sample is compared with the limit levels. Based on the results of this comparison, three decisions can be made: (i) the lot can be accepted, (ii) the lot can be rejected; (iii) a new sample is taken and inspected to make a decision. When only one sample is inspected at each sampling stage, the procedure is named as item sequential sampling. When only two decisions, accept and reject is defined after the inspection of the first sample, the sampling is named as single sampling plan. Single, double and sequential sampling plans with fuzzy parameters have also been studied in the literature. Single sampling plans with fuzzy parameters are investigated by Ohta and Ichihashi (1988), Kanagawa and Ohta (1990), Tamaki, Kanagawa and Ohta (1991), and Grzegorzewski (1998), Grzegorzewski et al. (2001), Jamkhaneh et al. (2010), Jamkhaneh et al. (2011). Sequential sampling plans with fuzzy parameters are discussed by Jamkhaneh and Gildeh (2010).

7.6   Design of Acceptance Sampling Plans Under Fuzzy Environment

Acceptance sampling procedures can be applied to lots of items when testing reveals non-conformance or non-conformities regarding product functional attributes. It can also be applied to variables characterizing lots, thus revealing how far product quality levels are from specifications. These applications have the main purpose of sort outing a lot as accepted or rejected, given the quality levels required for it. Generally, there are two major assumptions made when creating sampling plans. The first is that the sampling parameters are crisp, such as the fraction of nonconformities which is the rate of the observed nonconformities in the inspected samples, and sample rate which is a compromise between the accuracy and cost of the inspection. The second is that these parameters are vague values, particularly in the case where they can only be expressed by linguistic variables. According to Literature Review (Kahraman and Kaya 2010; Ohta and Ichihashi 1988; Kanagawa and Ohta 1990) some of the acceptance sampling studies have concentrated on fuzzy parameters. Some of these are given with the illustrative examples in the following subsections.

7.6.1   Design of Single Sampling Plans Under Fuzzy Environment

The single attribute sampling plan provides a decision rule to accept or reject a lot based on the inspection results obtained from a single random sample. The procedure corresponds to taking a random sample from the lot with size n_1 and inspects each item. If the number of non-conformities or nonconforming items does not exceed the specified acceptance number c_1, the entire lot is accepted. Many different acceptance plans meet the requirements of both the producer and the consumer. However, the producer is also interested in keeping the average number of items inspected to a minimum, aiming to reduce the costs of sampling and inspection, and economic aspects of the sampling plans must also be considered in practical implementations (Duarte and Saraiva 2008).

Kahraman and Kaya (2010) analyzed single and double sampling plans by taking into account two fuzzy discrete distributions such as binomial and Poisson distribution. They developed a single sampling plan assuming that a sample whose size is a fuzzy number $\tilde{n}$ is taken and 100 % inspected. The fraction nonconforming of the sample is also a fuzzy number $\tilde{p}$. The acceptance number is determined as a fuzzy number $\tilde{c}$. The acceptance probability for this single sampling plan can be calculated as follows:

$$\tilde{P}_a = P(d \le \tilde{c}|\tilde{n},\tilde{c},\tilde{p}) = \sum_{d=0}^{\tilde{c}} \frac{\tilde{\lambda}^d e^{-\tilde{\lambda}}}{d!} \tag{7.10}$$

where $\quad \tilde{\lambda} = \tilde{n}\tilde{p}$.

$$P_a(\alpha) = \left[P_{al,d,\tilde{\lambda}}(\alpha), P_{ar,d,\tilde{\lambda}}(\alpha) \right] \tag{7.11}$$

$$
\begin{aligned}
P_{al,d,\tilde{\lambda}}(\alpha) &= \min\left\{ \sum_{d=0}^{c} \frac{\lambda^d e^{-\lambda}}{d!} \middle| \lambda \in \lambda(\alpha), n \in n(\alpha), c \in c(\alpha) \right\} \\
P_{ar,d,\tilde{\lambda}}(\alpha) &= \max\left\{ \sum_{d=0}^{c} \frac{\lambda^d e^{-\lambda}}{d!} \middle| \lambda \in \lambda(\alpha), n \in n(\alpha), c \in c(\alpha) \right\}
\end{aligned}
\tag{7.12}
$$

If the binomial distribution is used, acceptance probability can be calculated as follows:

$$\tilde{P}_a = \sum_{d=0}^{\tilde{c}} \binom{\tilde{n}}{d} \tilde{p}^d \tilde{q}^{\tilde{n}-d} \tag{7.13}$$

$$
\begin{aligned}
\tilde{P}_a &= \sum_{d=0}^{\tilde{c}} \binom{\tilde{n}}{d} \tilde{p}^d \tilde{q}^{\tilde{n}-d} \\
&= \left\{ \sum_{d=0}^{\tilde{c}} \binom{\tilde{n}}{d} \tilde{p}^d \tilde{q}^{\tilde{n}-d} \middle| p \in p(\alpha), q \in q(\alpha), n \in n(\alpha), c \in c(\alpha) \right\}
\end{aligned}
\tag{7.14}
$$

$$P_a(\alpha) = [P_{al}(\alpha), P_{ar}(\alpha)] \tag{7.15}$$

$$
\begin{aligned}
P_{al}(\alpha) &= \min\left\{ \sum_{d=0}^{c} \binom{n}{d} p^d q^{n-d} \middle| p \in p(\alpha), q \in q(\alpha), n \in n(\alpha), c \in c(\alpha) \right\}, a \\
P_{ar}(\alpha) &= \max\left\{ \sum_{d=0}^{c} \binom{n}{d} p^d q^{n-d} \middle| p \in p(\alpha), q \in q(\alpha), n \in n(\alpha), c \in c(\alpha) \right\}
\end{aligned}
\tag{7.16}
$$

AOQ values for fuzzy single sampling can be calculated as follows:

$$A\tilde{O}Q \cong \tilde{P}_a\tilde{p} \tag{7.17}$$

$$AOQ(\alpha) = [AOQ_l(\alpha), AOQ_r(\alpha)] \tag{7.18}$$

$$AOQ_l(\alpha) = \min\{P_a p | p \in p(\alpha), P_a \in P_a(\alpha)\},$$
$$AOQ_r(\alpha) = \max\{P_a p | p \in p(\alpha), P_a \in P_a(\alpha)\}$$
(7.19)

ATI curve can also be calculated as follows:

$$A\tilde{T}I = \tilde{n} + \left(1 - \tilde{P}_a\right)\left(\tilde{N} - \tilde{n}\right)$$
(7.20)

$$ATI(\alpha) = [ATI_l(\alpha), ATI_r(\alpha)]$$
(7.21)

$$ATI_l(\alpha) = \min\{n + (1 - P_a)(N - n) | p \in p(\alpha), P_a \in P_a(\alpha), p \in N(\alpha), N \in N(\alpha)\},$$
$$ATI_r(\alpha) = \max\{n + (1 - P_a)(N - n) | p \in p(\alpha), P_a \in P_a(\alpha), p \in N(\alpha), N \in N(\alpha)\}$$
(7.22)

Numerical Example-1

Suppose that a product is shipped in lots of size "Approximately 5000". The receiving inspection procedure used is a single sampling plan with a sample size of "Approximately 50" and an acceptance number of "Approximately 2". If fraction of nonconforming for the incoming lots is "Approximately 0.05", calculate the acceptance probability of the lot. Based on Eq. (7.16), the acceptance probability of the sampling plan is calculated as $\tilde{P}_a = P(d \leq \tilde{2}) = \text{TFN}(0.190, 0.544, 0.864)$ and its membership function is shown in Fig. 7.4.

AOQ is calculated as $A\tilde{O}Q = \text{TFN}(0.008, 0.027, 0.052)$ by using Eq. (7.19). ATI is also calculated as $ATI = \text{TFN}(707.163, 2308.125, 4140.47)$ by using Eq. (7.22) and its membership function is illustrated in Fig. 7.5.

Kanawaga and Ohta (1990) presented a design procedure for the single sampling attribute plan based on the fuzzy sets theory. They improved the fuzzy design

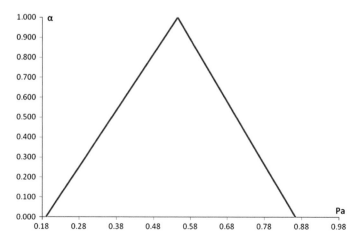

Fig. 7.4 Membership function of acceptance probability

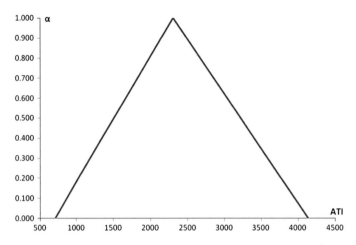

Fig. 7.5 Membership function of ATI for single sampling

procedure proposed by Ohta and Ichihashi (1988) with getting rid of the imbalance between the producer's and consumer's risks and the case in which a large sample size is (needlessly) required, both of which often arise in traditional crisp formulation, by means of the orthodox formulation as fuzzy mathematical programming with several objective function. They proposed the following formulation for the fuzzy design of the single sampling attribute plan.

$$P(p_1) \gtrsim 1 - \alpha, \quad P(p_2) \lesssim \beta, \tag{7.23}$$

$$n \to 0 \tag{7.24}$$

where the symbols $\gtrsim$ and $\lesssim$ stand for fuzzy inequlities. The membership functions $\mu_A(\alpha^*)$ and $\mu_B(\beta^*)$ in this case are shown in Fig. 7.6. The membership function $\mu_n(n)$ which represents the grade of satisfaction for the sample size must monotonically decrease as n increases as shown in Fig. 7.7. The fuzzy formulation can be written as the following fuzzy mathematical programming problem:

Problem 1 Find (n, c) so that

$$\min\{\mu_A(\alpha^*), \mu_B(\beta^*), \mu_n(n)\}$$

is maximized.

These membership functions in Fig. 7.6 are as follows:

$$\mu_A(\alpha^*) = \begin{cases} 1 & (\alpha^* \leq \alpha) \\ \frac{\alpha_u - 1 + P(p_1)}{\alpha_u - \alpha} & (\alpha \leq \alpha^* \leq \alpha_u) \\ 0 & (\alpha_u \leq \alpha^*) \end{cases} \tag{7.25}$$

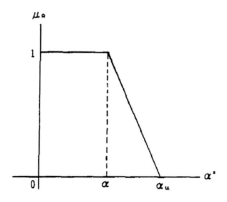
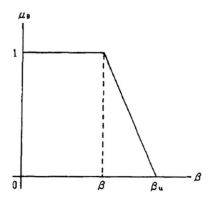

Fig. 7.6 Membership functions μ_A and μ_B

Fig. 7.7 Membership
function μ_n which represents
the grade of satisfaction for
the sample size n

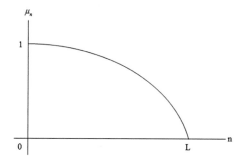

$$\mu_B(\beta^*) = \begin{cases} 1 & (\beta^* \leq \beta) \\ \frac{\beta_u - 1 + P(p_2)}{\beta_u - \beta} & (\beta \leq \beta^* \leq \beta_u) \\ 0 & (\beta_u \leq \beta^*) \end{cases} \qquad (7.26)$$

where

$$P(p) = \sum_{k=0}^{c} \binom{n}{k} p^k (1-p)^{n-k} \qquad (7.27)$$

Figure 7.8 shows the graphs of the membership functions $\mu_A(\alpha^*)$ and $\mu_B(\beta^*)$ with respect to the sample size n. In this figure, $n_{\alpha u}$ and $n_{\beta u}$ are real numbers which satisfy respectively.

$$2\lambda(n_{\alpha_u}, c, p_1) = 2g(n_{\alpha_u}, c)h(p_1) = X_{1-\alpha_u}^2(2c+2), \qquad (7.28)$$

$$2\lambda(n_{\beta_u}, c, p_2) = 2g(n_{\beta_u}, c)h(p_2) = X_{1-\beta_u}^2(2c+2), \qquad (7.29)$$

Fig. 7.8 Membership
functions $\mu_A(\alpha^*)$ and $\mu_B(\beta^*)$
with respect to sample size n

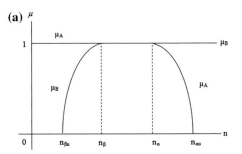

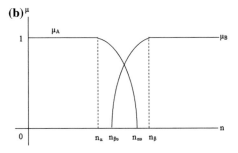

Let c_u be the minimum integer which satisfies $\geq R(c_u; \alpha, \beta)$. If the membership
functions μ_A and μ_B were decided as in Fig. 7.6, that is, in Fig. 7.8, it is found that
acceptance number c in Problem 1 is less than or equal to c_u. Because when c is
greater than or equal to c_u, it is $r \geq R(c_u; \alpha, \beta)$, then the membership functions
μ_A and μ_B are shown in Fig. 7.8a. Accordingly, which maximizes min $\{\mu_A, \mu_B, \mu_n\}$
depends on only the intersection of μ_A and μ_B. The grade of max min $\{\mu_A, \mu_B, \mu_n\}$
decreases with c, because μ_n is monotonically decreasing. So c is found to be less
than or equal to c_u. Let the sample size n expand to a real number. Setting $n^* \in R$,
which satisfies

$$\mu_B(\beta^*(n^*, c)) = \mu_n(n) \tag{7.30}$$

It is obvious that when c is equal to c_u, n^* belongs to interval $[n_{\beta_u}, n_\beta]$. If n^* is
found, the integer solution n is either $[n^*]$ or $[n^*] + 1$. Note that when c is less than
c_u, the relation is changed to $n_\beta \geq n_\alpha$. In the both cases, the integer solution n will be
found by means of searching in the integer interval $[[n_{\beta_u}] - 1, [n_{\beta_u}] + 1]$. Finally
sample size n is selected as

$$n_i = \frac{\max}{i} \{\mu_0(n_0), \mu_1(n_1), \mu_2(n_2), \ldots, \mu_i(n_i), \ldots\} \tag{7.31}$$

and acceptance number is selected as $c_u - i$.

Table 7.1 The grade of fuzzy union set with respect to n on each c

c:	Grade of fuzzy product set with respect to n on each c
4:	Max min $\{\mu_B, \mu_n\}$ in $[73, 89] = \mu_n(79) = 0.4846$
4-1:	Max min $\{\mu_A, \mu_B, \mu_n\}$ in $[60, 73] = \mu_n(66) = 0.5310$
4-2:	Max min $\{\mu_A, \mu_B, \mu_n\}$ in $[46, 58] = \mu_n(48) = 0.1853$
4-3:	Fuzzy product set does not exist

Numerical Example-2

It is set up the membership functions μ_A *and* μ_B as follows Kahraman and Kaya (2010):

$$\alpha = 5\,\%, \quad \alpha_u = 8\,\%, \quad \beta = 10\,\%, \quad \beta_u = 20\,\%.$$

For the membership function functions $\mu_n(n)$, it will be accepted the following function:

$$\mu_n(n) = 1 - \left(\frac{n}{L}\right)^m$$

where L is the tolerance limit of the sample size so that n should be smaller than L. it is better to select L to be less than N/10 for use of the binomial distribution. m is the shape parameter of the membership function, and m is selected so that $0 \leq m \leq 1$. In the case where L = 300, m = 0.5, $p_1 = 0.02$, $p_2 = 0.09$, the solving procedure of Problem 1 is as follows:

$$r = (h(p_2)/h(p_1)) = 4.665$$

So that $c_u = 4$. Then the grade of fuzzy union set with respect to n on each c is shown in Table 7.1. After all it is c = 3, n = 66, $\alpha^* = 0.04338$, $\beta^* = 0.1441$.

7.6.2   Design of Double Sampling Plans Under Fuzzy Environment

In double sampling by attributes, an initial sample is taken, and a decision to accept or reject the lot is reached on the basis of this first sample if the number of nonconforming units is either quite small or quite large. A second sample is taken if the results of the first sample are not decisive. Since it is necessary to draw and inspect the second sample only in borderline cases, the average number of pieces inspected per lot is generally smaller with double sampling. It has been demonstrated to be simple to use in a wide variety of conditions, economical in total cost, and acceptable psychologically to both producer and consumer (Juran 1998).

Kahraman and Kaya (2010) used a double sampling plan with fuzzy parameters $(\tilde{n}_1, \tilde{c}_1, \tilde{n}_2, \tilde{c}_2)a$. $\tilde{N}$ and $\tilde{p}$ are also fuzzy. If the Poisson distribution is used, the acceptance probability of double sampling can be calculated as follows:

$$P_a = P(d_1 \le \tilde{c}_1) + P(\tilde{c}_1 < d_1 \le \tilde{c}_2)P(d_1 + d_2 \le \tilde{c}_2) \tag{7.32}$$

$$\tilde{P}_a = \sum_{d_1=0}^{\tilde{c}_1} \frac{\lambda^{d_1} e^{-\tilde{n}_1 \tilde{p}}}{d_1!} + \sum_{d_1 > \tilde{c}_1}^{\tilde{c}_2} \left(\frac{\lambda^{d_1} e^{-\tilde{n}_1 \tilde{p}}}{d_1!} \times \sum_{d_2=0}^{\tilde{c}_2-d_1} \frac{\lambda^{d_2} e^{-\tilde{n}_2 \tilde{p}}}{d_2!} \right) \tag{7.33}$$

$$P_a(\alpha) = \left[P_{al,d;\tilde{\lambda}}(\alpha), P_{ar,d;\tilde{\lambda}}(\alpha) \right] \tag{7.34}$$

$$
\begin{aligned}
P_{al,d;\tilde{\lambda}}(\alpha) &= \min \left\{ \sum_{d_1=0}^{c_1} \frac{\lambda^{d_1} e^{-n_1 p}}{d_1!} + \sum_{d_1 > c_1}^{c_2} \left(\frac{\lambda^{d_1} e^{-n_1 p}}{d_1!} \times \sum_{d_2=0}^{c_2-d_1} \frac{\lambda^{d_2} e^{-n_2 p}}{d_2!} \right) \right\} \\
P_{ar,d;\tilde{\lambda}}(\alpha) &= \max \left\{ \sum_{d_1=0}^{c_1} \frac{\lambda^{d_1} e^{-n_1 p}}{d_1!} + \sum_{d_1 > c_1}^{c_2} \left(\frac{\lambda^{d_1} e^{-n_1 p}}{d_1!} \times \sum_{d_2=0}^{c_2-d_1} \frac{\lambda^{d_2} e^{-n_2 p}}{d_2!} \right) \right\}
\end{aligned}
\tag{7.35}
$$

where $p \in p(\alpha), n \in n(\alpha)$, and $c \in c(\alpha)$.

If the binomial distribution is used, acceptance probability can be calculated as follows:

$$P_a = \sum_{d_1=0}^{\tilde{c}_1} \binom{\tilde{n}_1}{d_1} \tilde{p}^{d_1} (1-\tilde{p})^{\tilde{n}_1-d_1} + \sum_{d_1 > \tilde{c}_1}^{\tilde{c}_2} \left(\binom{\tilde{n}_1}{d_1} \tilde{p}^{d_1} (1-\tilde{p})^{\tilde{n}_1-d_1} \times \sum_{d_2=0}^{\tilde{c}_2-d_1} \binom{\tilde{n}_2}{d_2} \tilde{p}^{d_2} (1-\tilde{p})^{\tilde{n}_2-d_2} \right) \tag{7.36}$$

$$
\begin{aligned}
P_{al}(\alpha) &= \min \left\{ \sum_{d_1=0}^{c_1} \binom{n_1}{d_1} p^{d_1} (1-p)^{n_1-d_1} + \sum_{d_1 > c_1}^{c_2} \left(\binom{n_1}{d_1} p^{d_1} (1-p)^{n_1-d_1} \times \sum_{d_2=0}^{c_2-d_1} \binom{n_2}{d_2} p^{d_2} (1-p)^{n_2-d_2} \right) \right\}, \\
P_{ar}(\alpha) &= \max \left\{ \sum_{d_1=0}^{c_1} \binom{n_1}{d_1} p^{d_1} (1-p)^{n_1-d_1} + \sum_{d_1 > c_1}^{c_2} \left(\binom{n_1}{d_1} p^{d_1} (1-p)^{n_1-d_1} \times \sum_{d_2=0}^{c_2-d_1} \binom{n_2}{d_2} p^{d_2} (1-p)^{n_2-d_2} \right) \right\}
\end{aligned}
\tag{7.37}
$$

where $p \in p(\alpha), q \in q(\alpha), n_1 \in n_1(\alpha), c_1 \in c_1(\alpha), n_2 \in n_2(\alpha)$, and $c_2 \in c_2(\alpha)$.

AOQ values for fuzzy double sampling can be calculated as in Eqs. (7.17–7.19). ASN for double sampling can be calculated as follows:

$$
\begin{aligned}
A\tilde{S}N &= \tilde{n}_1 \tilde{P}_I + (\tilde{n}_1 + \tilde{n}_2)(1 - \tilde{P}_I) \\
&= \tilde{n}_1 + \tilde{n}_2 (1 - \tilde{P}_I)
\end{aligned}
\tag{7.38}
$$

$$ASN(\alpha) = [ASN_l(\alpha), ASN_r(\alpha)] \tag{7.39}$$

$$
\begin{aligned}
ASN_l(\alpha) &= \min\{n_1 + n_2(1 - P_I) | p \in p(\alpha), n_1 \in n_1(\alpha), n_2 \in n_2(\alpha), P_I \in P_I(\alpha)\}, \\
ASN_r(\alpha) &= \max\{n_1 + n_2(1 - P_I) | p \in p(\alpha), n_1 \in n_1(\alpha), n_2 \in n_2(\alpha), P_I \in P_I(\alpha)\}
\end{aligned}
\tag{7.40}
$$

ATI curve for fuzzy double sampling can also be calculated as follows:

$$A\tilde{T}I = A\tilde{S}N + (\tilde{N} - \tilde{n}_1)P(d_1 > \tilde{c}_2) + (\tilde{N} - \tilde{n}_1 - \tilde{n}_2)P(d_1 + d_2 > \tilde{c}_2) \quad (7.41)$$

$$ATI(\alpha) = [ATI_l(\alpha), ATI_r(\alpha)] \quad (7.42)$$

$$\begin{aligned} ATI_l(\alpha) &= \min\{ASN + (N - n_1)P(d_1 > c_2) + (N - n_1 - n_2)P(d_1 + d_2 > c_2)\}, \\ ATI_r(\alpha) &= \max\{ASN + (N - n_1)P(d_1 > c_2) + (N - n_1 - n_2)P(d_1 + d_2 > c_2)\} \end{aligned}$$
$$(7.43)$$

where $p \in p(\alpha), ASN \in ASN(\alpha), n_1 \in n_1(\alpha), N \in N(\alpha), n_2 \in n_2(\alpha)$, and $c_2 \in c_2(\alpha)$.

Numerical Example-3

Let us reconsider Numerical Example-1 for the case of fuzzy double sampling. The sample sizes are determined as "Approximately 75" and "Approximately 300" for the first and second samples, respectively. Also the acceptance numbers are determined as "Approximately 0" and "Approximately 3" for the first and second samples, respectively. Based on Eq. (7.35), acceptance probability of the double sampling plan is calculated as follows:

$$\begin{aligned} \tilde{P}_a &= P(d_1 \leq \tilde{0}) + [P(d_1 = \tilde{1}) \times P(d_2 \leq \tilde{2})] + [P(d_1 = \tilde{2}) \times P(d_2 \leq \tilde{1})] + [P(d_1 = \tilde{3}) \times P(d_2 \leq \tilde{0})] \\ &= (0.0105, 0.0235, 0.2052) + [(0.0105, 0.0882, 0.227) \times (0, 0, 0.0024)] \\ &\quad + [(0.0477, 0.1654, 0.224) \times (0, 0, 0.0005)] + [(0.1088, 0.2067, 0.227) \times (0, 0, 0.0001)] \\ &= (0.0105, 0.0235, 0.2052) + (0, 0, 0.0005) + (0, 0, 0.0001) + (0, 0, 0) \\ \tilde{P}_a &= (0.0105, 0.0235, 0.2058). \end{aligned}$$

Its membership function is shown in Fig. 7.9.

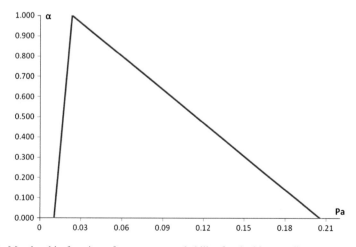

Fig. 7.9 Membership function of acceptance probability for double sampling

ASN is calculated as ASN = TFN(74.00, 213.08, 320.24) by using Eqs. (7.38–7.40). Also AOQ is calculated as AOQ = TFN(0.00042, 0.001175, 0.01235).

Wang and Chen (1997) formulated the problem of determining the Dodge-Romig LTPD double sampling plan under the fuzzy environment satisfies the consumer's risk closely around β using fuzzy mathematical programming. They proposed a model to minimize the ATI at the process average p_m subject to satisfying the consumer's risk closely around b, the Dodge-Romig LTPD double sampling plan finds a non-fuzzy non-negative integer pair (n_1, n_2, c_1, c_2) that minimizes:

$$I(p_m, n_1, n_2, c_1, c_2) = n_1 + n_2 * (1 - G(c_1, n_1 p_m)) + (N - n_1 - n_2)*$$
$$\left[1 - G(c_1, n_1 p_m) - \sum_{j=1}^{c_2 - c_1} g(c_1 + j, n_1 p_m) * G(c_2 - c_1 - j, n_2 p_m) \right] \tag{7.44}$$

subject to

$$K(p_n, n_1, n_2, c_1, c_2) \lesssim \beta \tag{7.45}$$

$$n_1, n_2, c_1, c_2 \geq 0, \text{ integer} \tag{7.46}$$

where

$$K(p_n, n_1, n_2, c_1, c_2) = G(c_1, n_1 p_n) + \sum_{j=1}^{c_2 - c_1} g(c_1 + j, n_1 p_m) * G(c_2 - c_1 - j, n_2 p_n)$$
$$\tag{7.47}$$

where $g(x, np)$ represents the probability of the Poisson distribution for the randım variable x with parameter np, i.e. $g(x, np) = e^{-np}(np)^x / x!$ and $G(x, np)$ is its cumulative distribution. $I(p_m, n_1, n_2, c_1, c_2)$ is the ATI. The symbol $\lesssim$ stands for fuzzy inequality.

The symmetry of the decision model in a fuzzy environment rests essentially on the assumption that the objective function as well as the constraint can be fuzzy sets, and that the degree of membership of solutions to the objective function and to the constraint could be considered comparable. The above model shown using Eqs. (7.44–7.46) is a non-linear integer fuzzy mathematical programming problem. Because the objective function in Eq. (7.44) is a crisp set and the constraint in Eq. (7.45) is a fuzzy set, it is the optimization problem is a non-symmetrical fuzzy model. So Wang and Chen (1997) used the studies of Zimmermann (1985) and Chakraborty (1988, 1992) with the minimum operator to aggregate the membership functions of fuzzy sets, they obtained the following model for Dodge-Romig LTPD double sampling plan problem:

$$Maximize \left\{ \begin{array}{l} \min n_1, n_2, c_1, c_2 \geq 0, \ integer \\[2mm] \left[\dfrac{\inf\limits_{R_1} I - I(p_m, n_1, n_2, c_1, c_2)}{\inf\limits_{R_1} I - \inf\limits_{S(R)} I} = \lambda_1, \dfrac{\beta_u - K(p_n, n_1, n_2, c_1, c_2)}{\beta_u - \beta} = \lambda_2 \right] = \lambda \end{array} \right\}$$

$$(7.48)$$

Subject to

$$\inf\limits_{S(R)} I \leq I(p_m, n_1, n_2, c_1, c_2) \leq \inf\limits_{R_1} I \tag{7.49}$$

$$\beta \leq K(p_n, n_1, n_2, c_1, c_2) \leq \beta_u \tag{7.50}$$

$$0 \leq \lambda \leq 1 \tag{7.51}$$

$$n_1, n_2, c_1, c_2 \geq 0, \ integer \tag{7.52}$$

where R is a fuzzy feasible region, $S(R)$ support of R, and $R_{1\_}$ − level cut of R for $\alpha = 1$.

This problem can be rewritten in the equivalent optimization problem to find $n_1, n_2, c_1, c_2, \lambda$ that maximize λ and subject to

$$\lambda \leq \dfrac{\inf\limits_{R_1} I - I(p_m, n_1, n_2, c_1, c_2)}{\inf\limits_{R_1} I - \inf\limits_{S(R)} I} \tag{7.53}$$

$$\lambda \leq \dfrac{\beta_u - K(p_n, n_1, n_2, c_1, c_2)}{\beta_u - \beta} \tag{7.54}$$

and inequalities (7.49) through (7.52).Let $I_1 = \inf\limits_{R_1} I$ and $I_0 = \inf\limits_{S(R)} I$. Then the above optimization problem can be expressed as: find $n_1, n_2, c_1, c_2, \lambda$ that maximize λ and subject to

$$R^{(I_1 - I_0)} + I(p_m, n_1, n_2, c_1, c_2) \leq I_1 \tag{7.55}$$

$$R^{(\beta_u - \beta_0)} + K(p_n, n_1, n_2, c_1, c_2) \leq -_U \tag{7.56}$$

$$I_0 \leq I(p_m, n_1, n_2, c_1, c_2) \leq I_1 \tag{7.57}$$

and inequalities (7.50), (7.51) and (7.52).

Numerical Example-4

It is considered the example given in Hald (1981): $N = 2000$, $p_m = 0.02$, $p_n = 0.10$, and $\beta = 0.10$. The optimum LTPD double sampling plan$(n_1^*, n_2^*, c_1^*, c_2^*)$ is to be found for $\beta_u = 0.15$. For a LTPD double sampling plan, it is obtained $(n_1^a, n_2^a, c_1^a, c_2^a) = (55, 132, 2, 10)$ with $I_1 = 70.30$ and $(n_1^b, n_2^b, c_1^b, c_2^b) = (36, 111, 1, 8)$ with $I_0 = 58.45$. The DSP giving the largest value of λ for each c_1 near $\{c_1^a 1, \ c_2^a = 2\}$ is found. It is concluded that the optimum LTPD double sampling plan is; $(n_1^*, n_2^*, c_1^*, c_2^*) = (40, 104, 1, 8)$ with $\lambda^* = 0.52 I(p_m, n_1^*, n_2^*, c_1^*, c_2^*) = 64.12$ and $K(p_n, n_1^*, n_2^*, c_1^*, c_2^*) = 0.1235$.

In this particular example, the difference between the solution of the traditional Dodge-Romig LTPD double sampling plan and their is that the decision maker takes an additional consumer's risk of 2.35 % for a lot being rejected against a saving of inspection effort per lot.

7.7 Proposed Fuzzy Multi-objective Mathematical Models for Design of Acceptance Sampling Plans

In this section, multi-objective mathematical models for designing of single and double sampling by attributes are developed and the optimal results are obtained by considering the various constraints under fuzziness. As a result it is obtained that the lower sample sizes in developed single and double sampling plans under fuzzy environment.

7.7.1 Proposed Fuzzy Multiobjective Models for Design of Single Acceptance Sampling Plan

The case in which a large sample size is needlessly required often arises in conventional design. The sample size n can be reduced as desired by relaxing the conditions on the consumer's risks. Hence, the tradeoff between the reduction of sample size and the relaxation of the conditions becomes a serious problem. So we developed a design procedure based on fuzzy multi-objective mathematical model for single sampling plans. In practical applications, LTPD cannot be known precisely. Hence the following model is developed to find the most appropriate sample size n with minimizing of ATI and AOQ. Also in this model, LTPD and consumer's risks β are defined as fuzzy numbers. The closed form of the model is given following equations:

Objective function

Min ATI

Min AOQ

Subject to

$$P_a(n, c; \widetilde{LTPD}) \leq \tilde{\beta} \tag{7.58}$$

$$n \geq c \tag{7.59}$$

$$n > 0, \ \text{integer}; \ c \geq 0, \ \text{integer} \tag{7.60}$$

The open form of the model is given by using Eqs. (7.61–7.64).
Objective Function
Min ATI

$$n + (N - n) \times \left[1 - \sum_{x=0}^{c} \frac{e^{-n \times \widetilde{LTPD}} \times (n \times \widetilde{LTPD})^x}{x!} \right] \tag{7.61}$$

Min AOQ

$$\frac{\widetilde{LTPD}}{N} \times (N - n) \times \left[\sum_{x=0}^{c} \frac{e^{-n \times \widetilde{LTPD}} \times (n \times \widetilde{LTPD})^x}{x!} \right] \tag{7.62}$$

Subject to

$$\sum_{x=0}^{c} \frac{e^{-n \times \widetilde{LTPD}} \times (n \times \widetilde{LTPD})^x}{x!} \leq \tilde{\beta} \tag{7.63}$$

$$n \geq c$$

$$n > 0, \ \text{integer}; \ c \geq 0, \ \text{integer} \tag{7.64}$$

Numerical Example-5
Developed fuzzy multi-objective mathematical model for single sampling plan shown in Eqs. (7.61–7.64) is solved for N = 500, $\widetilde{LTPD} = \text{TFN}(0.02, 0.03, 0.04)$ and. The obtained results for *n*, *ATI* and *AOQ* are given in Table 7.2.

According to Table 7.2, when $\widetilde{LTPD}$ and $\tilde{\beta}$ re defined as fuzzy numbers, the smaller values of *n*, *ATI* and *AOQ* are obtained. Table 7.3 gives the comparison of

Table 7.2 The results of n, ATI, and AOQ for single sampling plan given values for $\widetilde{LTPD}$ and $\tilde{\beta}$ and N = 500

N	$\widetilde{LTPD}$	Single sampling plan	Acceptance number (c)				
			1	2	3	4	5
500	TFN(0.02, 0.03, 0.04)	n	34	66	100	136	174
		ATI	104.158	131.915	160.017	190.611	223.688
		AOQ	0.0158	0.0147	0.0136	0.0124	0.011

Table 7.3 The comparison of the values of *n*, *ATI* and *AOQ* for the crisp values of LTPD and β = 0.10 and the fuzzy values of $\widetilde{LTPD}$ and $\tilde{\beta}$

Parameters		Single sampling plan	LTPD			$\widetilde{LTPD}$ = TFN(0.02, 0.03, 0.04)
N	C		0.02	0.03	0.04	
500	1	n	195	130	98	20
		ATI	469.748	463.301	460.757	40.422
		AOQ	0.0012	0.0022	0.0031	0.0183
	2	n	267	178	134	44
		ATI	476,983	468.191	464.342	72.644
		AOQ	0.0009	0.0019	0.0028	0.0171
	3	n	335	223	168	71
		ATI	483,697	473.459	467.603	97.211
		AOQ	0.0006	0.0016	0.0025	0.0161
	4	n	400	267	200	101
		ATI	490.036	476.919	470.110	125.551
		AOQ	0.0004	0.0014	0.0024	0.0148
	5	n	464	311	232	132
		ATI	496.410	481.654	473.276	154.429
		AOQ	0.0001	0.0011	0.0021	0.0138

the values of *n, ATI* and *AQO* for the crisp values of LTPD and β = 0.10 and fuzzy values of $\widetilde{LTPD}$ and $\tilde{\beta}$.

7.7.2 Proposed Fuzzy Multiobjective Models for Design of Double Acceptance Sampling Plan

In the presented model for double sampling plan, the decision maker specifies consumer's risk β and *LTPD* as fuzzy numbers to find the most appropriate sample sizes n_1 and n_2 with minimizing *ATI* and *AOQ*. The closed form of the model is given following equations:

Objective function

Min ATI

Min AOQ

Subject to

$$P_a\left(n_1, n_2; c_1, c_2; \widetilde{LTPD}\right) \le \tilde{\beta} \tag{7.65}$$

$$n_1, n_2 > 0, \text{ integer}; c_1, c_2 \ge 0, \text{ integer} \tag{7.66}$$

The open form of the model is given by using Eqs. (7.67–7.70).

Objective function

Min ATI

$$n_1 + n_2 \times \left[\sum_{x=0}^{c_2} \frac{e^{-n_1 \times \widetilde{LTPD}} \times (n_1 \times \widetilde{LTPD})^x}{x!} - \sum_{x=0}^{c_1} \frac{e^{-n_1 \times \widetilde{LTPD}} \times (n_1 \times \widetilde{LTPD})^x}{x!} \right]$$

$$+ (N - n_1) \times \left[1 - \sum_{x=0}^{c_2} \frac{e^{-n_1 \times \widetilde{LTPD}} \times (n_1 \times \widetilde{LTPD})^x}{x!} \right] + (N - n_1 - n_2) \times \left[\sum_{x=0}^{c_2} \frac{e^{-n_1 \times \widetilde{LTPD}} \times (n_1 \times \widetilde{LTPD})^x}{x!} \right]$$

$$- \left(\begin{array}{l} \sum_{x=0}^{c_1} \frac{e^{-n_1 \times \widetilde{LTPD}} \times (n_1 \times \widetilde{LTPD})^x}{x!} + \frac{e^{-n_1 \times \widetilde{LTPD}} \times (n_1 \times \widetilde{LTPD})^{(c_1+1)}}{(c_1+1)!} \times \sum_{x=0}^{c_2-c_1-1} \frac{e^{-n_2 \times \widetilde{LTPD}} \times (n_2 \times \widetilde{LTPD})^x}{x!} \\[2ex] + \frac{e^{-n_1 \times \widetilde{LTPD}} \times (n_1 \times \widetilde{LTPD})^{(c_1+2)}}{(c_1+2)!} \times \sum_{x=0}^{c_2-c_1-2} \frac{e^{-n_2 \times \widetilde{LTPD}} \times (n_2 \times \widetilde{LTPD})^x}{x!} + \cdots + \frac{e^{-n_1 \times \widetilde{LTPD}} \times (n_1 \times \widetilde{LTPD})^{c_2}}{c_2!} \\[2ex] \times e^{-n_2 \times \widetilde{LTPD}} \end{array} \right)$$

$$(7.67)$$

Min AOQ

$$\frac{\widetilde{LTPD}}{N} \times \left[\begin{array}{l} \sum_{x=0}^{c_1} \frac{e^{-n_1 \times \widetilde{LTPD}} \times (n_1 \times \widetilde{LTPD})^x}{x!} \times (N - n_1) + (N - n_1 - n_2) \\[2ex] \times \sum_{x=0}^{c_2} \frac{e^{-n_1 \times \widetilde{LTPD}} \times (n_1 \times \widetilde{LTPD})^x}{x!} - \left(\begin{array}{l} \sum_{x=0}^{c_1} \frac{e^{-n_1 \times \widetilde{LTPD}} \times (n_1 \times \widetilde{LTPD})^x}{x!} + \frac{e^{-n_1 \times \widetilde{LTPD}} \times (n_1 \times \widetilde{LTPD})^{(c_1+1)}}{(c_1+1)!} \\[2ex] \times \sum_{x=0}^{c_2-c_1-1} \frac{e^{-n_2 \times \widetilde{LTPD}} \times (n_2 \times \widetilde{LTPD})^x}{x!} + \frac{e^{-n_1 \times \widetilde{LTPD}} \times (n_1 \times \widetilde{LTPD})^{(c_1+2)}}{(c_1+2)!} \\[2ex] \times \sum_{x=0}^{c_2-c_1-2} \frac{e^{-n_2 \times \widetilde{LTPD}} \times (n_2 \times \widetilde{LTPD})^x}{x!} + \cdots + \frac{e^{-n_1 \times \widetilde{LTPD}} \times (n_1 \times \widetilde{LTPD})^{c_2}}{c_2!} \\[2ex] \times e^{-n_2 \times \widetilde{LTPD}} \end{array} \right) \end{array} \right]$$

$$(7.68)$$

Subject to

$$\left[\begin{array}{l} \sum_{x=0}^{c_2} \frac{e^{-n_1 \times \widetilde{LTPD}} \times (n_1 \times \widetilde{LTPD})^x}{x!} \\[2ex] - \left(\begin{array}{l} \sum_{x=0}^{c_1} \frac{e^{-n_1 \times \widetilde{LTPD}} \times (n_1 \times \widetilde{LTPD})^x}{x!} + \frac{e^{-n_1 \times \widetilde{LTPD}} \times (n_1 \times \widetilde{LTPD})^{(c_1+1)}}{(c_1+1)!} \\[2ex] \times \sum_{x=0}^{c_2-c_1-1} \frac{e^{-n_2 \times \widetilde{LTPD}} \times (n_2 \times \widetilde{LTPD})^x}{x!} + \frac{e^{-n_1 \times \widetilde{LTPD}} \times (n_1 \times \widetilde{LTPD})^{(c_1+2)}}{(c_1+2)!} \\[2ex] \times \sum_{x=0}^{c_2-c_1-2} \frac{e^{-n_2 \times \widetilde{LTPD}} \times (n_2 \times \widetilde{LTPD})^x}{x!} + \cdots + \frac{e^{-n_1 \times \widetilde{LTPD}} \times (n_1 \times \widetilde{LTPD})^{c_2}}{c_2!} \\[2ex] \times e^{-n_2 \times \widetilde{LTPD}} \end{array} \right) \end{array} \right] \leq \widetilde{\beta}$$

$$(7.69)$$

$$n_1, n_2 > 0, \text{ integer}; \ c_1, c_2 \geq 0, \text{ integer} \qquad (7.70)$$

Table 7.4 The results of n for n_1, n_2ATI and AOQ for double sampling plan given values for $\widetilde{LTPD}$ and $\tilde{\beta}$ and N = 500

$\widetilde{LTPD}$	Double sampling plan	Acceptance numbers (c_1 and c_2)				
		0–1	1–2	2–3	3–4	4–5
TFN(0.02, 0.03, 0.04)	n_1	14	29	49	71	94
	n_2	14	29	49	71	94
	ATI	112.956	135.238	175.544	216.570	257.793
	AOQ	0.0218	0.0246	0.0243	0.0234	0.0221

Numerical Example-6

Developed fuzzy multi-objective mathematical model for double sampling plan shown in Eqs. (7.67–7.70) is solved for

N = 500, $\widetilde{LTPD}$ = TFN(0.02, 0.03, 0.04) and $\tilde{\beta}$ = TFN(0.10, 0.15, 0.20). The obtained results for n_1, n_2ATI and AOQ are given in Table 7.4.

According to Table 7.4, when $\widetilde{LTPD}$ and $\tilde{\beta}$ are defined as fuzzy numbers, the smaller values of n_1, n_2 and ATI are obtained. Table 7.5 shows the comparison of

Table 7.5 The comparison of the values of n_1, n_2, ATI and AQO for the crisp values of LTPD and $\beta = 0.10$ and fuzzy values of $\widetilde{LTPD}$ and $\tilde{\beta}$

Parameters		Double sampling plan	LTPD			$\widetilde{LTPD}$
c_1	c_2		0.02	0.03	0.04	TFN(0.02, 0.03, 0.04)
0	1	n_1	16	11	8	14
		n_2	16	11	8	14
		ATI	89.294	80.528	70.926	112.956
		AOQ	0.0151	0.0230	0.0310	0.0218
1	2	n_1	36	24	17	29
		n_2	36	24	17	29
		ATI	125.790	100.756	88.238	135.238
		AOQ	0.0166	0.0257	0.0347	0.0246
2	3	n_1	62	41	31	49
		n_2	62	41	31	49
		ATI	176.329	132.972	113.837	175.544
		AOQ	0.0162	0.0256	0.0349	0.0243
3	4	n_1	92	61	46	71
		n_2	92	61	46	71
		ATI	234.181	171.887	142.884	216.570
		AOQ	0.0152	0.0247	0.0342	0.0234
4	5	n_1	124	83	62	94
		n_2	124	83	62	94
		ATI	294.655	214.481	172.498	257.793
		AOQ	0.0140	0.0235	0.0331	0.0221

the values of n_1, n_2, ATI and AQO for the crisp values of LTPD and $\beta = 0.10$ and fuzzy values of $\widetilde{LTPD}$ and $\tilde{\beta}$.

7.8 Conclusion

The complexity of industrial manufacturing is growing and the need for higher efficiency, greater flexibility; better product quality and lower cost have changed the face of manufacturing practice. Statistical Quality Control is a tool for developing required resolution plans against problematic areas of manufacturing practice. One of the most important subjects of the Statistical Quality Control is acceptance sampling. Proper design of an acceptance sampling planning usually depends on knowing the true level of quality required by customers. However, it is sometimes not possible to determine this quality level with certain values. Especially in production, it is not easy to determine the parameters of acceptance sampling such as proportion of defect items, sample size, acceptable defect items.

A lot or batch of items can be inspected in several ways including the use of single, double, multiple, sequential sampling. In this chapter, the parameters used in acceptance sampling are defined with the help of linguistic variables and fuzzy set theory has successfully been applied to acceptance sampling to eliminate uncertainty and lack of knowledge mentioned above. We propose fuzzy multi-objective mathematical models for single and double sampling schemes. As a result it is obtained that the lower sample sizes in developed single and double sampling plans under fuzzy environment. For further studies, multi-objective mathematical models for multiple and sequential sampling schemes can be developed under fuzzy environment. Also decision trees that identify the causes of the non conformities of a rejected sample and indicate the appropriate interventions in the manufacturing process are worthwhile to study for acceptance sampling.

References

Ajorlou, S., Ajorlou, A.: A Fuzzy-based design procedure for a single-stage sampling plan. FUZZ-IEEE 2009, Korea, Aug 20–24 2009

Aslam, M.: Economic reliability acceptance sampling plan for generalized rayleigh distribution. J. Stat. **15**, 26–35 (2008)

Aslam, M., Jun, C.H., Ahmad, M.: New acceptance sampling plans based on life tests for Birnbaum-Saunders distributions. J. Stat. Comput. Simul. (2010). doi:10.1080/00949650903418883

Baklizi, A.: Acceptance sampling based on truncated life tests in the pareto distribution of the second kind. Adv. Appl. Stat **3**(1), 33–48 (2003)

Borget, I., Laville, I., Paci, A., Michiels, S.: Application of an acceptance sampling plan for post-production quality control of chemotherapic batches in an hospital pharmacy. J. Pharm. Biopharm. **64**, 92–98 (2006)

Chakraborty, T.K.: A single sampling attribute plan of given strength based on fuzzy goalProgramming. Opsearch **25**, 259–271 (1988)

Chakraborty, T.K.: A class of single sampling plans based on fuzzy optimization. Opsearch **29**(1), 11–20 (1992)

Cheng, T., Chen, Y.: A GA mechanism for optimizing the design of attribute double sampling. Autom. Constr. **16**, 345–353 (2007)

Cochran, W.G.: Sampling Techniques. Wiley, New York, pp. 428, (1977)

Degroot, M.H.: Probability and Statistics. Addison-Wesley, MA, pp. 723, (1986)

Duarte, B.P.M., Saraiva, P.M.: An optimization-based approach for designing attribute acceptance sampling plans. Int. J. Qual. Reliab. Manage **25**(8), 824–841 (2008)

Fallahnezhad, M.S., Niaki, S.T.A.: A new acceptance sampling design using bayesian modeling and backwards induction. Int. J. Eng. **25**, 45–54 (2012)

Fernández, A.J., Pérez-González, C.J.: Optimal acceptance sampling plans for log-location–scale lifetime models using average risks. Comput. Stat. Data Anal. **56**, 719–731 (2012)

Fountoulaki, A., Karacapilidis, N., Manatakis, M.: Exploiting machine learning techniques for the enhancement of acceptance sampling. World Acad. Sci., Eng. Technol. **17**, (2008)

Grzegorzewski, P.: A soft design of acceptance sampling by attributes. In: Proceedings of the VIth International Workshop on Intelligent Statistical Quality Control Wurzburg, pp. 29–38. Sept 14–16 1998

Grzegorzewski, P.: A Soft Design of Acceptance Sampling Plans by Variables, in: Technologies for Contrueting Intelligent Systems, vol. 2, pp. 275–286. Springer, Berlin, (2002)

Grzegorzewski, P.: Acceptance sampling plans by attributes with fuzzy risks and quality levels. In: Wilrich P., Lenz H. J (eds.) Frontiers in Frontiers in Statistical Quality Control, vol. 6, pp. 36–46. Springer, Heidelberg, (2001)

Hald, A.: Statistical Theory of Sampling Inspection by Attributes, vol. 101. Academic Press, New York, (1981)

Hryniewicz, O.: User-preferred solutions of fuzzy optimization problems—an application in choosing user-preferred inspection intervals. Fuzzy Sets Syst. **137**, 101–111 (2003)

Hryniewicz, O.: Statistical Acceptance Sampling with Uncertain Information from a Sample and Fuzzy Quality Criteria Working Paper of SRI PAS, Warsow, (in polish), (1992)

Hryniewisz, O.: Statistics with fuzzy data in statistical quality control. Soft. Comput. **12**, 229–234 (2008)

Hsieh, C.-C., Lu, Y.-T.: Risk-embedded Bayesian acceptance sampling plans via conditiona value-at-risk with Type II censoring. Comput. Ind. Eng. **65**, 551–560 (2013)

Jamkhaneh, E.B., Gildeh, B.S.: Sequential sampling plan by variable with fuzzy parameters. J. Math. Comput. Sci. **1**(4), 392–401 (2010)

Jamkhaneh, E.B., Sadeghpour-Gildeh, B., Yari, G.: acceptance single sampling plan with fuzzy parameter with the using of Poisson distribution. World Acad. Sci., Eng. Technol. **25** (2009)

Jamkhaneh, E.B., Gildeh, B.S.: Yari, G., Acceptance single sampling plan by using of poisson distribution. J. Math. Comput. Sci. **1**(1), 6–13 (2010)

Jamkhaneh, E.B., Gildeh, B.S., Yari, G.: Acceptance single sampling plan with fuzzy parameter. Iran. J. Fuzzy Syst. **8**, 47–55 (2011)

Jozani, M.J., Mirkamali, S.J.: Improved attribute acceptance sampling plans based on maxima nomination sampling. J. Stat. Plan. Infer. **140**, 2448–2460 (2010)

Juran, J.M., Godfrey, A.B.: Juran's Quality Handbook. McGraw-Hill, (1998)

Kahraman, C., Kaya, İ.: fuzzy acceptance sampling plans. Stud. Fuzziness Soft Comput. **252**, 457–481 (2010)

Kanagawa, A., Ohta, H.: A design for single sampling attribute plan based on fuzzy sets theory. Fuzzy Sets Syst. **37**(2), 173–181 (1990)

Krätschmer, V.: Sampling inspections by attributes which are based on soft quality standards. In: Proceedings of EUSFLAT'2005, Barcelona, pp. 611–614. Sept 2005

Kuo, Y.: Optimal adaptive control policy for joint machine maintenance and product quality control. Eur. J. Oper. Res. **171**, 586–597 (2006)

Nezhad, M.S.F., Niaki, S.T.A., Mehrizi, H.A., Hossein, M.: A new acceptance sampling plan based on cumulative sums of conforming run-lengths. Int. J. Ind. Syst. Eng. **4**, 256–264 (2011)

Ohta, H.: And Ichihashi, H.: Determination of single-sampling-attribute plans based on membership functions. Int. J. Prod. Res. **26**(9), 1477–1485 (1988)

Pearn, W.L., Chien-Wei, W.: An effective decision making method for product acceptance. Omega **35**, 12–21 (2007)

Sampath, S.: Hybrid single sampling plan. World Appl. Sci. J. **6**(12), 1685–1690 (2009)

Sampath, S., Deepa, S.P.: Determination of optimal double sampling plan using genetic algorithm. Pak. J. Stat. Oper. Res. **8**(2), 195 (2012)

Schilling, E.G.: Acceptance sampling in quality control. Marcel Dekker, New York, p. 800 (1982)

Tamaki, F., Kanagawa, A., Ohta, H.: A fuzzy design of sampling inspection plans by attributes. Jpn J Fuzzy Theory Syst. **3**, 315–327 (1991)

Tsai, T.R., Chiang, J.Y.: Acceptance sampling procedures with intermittent inspections under progressive censoring. ICIC Express Lett. **3**(2), 189–194 (2009)

Turanoglu, E., Kaya, I., Kahraman, C.: Fuzzy acceptance sampling and characteristic curves. The Int. J. Comput. Intell. Syst. **5**(1), 13–29 (2012)

Vasudevan, D., Selladurai, V., Nagaraj, P.: Determination of closed form solution for acceptance sampling using ANN. Qual. Assur. **11**, 43–61 (2012)

Von Mises, R.: Probability, statistics and truth, 2nd edn. Macmillan, New York (1957)

Wang, R.-C., Chen, C.-H.: The Dodge-Romig double sampling plans based on fuzzy Optimization. Int. J. Qual. Sci. **2**(1), 52–62 (1997)

Wang, J., Chankong, V.: Neurally-inspired stochastic algorithm for determining multi-stage multi-attribute sampling inspection plans. J. Intell. Manuf. **2**(5), 327–336, (1991)

Wang, J.F., Haining, R., Cao, Z.D.: Sampling surveying to estimate the mean of a heterogeneous surface: reducing the error variance through zoning. Int. J. Geogr. Inf. Sci. **24**(4), 523–543 (2010)

Zimmermann, H.-J.: Fuzzy Set Theory and its Applications. Kluwer-Nijhoff Publishing Company, Boston, MA (1985)

Chapter 8
The Role of Computational Intelligence in Experimental Design: A Literature Review

Erkan Işıklı and Seda Yanık

Abstract Experimental design (DOE) is a well-developed methodology that has been frequently adopted for different purposes in a wide range of fields such as control theory, optimization, and intelligent decision making. The main objective of DOE is to best select experiments to estimate a set of parameters while consuming as little resources as possible. The enrichment of literature on computational intelligence has supported DOE to extend its sphere of influence in the past two decades. Specifically, the most significant progress has been observed in the area of optimal experimentation, which deals with the calculation of the best scheme of measurements so that the information provided by the data collected is maximized. Nevertheless, determining the design that captures the true relationship between the response and control variables is the most fundamental objective. When deciding whether a design is better (or worse) than another one, usually a criterion is utilized to make an objective distinction. There is a wide range of optimality criteria available in the literature that has been proposed to solve theoretical or practical problems stemming from the challenging nature of optimal experimentation. This study focuses on the most recent applications of DOE related to heuristic optimization, fuzzy approach, and artificial intelligence with a special emphasis on the optimal experimental design and optimality criteria.

Keywords Optimal experimental design · Optimality criteria · Neural networks · Heuristic optimization · Fuzzy

E. Işıklı (✉) · S. Yanık
Department of Industrial Engineering, Istanbul Technical University,
34367 Macka, Istanbul, Turkey
e-mail: isiklie@itu.edu.tr

© Springer International Publishing Switzerland 2016
C. Kahraman and S. Yanık (eds.), *Intelligent Decision Making
in Quality Management*, Intelligent Systems Reference Library 97,
DOI 10.1007/978-3-319-24499-0_8

8.1 Introduction

Experimental design (DOE) aims to quantify the cause and effect relationship between the inputs (process variables) and outputs (responses) of a process as economically as possible. The process of interest could belong in any field where variance reduction or quality improvement is one of the main objectives. Since its introduction by R.A. Fisher in the 1930s, DOE has attracted much attention and has been applied in various areas ranging from manufacturing to biochemistry; service industry to quality control; and biomedical sciences to marketing. As the number of cases in which DOE approach has been adopted increased, interesting challenges, mostly related with the main assumptions of DOE or applicability and efficiency of traditional designs, have arisen. In order to deal with these challenges, newer designs such as Box–Wilson Central Composite Design, Doehlert Design, Box–Behnken Design, Plackett–Burman Design, Split Plot Design, and Rechtschaffner Design have been introduced. However, Box Behnken and Central Composite designs may not have performed well in case the process behaviour is more complex than a second order (Rollins and Bhandari 2004). Thus, the significant acceleration in the rise of new experimentation techniques was observed with the introduction of computer-aided designs in which one or more optimality criteria are used to construct optimal experimental designs. Lately, the ease of computation has propelled the use of computational intelligence based methods in optimal experimental design. In this study, four streams of the related literature are reviewed: optimization methods, heuristics, fuzzy techniques, and artificial intelligence. However, we should note that these streams are not clear-cut and it is highly likely to come across studies combining methods from different research streams when reviewing the DOE literature. The remainder of this chapter is organized as follows: Sect. 8.2 overviews the basic terminology of DOE and provides some insights on the extent of the use of DOE by reviewing its most recent applications, Sects. 8.3 and 8.4 focus on heuristic optimization methods, and artificial intelligence and fuzzy methods employed in DOE, respectively, Sect. 8.5 concludes with a discussion of potential research avenues.

8.2 The Fundamentals of Experimental Design

Experimental design (DOE) aims to reduce the experimental cost while observing how a response variable (output) is influenced by alternating one or more process variables (inputs). Traditional approaches in DOE include full factorial designs, fractional factorial designs, mixture designs, Taguchi designs, central composite designs (Box–Wilson, Box–Behnken, etc.), and Latin hypercube designs. Lundstedt et al. (1998) provided an insightful review on DOE with a special emphasis on the screening methods along with central composite designs and the Doehlert design. Anderson-Cook et al. (2009) paid particular attention to robust parameter designs, split-plot designs, mixture experiment designs, and designs for generalized linear models. The authors underlined the importance of investing more in the analysis stage, before data collection, to obtain better results. Resolving issues regarding a design could drastically increase the related cost. Thus, attention should be paid in the earlier stages of the experimentation. Recently, a remarkable progress in optimal experimentation has been observed, especially due to new algorithmic approaches and a significant decrease in computation times; however, this stream of the literature is still developing and needs more attention even though its roots date back to 1920s.

Optimal experimental designs (sometimes also called computer-aided designs) are generated by an optimization algorithm that uses a design criterion to measure the quality of the experiment. There are several optimality criteria proposed in the literature which can be mainly classified into two groups: information-based criteria and distance-based criteria. The former are based on the Fisher information matrix, $X^T X$, whereas the latter are based on the distance $d(y, A)$ from a point (y) in the n-dimensional Euclidean space (R^n) to a subset (A) of R^n. These criteria play a vital role in optimal experimentation as they help experimenters choose between alternative designs—by calculating their efficiencies—without wasting too much resource, time, effort, and money. However, experimenters should also take into account the robustness of these candidate designs and the effect of missing data to make better conclusions (Anderson-Cook et al. 2009).

To provide the reader a background to better understand the details of various designs to be discussed in subsequent sections, we briefly cover the information-based criteria (also known as the alphabet criteria) below. Interested readers should refer to Das (2002) or Pukelsheim (1993) for a thorough review on this topic.

- A-optimality minimizes the trace of $(X^T X)^{-1}$, which is equivalent to minimizing the average variance of the parameter estimates. It is vulnerable to changes in the coding of the design variable(s) (Anderson-Cook et al. 2009).
- C-optimality minimizes the variance of the best linear unbiased estimator of a predetermined combination of model parameters (Harman and Jurik, 2008).
- D-optimality maximizes $det(X^T X)$, which is equivalent to minimizing the inverse Fisher information matrix. This way, the volume of the confidence ellipsoid around the parameter vector is minimized. The higher the D-optimality criterion the smaller the confidence region for the parameter estimates (Balsa-Canto et al. 2007).

- E-optimality maximizes the minimum eigenvalue of $X^T X$, which implies the minimization of the maximum variance of all possible normalized linear combinations of parameter estimates. Modified E-optimality, which minimizes the ratio of the largest eigenvalue of $X^T X$ to the smallest one, represents the relationship between the longest and shortest semi-axes of the information hyper–ellipsoid (Balsa–Canto et al. 2007).
- T-optimality maximizes the trace of $X^T X$.
- G-optimality minimizes the maximum entry in the diagonal of $X(X^T X)^{-1} X^T$, which corresponds to the maximum variance of any predicted value over the design space.
- I-optimality (also known as Q-optimality, V-optimality, or I_V-optimality) minimizes the (normalized) average prediction variance over the region of interest.
- L-optimality, a modified version of A–optimality, minimizes the average variance of the parameter estimates (Wit et al. 2005).
- V-optimality minimizes the average prediction variance over a set of m specific points.

Fraleigh et al. (2003) mentioned that there were two particular optimal designs of interest: variance optimal and model discrimination designs. There are a range of variance optimal designs including A-, D-, E-, G- and Q-optimal approaches. D-optimal experimental design, which was developed to determine the experimental conditions that minimize the volume of the uncertainty region for the parameter estimates, has been very popular. T-optimal experimental designs are used to decide which experimental conditions to use so that one can discriminate between alternative models, and is based on the prediction error. The objective of T-optimal design is to maximize the sum of squares lack of fit between the observations and the model predictions. López-Fidalgo et al. (2007) proposed an extension to the conventional T-optimality criterion that considers the case of non–normal parametric regression models. Their criterion was further modified by Otsu (2008) to also cover the case of semi-parametric models as an assumption on the distribution of residuals may be restrictive in some cases. Fang et al. (2008) explored five different approaches to derive the lower bounds of the most common criteria employed in DOE.

As Anderson-Cook et al. (2009) informed, optimality criteria should not be the only aspects to consider for the estimation and/or prediction. Collecting data reasonably, estimating/interpreting model parameters carefully, and having Plan B are equally important aspects in DOE. Thus, creating a design that balances the pros and cons of each such aspect should be the first priority of an experimenter, which would result in a near optimal design in many occasions. Imhof et al. (2004) also discussed the pitfalls of an optimal experimental design methodology when some of the observations may not be available at the end of the experiment and showed how inefficient the experimentation could be if the anticipated missingness pattern was not accounted for at the design stage.

DOE is an efficient procedure for planning experiments such that the data obtained can be analyzed to yield valid and objective conclusions. Well-chosen experimental designs maximize the amount of information that can be obtained with a given amount of the experimental effort. The main goal of DOE is to plan a process in an optimal way with a single or multiple underlying objectives such as cost minimization, effective resource consumption, and reduced environment pollution. Therefore, it is natural that DOE can be viewed as an optimization technique (Siomina and Ahlinder 2008).

DOE is often used to select the significant factors that affect the output. Fraleigh et al. (2003) adopted DOE for this purpose in a sensor subsystem to ensure an effective real time optimization (RTO) system. The authors suggested a procedure that combines a modified D-optimal and a modified T-optimal design that fits the RTO problem geometry well and illustrated its use via a simulation study.

Rollins and Bhandari (2004) adopted DOE to determine the design points (to generate data) for sequential step tests in a new multiple input, multiple output (MIMO) constrained discrete-time modelling (DTM) approach for dynamic block-oriented processes. Their approach is essentially innovative as DOE provides the efficient information to estimate ultimate response and dynamic response behaviour. Similarly, Patana and Bogacka (2007) attempted to use DOE to properly design the data collection process and to avoid the noise in the parameter estimates for multi-response dynamic systems when one of its basic assumptions is violated: uncorrelated error terms.

Siomina and Ahlinder (2008) stressed one of the most important reasons to use DOE in practical applications: reducing the cost of experimental time and effort. The authors presented a lean optimization algorithm that sequentially uses supersaturated experimental designs for the optimization of a multi-parameter system in which the maximum number of experiments cannot exceed the number of factors. Their algorithm was proven to be computationally efficient and to significantly outperform the well-known Efficient Global Optimization (EGO) algorithm (Jones et al. 1998). EGO algorithm first fits a response surface to data collected by evaluating the objective function at a few points and then balances between finding the optimum point of the surface and improving the approximation by sampling where the prediction error may be high (Siomina and Ahlinder 2008).

Myers et al. (2004) observed that the response surface framework had become the standard approach for much of the experimentation carried out in industrial research, development, manufacturing, and technology commercialization. The Response Surface Methodology (RSM) has been originally designed to approximate an unknown or complex relationship between design variables and design functions by fitting a simpler model to a (relatively small) number of experimental points. In RSM, the direction of improvement is determined using the path of the steepest descent/ascent (for a minimization/maximization problem) based on the estimated first-order model or using ridge trace analysis for the second-order model (Siomina and Ahlinder 2008). Anderson-Cook et al. (2009) provided an insightful discussion on good response surface designs considering qualitative and quantitative characteristics.

RSM is a widely used technology also for rational experimental design and process optimization in the absence of mechanistic information. RSM initiates from design of experiments (DOE) to determine the factors' values for conducting experiments and collecting data. The data are then used to develop an empirical model that relates the process response to the factors. Subsequently, the model facilitates to search for better process response, which is validated through experiment(s). The above procedure iterates until an optimal process is identified or the limit on experimental resources is reached (Chi et al. 2012).

In traditional RSM, the first-order or second-order polynomial function is adopted for empirical modelling. However, the restrictive functional form of polynomials has long been recognized as ineffective in modelling complex processes. The non-traditional RSM is a stage-wise heuristic that searches for the input combination that maximizes the output (Kleijnen et al. 2004). Progress in adopting more flexible models in RSM includes artificial neural networks (ANN), support vector regression (SVR), and more recently Gaussian process regression (GPR). GPR, also known as kriging model with a slightly different formulation, has been accepted as a powerful modelling tool in various fields, especially in process systems engineering (Chi et al. 2012). The next two sections delve deeper into these topics.

8.3   The Use of Heuristic Optimization Methods in DOE

Lundstedt et al. (1998) compared the theoretical and practical aspects of two optimization approaches (simplex method and response surface methodology) in experimental design (DOE). It is possible to reach the optimal set of parameters using Response Surface Methodology (RSM); however, the experiments are performed one by one in simplex optimization and the global optimum is not guaranteed. Coles et al. (2011) emphasized the need for a comprehensive approach that compares the quality of the optimal experimental design and the computational efficiency of the algorithm used for parameter estimation. They claimed that it would not always be possible to find a unique algorithm that could perform well for different types of objective functions. This is one of the most important challenges in DOE: the trade-off between the optimum use of resources and the computational efficiency. Such challenges have usually been approached by using both linear and non-linear programming techniques. However, traditional algorithms may not work at some instances. This is where heuristic approach comes into play. A detailed discussion on the use of heuristic techniques is made in this section after providing a concise review on how, in general, optimization techniques have been employed in DOE.

On the linear side, Joutard (2007) proposed a large deviations principle for the least-squares estimator in a linear model and used its results to find optimal experimental designs. The author demonstrated the performance of this principle by estimating the whole parameter vector in a Gaussian linear model and one component of the parameter vector in an arbitrary linear model in which certain assumptions on the distribution of errors were made. Harman and Jurík (2008) formulated the approximate C-optimal design for a linear regression model with uncorrelated observations and a finite experimental domain as a specific linear programming problem. The authors stated that the proposed algorithm can also be applied to *difficult* problems with singular C-optimal designs and relatively high dimension of β; however, computing optimal designs with respect to other well-known criteria cannot be reduced to a linear programming problem. The algorithm the authors proposed (called SAC), which is based on the simplex method, identifies the design support points for a C-optimal design. It can also be applied to C-optimal design problems with a large experimental domain without significant loss of efficiency.

Research on the non-linear side has been more diversified. The related literature has mostly focused on the construction of D-optimal designs to estimate some fixed parameters. Recently, Loeza–Serrano and Donev (2014) drew attention to the lack of research on the estimation of variance components (or variance ratios) contributed to the literature by proposing a new algorithm for the construction of A- and D-optimal designs at such instances. Parameter estimation can get tedious for non-linear models in the sense of experimental effort and computational effectiveness. Sequential DOE has been proven to be very helpful in such cases to substantially reduce experimental cost. The experiments excluding the first one are run using the information on preceding experiments in order to optimize the design. Harman and Filová (2014) used a quadratic approximation of D-optimality criterion (DQ criterion) in the method they proposed when computing efficient exact experimental designs for linear regression models. They asserted that the main advantage of their method can be realized in case there are general linear constraints such as cost constraints on permissible designs. Bruwer and MacGregor (2006) extended the open-loop D-optimal design formulation of Koung and MacGregor (1994) for robust multi-variable identification. Their design formulations enable effective and efficient identification of robust models. The authors regarded that their design formulations also performed better in the presence of constraints using a two-input, two-output system as a case study. Even though the designs they proposed resulted in highly correlated physical input sequences as in the unconstrained case, the authors maintained that the designs would overcome this when highly unbalanced replications were used among the support-points to emphasize excitation of these directions. Similarly, Ucinski and Bogacka (2007) studied optimal experimental designs in the presence of constraints aiming to develop a theoretical background along with numerical algorithms for model discrimination design. The authors applied their numerical procedure in a chemical kinetic model discrimination problem in which some of the experimental conditions were allowed to continually vary during the experimental run.

Sagnol (2011) proposed an extension to Elving's theorem in the case of multi-response experiments and concluded that it would be possible to use second-order cone programming to compute the C-, A-, T-, and D-optimality criteria when a finite number of experiments was to be run. The author also provided a way to avoid the complexity in the multi-response C-optimal designs.

DOE has often been employed in adaptive and optimal control. Pronzato (2008) underlined the role of DOE in the asymptotic behaviour of the parameters estimated. The author pointed out the strength of DOE as a tool to establish links between optimization, estimation, prediction and control problems. Pronzato (2008) presented a comprehensive review on the relationship between sequential design and adaptive control, the mathematical foundations of optimal experimental design when estimating parameters of dynamical models.

Mandal and Torsney (2006) proposed a way that makes it possible to calculate a probability distribution by first discretizing the (continuous) sample space and then using these disjoint clusters of points at each iteration until the algorithm converges.

According to Coles et al. (2011), the non-linear nature of most of the design criteria add too much complexity to the design algorithms. The authors also questioned the heuristic nature of many design algorithms and the lack of their convergence properties in the related literature. Thus, the design criteria should not be determined without taking the design algorithm into account as the individual choice for both the former and the latter could alter the final result. Goujot et al. (2012) proposed a method that does not require the use of a global optimization algorithm. In a similar study, a new method that blends results obtained from initial experimental design, empirical modelling, and model-based optimization to determine the most promising experiments that would be used as an input at the subsequent stage was introduced (Chi et al. 2012). The authors claimed that their approach could be used as an alternative to RSM, especially in case prediction uncertainty should be taken into account. The problem they were interested in can be classified as a multi-objective optimization problem. Balsa-Canto et al. (2007) formulated the optimal experimental design problem as a general dynamic optimization problem where the objective is to find those experimental variables that could be manipulated in order to achieve maximum information content (or minimum experimental cost), as measured by the Fisher information matrix. They illustrated their approach in the estimation of the thiamine degradation kinetic parameters during the thermal processing of canned tuna. Based on their results, the authors concluded that optimal dynamic experiments could both improve identifiability essentially and reduce the experimental effort. The authors employed a metaheuristic approach called scatter search method (SSm), which could guarantee convergence to the global solution, when simultaneously computing the system dynamics and the local parametric sensitivities.

Coles et al. (2011) presented an empirical formula for designing Bayesian experimental designs when D-optimality is employed. The authors considered the case of linearized experimental design and claimed that their approach can be generalized for both the case of non-linear experimental design and the case of Bayesian experimental design. They concluded that the choice of the design algorithm should be made by considering different aspects of the problem such as the experimental quality and the importance of computational efficiency.

Myunga et al. (2013) provided a thorough review on the use of Adaptive Design Optimization (ADO) in the construction of optimal experimental designs. ADO is a Bayesian statistical framework that can be employed to conduct maximally informative and highly efficient experiments. The authors compared the practicality of ADO and the traditional, non-adaptive heuristic approach to DOE and claimed that ADO combined with modern statistical computing techniques had high potential to lead the experimenter to better statistical inference while keeping the related cost at a minimum.

Even though DOE has been applied in a wide variety of areas, some problems are intrinsically ill-conditioned and/or very large and their solutions require the use of alternative methods such as metaheuristics that can reduce computation time while guaranteeing robustness in many occasions.

Kleijnen et al. (2004) focused on the non-traditional RSM, which searches for the input combination maximizing the output of a real system or its simulation. It is a heuristic that locally fits first-order polynomials, and estimates the corresponding steepest ascent paths. The authors proposed novel techniques that combined mathematical statistics with mathematical programming to solve issues stemming from the scale-dependence of the steepest ascent and the intuitive selection of its step size. One of the techniques, called adapted steepest ascent (ASA), accounts for the covariances between the components of the estimated local gradient. It is scale-independent; however, the step-size problem can only be solved tentatively. The other technique follows the steepest ascent direction using a step size inspired by ASA. Monte Carlo experiments showed that ASA would more likely lead to a better search direction than the steepest ascent would.

Box and Draper (1969) developed a heuristic approach called Evolutionary Operation, which iteratively builds a response surface around the optimum from the previous iteration. Torczon and Trosset (1998) defined and experimented with the use of merit functions chosen to simultaneously improve both the solution to the optimization problem and approximation quality. They used the distance between a possible new candidate point and an already evaluated point as a measure for the error of the metamodel (Bonte et al. 2010). A number of heuristic move-limit strategies have been developed for approximate design optimization. These methods vary the bounds of design variables in approximation iterations and differ from each other by different bound-adjustment strategies (Siomina and Ahlinder, 2008).

Alonso et al. (2011) proposed using simulated annealing to find the right permutations of levels of each factor in order to obtain uncorrelated main effects with a minimum number of runs. Factorial experiments are used in many scientific fields. As the number of factors increases, the number of runs required for a complete replica of the design grows exponentially. Usually, only a fraction of the full factorial is used. This is called a fractional factorial design. The key issue is to choose an appropriate fraction that verifies the desired properties, especially the orthogonal property.

When characterizing orthogonal fractional factorial, the following notation is used: $s_1^{k1}; s_2^{k2};\ldots; s_h^{kh}(n)$ where n is the number of runs, s_i is the number of levels of the factors, and k_i is the number of factors with s_i levels. Let matrix d of dimension

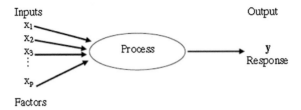

Fig. 8.1 Model of experimental design

$n \times p$ be built by factors as columns and runs as rows, with $p = k_1 + k_2 + \ldots + k_h$. In the experimental design literature, d is known as the design matrix. An illustration of a typical DOE model is given in Fig. 8.1.

It is well known that the multiple linear regression model which represents an experimental design can be written as in Eq. (8.1).

$$y = \mu + X_1 \beta + \varepsilon \tag{8.1}$$

where μ is the grand mean, y denotes the matrix of the response values, β denotes the matrix of the main effects coefficients, X_1 denotes the matrix of contrast coefficients for the vector of main effects, and ε denotes the vector of random errors.

If X^T is the transpose matrix of X, the correlations matrix is $X^T X$. The correlation matrix is an indicator of a good design. If the correlation matrix is diagonal, the computations will be simple and the estimators of all the regression coefficients are uncorrelated.

When orthogonal designs are not possible due to excessive runs and restricted budgets, it would be desirable to obtain a design as close as possible to an orthogonal one, with just a few runs. Such designs are called nearly-orthogonal, and generated by using several criteria. Alonso et al. (2011) employed a criterion based on Addelman frequencies that works with the design matrix. They applied simulated annealing to fractional factorial designs using the Addelman proportional frequencies criterion in order to obtain orthogonal designs.

Bates et al. (2003) used Genetic Algorithms (GA) to find the optimum points in the Audze–Eglais experimental design, which is achieved by distributing experimental points as uniformly as possible within the design domain. A uniform distribution results in the minimization of the potential energy of the points of a DOE. The potential energy is formulated in Eq. (8.2).

$$\min U = \min \sum_{p-1}^{P} \sum_{q-p+1}^{P} \frac{1}{L_{pq}^2} \tag{8.2}$$

where U is the potential energy and L_{pq} is the distance between the points p and q.

An example of Audze–Eglais Uniform Latin Hypercube (AELH) for two design variables and three points are given in Fig. 8.2.

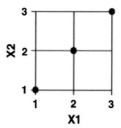

Fig. 8.2   An illustration of Audze-Eglais Latin Hypercube

Various experimental design combinations can be evaluated and the one with minimum objective function (i.e., Eq. (8.2) is minimized) is the AELH experimental design. Bates et al. (2003) carried out the search for the best DOE by minimizing the objective function in Eq. (8.2) using the Genetic Algorithm (GA). The fitness function of the GA is given in Eq. (8.2). For the encoding of two alternatives, the node numbers and the coordinates of the points are evaluated, and coordinates are chosen since it results in shorter length of chromosomes. Various numerical studies have been conducted varying the number of design variables and the number of points. The results indicate that the method works well and an improvement over previous results of Audze–Eglais Uniform Latin Hypercube experimental design and of random sampling Latin Hypercube experimental design has been achieved.

Chen and Zhang (2003) employed a GA for 2^{k-p} fractional factorial design. They used a MD-optimality criterion for optimizing the fractional factorial design. To select the optimal follow-up design using MD-optimality, traditionally the procedure below is used:

(1) Identify potential regression models that can describe the response values in the initial experiment by using Bayesian analysis, and define all the factors appearing in these models as active factors.
(2) Choose a set of runs (follow-up design) from all the experimental combinations of the active factors such that the best model can be discriminated from the potential regression models. Note that the effects of the factors and interactions included in the model are the significant effects in the experiment. Therefore, the confounded effects produced in the initial experiment are separated.

There is a weakness in this approach as the number of follow-up designs that needs to be examined significantly increases when the number of active factors increases, or the number of runs included in a follow-up design increases. Thus, Chen and Zhang (2003) developed a heuristic method based on an effective evolutionary algorithm and genetic algorithms (GA) for finding the optimal follow-up design. This heuristic is denoted as GA for maximum model-discrimination design (GAMMDD). In this GA, the encoding of a solution is represented as a follow-up design, U_i, which is described as a $n_1 \times k$ matrix, where n_1 is the number of experimental runs in the follow-up design, and k is the number of active factors. The fitness value is specified as the model-discrimination (MD) value of a design, since the problem is to find a follow-up design that can identify a model with maximum model-discrimination value. Let X be a $n_1 \times k$ follow-up design matrix for k factors $f_1, f_2, \ldots, f_k$; and y denote the predicted vector under X, then the MD value for the design is calculated using Eq. (8.3).

$$MD = \frac{1}{2} \sum_{0 \leq i \neq j \leq m} p(M_i|y)p(M_j|y), \tag{8.3}$$

where $p(M_i|y)$ is the posterior probability of the model M_i, given y and considering a regression model M_i as in Eq. (8.4).

$$y = X_i \beta_i + \varepsilon_i \tag{8.4}$$

The computational results in their research show that the performance of GAMMDD was significantly better to that of the exchange algorithm, and would be able to enhance the strength of traditional two-step approach.

Lejeune (2003) implemented a one-exchange algorithm and used a generalized simulated annealing for the construction of D-optimal designs. The proposed method does not require to construct or to enumerate each point of the candidate set, whose size grows exponentially with the number of variables. In order to handle more complex problems, their procedure generates guided starting designs.

The focus is given to the D-optimality criterion, which requires the maximization of the determinant of the information matrix, $|X^T X|$, or, equivalently, its D-efficiency level formulated using Eq. (8.5)

$$D_{eff} = 100 \left(\frac{|X^T X|^{\left(\frac{1}{P}\right)}}{N} \right) \tag{8.5}$$

where P is the number of parameters, N is the number of experiments in the model. When a linear regression model is considered, $y = X_i \beta_i + \varepsilon_i$, any increase in the determinant of $X^T X$ reduces the error variances of the estimates.

The integrated algorithmic process presented to find the D-optimal designs has the following characteristics:

- The proposed algorithm selects a new point of the candidate space randomly and does not require such maximization operations.
- This is an important aspect in simulated annealing, which has also the advantage of preventing from premature convergence towards local optima and giving the possibility to escape from a sequence of local optima. In addition, the method does not involve the construction or enumeration of each point of the candidate set and is time-saving.
- The exchange algorithm is a one-exchange procedure.
- The algorithmic process includes a procedure for constructing guided starting designs. This procedure is implemented with, in mind, the objective of applying the algorithmic process for more complex models.

This procedure resulted in a highly D-efficient algorithmic process that could be applied for more complex models than those treated in the literature. The latter objective requires that the computing time does not rise exponentially with the number of factors. The time-saving property constitutes the third characteristic of the algorithmic process proposed.

Sanchez et al. (2012) focused on finding an experimental design that balances different competing criteria which is a multi-objective optimization problem. They tackled the problem by looking for the Pareto-optimal front in the competing criteria. They reported various criteria used in the literature such as A-, E-, and D-optimality criteria related to the joint estimation of the coefficients, or the I- and G-optimality criteria related to the prediction variance.

A design is said to be D-optimal when it achieves the maximum value of D in Eq. (8.5), which means the minimum volume of the joint confidence region, so the most precise joint estimation of the coefficients.

A- and E-optimality criteria are related to the shape of the confidence region (the more spherical the region, the less correlated the estimates). When the estimates are jointly considered, the $(1 - \alpha) \times 100\%$ joint confidence ellipsoid for the coefficients is determined by the set of vectors β such that

$$(\beta - b)'X^TX(\beta * b) \leq P\widehat{\sigma}^2 F_{\alpha,P,N-P} \tag{8.6}$$

where P is the number of estimated coefficients, N denotes the number of experiments in the design, $\widehat{\sigma}^2$ is the variance of the residuals, (an estimate of σ^2) and $F_{\alpha,P,N-P}$ is the corresponding upper percentage point of an F distribution with P and $N - P$ degrees of freedom.

When using the I- and G-optimality criteria, the variance of the prediction is taken into account through the prediction variance. The variance of the response predicted for a given point x in the experimental domain, is given by Eq. (8.7) and the G-optimality criterion is shown in Eq. (8.8)

$$Var(\widehat{y}(x)) = x'_{(m)}(X^TX)^{-1}x_{(m)}\widehat{\sigma}^2 = d(x)\widehat{\sigma}^2 \tag{8.7}$$

$$G = Nd_{max} = Nmax_x(d(x)) \tag{8.8}$$

A design is said to be G-optimal when it achieves the minimum value of G in Eq. (8.9), whereas I-optimality criterion uses the average value of $Nd(x)$ obtained by integrating it over the domain.

Sanchez et al. (2012) employed an evolutionary algorithm to compute the Pareto-optimal front for a given problem. The input for the algorithm is the number of factors (k), domain, model to be fit (that determines the number of coefficients, P) and number of experiments (N, $N \geq P$) to do so, and also the criteria to be taken into account. The evolutionary algorithm is designed such that each individual in the population is an experimental design ($N \times k$ design matrix), codified according to the search space and such that $\det(X^TX) \geq 0.01$. Every design is evaluated in terms of the criteria, so that the fitness associated to each individual is a vector.

The applicability and interpretability of the proposed approach was shown by an application to determine sulfathiazole in milk (substance that has a maximum residue limit established by the European Union) by using molecular fluorescence spectroscopy. Numerical results are presented and the results show that the proposed algorithmic approach makes it possible to address the computation of ad hoc experimental designs with the property of being optimal in one or several criteria stated by the user.

Fuerle and Sienz (2011) presented a procedure that creates Optimal Latin Hypercubes (OLH) for constrained design spaces. OLH in a constrained design space may result in infeasible points of experimental designs. Instead of omitting these infeasible points, a better mapping of the feasible space is generated using the same number of points by using permutation genetic algorithm. In the search procedure, the objective was set so that the Audze-Eglais potential energy of the points as shown in Eq. (8.2) is minimized.

8.4 The Use of Experimental Design in Artificial Intelligence and Fuzzy Methods

Experimental design (DOE) has been one of the most important tools to verify interactions and interrelations between parameters in the design of intelligent systems. Among these systems, artificial neural networks and fuzzy inference systems have been the most prominent ones to search for representations of the domain knowledge, reasoning on uncertainty, automatic learning and adaptation. Neuro-fuzzy system is an approach that can learn from the environment and then reason about its state. A neuro-fuzzy system is based on a fuzzy inference system, which is trained by a learning algorithm derived from artificial neural network theory.

The design of a neuro fuzzy system requires the tuning and configuration of the topology and many parameters. Setting the parameters such as the membership functions, number and shape of each input variable, learning rates is a difficult task. Zanchettin et al. (2010) used DOE for parameter estimation of two neuro-fuzzy systems—Adaptive Neuro Fuzzy Inference System (ANFIS) and Evolving Fuzzy Neural Networks (EFuNNs). A depiction of two intelligent systems (ANFIS and EFuNNs) is provided in Fig. 8.3.

The ANFIS architecture consists of a five-layer structure. In the first layer, the node output is the degree to which the given input satisfies the linguistic label associated to the membership functions named as premise parameters. In the second layer, each node function computes the firing strength of the associated rule. In the third layer, each node i calculates the ratio of the ith rule firing strength for the sum of firing strength of all rules. The fourth layer is the product of the normalized firing level and the individual rule output of the corresponding rule. Parameters in this layer are referred to as consequent parameters. EFuNNs also have a five-layer structure. Each input variable is represented by a group of spatially arranged

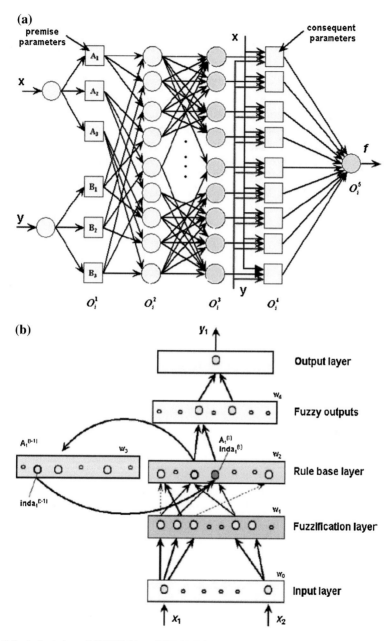

Fig. 8.3 A depiction of ANFIS (**a**) and EFuNN (**b**) (Zanchettin et al. 2010)

neurons to represent a fuzzy quantization of this variable. Fuzzy quantization in variable space is represented in the second layer of nodes. Different membership functions can be attached to these neurons (triangular, Gaussian, etc.). The experiments for setting the parameters of the two intelligent systems are performed with four different prediction and classification problem datasets. The results show that for ANFIS, number of input membership functions and the shape of the output membership functions are usually the factors with the largest influence on the system's error measure. For the EFuNN, the membership function shape and the interaction between membership function shape and the number usually have the largest effect (Zanchettin et al. 2010).

Breban et al. (2013) used DOE for choosing the optimized parameters and determining the influence of the parameters of a fuzzy-logic supervision system in an embedded electrical power system. The fuzzy logic supervision system was developed to minimize the DC-link voltage variations, and to increase the system efficiency by reducing the dissipated power. In the experimental design step, first, the parameters and their variation range are chosen. Second task is to find the optimal ones and to test the system response to their changes. The most influential parameters are determined by testing the system response to each parameter extremity range modification. Breban et al. (2013) chose eight parameters, each with two extremity range values. Then, the influence of each parameter is tested on each optimization factor. For each parameter optimization, the low extremity range value becomes -1, and the high extremity range value becomes $+1$. This assumption creates a matrix, called *test matrix*. Using relation (8.9), the influence E of each indicator is calculated as follows:

$$E = \frac{1}{n}M^t F \tag{8.9}$$

where n is the number of tests, M^t, the transpose of the test matrix and F, the indicators matrix of the parameters.

Basu et al. (2014) analyzed the process parameters of soap manufacturing industries. The process capability was determined using Fuzzy Inference System rule editor based on a set of justified "if-then" statements as applicable for the process. The data was collected in linguistic form to derive its process capability, using a set of justified rules and the effect of each factor was determined using DOE and ANOVA for improving the soap quality from the perspective of its softness. This article concludes that integrating fuzzy inference systems with DOE provides better results compared to those retrieved from DOE and Fuzzy Inference system in isolation.

Plumb et al. (2002) investigated the effect of experimental design strategy on the modelling of a film coating formulation by artificial neural networks (ANNs). Three different DOE approaches: (i) Box–Behnken, (ii) central composite and (iii) pseudo-random designs were used to train a multilayer perceptron (MLP). The structure of the ANN was optimized by training networks containing 3, 4, 5, 6, 7, or 9 nodes in the hidden layer. The predictive ability of each architecture was assessed by comparing the deviations mean square and R^2 from ANOVA analysis of the linear regression of predicted against observed property values. The architecture with the lowest deviations mean square and highest R^2 was considered to be the most predictive one. Over-training was minimized by attenuated training.

Specifically, the onset of over-training was detected by setting a test error weight (W_T) calculated by Eq. (8.10):

$$W_T = \frac{N_{Test}}{(N_{Test} + N_{Train})} \tag{8.10}$$

where N_{Test} and N_{Train} are the number of records in the training and test sets, respectively.

As a result, ANN comprising six input and two output nodes separated by a single hidden layer of five nodes. The Box–Behnken and central composite models showed a poor predictive ability which is related to the high curvature of the response surfaces. In contrast, the pseudo-random design mapped the interior of the design space allowing improved interpolation and predictive ability. It was concluded that Box–Behnken and central composite experimental designs were not appropriate for ANN modelling of highly curved responses.

Alam et al. (2004) presented a case study which also investigated the experimental design on the development of artificial neural networks as simulation metamodels. The simulation model used in the study is a deterministic systems dynamics model. Six different DOE approaches which are the traditional full factorial design, random sampling design, central composite design, modified Latin Hypercube design and designs supplemented with domain knowledge are compared for developing the neural network metamodels. Various performance measures were used to evaluate the networks. The relative prediction error (RPE) which is commonly used for metamodels of deterministic simulations was used as a performance measure, which was defined as in Eq. (8.11)

$$RPE = \frac{\widehat{Y}_r}{Y_r} \tag{8.11}$$

where Y_r is the known target value (simulation response) from the independent test data set, and $\widehat{Y}_r$ is the corresponding network output or prediction. Another measure of performance is the mean squared error of prediction (MSEP), defined as Eq. (8.12)

$$MSEP = \frac{1}{N}\sum(Y_r - \widehat{Y}_r)^2 \tag{8.12}$$

The mean absolute percentage deviation (MAPD), which is used as the third performance measure is defined as Eq. (8.13)

$$MAPD = \frac{1}{N}\sum\left|\left[\widehat{Y}_r - Y_r\right]/Y_r\right| \tag{8.13}$$

The neural network developed from the modified Latin Hypercube design supplemented with domain knowledge produced the best performance, outperforming networks developed from other designs of the same size.

Chang (2008) presented a case for the use of the Taguchi method for product design. Specifically, the aim was to optimize the parameter robust product design in terms of production time, cost, and quality as continuous control factors. They employed a four-stage approach based on artificial neural networks (ANN), desirability functions, and a simulated annealing (SA) algorithm to resolve the problems of dynamic parameter design with multiple responses. An ANN was employed to build a system's response function model. Desirability functions were used to evaluate the performance measures of multiple responses. AnSA algorithm was applied to obtain the best factor settings through the response function model.

Chang and Low (2008) also used Taguchi experiments to minimize various measures simultaneously (i.e., cost of the filter, its power loss, the total demand distortion of harmonic currents and the total harmonic distortion of voltages at each bus) of large-scale passive harmonic filters. Using the results of the Taguchi experiments as the learning data for an artificial neural network (ANN) model, an ANN was developed to predict the parameters at discrete levels. Then, the discrete levels were transformed into continuous scale using a genetic algorithm. Besides, the multiple objectives of the problem were tackled using the membership functions of fuzzy logic theory which were adopted in the algorithm for determining the weight of each single objective. The proposed approach significantly improves the performance of the harmonic filters when compared with the original design.

Balestrassi et al. (2009) applied DOE to find the optimal parameters of an Artificial Neural Network (ANN) in a problem of nonlinear time series forecasting. They presented a case study for six time series representing the electricity load for industrial consumers of a production company in Brazil. They employed an approach based on factorial DOE using screening, Taguchi, fractional and full factorial designs to set the parameters of a feed-forward multilayer perceptron neural network. The approach used classical factorial designs to sequentially define the main ANN parameters that a minimum prediction error could be reached. The main factors and interactions were identified using this approach and results suggest that ANNs using DOE can perform better comparably to the existent nonlinear autoregressive models.

Tansel et al. (2011) proposed using Taguchi Method and Genetically Optimized Neural Networks (GONNS) to estimate optimal cutting conditions for the milling of titanium alloy with PVD coated inserts. Taguchi method was used to determine the test conditions, the optimal cutting condition and influences of the cutting speed, feed rate and cutting depth on the surface roughness. GONNS was used to minimize or maximize one of the output parameters while the others were kept within a specified range.

Salmasnia et al. (2012) used DOE for data gathering to find the most valuable information used in a multiple response optimization problem. The multiple response optimization problem aims to find optimal inputs (design variables) to the system that yields in desirable values for stochastic outputs (responses).

Specifically, the problem of correlated multiple responses where relationship among response and design variables is highly nonlinear and the assumption that variance of each response is constant over the feasible region was tackled with a neuro-fuzzy (i.e. ANFIS) and principal component analysis derived desirability function. The resulting desirability functions were used to form a fitness function for optimization in GA. Effectiveness of the proposed method was presented through a numerical example.

Richard et al. (2012) proposed an alternative method to the classical response surface technique where the response surface was chosen as a support vector machine (SVM). An adaptive experimental design was used for the training of the SVM. As a result, the design can rotate according to the direction of the gradient of the SVM approximation leading to realistic samples. Furthermore, the precision of the probability of failure computation was improved since a closed form of expression of the Hessian matrix could be derived from the SVM approximation. This method was tested through a case study showing that high-dimensional problems can be solved with a fairly low computational cost and a good precision.

Hametner et al. (2013) dealt with the model based design of experiments for the identification of nonlinear dynamic systems. The aim of designing experiments was to generate informative data and to reduce the experimentation effort as much as possible as well as to comply with constraints on the system inputs and the system output. Two different modelling approaches, namely multilayer perceptron networks and local model networks were employed and the experimental design was based on the optimization of the Fisher information matrix of the associated model architecture. Deterministic data driven models with a stochastic component at the output were considered. The parameters of the considered models were denoted by θ. The measured output $y(k)$ at the time k was given by the model output $\hat{y}(k, \theta)$ plus some error $e(k)$. Then, the Fisher information matrix was formulated as in Eq. (8.14)

$$\tau = \frac{1}{\sigma^2} \sum_{k=1}^{N} \frac{\partial \hat{y}(k, \theta)}{\partial \theta} \frac{\partial \hat{y}(k, \theta)'}{\partial \theta} \tag{8.14}$$

The effects of the Fisher information matrix in the static and the dynamic configurations were discussed. Finally, the effectiveness of the proposed method was tested on a complex nonlinear dynamic engine simulation model. The presented model architectures for model based experiment design were compared.

Lotfi and Howarth (1997) proposed a novel technique named as the Experimental Design with Fuzzy Levels (EDFLs), which assigns a membership function for each level of variable factors. Traditionally, variable factors can be expressed with some linguistic terms such as low and high and they are converted into crisp values such as -1, 0, and $+1$. If some of the factor levels are not measurable, their values should be represented by equivalent fuzzy terms so that their importance is included in the system response. Using the fuzzy levels of factors, a set of fuzzy rules was used to represent the design matrix and observed responses. In this study, a number of examples were presented to clarify the proposed idea and the results were compared with the conventional Taguchi methodology. In their study, they used a L_{18} orthogonal array EDFL for the application of the solder paste printing stage of surface mount printed circuit board assembly. For this case study, they provided a model for the process and optimized the selection of variable factors.

8.5   Conclusion

DOE is concerned with the selection of experimental settings that provide maximum information for the least experimental cost and can prove essential to successful modelling in an operating process application. According to the experimenter's objectives, DOE can dictate which variables should be measured, at which settings, and how many replicate measurements are needed to provide the required information (Fraleigh et al. 2003). The related literature offers very good examples for standard designs in case of fitting a first-order model; however, the choice of a response surface design for fitting a response surface design can be extremely challenging. Specifically, parameter estimation may not always easy for non-linear models regarding experimental effort and computational effectiveness. Thus, the need for more flexible and/or specific designs is still viable (Anderson-Cook et al. 2009). Response surface methodology has seen the most significant progress in DOE-oriented research due to recent advances in metaheuristics and fuzzy techniques.

Coles et al. (2011) emphasized the need for a holistic approach that compares the quality of the optimal experimental design and the computational efficiency of the algorithm used for parameter estimation. They claimed that it would not always be possible to find a unique algorithm that could perform well for different types of objective functions. This is one of the most important challenges in DOE: maximizing the information to retrieve with scarce resources.

The number of avenues for future research is enormous. Bayesian techniques have been slightly touched in the literature. Active learning and nonlinear feedback control (NFC) are also available for further development according to Pronzato (2008). Computationally faster algorithms are still necessary especially for recently developed optimality criteria (Otsu, 2008). The derivation of lower (or upper) bounds or convergence properties of some algorithms should also be studied in more detail.

Another use of DOE is for tuning the parameters of artificial intelligence techniques such as neural networks, support vector machines or fuzzy inference systems. The literature shows that commonly traditional DOE methods are used to this aim. More sophisticated experimental design techniques (i.e. optimal DOE) for tuning the parameters of such systems present a new potential stream of research.

References

Alam, F.M., McNaught, K.R., Ringrose, T.J.: A comparison of experimental designs in the development of a neural network simulation metamodel. Simul. Model. Pract. Theory **12**, 559–578 (2004)

Alonso, M.C., Bousbaine, A., Llovet, J., Malpica, J.A.: Obtaining industrial experimental designs using a heuristic technique. Expert Syst. Appl. **38**, 10094–10098 (2011)

Anderson-Cook, C.M., Borror, C.M., Montgomery, D.C.: Response surface design evaluation and comparison. J. Stat. Plann. Infer. **139**, 629–641 (2009)

Balestrassi, P.P., Popova, E., Paiva, A.P., Lima, J.W.M.: Design of experiments on neural network's training for nonlinear time series forecasting. Neurocomputing **72**, 1160–1178 (2009)

Balsa-Canto, E., Rodriguez-Fernandez, M., Banga, J.R.: Optimal design of dynamic experiments for improved estimation of kinetic parameters of thermal degradation. J. Food Eng. **82**, 178–188 (2007)

Basu, S., Dan, P.K., Thakur, A.: Experimental design in soap manufacturing for optimization of fuzzified process capability index. J. Manufact. Syst. **33**, 323–334 (2014)

Bates, S.J., Sienz, J., Langley, D.S.: Formulation of the Audze–Eglais Uniform latin hypercube design of experiments. Adv. Eng. Softw. **34**, 493–506 (2003)

Bonte, M.H.A., Fourment, L., Do, T., van den Boogaard, A.H., Huétink, J.: Optimization of forging processes using finite element simulations: a comparison of sequential approximate optimization and other algorithms. Struct. Multidisc Optim. **42**, 797–810 (2010)

Box, G.E.P., Draper, N.R.: Evolutionary Operation: A Statistical Method for Process Management. Wiley, Toronto (1969)

Breban, S., Saudemont, C., Vieillard, S., Robyns, B.: Experimental design and genetic algorithm optimization of a fuzzy-logic supervisor for embedded electrical power systems. Math. Comput. Simul. **91**, 91–107 (2013)

Bruwer, M.J., MacGregor, J.F.: Robust multi-variable identification: Optimal experimental design with constraints. J. Process Control **16**, 581–600 (2006)

Chang, H.H.: A data mining approach to dynamic multiple responses in Taguchi experimental design. Expert Syst. Appl. **35**, 1095–1103 (2008)

Chang, Y.P., Low, C.: Optimization of a passive harmonic filter based on the neural-genetic algorithm with fuzzy logic for a steel manufacturing plant. Expert Syst. Appl. **34**, 2059–2070 (2008)

Chen, C.L., Lin, R.H., Zhang, J.: Genetic algorithms for MD-optimal follow-up designs. Comput. Oper. Res. **30**, 233–252 (2003)

Chi, G., Hua, S., Yang, Y., Chen, T.: Response surface methodology with prediction uncertainty: a multi–objective optimisation approach. Chem. Eng. Res. Des. **90**, 1235–1244 (2012)

Coles, D., Curtis, A., Coles, D., Curtis, A.: A free lunch in linearized experimental design? Comput. Geosci. **37**, 1026–1034 (2011)

Das, A.: An Introduction to optimality criteria and some results on optimal block design. In: Design Workshop Lecture Notes, pp. 1–21. ISI, Kolkata 25–29 Nov 2002

Fang, K.T., Tang, Y., Yin, J.: Lower bounds of various criteria in experimental designs. J. Stat. Plann Infer **138**, 184–195 (2008)

Fraleigh, L.M., Guay, M., Forbes, J.F.: Sensor selection for model–based real–time optimization: relating design of experiments and design cost. J. Process Control **13**, 667–678 (2003)

Fuerle, F., Sienz, J.: Formulation of the Audze–Eglais uniform Latin hypercube design of experiments for constrained design spaces. Adv. Eng. Soft. **42**, 680–689 (2011)

Goujot, D., Meyer, X., Courtois, F.: Identification of a rice drying model with an improved sequential optimal design of experiments. J. Process Control **22**, 95–107 (2012)

Hametner, C., Stadlbauer, M., Deregnaucourt, M., Jakubek, S., Winsel, T.: Optimal experiment design based on local model networks and multi–layerperceptron networks. Eng. Appl. Artif. Intell. **26**, 251–261 (2013)

Harman, R., Filová, L.: Computing efficient exact designs of experiments using integer quadratic programming. Comput. Stat. Data Anal. **71**, 1159–1167 (2014)

Harman, R., Jurík, T.: Computing c–optimal experimental designs using the simplex method of linear programming. Comput. Stat. Data Anal. **53**, 247–254 (2008)

Imhof, L.A., Song, D., Wong, W.K.: Optimal design of experiments with anticipated pattern of missing observations. J. Theor. Biol. **228**, 251–260 (2004)

Jones, D.R., Schonlau, M., Welch, W.J.: Efficient global optimization of expensive black-box functions. J. Glob. Optim. **13**, 455–492 (1998)

Joutard, C.: Applications of large deviations to optimal experimental designs. Stat. Probab. Lett. **77**, 231–238 (2007)

Kleijnen, J.P.C., den Hertog, D., Angün, E.: Response surface methodology's steepest ascent and step size revisited. Eur. J. Oper. Res. **159**, 121–131 (2004)

Koung, C.W., MacGregor, J.F.: Identification for robust multivariable control: the design of experiments. Automatica **30**(10), 1541–1554 (1994)

Lejeune, M.A.: Heuristic optimization of experimental designs. Eur. J. Oper. Res. **147**, 484–498 (2003)

Loeza-Serrano, S., Donev, A.N.: Construction of experimental designs for estimating variance components. Comput. Stat. Data Anal. **71**, 1168–1177 (2014)

Lopez-Fidalgo, J., Trandafir, C., Tommasi, C.H.: An optimal experimental design criterion for discriminating between non–normal models. J. Roy. Stat. Soc. B **69**, 231–242 (2007)

Lotfi, A., Howarth, M.: Experimental design with fuzzy levels. J. Intell. Manuf. **8**, 525–532 (1997)

Lundstedt, T., Seifert, E., Abramo, L., Thelin, B., Nystrom, A., Pettersen, J., Bergman, R.: Experimental design and optimization. Chemometr. Intell. Lab. Syst. **42**, 3–40 (1998)

Mandal, S., Torsney, B.: Construction of optimal designs using a clustering approach. J. Stat. Plann. Infer. **136**, 1120–1134 (2006)

Myers, R.H., Montgomery, D.C., Vining, G.G., Borror, C.M., Kowalski, S.M.: Response surface methodology: A retrospective and literature survey, J Qual Technol, **36**(1), 53–77 (2004)

Myunga, J.I., Cavagnaro, D.R., Pitt, M.A.: A tutorial on adaptive design optimization. J. Math. Psychol. **57**(3–4), 53–67 (2013)

Otsu, T.: Optimal experimental design criterion for discriminating semiparametric models. J. Stat. Plann. Infer. **138**, 4141–4150 (2008)

Patana, M., Bogacka, B.: Optimum experimental designs for dynamic systems in the presence of correlated errors. Comput. Stat. Data Anal. **51**, 5644–5661 (2007)

Plumb, A.P., Rowe, R.C., York, P., Doherty, C.: The effect of experimental design on the modeling of a tablet coating formulation using artificial neural networks. Eur. J. Pharm. Sci. **16**, 281–288 (2002)

Pronzato, L.: Optimal experimental design and some related control problems. Automatica **44**, 303–325 (2008)

Pukelsheim, F.: Optimal Design of Experiments. Wiley, New York (1993)

Richard, B., Cremona, C., Adelaide, L.: A response surface method based on support vector machines trainedwith an adaptive experimental design. Struct. Saf. **39**, 14–21 (2012)

Rollins, D.K., Bhandari, N.: Constrained MIMO dynamic discrete-time modeling exploiting optimal experimental design. J. Process Control **14**, 671–683 (2004)

Sagnol, G.: Computing optimal designs of multiresponse experiments reduces to second–order cone programming. J. Stat. Plann. Infer. **141**, 1684–1708 (2011)

Salmasnia, A., Kazemzadeh, B., Tabrizi, M.M.: A novel approach for optimization of correlated multiple responses based ondesirability function and fuzzy logics. Neurocomputing **91**, 56–66 (2012)

Sánchez, M.S., Sarabia, L.A., Ortiz, M.C.: On the construction of experimental designs for a given task by jointly optimizing several quality criteria: Pareto-optimal experimental designs. Analytica Chimica Acta **754**, 39–46 (2012)

Siomina, I., Ahlinder, S.: Lean optimization using supersaturated experimental design. Appl. Numer. Math. **58**, 1–15 (2008)

Tansel, I.N., Gülmez, S., Demetgul, M., Aykut, S.: Taguchi Method-GONNS integration: Complete procedure covering from experimental design to complex optimization. Expert Syst. Appl. **38**, 4780–4789 (2011)

Torczon, V., Trosset M.W.: Using approximations to accelerate industrial design optimization. In: Grandhi, R.V., Canfield, R.A. (eds.), Proceedings of the 7th AAIA/NASA/USAF/ISSMO Symposium on Multidisciplinary Analysis and Optimization: A Collection of Technical Papers, Part 2, pp. 738–748. American Institute of Aeronautics and Astronautics, Virginia, USA (1998)

Ucinski, D., Bogacka, B.: A constrained optimum experimental design problem for model discrimination with a continuously varying factor. J. Stat. Plan Infer. **137**, 4048–4065 (2007)

Wit, E., Nobile, A., Khanin, R.: Near-optimal designs for dual channel microarray studies. Appl. Stat. **54**, 817–830 (2005)

Zanchettin, C., Minku, L.L., Ludermir, T.B.: Design of experiments in neuro-fuzzy systems. Int. J. Comput. Intell. Appl. **9**(2), 137–152 (2010)

Chapter 9
Multivariate Statistical and Computational Intelligence Techniques for Quality Monitoring of Production Systems

Tibor Kulcsár, Barbara Farsang, Sándor Németh and János Abonyi

Abstract The ISO 9001:2008 quality management standard states that organizations shall plan and implement monitoring, measurement, analysis and improvement processes to demonstrate conformity to product requirements. According to the standard, detailed analysis of data is required for this purpose. The analysis of data should also provide information related to characteristics and trends of processes and products, including opportunities for preventive action. The preliminary aim of this chapter is to show how intelligent techniques can be used to design data–driven tools that are able to support the organization to continuously improve the effectiveness of their production according to the Plan—Do—Check—Act (PDCA) methodology. The chapter focuses on the application of data mining and multivariate statistical tools for process monitoring and quality control. Classical multivariate tools such as PLS and PCA are presented along with their nonlinear variants. Special attention is given to software sensors used to estimate product quality. Practical application examples taken from chemical and oil and gas industries illustrate the applicability of the discussed techniques.

Keywords Multivariate statistics · Computational intelligence · Quality monitoring · Production systems · PDCA

9.1 Introduction

The modern definition of quality states that "quality is inversely proportional to variability". This definition implies that if variability in the important characteristics of a production system decreases, then the quality of the product increases. Statistical process control (SPC) provides techniques to assure and improve the

T. Kulcsár · B. Farsang · S. Németh · J. Abonyi (✉)
Department of Process Engineering, University of Pannonia, Veszprém 158 Hungary
e-mail: janos@abonyilab.com

© Springer International Publishing Switzerland 2016   237
C. Kahraman and S. Yanık (eds.), *Intelligent Decision Making
in Quality Management*, Intelligent Systems Reference Library 97,
DOI 10.1007/978-3-319-24499-0_9

quality of products by reducing the variance of process variables. The role of these tools is illustrated in Fig. 9.1, which presents a manufacturing process. The control chart of SPC is a very useful process monitoring technique, when unusual sources of variability are present and important process variables will plot outside the control limits. In these cases some investigation of the process should be made and corrective action to remove these unusual sources of variability should be taken. Systematic use of a control chart is an excellent way to reduce variability (Montgomery 2009).

As new products are required to be introduced to the market over a short time scale to ensure competitive advantage, the development of process monitoring models of multi-product manufacturing environment necessitates the use of empirical based techniques as opposed to first-principles models since phenomenological model development is unrealizable in the time available. Hence, the mountains of data, that computer-controlled plants generate, must be used by the operator support systems to distinguish normal from abnormal operating conditions. Detection and diagnosis of faults and control of product quality are the pivotal tasks of plant operators. The aim of multivariate statistical based approaches is to reduce the dimensionality of the correlated process data by projecting them down onto a lower dimensional latent variable space where the operation can be easily visualized and hidden functional relationships among process and quality variables can be detected.

In modern production systems huge amount of process operational data are recorded. These data definitely have the potential to provide information for product and process design, monitoring and control (Yamashita 2000). This is especially important in many practical applications where first-principles modeling of complex "data rich and knowledge poor" systems are not possible (Zhang et al. 1997). The term knowledge discovery in databases (KDD) refers to the overall process of

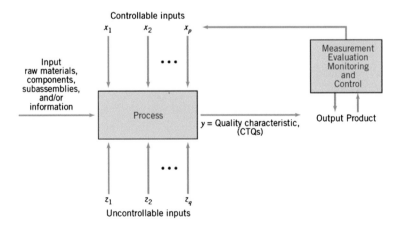

Fig. 9.1 Scheme of a production process where statistical process control (SPC) can be applied to improve the quality characteristic by adjusting and monitoring important process variables (Montgomery 2009)

discovering knowledge from data. KDD has evolved from the intersection of research fields such as machine learning, pattern recognition, databases, statistics, artificial intelligence, and more recently it gets new inspirations from soft computing. KDD methods have been successfully applied in the analysis of process systems, and the results have been used for process design, process improvement, operator training and so on (Wang 1999).

Application of knowledge discovery and data mining for quality development requires sophisticated methodology. Deming recommended the four steps (Plan, Do, Check, Act) based PDCA cycle as model to guide improvement. In the *Plan* step, we propose a change in the system that is aimed at improvement. In *Do*, we carry out the change, usually on a small or pilot scale to ensure that to learn the results that will be obtained. *Check* consists of analyzing the results of the change to determine what has been learned about the changes that we carried out. In *Act*, we either adopt the change or, if it was unsuccessful, abandon it. The process is almost always iterative, and may require several cycles for solving complex problems. It is interesting to note that the concept of PDCA is also applied in data mining. CRISP-DM stands for Cross Industry Standard Process for Data Mining (CRISP-DM 2000) (see Fig. 9.2). It is a data mining process model that describes commonly used approaches that expert data miners use to tackle problems.

Fig. 9.2 The CRISP-DM methodology as continuous data-driven improvement process (CRISP-DM 2000)

Plan:

Business understanding: This initial phase focuses on understanding the project objectives and requirements from a business perspective, then converting this knowledge into a data mining problem definition and a preliminary plan designed to achieve the objectives.

Data understanding: The data understanding phase starts with initial data collection and proceeds with activities that identify data quality problems, discover first insights into the data, and/or detect interesting subsets to form hypotheses regarding hidden information.

Do:

Data preparation: The data preparation phase covers all activities needed to construct the final dataset from the initial raw data.

Modeling: In this phase, various modelling techniques are selected and applied, and their parameters are calibrated to optimal values. Typically, there are several techniques for the same data mining problem. Some techniques have specific requirements on the form of data. Therefore, going back to the data preparation phase is often necessary.

Check:

Evaluation: At this stage, model (or models) is built that appears to have high quality from a data analysis perspective. A key objective is to determine if there is some important business issue that has not been sufficiently considered. At the end of this phase, a decision on the use of the data mining results should be reached.

Act:

Deployment: Creation of the model is generally not the end of the project. Even if the purpose of the model is to increase knowledge, the knowledge gained should be organized and presented in a way that the customer can use it. It often involves applying "dynamic" models within an organization's decision making processes— for real-time control. Depending on the requirements, the deployment phase can be as simple as generating a report or as complex as implementing a repeatable data mining process across the enterprise.

The previously presented data mining procedure should be embedded into the whole quality development process. As we mentioned, most of quality management methodologies are based on intensive analysis of data. Among the wide ranges of methodologies, we suggest the application of DMAIC (Define, Measure, Analyze, Improve, and Control) process (see Fig. 9.3). DMAIC is a structured problem-solving procedure extensively used in quality and process improvement.

Among the wide range of data mining tools, in this chapter we focus on multivariate statistical tools that are extensively applied in process monitoring and quality development.

Process monitoring based on multivariate statistical analysis of process data has recently been investigated by a number of researchers (MacGregor and Kourti 1995). The aim of these approaches is to reduce the dimensionality of the correlated process data by projecting them down onto a lower dimensional latent variable

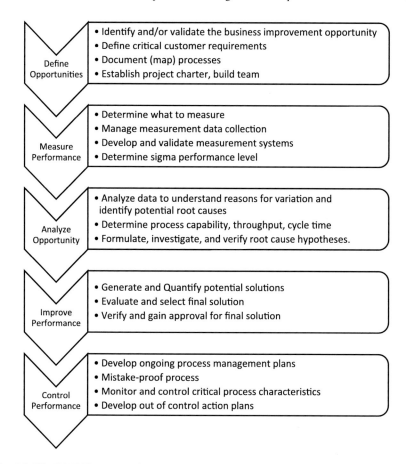

Fig. 9.3 The DMAIC process of quality development

space where the operation can be easily visualized. These approaches use the techniques of principal component analysis (PCA) or Partial Least Squares (PLS). Beside process performance monitoring, these tools can also be used for system identification (MacGregor and Kourti 1995), ensuring consistent production (Martin et al. 1996) and product design (Lakshminarayanan et al. 2000). Data analysis based formulation of new products was first reported by Moteki and Arai (Moteki and Arai 1986), who used PCA to analyze data from a polymer production. Jaeckle and MacGregor (1998) used PLS and principal component regression (PCR) to investigate the product design problem. Their methodology was illustrated using simulated data from a high-pressure tubular low-density polyethylene process. Borosy (1998) used artificial neural networks to analyze data from the rubber industry.

The large number of examples taken from polymer industry is not surprising. Formulated products (plastics, polymer composites) are generally produced from many ingredients, and high number of interactions between the components and the processing conditions has an effect on the final product quality. When these effects are detected, significant economic benefits can be realized. The major aims of monitoring plant performance are the reduction of off-specification production, the identification of important process disturbances and the early warning of process malfunctions or plant faults. Furthermore, when a reliable model is available that is able to estimate the quality of the product; it can be inverted to obtain the suitable operating conditions required for achieving the target product quality (Lakshminarayanan et al. 2000).

When we attempted to use standard data mining, KDD, and multivariate statistical tools for industrial problems such as extracting knowledge from large amount of data, we realized that production systems are typically ill-defined, difficult to model and they have large-scale solution spaces. In these cases, precise models are impractical, too expensive, or non-existent. Furthermore, the relevant available information is usually in the form of empirical prior knowledge and input-output data representing instances of the system's behaviour. Therefore, we need an approximate reasoning system capable of handling such imperfect information (Abonyi and Feil 2005). Computational Intelligence (CI) and Soft Computing (SC) are recently coined terms describing the use of many emerging computing disciplines. According to Zadeh (1994): "... in contrast to traditional, hard computing, soft computing is tolerant of imprecision, uncertainty, and partial truth." In this context Fuzzy Logic (FL), Probabilistic Reasoning (PR), Neural Networks (NNs), and Genetic Algorithms (GAs) are considered as main components of SC.

Most of the SC based models can be effectively used in data mining and lend themselves to transform into other traditional data mining or advanced SC-based model structures that allow information transfer between different models. For example, in Sethi (1990) a decision tree was mapped into a feed forward neural network. A variation of this method is given in Ivanova and Kubat (1995) where the decision tree was used only for the discretization of input domains. Another example is that as radial basis functions (RBF) are functionally equivalent to fuzzy inference systems (Jang and Sun 1993), tools developed for the identification of RBFs can also be used to design fuzzy models. The KDD process also includes the interpretation of the mined patterns. This step involves the visualization of the extracted patterns/models, or visualization of the data given the extracted models. Among the wide range of SC tools (Pal 1999), the Self-Organizing Map (SOM) is the most applicable for this purpose (Kohonen 1990). The main objective of this chapter is to propose an SOM based methodology that can be effectively used for the analysis of operational process data and product quality.

Nowadays, more and more articles deal with SOM-based data analysis (Astudillo and Oommen 2014; Poggy et al. 2013; Ghosh et al. 2014) that is a new, powerful software tool for the visualization of high-dimensional data. The SOM algorithm performs a topology preserving mapping from high dimensional space onto a two dimensional grid of neurons so that the relative distances between data points are preserved (Valova et al. 2013). The net roughly approximates the probability density function of the data and, thus, serves as a clustering tool

(Kohonen 1990). It also has the capability to generalize, i.e. the network can interpolate between previously encountered inputs. Since SOM is a special clustering tool that provides compact representation of the data distribution, it has been widely applied in the visualization of high-dimensional data (Kohonen 1990). The SOM facilitates visual understanding of processes so that several variables and their interactions may be inspected simultaneously. For instance, Kassalin used SOM to monitor the state of a power transformer and to indicate when the process was entering a non-desired state represented by a "forbidden" area on the map (Kassalin et al. 1992). Tryba and Goser (1991) applied the SOM in monitoring of a distillation process and discussed its use in chemical process control in general. Alander (1991) and Harris and Kohonen (1993) have used SOM in fault detection. Since the model is trained using measurement vectors describing normal operation only, a faulty situation can be detected by monitoring the quantization error (distance between the input vector and the best matching unit (BMU)), as large error indicates that the process is out of normal operation space. SOM can also be used for prediction, where SOM is used to partition the input space of piecewise linear models. This partitioning is obtained by the Voronoi diagram of the neurons (also called codebook) of SOM. The application of Voronoi diagrams of SOM has already been suggested in the context of time series prediction (Principe et al. 1998).

Based on the aforementioned beneficial properties of SOM, a new approach for process analysis and product quality estimation is proposed in this chapter. This approach is applied in an industrial polyethylene plant, where medium and high-density polyethylene (MDPE and HDPE) grades are manufactured in a low-pressure catalytic process, a slurry polymerization technology under license from Phillips Petroleum Company. The main properties of polymer products (Melt Index (MI) and density) are controlled by the reactor temperature, monomer, comonomer and chain-transfer agent concentration. The detailed application study demonstrates that SOM is very effective in the detection of typical operating conditions related to different products, and can be used to predict the product quality (MI and density) based on measured and calculated process variables.

The chapter is organized as follows. In Sect. 9.2, multivariate techniques for process monitoring and product quality estimation are overviewed. In Sect. 9.3, case studies are presented where the proposed methodologies are applied in real-life quality development problems of chemical industry. Finally, conclusions are given in Sect. 9.4.

9.2   Multivariate Techniques for Quality Development

Measurements on process variables $z_k = \left[k_{k,1}, \ldots, z_{k,m} \right]^T$ such as temperatures, pressure, flow rates are available every second. Final product quality variables $y_k = \left[y_{k,1}, \ldots, y_{k,n} \right]^T$, such as polymer molecular weights or melt index are available in much less frequent basis. All such data should be used to extract information in any effective scheme for monitoring and diagnosis operating performance. However, all of these variables are not independent of one another. Only a few underlying events are driving a process at any time, and all of these measurements are simply different reflections of these same underlying events. When the quality properties are not correlated, it is customary to build a model that relates z_k to each $y_{k,i}$ separately: $y_{k,i} = f_i(z_k)$. This approach is satisfactory in general if the model is just being used for calibration, inferential control or prediction. For monitoring purposes; however, since quality is a multivariate property, it is important to fit all the variables from the y-space in a single model in order to obtain a single low-dimensional monitoring space. Hence, the following multivariate models are introduced to model the joint distribution of the process and quality variables $x_k = \left[x_{k,1}, \ldots, x_{k,l}\right]^T = \left[y_k^T, z_k^T\right]^T$, where $l = n + m$.

9.2.1 Principal Component Analysis and Partial Least Squares

PCA and PLS are the most common algorithms used for the analysis of multivariate processes data (Jolliffe 2008). In the literature several papers deal with the application of these methods (Kano and Nakagawa 2008; Höskuldsson 1995; Godoy et al. 2014). Analysis of chemical and spectroscopic data mostly requires the utilization of these models. Results of the related research work are mostly published in Journal of Chemometrics (Kettaneh et al. 2003; Janné et al. 2001) and Chemometrics and Intelligent Laboratory Systems (Godoy et al. 2014; Nelson et al. 2006). Multivariate statistical methods can also be used in production systems to estimate unmeasured process and product quality variables (soft sensors) and fault detection. Chen et al. (1998) demonstrated how PLS and PCA are used for on-line quality improvement in two case studies: a binary distillation column and Tennessee Eastman process (Chen et al. 1998). Kresta (1992) showed how PLS and PCA are used to increase process operating performance in case of fluidized bed reactor and extractive distillation column (Kresta et al. 1992).

Fault detection and isolation algorithms detect outliers and isolate (root) causes of faults. PLS and PCA models can evaluate the consistency of multivariate data, characterize normal operation, and generate informative symptoms of deviations (Chiang et al. 2001; Wise and Gallagher 1996; Hu et al. 1995). It should be noted that these outliers may significantly reduce model accuracy when they are involved in the identification of PLS and PCA models. Therefore, data preprocessing and cleaning are important steps of model building (Wang and Srinivasan 2009; Fujiwara et al. 2012).

PLS and PCA are similar in that they are both factor analysis methods, and they both reduce the dimensionality of the variable space. This is done by representing the data matrix (X) with a few orthogonal variables that explain most of the variance.

The main difference between PLS and PCA is that PLS can be referred to as a supervised technique that maximizes the covariance between the response (Y) and input variables (X) in as few factors as possible while PCA simply aims to maximize the covariance of X (Jolliffe 2008).

Mathematically, PCA reduces the data matrix using eigenvector decomposition of the covariance matrix of the data matrix. Essentially, the data matrix is broken down into principal components (PCs), represented by pairs of scores (t) and loadings (p) (Jolliffe 2008). The loading vectors are equivalent to the eigenvectors of the covariance matrix of X, and the corresponding eigenvalues (λ) represent the variance of each corresponding PC. Suppose X is composed of n samples on q variables. The first PC is defined as $t_1 = Xp_1$ and explains the greatest amount of variance, while the second PC is defined as $t_2 = Xp_2$ having the next greatest amount of variance, and so on. Up to q PCs can be defined, but only the first few (M) are significant in explaining the main variability of the system (Jolliffe 2008). Selection of optimal number of PCs can be accomplished in various ways.

Partial Least Squares (PLS) regression combines principal component analysis and multivariate regression. PLS captures variance and correlates X and Y (Vinzi 2010). The first latent variable (LV) $t_1 = Xw_1$ is a linear combination of the X variables that maximizes the covariance of X and Y, where w_1 is the first eigenvector (called weight vector) of the covariance matrix. The columns of X are then regressed on t_1 to give a regression vector p_1. The original X matrix is then deflated as follows: $X_2 = X - t_1 p_1^T$. X_2 is the resultant data matrix after removing the element of the original data matrix (X) that was most correlated with Y. The second LV is then computed from $X_2, t_2 = Xw_2$, where w_2 is the first weight vector of the covariance matrix of X_2 and Y. These steps are repeated until q number of LVs is computed. As with PCA, the optimum number of LVs may be chosen via cross-validation methods.

PCA and PLS are widely applied tools of quality development. PCA can be used in monitoring of groundwater (Sánchez-Martos et al. 2001), essential oil (Ochocka et al. 1992), pig meat (Karlsson 1992), and soil quality (Garcia-Ruiz et al. 2008).

PLS is essentially a regression tool and may be used to relate process variables to product quality attributes. PCA can also be used as a regression tool in that the significant PCs may be used to generate a regression model that relates process variables to product quality attributes (when PCA is used in such a way, it is referred to as principal component regression—PCR) (Vinzi 2010). Application examples can be found from biotechnology (measure fruit and vegetable or vegetable oil quality) (Nicolai et al. 2007; Pereira et al. 2008), chemical industry (predict gasoline properties (Bao and Dai 2009), prediction of crude oil quality (Abbas et al. 2012), quality improvement of batch processes (Ge 2014), food industry (food quality improvement (Steenkamp and van Trijp 1996).

PLS can also be used for the visualization of the data. We apply the algorithm developed in Ergon (2004) for the two-dimensional visualization of the PLS model.

Two components that are informative for visualization may be obtained in several ways. One example is principal components of predictions (PCP), where in the scalar response case $\widehat{y} = X\widehat{b}$ normalization is used as one component, while residuals of X not contributing to y are suggested for use as the second component (Ergon 2004).

The basic idea behind the applied mapping is illustrated in Fig. 9.2. The estimator $\widehat{b}$ is found in the space spanned by loading weight vectors in $\widehat{W} = [\widehat{w}_1, \widehat{w}_2, \ldots, \widehat{w}_A]$ i.e. it is a linear combination of these vectors. It is, however, also found in the plane defined by $\widehat{w}_1$ and a vector $\widetilde{w}_2$ orthogonal to $\widehat{w}_1$, which is a linear combination of the vectors $\widehat{w}_2, \widehat{w}_3, \ldots, \widehat{w}_A$.

The matrix $\widetilde{W} = [\widehat{w}_1, \widetilde{w}_2]$ is thus the loading weight matrix in a two component PLS solution (2PLS) giving exactly the same estimator $\widehat{b}$ as the original solution using any number of components. What matters in the original PLS model is not the matrix $\widehat{W}$ as such, but the space spanned by $\widehat{w}_1, \widehat{w}_2, \ldots, \widehat{w}_A$. In the 2PLS model, this represents the plane spanned by $\widehat{w}_1$ and $\widetilde{w}_2$ that is essential. Note that all samples in X (row vectors) in the original PLS model are projected onto the space spanned by $\widehat{w}_1, \widehat{w}_2, \ldots, \widehat{w}_A$.

Samples may thus be further projected onto the plane spanned by $\widehat{\omega}_1$ and $\widetilde{\omega}_2$, and form a single score plot containing all y-relevant information. When for some reasons, for example, $\widehat{w}_2$ is more informative than $\widehat{w}_1$, a plane through $\widehat{w}_2$ and $\widehat{b}$ may be a better alternative. It will in any case result in a 2PLS model that gives the estimator $\widehat{b}$, as will in fact all planes through $\widehat{b}$ that are at the same time subspaces of the column space of $\widehat{W}$ (Ergon 2004).

9.2.2   Self-organizing Map

Cluster analysis organizes data into groups according to similarities among them. In metric spaces, similarity is defined by means of distance based upon the length from a data vector to some prototypical object of the cluster. The prototypes are usually not known beforehand, and are sought by the clustering algorithm simultaneously with the partitioning of the data. In this chapter, the clustering of the operational data is considered. Hence, the data are the measured input and output process variables, parameters of the operating conditions, and laboratory measurements of the product quality. Each observation consists of l measured variables, grouped into an l-dimensional column vector $x_i = [x_{i,1}, \ldots, x_{i,l}]^T$. A set of N observations is denoted by X and represented as a matrix $X = [x_1, \ldots, x_N]$. In pattern recognition terminology, the columns of X are called patterns or objects, the rows are called the features or attributes, and X is called the pattern matrix. The objective of clustering is to divide the data set X into c clusters.

The SOM algorithm is a kind of clustering algorithm which a performs a topology preserving mapping from high dimensional space onto map units so that relative distances between data points are preserved. The map units, or neurons, form usually a two dimensional regular lattice. Each neuron, i, of the SOM is represented by an l-dimensional weight, or model vector $\boldsymbol{m}_i = \left[m_{i,1}, \ldots, m_{i,l}\right]^T$. These weight vectors of the SOM form a codebook and can be considered as cluster prototypes. The neurons of the map are connected to adjacent neurons by a neighbourhood relation, which dictates the topology of the map. The number of neurons determines the granularity of the mapping, which affects the accuracy and the generalization capability of the SOM.

SOM is a vector quantizer, where the weights play the role of the codebook vectors. This means that each weight vector represents a local neighbourhood of the space, also called Voronoi cell. The response of a SOM to an input $\boldsymbol{x}_k = \left[x_{k,1}, \ldots, x_{k,l}\right]^T$ is determined by the reference vector (weight) $\boldsymbol{m}_{i^0}$ which produces the best match of the input

$$i_k^0 = arg\left(min_i \|\boldsymbol{m}_i - \boldsymbol{x}_k\|\right) \tag{9.1}$$

where i_k^0 represents the index of the Best Matching Unit (BMU) of the kth input.

During the iterative training of SOM, the SOM forms an elastic net that folds onto the "cloud" formed by the data. The net tends to approximate the probability density of the data; the codebook vectors tend to drift there where the data are dense, while there are only a few codebook vectors where the data are sparse.

The training of SOM can be accomplished generally with a competitive learning rule as

$$\boldsymbol{m}_i^{(t+1)} = \boldsymbol{m}_i^{(t)} + \eta \Lambda_{i_k^0,i}\left(\boldsymbol{x}_k - \boldsymbol{m}_i^{(t)}\right) \tag{9.2}$$

where $\Lambda_{i_k^0,i}$ is a spatial neighbourhood function and η is the learning rate, and the (t) upper index denotes the iteration step. Usually, the neighbourhood function is

$$\Lambda_{i_k^0,i} = \exp\left(-\frac{\left\|r_i - r_{i_k^0}\right\|^2}{2\sigma^{2,(t)}}\right) \tag{9.3}$$

where $\left\|r_i - r_{i_k^0}\right\|$ represents the Euclidean distance in the low dimensional output space between the ith vector and the winner neuron (BMU).

There are two phases during learning. First, the algorithm should cover the full input data space and establish neighbourhood relations that preserve the input data structure. This requires competition among the majority of the weights and a large learning rate such that the weights can orient themselves to preserve local relation-ships. Hence, in the first phase relatively large initial σ^2 is used. The second phase of learning is the convergence phase where the local detail of the input space is preserved. Hence the neighbourhood function should cover just one unit and the learning rate should also be small. In order to achieve these properties, both the neighbourhood function and the learning rate should be scheduled during learning (Kohonen 1990).

SOM is increasingly applied in quality development (Pölzlbauer 2004). Thanks to the robustness of the method SOM is applied water management (Kalteh et al. 2008; Juntunen et al. 2013) for soil and sediment quality estimation (Olawoyin et al. 2013), in pulp and paper processes (Alhoniemi et al. 1999), and in biotechnology (Mele and Crowley 2008).

In addition, SOM is capable of detection of faults. Since SOM is a gradient based iterative technique, it is less sensitive if outliers are in data sets. Since this technique is a mapping, the performance of whole procedure is not influenced by outliers because they grouped or they are on the edge of the map.

When a cell contains outliers the performance of the local model may decrease. Since this cell represents the edge of the normal operating region, outliers do not influence the global modelling performance. Hence, SOM is much less sensitive to outliers than PCA or PLS. Therefore, SOM is excellent for fault detection, because cells contain outliers and data related to malfunction of the process can be easily identified (Fustes et al. 2013; Munoz and Muruzábal 1998).

9.2.2.1 SOM for Piecewise Linear Regression

The goal of this section is to develop a data-driven algorithm for the identification of a model in the form of $y_k = f(z_k)$, where z_k represents the model inputs (process variables) and y_k contains the product quality. In general, it may not be easy to find a global nonlinear model that is universally applicable to describe the relationships between the inputs and the outputs on the whole operating domain of the process. In that case, it would certainly be worthwhile to build local linear models for specific operating points of the process and combine these into a global model. This can be done by combining a number of local models, where each local model has a predefined operating region where the local model is valid. This results in the so-called operating regime based model. The applications and the possible identification of operating regime based modelling to the identification of dynamic systems are recent and rich (Murray-Smith 1997).

The operating regime based model is formulated as

$$y_k^T = \sum_{i=1}^{s} \omega_i(z_k) \left[z_k^T, 1 \right] \boldsymbol{\Theta}_i \tag{9.4}$$

where $\omega_i(z_k)$ describes the operating regime of the ith local linear model de-fined by the $\boldsymbol{\Theta}_i$ parameter matrix (or vector if y_k^T is a scalar). The piecewise linear models are special case of operating regime-based models. If we denote the input space of the model by $T : z \in T \subset \Re^m$, the piecewise linear model consists of a set of operating ranges $T_1, T_2, \ldots, T_s$ which satisfy $T_1 \cup T_2 \cup \ldots \cup T_s = T$ and $T_j \cap T_i = \varnothing; \forall i \neq j$.

Hence, the model can be formulated as

$$\text{If } z_k \in T_i \text{ then } y_k^T = \left[z_k^T, 1\right] \boldsymbol{\Theta}_i \tag{9.5}$$

where $\boldsymbol{\Theta}_i$ denotes the parameter estimate vector used in the ith local model.

The identification of these models can be divided into two tasks: structure identification that generates the operating ranges and parameter identification of the local models. As the simultaneous combination of these steps results in complex nonlinear optimization problem, several heuristic, mainly iterative algorithms have been worked out for this purpose (Murray-Smith 1997).

When SOM is used to represent nonlinear systems, it is trained based on the N input-output data pairs arranged in the $X = [x_1, \ldots, x_N]^T$ pattern matrix as $x_k = \left[y_k^T, z_k^T\right]^T$.

The SOM can be directly used for prediction of the output, y_k of the process given the input vector z_k. Regression is accomplished by searching for the BMU using the known vector components. As the output of the system is unknown, the BMU is determined as

$$i_k^0 = arg\left(min_i \left\|m_i^* - z_k\right\|\right) \tag{9.6}$$

where $m_i^* = [m_{i,n+1}, \ldots, m_{i,l}]$.

The output of the model can be defined as the unknown component of the BMU, $y_k^T = m_{i^0}^+ = \left[m_{i^0,1}, \ldots, m_{i^0,n}\right]$, which results in a piecewise constant model.

The accuracy of this model can be increased by building local models for data in the Voronoi cells of the SOM,

$$y_k^T = \left[z_k - m_{i_k^*}^*\right]^T \boldsymbol{\Theta}_{i_k^0} + \left(m_{i^0}^+\right)^T \tag{9.7}$$

or

$$y_k^T = \left[z_k^T, 1\right] \boldsymbol{\Theta}_{i_k^0} \tag{9.8}$$

where the $\boldsymbol{\Theta}_{i_k^0}$ parameter matrix of the local model is calculated by least squares method based on the local data set on the operating regime T_i only, where T_i is the ith Voronoi cell of the Voronoi diagram of the codebook of the SOM, $M = \{m_1^*, \ldots, m_c^*\}$. $Vor(M)$ is defined as the subdivision of T into c cells $T_i, i = 1, \ldots, c$, with the property that a point z_k lies in the cell corresponding to the site $m_{i_k^*}^*$ if and only if $i_k^0 = arg(min_i \|m_i - x_k\|)$. Thus, each cell of the diagram is the intersection of a number of half-planes.

When the process is nonlinear there is a need for local linear approximation of the operating regime of the system. Sliced Inverse Regression (SIR) and the related techniques are suitable for the extraction and characterization of local linear subspaces (Li 2012; Lue 2009; Kuentz and Saracco 2010). In this context these techniques are similar to SOM as SOM also defines local operating regimes that can be also considered Voronoi cells of SOM. As this section showed, these clusters can be used to build local linear models. In our case least squares regression is used to build local models. However, local models of the clusters can also be defined by local PCA (similar to SIR) or sub-PLS models. This approach also illustrates that local linear modelling and clustering can be effectively combined to get accurate and interpretable models (Kenesei and Abonyi 2013; Abonyi et al. 2002).

9.2.2.2 SOM for Classification of Product Grades and Operating Conditions

The SOM can be used for classification purposes by assigning a class for each codebook vector and deciding the class of a sample vector based on the class of its BMU. The rule-based classifier consists of rules that describe N_c number of classes, given n data points. The rule antecedent defines the operating region of the rule in the l-dimensional feature space and the rule consequent is a class label from the set $g_i = \{1, \ldots, N_c\}$.

$$\text{If } x_k \in T_i \text{ then } class \text{ is } g \quad (9.9)$$

The interpretability of classifier depends on the number of utilized features. For the selection of the most relevant features, we modify the Fischer interclass separability method which is based on statistical properties of labeled data. The importance of a feature is measured by leaving out a feature and calculating a cost function for the reduced model. The feature selection is made iteratively by leaving out the less needed feature.

9.3 Application Examples

In this chapter two examples are given to demonstrate the applicability of multivariate data-driven tools. In the first example, PLS is applied to visualize the production of a fuel mixing process and estimate the product quality. The second example is similar in the application point of view, SOM is applied to monitor product quality of a polymerization process.

9.3.1 Online NIR—PLS Example

Present research focuses on two tasks. Datasets collected at the Dune Refinery of MOL Ltd (Hungary) are analyzed. The first task is the development of a prediction model that can estimate product properties based on spectra taken by online NIR

Table 9.1 Effect of the number of latent variables to the performance of the model (correlation between the estimated and measured variables are shown)

Property	Latent dimensions					
	2	6	12	18	24	48
Density	0.776	0.988	0.993	0.993	0.993	0.989
T90	0.432	0.654	0.849	0.895	0.868	0.796
CFPP0	0.657	0.942	0.947	0.953	0.921	0.888
CFPP	0.516	0.755	0.769	0.728	0.703	0.610
Cloud Pt	0.668	0.924	0.950	0.958	0.955	0.943
Flash Pt	0.408	0.596	0.878	0.901	0.895	0.854

analyzers. The second task is the development a monitoring tool based on the visualization of the same spectra.

The prediction performance of the models is measured by the correlation coefficient defined as:

$$R(i,j) = \frac{C(i.j)}{\sqrt{C(i,i)C(j,j)}} \tag{9.10}$$

where C is the covariance matrix and it is calculated as $C = \mathrm{cov}(y, \hat{y})$. Table 9.1 shows that the number of the available samples, N, differs for each properties. Among the 651 spectra, only 560 were different and in most of the cases, only a fragment of the properties were measured. Firstly the effect of dimensionality of latent space of the PLS model was analyzed (from 2 to 48 dimensions). To perform an adequate comparison, leave-one-out and 10-fold cross validation technique was applied. As it is shown in this table, the accuracy of the model increases rapidly by increasing the dimensionality of the latent space from 2 to 6 dimensions; however, it reaches a maximum since when the complexity of the model is higher than the complexity of the modelled system.

In Sect. 9.2.2, a special method is presented that can map the PLS latent space into two dimensional space by orthogonal signal correction. This method is compared with Principal Component Analysis and Topological Near-Infrared Modeling (CRISP-DM 2000; Abonyi and Feil 2005) (TOPNIR) developed specifically to visualize NIRspectra and building topological prediction models with the help of resulted maps. As shown in Fig. 9.4, this technique is effective in visualization of high dimensional spectral space. This plot gives information about how summer and winter fuel samples are clustered.

9.3.2 Application in Polyethylene Production

To illustrate the proposed approach, the monitoring of a medium and high-density polyethylene (MDPE, HDPE) plant of the TVK Ltd. in Hungary is considered. HDPE is versatile plastic used for household goods, packaging, car parts and pipe,

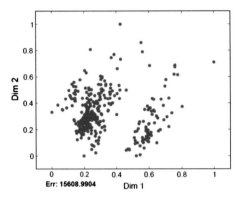

Fig. 9.4 Visualization of DS1 using PLS (CFPP0)

and TVK Ltd. is the largest Hungarian polymer production company (www.tvk.hu). A brief explanation of the Phillips license based low-pressure catalytic process is provided in the following section.

Figure 9.5 represents the Phillips Petroleum Co. suspension ethylene polymerization process. The polymer particles are suspended in an inert hydrocarbon. The melting point of high-density polyethylene is approximately 135 °C. Therefore, slurry polymerization takes place at a temperature below 135 °C; the polymer formed is in the solid state. The Phillips process takes place at a temperature between 85 and 110 °C. The catalyst and the inert solvent are introduced into the loop reactor where ethylene and an olefin (hexene) are circulating. The inert solvent

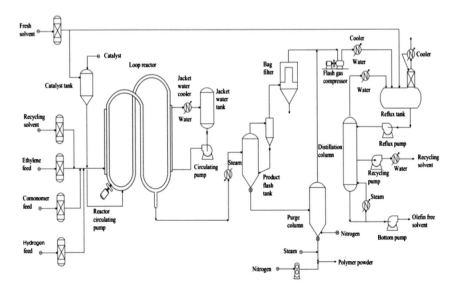

Fig. 9.5 Scheme of the Phillips loop reactor process (Nagy 1997)

(isobuthane) is used to dissipate heat as the reaction is highly exothermic. A cooling jacket is also used to dissipate heat. The reactor consists of a folded loop containing four long runs of pipe that are 1 m in diameter, connected by short horizontal lengths of 5 m. The slurry of HDPE and catalyst particles circulates through the loop at a velocity between 5 and 12 m/s. The reason for the high velocity is due to the fact that at lower velocities, the slurry will deposit on the walls of the reactor causing fouling. The concentration of polymer products in the slurry is 25–40 % by weight. Ethylene, olefin comonomer (if used), an inert solvent, and catalyst components are continuously charged into the reactor at a total pressure of 450 psig. The polymer is concentrated in settling legs to about 60–70 % by weight slurry and continuously removed. The solvent is recovered by hot flashing. The polymer is dried and pelletized. The conversion of ethylene to polyethylene is very high (95–98 %), eliminating ethylene recovery. The molecular weight of high-density polyethylene is controlled by the temperature of catalyst preparation (Nagy 1997). The main properties of polymer products (Melt Index (MI) and density) are controlled by the reactor temperature, monomer, comonomer and chain-transfer agent concentration.

9.3.2.1 Problem Description

An interesting problem with the process is that it is required to produce about ten product grades according to market demand. Hence, there is a clear need to minimize the time of changeover because off-specification product may be produced during transition. The difficulty of the problem comes from the fact that there are more than ten process variables to consider. Measurements are available in every 15 s on process variables z_k, which are the $z_{k,1}$ reactor temperature (T), $z_{k,2}$ ethylene concentration in the loop reactor $(C2)$, $z_{k,3}$ hexene concentration $(C6)$, $z_{k,4}$ the ratio of the hexene and ethylene inlet flowrate $(C6/C2in)$, $z_{k,5}$ the flowrate of the isobutane solvent $(C4)$, $z_{k,6}$ the hydrogen concentration (H_2), $z_{k,7}$ the density of the slurry in the reactor (roz), $z_{k,8}$ polymer production intensity (PE), and $z_{k,9}$ the flowrate of the catalyzer (KAT). The product quality y_k is only determined later, in another process. The interval between the product samples is between half an hour and 5 h. The $y_{k,1}$ melt index (MI) and the $y_{k,2}$ density of the polymer (ro) are monitored by off-line laboratory analysis after drying and extrusion of the polymer that causes 1 h time-delay.

Since, it would be useful to know if the product is good before testing it, the monitoring of the process would help in the early detection of poor-quality product. There are other reasons why monitoring the process is advantageous. Only a few properties of the product are measured and sometimes these are not sufficient to entirely define the product quality. For example, if only rheological properties of polymer are measured (melt index), any variation in end-use application that arise due to variation of chemical structure (branching, composition, etc.) will not be captured by following only these product properties. In these cases, the process data may contain more information about events with special causes that may affect the product quality (Jeackle and MacGregor 1998).

9.3.2.2 SOM Based Historical Analysis of the Process

The modelling and monitoring of processes from data involves solving the problem of data gathering, preprocessing, model architecture selection, identification or adaptation and model validation. The process data analyzed in this chapter have been collected over 3 months of operation. The data have been extracted from the distributed control system (DCS) of the process. An SQL server has been installed to store and merge this data with the product quality database. According to the data warehousing methodology, the application relevant data have been extracted from this SQL database. As one of the objectives is to infer the values of product quality from process data obtained at different operating regions, a set of transition-free data is used that covers the whole range of specifications of the quality properties and the process variables over all the possible operating regions. The data were preprocessed by normalization performed on single variables. Scaling of variables is of special importance since the SOM algorithm uses Euclidean metric. In the current phase of the project, this data are processed by the modified version of the MATLAB SOM Toolbox (Vesanto et al. 2015). The whole methodology is illustrated in Fig. 9.6.

The SOM of the process has been applied to predict polymer properties from measured process variables and to interpret the behaviour of the process. The

Fig. 9.6 Scheme of the SOM based process analysis approach

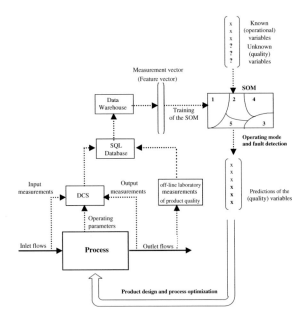

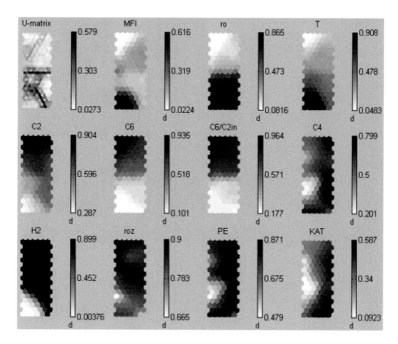

Fig. 9.7 Component planes of the polyethylene production map

constructed SOM (size 17 by 6 units) with eleven component planes is shown in Fig. 9.7. Based on the map the typical operating regions related to different product grades could be determined. Furthermore, the SOM is a good tool for hunting for correlation among the operating variables (Simula et al. 1999). For example, it can be easily seen that the melt index of the polymer (MFI) is highly correlated with reactor temperature (T).

Figure 9.8 shows the labels of the products and the distribution of the data marked by black hexagons with proportional size to the number of data in the operating regions of the clusters. This figure shows that the SOM is a useful tool for the visualization of multivariate data. A common procedure for reducing the dimensionality of the variable space is Principal Component Analysis (PCA) (MacGregor and Kourti 1995). For a comparison of the SOM with "standard" techniques, the historical data have been transformed into a two-dimensional space spanned by the first two principal components of the data. In Fig. 9.9, the grid of the transformed codebook of the SOM is shown to illustrate how the clusters approximate the density of the data. It is interesting to compare the SOM and the PCA model of the process (Fig. 9.11) as in both transformed spaces the regions of the different products are similar; the data points appear to cluster into four regions which corresponded to different product grades and operating conditions.

Since the distance preserving mapping property of the SOM, products that are close to each other on the map are similar. In the discrete two-dimensional output space of the SOM, the trajectory of the production can be effectively visualized by

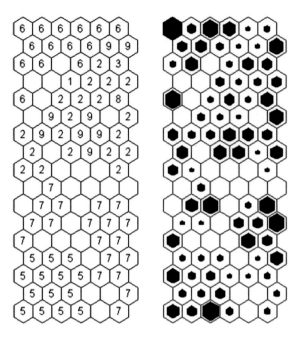

Fig. 9.8 Product labels (numbers) and distribution of the data

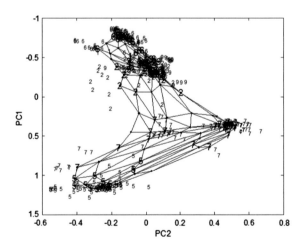

Fig. 9.9 PCA scores plots for three months of operation. The grid of the transformed codebook of the SOM is also shown to illustrate how the clusters approximate the density of the data

plotting the trajectory of the BMUs (Principe et al. 1998), which is especially useful in process monitoring and fault detection. Hence, the map can be effectively used for scheduling the different products by designing the trajectory of the production on the map of the products. An example for a grade transition is shown in Fig. 9.10, where the production of Product 6 is followed by the production of Product 7. The multivariate historical data of this transition depicted in Fig. 9.10 can be easily visualized by the PCA and the SOM model of the process as shown in Fig. 9.11.

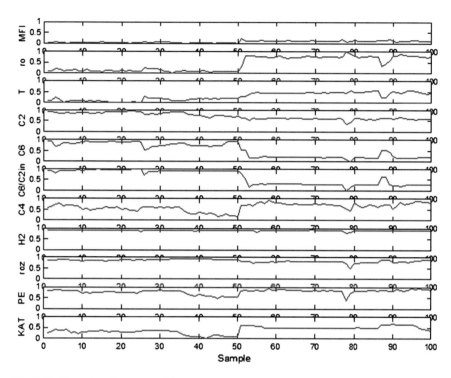

Fig. 9.10 Example of grade transition

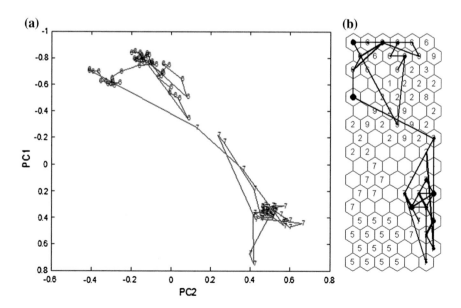

Fig. 9.11 Grade transition shown in Fig. 9.10 mapped into the two dimensional space by PCA
(**a**) and SOM (**b**)

The previous example has shown that the SOM results in a good representation of the operating regions of the products. Hence, it can be also used for classification. When the whole SOM is used as a rule-based classifier system with rules like

$$\text{If } x_k \in T_i \text{ then } class \text{ is } g_i \tag{9.11}$$

it gives 8 % classification error. This can be considered as a good result taking into account that the data is quite noisy and not too much effort was put to select the training data related to normal operating conditions.

9.3.2.3   SOM Based Product Quality Estimation

The SOM has been also used to estimate the product quality variables. Figure 9.12 shows the estimation error of the linear model and the SOM presented in the previous section. Although in this case the SOM is used as a piecewise constant model, it gives better results than the linear model. The performances of the models have been measured by the Root Mean Square Error (RMSE) of the models. In Table 9.2, it can be seen that the SOM is more accurate than the linear model. This is not surprising since the linear model has only two times ten (20) parameters and cannot capture the nonlinearity of the process. The good performance of the BMU-based piecewise constant model shows that the SOM gives a good approximation of the density of the data, hence it can be considered as a good

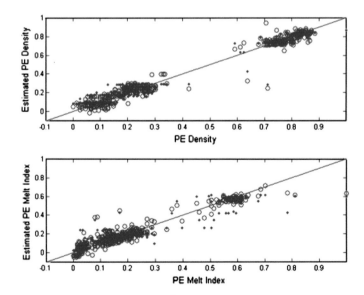

Fig. 9.12 Estimation performance of the SOM based piecewise constant model (.) and a multivariate linear model (o)

9 Multivariate Statistical and Computational Intelligence Techniques ...   259

Table 9.2 Root Mean Square Errors (RMSE) achieved by different models

	Density (ro)	Melt index (MFI)
Linear	0.0355	0.0372
SOM pw constant (BMU)	0.0330	0.0341
Linear (4 variables)	0.0368	0.0387
SOM pw linear (4 variables)	0.0251	0.0312

Table 9.3 Relative importance of process variables obtained by OLS

Importance	Density (ro)	Melt index (MFI)
1	'C6/C2in'	'T'
2	'C4'	'C6/C2in'
3	'H2'	'C4'
4	'roz'	'H2'
5	'PE'	'C2'
6	'KAT'	'roz'
7	'C6'	'KAT'
8	'C2'	'C6'
9	'T'	'PE'

nonparametric model. Because of the large number of the clusters (size 17 by 6 units), the Voronoi cell based multiple linear model approach cannot be used. The reason is that the identification of the most local models become badly conditioned due to the small number of data related to the operating region of the models. Hence, in this chapter an approach based on the reduction of the original SOM is introduced, where for the regression purpose a smaller SOM is identified.

This task requires the selection of the most important variables having effect to the product quality variables. This can be done by analyzing SOM of the process (Fig. 9.7) to detect similarities between the component planes that shows the correlation of the variables. Another possible approach is to use Orthogonal Least Squares (OLS) techniques for ordering of the process variables. The ordering obtained is shown in Table 9.3. This ordering can then be easily used to select a subset of the inputs in a forward-regression manner. It is interesting to see that ordering of the OLS gives similar results that we can obtain from the visual inspection of the SOM of the process. Based on this model reduction approach, two independent SOMs with 24 clusters on four process variables (the first four shown in Table 9.2.) have been identified to estimate the density and the melt index of the product. As shown in Table 9.3, these compact models give good estimations of the quality variables.

9.4 Conclusion

Quality development intensively applies process data. The iterative data mining methodology effectively supports the PDCA cycle based quality development. Since several process variables have to be monitored and unknown functional relationships among process and quality variables have to be explored, multivariate statistical techniques are the most widely applied tools. Beside the classical principal component analysis (PCA) and partial least squares regression models (PLS), we applied soft computing based tools to handle uncertainty and nonlinearity and complexity of the problem.

We demonstrate that PLS based model is able to simultaneously predict unmeasured material properties and monitor the state of a complex production process. Process monitoring is realized in orthogonal two dimensional plots. These plots can also be used for the effective identification of outliers.

Self-Organizing Map (SOM) is a soft-computing based approach and it is used for the extraction of knowledge from the historical data of production. Since SOM provides a compact representation of the data distribution, the typical operating conditions of the process are efficiently detected. It has been shown that efficient process monitoring can be performed in the two-dimensional projection of the process variables. For the estimation of the product quality variables multiple local linear models are introduced, where the operating regimes of the local linear models are obtained by the Voronoi diagram of the prototype vectors of the SOM. The important process variables having effect to the product quality have been selected by orthogonal least squares method. The approach has been demonstrated by means of the analysis of a polyethylene production plant. The results show that the SOM is very effective in the detection of typical operating conditions related to different product grades and can be used to predict the product quality (melt index and density) based on the process variables measured. The proposed method is attractive in comparison with other advanced process monitoring schemes such as Principal Component Analysis.

The interested reader might want to know under what conditions these methods can be employed and what kind of diagnostic tests are available. Some books that are dealing with only one technique in detail are suggested for them: Handbook of Partial Least Squares (Vinzi et al. 2010), Principal Component Analysis (Jolliffe 2008), Introduction to Statistical Quality Control (Montgomery 2009) and Self-Organizing Maps (Kohonen 2001).

More technical details and illustrative examples and MATLAB program codes related to the application of intelligent tools for fault detection and quality estimation can be found at the website of the authors: www.abonyilab.com.

Acknowledgments The work was supported by the frames of TÁMOP-4.2.2.C-11/1/KONV-2012-0004—National Research Center for Development and Market Introduction of Advanced Information and Communication Technologies and TÁMOP 4.2.4. A/2-11- 1-2012-0001 "National Excellence Program—Elaborating and operating an inland student and researcher personal support system". These projects were subsidized by the European Union and co-financed by the European Social Fund.

References

Abbas, O., et al.: PLS regression on spectroscopic data for the prediction of crude oil quality: API gravity and aliphatic/aromatic ratio. Fuel **98**, 5–14 (2012)

Abonyi, J., Feil, B.: Computational intelligence in data mining. Informatica **29**, 3–12 (2005)

Abonyi, J., Babuska, R., Szeifert, F.: Modified Gath-Geva fuzzy clustering for identification of Takagi-Sugeno fuzzy models. IEEE Trans. Syst. Man Cybern. B Cybern. **32**(5), 612–621 (2002)

Alander, J.T., et al.: Process error detection using self-organizing feature maps, Artif. Neural Netw. **2**, 1229–1232 (1991)

Alhoniemi, E., et al.: Process monitoring and modeling using the self-organizing map. Integr. Comput. Aided Eng. **6**, 3–14 (1999)

Astudillo, C.A., Oommen, B.J.: Self-organizing maps whose topologies can be learned with adaptive binary search trees using conditional rotations. Pattern Recogn. **47**, 96–113 (2014)

Bao, X., Dai, L.: Partial least squares with outlier detection in spectral analysis: a tool to predict gasoline properties. Fuel **88**(7), 1216–1222 (2009)

Borosy, A.P.: Quantitative composition-property modelling of rubber mixtures by utilizing artificial neural networks. Chemom. Intell. Lab. Syst. **47**, 227–238 (1998)

Chen, G., McAvoy, T.J., Piovoso, M.J.: A multivariate statistical controller for on-line quality improvement. J. Process Control **8**(2), 139–149 (1998)

Chiang, L.H., Russel, E.L., Braatz, R.D.: Fault Detection and Diagnosis in Industrial Systems. Springer, London (2001)

CRISP-DM Cross Industry Standard Process for Data Mining (2000). http://en.wikipedia.org/wiki/Cross_Industry_Standard_Process_for_Data_Mining

Ergon, R.: Informative PLS score-loading plots for process understanding and monitoring. J. Process Control **14**, 889–897 (2004)

Fujiwara, K., Sawada, H., Kano, M.: Input variable selection for PLS modeling using nearest correlation spectral clustering. Chemometr. Intell. Lab. Syst. **118**, 109–119 (2012)

Fustes, D., et al.: SOM ensemble for unsupervised outlier analysis. Application to outlier identification in the Gaia astronomical survey. Expert Syst. Appl. **40**(5), 1530–1541 (2013)

Garcia-Ruiz, R., et al.: Suitability of enzyme activities for the monitoring of soil quality improvement in organic agricultural systems. Soil Biol. Biochem. **40**(9), 2137–2145 (2008)

Ge, Z.: Two-level PLS model for quality prediction of multiphase batch processes. Chemometr. Intell. Lab. Syst. **130**, 29–36 (2014)

Ghosh, S., Roy, M., Ghosh, A.: Semi-supervised change detection using modified self-organizing feature map neural network. Appl. Soft Comput. **15**, 1–20 (2014)

Godoy, J.L., Vega, J.R., Marchetti, J.L.: Relationships between PCA and PLS-regression. Chemom. Intell. Lab. Syst. **130**, 182–191 (2014)

Harris, T., Kohonen, A.: S.O.M. based, machine health monitoring systems which enables diagnosis of faults not seen in the training set. Proc. Int. Conf. Neural Netw. (IJCNN'93) Nagoya, Japan **1**, 947–950 (1993)

Höskuldsson, A.: A combined theory for PCA and PLS. J. Chemom. **9**(2), 91–123 (1995). doi:10.1002/cem.1180090203

Hu, W., Storer, R., Georgakis, C.: Disturbance detection and isolation by dynamic principal component analysis. Chemometr. Intell. Lab. Syst. **30**(1), 179–196 (1995)

Ivanova, I., Kubat, M.: Initialization of neural networks by means of decision trees. Knowl. Based Syst. **8**, 333–344 (1995)

Jang, J.-S.R., Sun, C.T.: Functional equivalence between radial basis function networks and fuzzy inference systems. IEEE Trans. Neural Netw. **4**, 156–159 (1993)

Janné, K., et al.: Hierarchical principal component analysis (PCA) and projection to latent structure (PLS) technique on spectroscopic data as a data pretreatment for calibration. J. Chemom. **15**(4), 203–213 (2001). doi:10.1002/cem.677

Jeackle, C., MacGregor, J.: Product design through multivariate statistical analysis of process data. Am. Inst. Chem. Eng. J. **44**(5), 1105–1118 (1998)

Jolliffe, I.T.: Principal Component Analysis, 2nd edn. Springer Series in Statistics (2008)

Juntunen, P., et al.: Cluster analysis by self-organizing maps: an application to the modelling of water quality in a treatment process. Appl. Soft Comput. **13**(7), 3191–3196 (2013)

Kalteh, A.M., Hjorth, P., Berndtsson, R.: Review of the self-organizing map (SOM) approach in water resources: analysis, modelling and application. Environ. Model Softw. **23**(7), 835–845 (2008)

Kano, M., Nakagawa, Y.: Data-based process monitoring, process control, and quality improvement: recent developments and applications in steel industry. Comput. Chem. Eng. **32**(1–2), 12–24 (2008)

Karlsson, A.: The use of principal component analysis (PCA) for evaluating results from pig meat quality measurements. Meat Sci. **31**(4), 423–433 (1992)

Kassalin, M., Kangas, J., Simula, O.: Process state monitoring using self-organizing maps. Artif. Neural Netw. **2**, 1531–1534 (1992)

Kenesei, T., Abonyi, J.: Hinging hyperplane based regression tree identified by fuzzy clustering and its application. Appl. Soft Comput. J. **13**(2), 782–792 (2013)

Kettaneh, N., Berglund, A., Wold, S.: PCA and PLS with very large data sets. Comput. Stat. Data Anal. **48**, 69–85 (2003)

Kohonen, T.: The self-organizing map. Proc. IEEE **78**(9), 1464–1480 (1990)

Kohonen, T.: Self-Organizing Maps, 3rd edn. Springer Series in Information Sciences (2001)

Kresta, J.V.: The application of partial least squares to problems in chemical engineering, PhD Theis, McMaster University (1992).http://hdl.handle.net/11375/8576

Kresta, J.V., Macgregor, F.F., Marlin, T.E.: Multivariate statistical monitoring of process operating performance. Can. J. Chem. Eng. **69**(1), 35–47 (1992). doi:10.1002/cjce.5450690105

Kuentz, V., Saracco, J.: Cluster-based sliced inverse regression. J. Korean Stat. Soc. **39**, 251–267 (2010)

Lakshminarayanan, S., et al.: New product design via analysis of historical databases. Comput. Chem. Eng. **24**, 671–676 (2000)

Li, K.-C.: Sliced inverse regression for dimension reduction (2012). www.jstor.org/stable/2290563 . Accessed 19 Dec 2013

Lue, H.-H.: Sliced inverse regression for multivariate response regression. J. Stat. Plan. Inference **139**, 2656–2664 (2009)

MacGregor, J.F., Kourti, T.: Statistical process control of multivariate processes. Control Eng. Pract. **3**(3), 403–414 (1995)

Martin, E.B., et al.: Batch process monitoring for consistent production. Comput. Chem. Eng. **20**, S599–S605 (1996)

Mele, P.M., Crowley, D.E.: Application of self-organizing maps for assessing soil biological quality. Agric. Ecosyst. Environ. **126**(3–4), 139–152 (2008)

Montgomery D.C.: Introduction to Statistical Quality Control, John Wiley, New York (2009)

Moteki, Y., Arai, Y.: Operation planning and quality design of a polymer process. In: IFAC DYCORD, pp. 159–165 (1986)

Munoz, A., Muruzábal, J.: Self-organizing maps for outlier detection. Neurocomputing **18**, 33–60 (1998)

Murray-Smith, R., Johansen, T.A.: Multiple Model Approaches to Nonlinear Modeling and Control. Taylor & Francis, London (1997)

Nagy, G.: The polyethylene, Magyar Kémikusok Lapja (MKL). Hungary **52**(5), 233–242 (1997)

Nelson, P.R.C., MacGregor, J.F., Taylor, P.A.: The impact of missing measurements on PCA and PLS prediction and monitoring applications. Chemometr. Intell. Lab. Syst. **80**(1), 1–12 (2006)

Nicolai, B.M., et al.: Nondestructive measurement of fruit and vegetable quality by means of NIR spectroscopy: a review. Postharvest Biol. Technol. **46**(2), 99–118 (2007)

Ochocka, R.J., Wesolowski, M., Lamparczyk, H.: Thermoanalysis supported by principal component analysis (PCA) in quality assessment of essential oil samples. Thermochim. Acta **210**, 151–162 (1992)

Olawoyin, R., et al.: Application of artificial neural network (ANN)–self-organizing map (SOM) for the categorization of water, soil and sediment quality in petrochemical regions. Expert Syst. Appl. **40**(9), 3634–3648 (2013)

Pal, N.R.: Soft computing for feature analysis. Fuzzy Sets Syst. **103**, 201–221 (1999)

Pereira, A.F.C., et al.: NIR spectrometric determination of quality parameters in vegetable oils using iPLS and variable selection. Food Res. Int. **41**(4), 341–348 (2008)

Poggy, G., Cozzolino, D., Verdoliva, L.: Self-organizing maps for the design of multiple description vector quantizers. Neurocomputing **122**, 298–309 (2013)

Pölzlbauer, G.: Survey and comparison of quality measures for self-organizing maps. In: Fifth Workshop on Data Analysis (WDA) (2004). www.ifs.tuwien.ac.at/~poelzlbauer/publications/Poe04WDA.pdf. Accessed 17 Dec 2013

Principe, J.C., Wang, L., Motter, M.A.: Local dynamic modeling with self-organizing maps and applications to nonlinear system identification and control. Proc. IEEE **86**(11), 2241–2258 (1998)

Sánchez-Martos, F., Jiménez-Espinosa, R., Pulido-Bosch, A.: Mapping groundwater quality variables using PCA and geostatistics: a case study of Bajo Andarax, southeastern Spain. Hydrol. Sci. J.-des Sciences Hydrologiques **46**(2), 227–242 (2001)

Sethi, L.K.: Entropy nets: from decision trees to neural networks. Proc. IEEE **78**, 1605–1613 (1990)

Simula, O., et al.: Analysis and modeling of complex systems using the self-organizing map. In Neuro-Fuzzy Techniques for Intelligent Information Systems, pp. 3–22. Springer, New York (1999)

Steenkamp, J.B.E.M., van Trijp, H.C.M.: Quality guidance: a consumer-based approach to food quality improvement using partial least squares. Eur. Rev. Agric. Econ. **23**(2), 195–215 (1996). doi:10.1093/erae/23.2.195

Tryba, V., Goser, K.: Self-organizing feature maps for process control in chemistry. Artif. Neural Netw. 847–852 (1991)

Valova, I., et al.: Initialization Issues in Self-organizing Maps. Procedia Comput. Sci. **20**, 52–57 (2013)

Vinzi, V., et al.: Handbook of Partial Least Squares. In: Springer Handbooks of Computational Statistics (2010)

Wang, D., Srinivasan, R.: Eliminating the effect of multivariate outliers in pls-based models for inferring process quality. Comput. Aided Chem. Eng. **26**, 755–760 (2009)

Wang X.Z.: Data Mining and Knowledge Discovery for Process Monitoring and Control. Springer, New York (1999)

Wise, B.M., Gallagher, N.B.: The process chemometrics approach to process monitoring and fault detection. Journal of Process Control **6**(6), 329–348 (1996). doi:http://dx.doi.org/10.1016/0959-1524(96)00009-1

Vesanto, J., Himberg, J., Alhoniemi, E., Parhankangas, J.: SOM Tooolbox for MATLAB (2015). The Toolbox can be downloaded for free from http://www.cis.hut.fi/projects/somtoolbox

Yamashita, Y.: Supervised learning for the analysis of the process operational data. Comput. Chem. Eng. **24**, 471–474 (2000)

Zadeh, L.A.: Soft computing and fuzzy logic. Software, IEEE **11**(6), 48–56 (1994)

Zhang, J., Martin, E.B., Morris, A.J.: Process monitoring using non-linear statistical techniques. Chem. Eng. J. **67**, 181–189 (1997)

Chapter 10
Failure Mode and Effects Analysis Under Uncertainty: A Literature Review and Tutorial

Umut Asan and Ayberk Soyer

Abstract The multidimensional nature of risks as well as substantial uncertainties and subjectivities inherent in the risk assessment process led a growing number of researchers to develop alternative approaches for failure mode and effects analysis. The purpose of this chapter is to provide a comprehensive review of the multi-criteria approaches proposed for failure mode and effects analysis under uncertainty and offer a brief tutorial for those who are interested in these approaches.

Keywords Failure modes and effects analysis (FMEA) · Risk assessment · Multi-Criteria decision making (MCDM) approaches · Uncertainty

10.1 Introduction

Failure Modes and Effects Analysis (FMEA) is one of the first structured, systematic and proactive techniques used for failure analysis. The purpose of FMEA is to list out all possible failure modes (FMs) (i.e., the things that could go wrong in an organization); evaluate the causes of each FM and their subsequent effects on the performance of the system that is under consideration. By definition, FM refers to the termination of the ability of a system to perform a required function or its inability to perform within previously specified limits (ISO/IEC-15026-1 2013) and includes both known and/or potential failures, problems, or errors that may affect the customers and thus endanger the reputation of the entire organization. Since FMs are unavoidable for the majority of the systems, FMEA serves as an effective tool to ensure that potential threats to the system have been considered and addressed, and associated risks are minimized. The history of FMEA goes back to

U. Asan (✉) · A. Soyer
Industrial Engineering Department, Istanbul Technical University,
34357 Maçka/Istanbul, Turkey
e-mail: asanu@itu.edu.tr

© Springer International Publishing Switzerland 2016
C. Kahraman and S. Yanık (eds.), *Intelligent Decision Making
in Quality Management*, Intelligent Systems Reference Library 97,
DOI 10.1007/978-3-319-24499-0_10

the early 1950s and 1960s. In 1949, it was first used in the United States military as a reliability evaluation technique to determine the effect of system and equipment failures. In 1963, National Aeronautics and Space Administration (NASA) used FMEA during the Apollo missions to assure desired reliability of space systems (Chang et al. 1999). Later in 1974, the US Navy developed MIL-STD-1629, which discussed the proper use of the technique. In the late 1970s, Ford Motor Company introduced FMEA to the automotive industry and then the automotive industry collectively developed various standards in the 1990s. Over the years, FMEA became a universally used technique in many different industries, such as, aerospace, automotive, defense, medical, marine, nuclear power, semiconductor, etc., and it has been proven to be successful in any manufacturing or service industry (Chang et al. 1999, 2001; Chen 2007; Welborn 2010; Arabian-Hoseynabadi et al. 2010).

Commonly, there are four types of FMEA: (i) System FMEA, (ii) Design FMEA, (iii) Process FMEA, and (iv) Service FMEA. The general properties of FMEA types are shortly summarized below (Stamatis 2003; Carlson 2012):

(i) *System FMEA* (sometimes referred as *Concept FMEA*) is used to analyze systems and subsystems in the early concept and design stage. It focuses on potential FMs between the functions of the system caused by system-related deficiencies (such as, system integration, interaction between systems and/or subsystems, interaction with the external environment, etc. causing the system not to work as intended).
(ii) *Design FMEA* (sometimes referred as *Product FMEA*) is used to analyze products, early in the design phase to be able to identify potential design flaws. Therefore, it focuses on FMs caused by design-related deficiencies to improve the design and to ensure safety and reliability of the relevant product during its lifetime.
(iii) *Process FMEA* is used to analyze processes required to produce a product or service. It focuses on potential FMs caused by process-related or assembly-related deficiencies.
(iv) *Service FMEA* is used to analyze services before they reach to customer, and focuses on potential FMs caused by system-related or process-related deficiencies to maximize customer satisfaction.

Today, FMEA is one of the most widely utilized and powerful techniques, having several advantages for organizations that are trying to find the ways of improving quality and safety. Some of the major advantages of FMEA indicated in the literature include:

• Inclusion of people from different expertise areas in an organization, as each of them views the system from various perspectives, responsibilities, and concerns. By this means, it provides an opportunity to improve the communication and cooperation between the different functions of an organization, and the relationships with external factors, such as suppliers and customers (Kostina et al. 2012).

- It provides a simple analysis procedure which is easy to learn and implement, and makes even the evaluation of complex systems easy to do (Dhillon 2009; Mozaffari et al. 2013).
- It acts as a useful visibility tool for managers (Dhillon 2009; Braglia 2000) and serves as an excellent instrument for learning.
- It is a very structured, systematized and reliable method (Dhillon 2009; Mozaffari et al. 2013; Kostina et al. 2012) that helps to identify the connections between the FMs, the reasons of FMs, and the effects of FMs, as well.
- It permits a realistic appreciation of the conformity of products and services with the market and customer needs (Kerekes and Johanyák 1996); therefore, increases the safety and reliability of products/services, reduces warranty and service costs, shortens the development process, improves compliance with the deadlines (Bujna and Prístavka 2013), and eventually improves the customer satisfaction (Dhillon 2009).

As mentioned above, FMs (whether known or potential) are listed and prioritized in FMEA to prepare for them in the best way possible and to prevent problems from reaching the customer. To this end, FMEA uses Risk Priority Number (RPN) methodology to analyze the risks associated with each identified FM. This methodology consists of assessing the FMs with respect to their 'severity (S)', 'probability of occurrence (O)', and 'likelihood of detection (D)'. For each FM, an estimate is made of its S, O, and D on a numerical scale of 1 to 10, as described in Tables 10.1, 10.2, and 10.3 (Chin et al. 2009a, b; Pillay and Wang 2003; Seyed-Hosseini et al. 2006; Wang et al. 2009; Guimarães and Lapa 2007; Xu et al. 2002; Franceschini and Galetto 2001; Liu et al. 2013b; Chang and Cheng 2010). The S, O, and D ratings are then multiplied together to get the RPN. In equation form, $RPN = S \times O \times D$. The FMs with higher RPNs are assumed to be more important and should be given higher priorities (Wang et al. 2009; Liu et al. 2011).

Table 10.1 Severity scale for a FM

Rating	Effect	Severity of effect
10	Hazardous without warning	Very high severity ranking when a potential FM effects safe system operation without warning
9	Hazardous with warning	Very high severity ranking when a potential FM effects safe system operation with warning
8	Very high	System inoperable with destructive failure without compromising safety
7	High	System inoperable with equipment damage
6	Moderate	System inoperable with minor damage
5	Low	System inoperable without damage
4	Very low	System operable with significant degradation of performance
3	Minor	System operable with some degradation of performance
2	Very minor	System operable with minimal interference
1	None	No effect

Table 10.2 Probability of occurrence scale for a FM

Rating	Probability of occurrence	Failure probability
10	Extremely high: failure is almost inevitable	>1 in 2
9	Very high	1 in 3
8	Repeated failures	1 in 8
7	High	1 in 20
6	Moderately high	1 in 80
5	Moderate	1 in 400
4	Relatively low	1 in 2000
3	Low	1 in 15,000
2	Remote	1 in 150,000
1	Nearly impossible	<1 in 1,500,000

Table 10.3 Likelihood of detection scale for a FM

Rating	Detection	Likelihood of detection
10	Absolute uncertainty	Potential cause/mechanism and subsequent FM cannot be detected
9	Very remote	Very remote chance of detecting potential cause/mechanism and subsequent FM
8	Remote	Remote chance of detecting potential cause/mechanism and subsequent FM
7	Very low	Very low chance of detecting potential cause/mechanism and subsequent FM
6	Low	Low chance of detecting potential cause/mechanism and subsequent FM
5	Moderate	Moderate chance of detecting potential cause/mechanism and subsequent FM
4	Moderately high	Moderately high chance of detecting potential cause/mechanism and subsequent FM
3	High	High chance of detecting potential cause/mechanism and subsequent FM
2	Very high	Very high chance of detecting potential cause/mechanism and subsequent FM
1	Almost certain	Potential cause/mechanism and subsequent FM will be detected

Whether applied to a system, process, product, or service, FMEA, basically, consists of the implementation steps summarized in Fig. 10.1.

Despite its advantages mentioned above, there also exists several shortcomings of the FMEA methodology indicated in the literature (Bowles and Peláez 1995; Braglia 2000; Braglia et al. 2003, 2007; Chang 2009; Chang and Cheng 2010, 2011; Chang and Sun 2009; Chang and Wen 2010; Chang et al. 2001, 1999, 2010; Chen 2007; Chen and Ko 2009; Chin et al. 2009a, b; Franceschini and Galetto 2001; Gargama and Chaturvedi 2011; Geum et al. 2011; Kutlu and Ekmekçioğlu

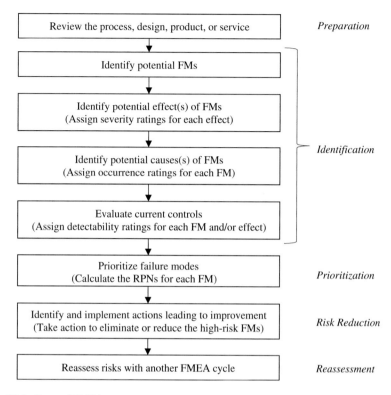

Fig. 10.1 Steps of FMEA process

2012; Liu et al. 2011, 2012, 2013b, c; Pillay and Wang 2003, Seyed-Hosseini et al. 2006, Sharma et al. 2008a, b; Wang et al. 2009; Xiao et al. 2011; Xu et al. 2002; Yang et al. 2008, 2011; Zammori and Gabbrielli 2012; Zhang and Chu 2011), some of which can be summarized as follows:

- Relative weights (importance) of the risk factors (i.e., S, O, and D) are not considered, while different weights will result in different priorities.
- Various sets of S, O, and D ratings may produce the same RPN value, although, relevant risk implications may be different.
- It is not possible to measure the amount of difference between the ranks in an ordinal scale, as the intervals of an ordinal scale are determined subjectively, and therefore, are not identical to each other. Because of this, mathematical operations cannot be performed on ordinal scales. However, in FMEA, RPN values are calculated by multiplying the numerical ratings of S, O, and D, which are measured on ordinal scales.
- The use of multiplication for RPN calculations, instead of other relationships, is questionable, as multiplication is very sensitive to the variations in the ratings of S, O, and D.

- It considers only three factors (S, O, and D), ignoring other significant factors such as, costs, production quantities, quality, etc.
- It considers only one FM at a time, interdependencies among various FMs and their effects are not taken into account.
- Since many of the points in the RPN scale, which ranges from 1 to 1000, cannot be formed from the product of S, O, and D (only 120 of the 1000 numbers can be generated), RPNs are not continuous. Furthermore, most of the unique points in the scale can be formed in several different ways (e.g., 60 can be formed from 24 different combinations of S, O, and D).
- The conversion of ratings for the three components of a FM is different. The relation between O and O's probability scale is non-linear, while the relation between D (S) and D's (S's) probability scale is linear.
- It is difficult, or even impossible, to give a precise and direct numerical evaluation for intangible quantities, such as S, O, and D.

In their review of literature on FMEA, Liu et al. (2013b) investigated the shortcomings of the FMEA methodology (which some of them are mentioned above) and discussed the approaches used in the FMEA literature. They proposed a framework for classifying the reviewed articles according to the FM prioritization method used, in which the relevant approaches were divided into five main categories: (i) Multi-Criteria Decision Making (MCDM) Approaches, (ii) Mathematical Programming (MP) Approaches, (iii) Artificial Intelligence (AI) Approaches, (iv) Integrated Approaches, and (v) Other Approaches.

Among others, some of the common approaches classified into these five categories are:

i. Evidence Theory (Yang et al. 2011), Analytical Hierarchy Process (AHP) (Hu et al. 2009), Analytical Network Process (ANP) (Zammori and Gabbrielli 2012), Grey Theory (Chang et al. 1999, 2001), and Intuitionistic Fuzzy Sets (Chang and Cheng 2010; Chang et al. 2010)
ii. Linear Programming (Chen and Ko 2009), Data Envelopment Analysis (DEA) (Chin et al. 2009a; Chang and Sun 2009), Fuzzy DEA (Garcia et al. 2005)
iii. Rule-base Systems and Fuzzy Rule-base Systems (Gargama and Chaturvedi 2011; Sharma et al. 2008a, b)
iv. Fuzzy Cognitive Maps (Pelaez and Bowles 1996), Fuzzy Evidential Reasoning and Grey Theory (Liu et al. 2011), Fuzzy AHP and Fuzzy TOPSIS (Kutlu and Ekmekçioğlu 2012), Intuitionistic Fuzzy Sets (IFS) and DEMATEL (Chang and Cheng 2010)
v. Monte Carlo Simulation (Bevilacqua et al. 2000), Minimum Cut Sets Theory (Xiao et al. 2011), Quality Function Deployment (QFD) (Braglia et al. 2007), and Probability Theory (Sant'Anna 2012).

According to Liu et al. (2013b), the categories including the most frequently used approaches for the prioritization of FMs, are AI and MCDM, respectively. Particularly, fuzzy rule-base system in AI category is the most used approach, followed by grey theory and AHP/ANP in the MCDM category. Fuzzy rule-based approaches, although applied extensively in the literature, have also been criticized, since they have some drawbacks that will be discussed in detail, in Sect. 10.2.

This chapter will be focusing on studies addressing the issues related to modelling, qualitative nature of risk assessment, as well as subjectivities and substantial uncertainties inherent in the assessment process. In other words, the approaches dealing with both complexity and uncertainty of the risk assessment process will be reviewed. The rest of this chapter is organized as follows. First, a comprehensive review of the literature is provided which is followed by illustrative examples for selected approaches. Then, the methodological differences of these alternative approaches are examined. Finally, conclusions and further research opportunities are presented.

10.2 Multi-criteria Risk Prioritization Under Uncertainty

Identifying and prioritizing potential failure modes and their effects generally requires dealing with uncertain information (including incomplete, vague and/or ambiguous information) as well as highly subjective judgments of experts. The uncertainties and subjectivities that arise here may stem from different sources, such as (1) lack of knowledge, limited attention and information processing capabilities (Asan et al. 2013); (2) vague assessment and grading criteria whose meaning, value, or boundaries vary considerably according to context or conditions; and (3) fragmented expert judgments. The last source, also known as inter-personal uncertainty (see Wu and Mendel 2009), can even emerge in situations where sufficient knowledge is available. This is related to the fact that FMEA is commonly performed in a group decision environment where experts may provide different judgments for the same risk factors because of their different expertise and backgrounds (Chin et al. 2009b; Song et al. 2014).

Thus, it becomes often unrealistic and impractical to acquire exact judgments in risk assessment when distinct interpretations are present and/or available data is incomplete or vague. Several authors have similarly reported that precision based methods suggested in the literature have largely or totally failed to address these certain sources of uncertainties. Below, the extensively criticized limitations of the conventional FMEA methods in dealing with uncertainties associated with the judgment process are summarized:

- They can't handle imprecise data and subjective judgments of domain experts, especially when the data set is small in size and its distribution is unknown.
- They can't cope with incomplete assessments and total ignorance.
- They can't deal with different types of assessment information simultaneously.
- They require prior information, such as, assumptions or pre-defined functions to deal with uncertainty.

- They ignore the level of confidence (belief degrees) experts are often willing to express in their subjective assessments.
- They ignore diversity in expert judgments.
- They lack a framework to analyze complex structures.
- They ignore other important factors (e.g., economical aspects).

Ideally, a complete theory and its accompanying tools used for identifying and prioritizing potential FMs and their effects should therefore address not only modeling issues, but also issues related to the qualitative nature of risk assessment as well as the analysis of subjectivities and substantial uncertainties inherent in the assessment process. To overcome some of the mentioned limitations of the conventional FMEA methods, many alternative approaches have been suggested in the literature. According to the literature review conducted by Liu et al. (2013b), the most frequently studied class of approaches was found to be artificial intelligence—in particular fuzzy rule-based approaches. There are several reasons why these approaches have been more preferred. First of all, they can handle ambiguous, qualitative as well as quantitative data in a consistent manner; second, they allow combining risk factors (i.e. FMs) in a more flexible and realistic manner; and finally, the risk assessment function can be customized according to the particular product, process or system under consideration (for more detail see Liu et al. 2013b). However, rule-based approaches have also significant limitations. For example, rule-based approaches require experts to design a sufficiently rich set of if-then rules and maintain them over time, which is often highly costly and time-consuming. Otherwise, an incomplete rule base will produce biased or even wrong inferences. Moreover, the rules with the same consequence but different antecedents cannot be distinguished from one another, which makes a complete prioritization or ranking of the failure modes impossible (Song et al. 2014). It is also hard to define proper membership functions for the risk factors and priority levels (Liu et al. 2013b). Thus, rule based approaches, which tend to be highly subjective, costly, and time consuming, should not necessarily be regarded as the best possible method.

An alternative class developed for FMEA under uncertainty consists of multi-criteria approaches. These approaches are able to handle both modelling issues (e.g., scaling, structuring, aggregation, weighting, etc.) and issues related to the analysis of subjectivities and substantial uncertainties inherent in the assessment process. A review of the literature indicates a growing interest in these approaches, especially in the past 5 years (see Fig. 10.2). Note that the source used for the review was only academic journal articles published in the past 15 years. According to the review, the most common theories and techniques employed in this class of approaches are grey relational analysis, aggregation operators, fuzzy technique for order preference by similarity to ideal solution (fuzzy TOPSIS), evidential reasoning (ER), intuitionistic fuzzy sets, type-1 fuzzy sets, 2-tuple fuzzy linguistic representation, fuzzy analytic hierarchy process (fuzzy AHP), rough set theory, fuzzy weighted geometric mean, fuzzy weighted least square, and possibility theory

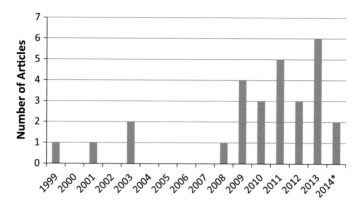

Fig. 10.2 Distribution of the reviewed articles (*: only the first quarter)

(see Table 10.4). Below, a summary of the reviewed papers based on this classification is provided.

Approaches Based on Grey Relational Analysis

Grey theory was initiated by Deng at the beginning of 1980s (Deng 1982, 1989). Like fuzzy set theory, grey theory also deals with making decisions with poor, incomplete, and multi-input information, and explores the behavior of a system using relational analysis (Deng 1982, 1989; Sharma et al. 2008b; Kuo et al. 2008; Geum et al. 2011; Chang et al. 2013). It provides a measure to analyze relationship between discrete quantitative and qualitative series (Chang et al. 2001). As one of the most applied techniques in FMEA, Grey Relational Analysis (GRA) is part of grey system theory, which can easily handle complicated interactions between multiple factors and variables. GRA provides a better distinction among decision alternatives (Kuo et al. 2008), and it gives the opportunity to assign different importance weights to S, O, and D. In one of the early studies, Chang et al. (2001) used GRA where they assigned different weights to the risk factors, and eliminated the need for a utility function and the conversion of ratings for the three components of FMs. They demonstrated the applicability of the proposed approach in an automobile PCB assembly case. Similarly, Chang et al. (1999), proposed a new approach for prioritizing the risks related to FMs. They adopted the fuzzy linguistic assessment to rate the risk factors, and applied grey theory to calculate the risk priority numbers (RPNs) of potential causes of each FM. In another study, Pillay and Wang (2003) presented an improved FMEA methodology utilizing fuzzy rule base and grey relation theory. In their illustrative application to an ocean going fishing vessel, they generated 35 fuzzy if-then rules in order to identify lacking safety features, and thus, to improve the operational safety of the vessel. Using this fuzzy rule base, risk factor ratings for each FM were integrated to obtain linguistic variables that were then used to rank FMs. As mentioned above, to address some of the limitations of traditional FMEA (i.e., identical RPNs and equally weighted risk factors) Sharma et al. (2008b), also proposed GRA to prioritize the causes of FMs.

Table 10.4 Classification of multi-criteria methods developed for FMEA under uncertainty

Approach	Literature	Total number*
Grey relational analysis	Chang et al. (1999), (2001), (2013); Pillay and Wang (2003); Geum et al. (2011); Liu et al. (2011)	6
Aggregation operators	Chang (2009); Chang and Wen (2010); Chang and Cheng (2011); Chang et al. (2012); Liu et al. (2014)	5
Fuzzy TOPSIS	Braglia et al. (2003); Kutlu and Ekmekçioğlu (2012); Song et al. (2013), (2014); Hadi-Vencheh and Aghajani (2013)	5
Evidential reasoning	Chin et al. (2009b); Liu et al. (2011); Yang et al. (2011); Liu et al. (2013a)	4
Intuitionistic fuzzy sets	Chang et al. (2010); Chang and Cheng (2010); Liu et al. (2014)	3
Ordinary fuzzy sets	Sharma et al. (2008b); Liu et al. (2012); Lin et al. (2014)	3
2-Tuple fuzzy linguistic representation	Chang and Wen (2010); Chang et al. (2012)	2
Fuzzy AHP	Hu et al. (2009); Kutlu and Ekmekçioğlu (2012)	2
Rough set theory	Song et al. (2014)	1
Fuzzy weighted geometric mean	Wang et al. (2009)	1
Fuzzy weighted least square	Zhang and Chu (2011)	1
Possibility theory	Mandal and Maiti (2014)	1

*Studies that involve more than one method are classified in more than one category in the table

Geum et al. (2011) developed a two-stage approach where a service-specific FMEA was constructed in the first stage; and GRA was applied to calculate the RPN of each FM in the second stage. To represent the service characteristics, they determined 19 sub-dimensions for the three risk factors of FMEA. When performing GRA in the second stage, risk scores for each sub-dimension were calculated firstly to establish S, O, and D scores, respectively; and then, overall RPN of each FM were obtained using these risk scores. Finally, in a recent study, Chang et al. (2013) integrated the GRA and the DEMATEL method to rank FMs according to the risks they represent for the organization, and presented an actual case of the TFT-LCD cell process. They argued that their new approach provides a lower duplication rate, generates more ideal rankings, and helps decision-makers to make more ideal determinations. Consequently, the major advantages of applying GRA to FMEA can be summarized as follows: (i) capability of dealing with incomplete information (ii) eliminating the need for a utility function, (iii) eliminating the need for the conversion of ratings for the three components of FMs, (iv) capability of assigning different importance weights to each risk factor, and (v) capability of providing a better distinction among FMs.

Approaches Based on Aggregation Operators

Aggregation operators weight values according to their ordering. In other words, these techniques are used to find optimal weights of the risk factors based on the ranks of the weighting vectors after an aggregation process (Chang et al. 2012). At the end, more accurate and reasonable ranking of the risk of failures may be obtained. In this category, Chang (2009) and Chang and Cheng (2011) proposed methodologies, which combine aggregation operators, such as, the ordered weighted geometric averaging and fuzzy ordered weighted averaging (OWA) operator, respectively, with the DEMATEL approach to evaluate the orderings of FMs. Findings suggest that it is more suitable to consider preferences in form of linguistic variables rather than numerical ones (Chang et al. 2012). Chang and Wen (2010) proposed a technique, where the OWA operator and 2-tuple fuzzy linguistic modelling is integrated, to prioritize failures in product design. They showed that the proposed approach, in comparison to the conventional RPN method, provides a more flexible structure for combining S, O, and D factors. Finally, Liu et al. (2014) proposed a new operator (intuitionistic fuzzy hybrid weighted Euclidean distance) that takes into account both subjective and objective weights of risk factors during the assessment process. The fragmented and uncertain assessments provided by a group of experts are treated as linguistic terms expressed in intuitionistic fuzzy numbers. The proposed operator allows reducing the impact of disproportionately large (or small) deviations on the results by assigning them low (or high) weights.

Approaches Based on Fuzzy TOPSIS

Another powerful method suggested to improve the conventional FMEA is TOPSIS —a multi-criteria decision making approach used to rank alternatives on the basis of the Euclidean distance of an alternative from both the positive and negative ideal solutions. Here, FMs are considered as the alternatives to be ranked with respect to the risk factors S, O, and D, which correspond to the criteria. Braglia et al. (2003) developed a fuzzy version of TOPSIS to provide a framework that allows dealing with imprecise quantities, such as those deriving from linguistic evaluations or subjective and qualitative assessments. By performing a sensitivity analysis of the fuzzy judgment weights and comparing results with the conventional method, they confirmed that the proposed approach gives a reasonable and robust final ranking of FMs. In another study, Kutlu and Ekmekçioğlu (2012) integrated fuzzy AHP with fuzzy TOPSIS in order to determine more realistic weights for the risk factors. Fuzzy AHP allows experts weighting the risk factors in linguistic variables. Song et al. (2013) also suggested a fuzzy weighted TOPSIS for FMEA under uncertainty. However, they developed a novel weighting approach where subjective weights derived from experts and objective weights obtained from an entropy-based method are integrated to avoid any underestimation or overestimation of the FMs. In another study of Song et al. (2014) a rough group TOPSIS method was proposed. The method integrates the strength of rough set theory in handling vagueness and the advantages of TOPSIS in modeling multi-criteria problems. Finally, Hadi-Vencheh and Aghajani (2013) proposed a fuzzy TOPSIS method based on α-level sets and the fuzzy extension principle. They formulated a new relative closeness coefficient in form of nonlinear programming (NLP) models and solved them in a series of linear programming models. Consequently, all these studies have shown that fuzzy TOPSIS is capable of: (1) assigning relative importance to risk factors, (2) introducing a potentially larger number of risk factors, and (3) using imprecise data in the form of fuzzy numbers.

Approaches Based on Evidential Reasoning

Evidential reasoning, as another popular approach in FMEA under uncertainty has been originally developed in the 1990s to support the solution of multi-attribute decision analysis problems with ignorance (see Yang and Singh (1994)). The recent ER approaches can model both quantitative and qualitative attributes using a distributed modelling framework, in which each attribute is characterized by a set of collectively exhaustive assessment grades (including incomplete information, complete ignorance and/or fuzzy uncertainty) with different degrees of belief (Wang et al. 2006). Experiences show that an expert may not always be fully confident in his assessments and may be willing to express beliefs to subsets of adjacent grades (Liu et al. 2011). A belief structure, in the FMEA context, captures the performance distribution of a subjective assessment of a FM. In one of the first studies in this group, Chin et al. (2009b) proposed a group-based ER approach, which can capture diversity in FMEA team members' opinions and prioritize FMs under different types of uncertainties, such as, incomplete assessment, ignorance and intervals. They calculate the overall belief structures and convert them into expected risk scores, which are finally ranked using the minimax regret approach. Inspired by the work of Chin et al. (2009b); Yang et al. (2011) adopted the modified Dempster–Shafer evidence theory to aggregate the different opinions about FMs, which may be inconsistent and uncertain. However, in the proposed model, the three risk factors are regarded as discrete random variables and all assessment grades are assumed to be crisp and independent of each other. The ER approach is further developed by Liu et al. (2011) to deal with risk evaluation problems which involve both probabilistic and fuzzy uncertainties. These are problems, where some of the assessment grades are difficult to be expressed as clearly distinctive crisp sets, but easier as overlapping fuzzy sets (Yang et al. 2006). The most recent study in this group, by Liu et al. (2013a), combined the fuzzy evidential reasoning (FER) approach with belief rule-based (BRB) methodology. The FER approach is used to capture and aggregate expert opinions, while the BRB methodology is used to model the uncertain causal relationships between risk factors and the risk level. A belief rule-base is a collection of expert knowledge that represents functional mappings between risk factors (antecedents) and risk levels (conclusions), possibly with uncertainty. According to Yang et al. (2008), BRB provides a more informative and realistic scheme than a simple if-then rule base on uncertain knowledge representation. To sum up, in comparison with the traditional FMEA and its variants, an ER approach to FMEA yields the following advantages (see also Chin et al. (2009b)): (1) the relative importance of risk factors are considered, (2) the diversity and uncertainty of experts' assessment information and related confidence values can be well reflected and modelled using belief structures, (3) FMs can be fully ranked and well distinguished from each other, (4) the expected risk score is a continuous number, and (5) risk factors are aggregated in a highly nonlinear manner.

Approaches Based on Intuitionistic Fuzzy Sets

Intuitionistic Fuzzy Set (IFS), which is an extension of fuzzy set, was introduced by Atanassov in 1983 (Atanassov 1986). In fuzzy set theory, the degree of non-membership is calculated by subtracting the degree of membership from one. However, this is not the case for IFSs. IFS adds an extra degree of uncertainty to classic fuzzy sets for modelling the hesitation and uncertainty about the degree of membership (Da Costa et al. 2010). Therefore, IFS can represent the imprecision of data in a more comprehensive manner than fuzzy sets (Xu 2011). An IFS A in a universe U, is defined as (Atanassov 1986):

$$A = \{(u, \mu_A(u), \nu_A(u)) | u \in U\} \; (A = (\mu_A, \nu_A) \text{ for short}) \qquad (10.1)$$

where the functions $\mu_A : U \rightarrow [0, 1]$ and $\nu_A : U \rightarrow [0, 1]$ define the grade of membership and the grade of non-membership of the each element of U to A, respectively. The functions $\mu_A(u)$ and $\nu_A(u)$ should satisfy the condition:

$$0 \leq \mu_A(u) + \nu_A(u) \leq 1 \quad (\forall u \in U) \qquad (10.2)$$

and

$$\pi_A(u) = (1 - \nu_A(u) - \mu_A(u)) \qquad (10.3)$$

where $\pi_A(u)$ denotes the uncertainty of u (also called as the hesitancy of u). Clearly, in the case of ordinary fuzzy sets, $\pi_A(u) = 0$ for $\forall u \in U$. For further detail, the reader should refer to Atanassov (1986).

As mentioned above, due to its capability to deal with uncertainty, IFS has recently been used in the FMEA literature. Chang et al. (2010) proposed a new approach utilizing the IFS ranking technique for reprioritization of FMs and presented an illustrative example of a silane supply system in a TFT-LCD process. According to Chang et al. (2010), their new approach reduces the occurrence of duplicate RPNs and provides more accurate information, and real situations are reflected in a more realistic and flexible manner. In another study, Chang and Cheng (2010), integrated the IFS and DEMATEL approach on risk assessment providing a more flexible structure for combining risk factors. They claim that, the proposed approach provides a more reasonable ranking where FMs are better distinguished. Finally, Liu et al. (2013c) developed a methodology using Intuitionistic Fuzzy Hybrid Weighted Euclidean Distance (IFHWED) operator. In this methodology, linguistic terms were used for the assessment of risk factors. In order to aggregate multiple experts' assessments into a group assessment, fuzzy weighted averaging operator was used, and then, IFHWED operator was applied to rank FMs, considering the weights of risk factors.

Approaches Based on Ordinary Fuzzy Sets

One of the prominent area of application of fuzzy set theory is in modeling where typically the available information contains various kinds of uncertainty due to internal and external disturbances and limitation of human knowledge and understanding (Liu and Lin 2010). As experts from different expertise areas and skill levels are included in FMEA process (Chin et al. 2009a, b), there usually exists an imprecise information to be treated as an input of this process. Additionally, complexity of the systems/products under investigation also increases the imprecision and uncertainty in FMEA. Therefore, as an effective tool providing a means for representing the uncertainty, fuzzy set theory has been extensively employed in FMEA literature. Bowles and Peláez (1995)'s study, in which the risk factors used in FMEA (i.e., S, O, and D) were represented as members of a fuzzy set, was the first study using the fuzzy sets theory for criticality analysis. In this study, linguistic variables were used to assess the S, O, and D of FMs. Following the determination of the degree of membership of each FM assessment to the corresponding fuzzy sets, which were identified as a guide for ranking S, O, and D; these fuzzy inputs were then evaluated using a linguistic rule base and fuzzy logic operations. Finally, the results were defuzzified and all FMs were ranked according to their criticality levels. Sharma et al. (2008b) established a framework based on fuzzy methodology and grey relation analysis to evaluate and assess system failure behavior, and presented a case from a process industry to demonstrate the applicability of the proposed framework. They concluded that their framework provides an effective way to combine expert knowledge and experience as well as to deal with uncertainty and imprecision in a more realistic manner. In an another study, Liu et al. (2012) used linguistic variables to assess the ratings and weights of the risk factors S, O, and D, and proposed a new risk priority model that extends VIKOR method to determine the risk priorities of FMs. They applied this model to the assessment of risks in general anesthesia process and claimed that they address some of the shortcomings of the traditional FMEA. Finally, in a very recent study, Lin et al. (2014) proposed an assessment model for human reliability in the risk assessment of medical devices, which applies the fuzzy linguistic theory to deal with the subjective assessments of experts. They noted that their proposed model, differing from the qualitative and quantitative methods used in human reliability analysis, considers some critical aspects, such as, context related factors, organizational factors, and errors in FMEA team members' assessments. Consequently, fuzzy set theory yields the following advantages over traditional FMEA: (i) qualitative, as well as quantitative, data can be used in the assessment, (ii) risk factors of FMs can be directly assessed using the linguistic terms, and (iii) S, O, and D can be combined in a more flexible manner.

Approaches Based on 2-Tuple Fuzzy Linguistic Representation

As indicated above, several authors have applied the fuzzy linguistic approach to FMEA problems with uncertain data. In these studies, the FMs are evaluated with respect to S, O, and D using a linguistic domain treated as discrete. However, operations (most notably multiplication) on fuzzy numbers produce results that usually do not exactly match any of the initial linguistic terms. To resolve this issue, an approximation process is used to express the results in the discrete initial expression domain that, however, leads to loss of information and hence lack of precision in the final results (Herrera and Martínez 2000). To overcome this critical shortcoming, Chang and Wen (2010) have suggested an FMEA model based on the 2-tuple fuzzy linguistic representation developed by Herrera and Martínez (2000). This model represents the crisp or linguistic information with a pair of values, called 2-tuple, which is composed by a linguistic term and a numeric value assessed in [−0.5, 0.5). In this way, any information obtained in an aggregation process can be represented on its domain. In their case study, Chang and Wen (2010) showed that their fuzzy linguistic representation model combined with the OWA operator effectively solves the problem of measurement scales (i.e., information loss in aggregation). In another study, Chang et al. (2012) have integrated 2-tuple fuzzy linguistic representation and the Linguistic Ordered Weighted Geometric Averaging (LOWGA) operator in process FMEA. This approach, as in Chang and Wen (2010), provides reasonable rankings for cases including FMs having the same RPN.

Approaches Based on Fuzzy AHP

Some recent studies have suggested using fuzzy AHP to explicitly accommodate the inherent uncertainty and complexity associated with risk assessment. Fuzzy AHP involves several concepts and techniques, such as, hierarchical structuring, pairwise comparison, prioritization principles for deriving weights, consistency considerations, and priority synthesis (see Saaty 1988). The hierarchical models developed in these studies typically consist of a goal (risk assessment), criteria (risk factors) and alternatives (FMs). Hu et al. (2009), for example, suggested a hierarchical risk assessment model to evaluate the risk of green components. They used triangular fuzzy numbers to express the comparative judgments of decision-makers. The resulting global priority values, i.e., the green component RPNs, are used to identify high-risk components and provide insight to the incoming quality control staff for improving the efficiency of inspection and mitigating risk. Kutlu and Ekmekçioğlu (2012) have also suggested applying fuzzy AHP to determine the weight vector of the three risk factors (S, O, and D). However, differently from the former study, they preferred Chen (2000)'s fuzzy TOPSIS to prioritize the final risk scores of the FMs.

Approaches Based on Rough Set Theory

Rough set theory, proposed by Pawlak in the early 1980s, is a formal approximation of the classical set theory that can handle imprecise and subjective judgments without any assumption and additional information (e.g., membership functions). In fact, predefined fuzzy membership functions or crisp rating scales in FMEA allow only judgments in form of point or fixed interval values and, hence, do not fully reflect the subjectivity and preference differences of experts. However, the rough set approach to FMEA, proposed by Song et al. (2014), provides a more rational risk evaluation framework where flexible intervals (i.e., rough intervals) are used to represent the inter-personal uncertainty. Here, a larger rough interval indicates a higher inconsistency among the experts. In this respect, the proposed rough FMEA not only provides an improved representation of the subjectivity and uncertainty in the evaluations, but also maintains the objectivity of original information (Zhai et al. 2007).

Approaches Based on Fuzzy Weighted Geometric Mean

As discussed in previous sections, traditional FMEA has limitations in terms of acquiring precise assessment information on the three components of FMs. In response to this limitation, it has been suggested in the literature to evaluate these risk factors by using linguistic scales and to use fuzzy FMEA, which utilizes a fuzzy rule-based reasoning approach to obtain RPN ratings. However, building up a complete and accurate rule base is a tedious and time-consuming task, particularly for the complex systems/products. Additionally, relative importance weights of the risk factors are not taken into consideration in traditional FMEA. Therefore, to overcome these limitations, Wang et al. (2009) suggested using Fuzzy Weighted Geometric Mean (FWGM) method for the calculation of FRPNs to prioritize FMs. In their study, Wang et al. (2009), firstly, evaluated the risk factors and their importance weights in a linguistic manner; then computed FRPNs applying an alpha cut based linear programming approach; and finally, defuzzified FRPNs using centroid defuzzification method for the final ranking of FMs. According to the authors, besides the above-mentioned advantages, the proposed methodology has the potential to fully prioritize FMs and hence to distinguish each FM from one another, and is not limited to the risk factors, S, O, and D.

Approaches Based on Fuzzy Weighted Least Square

As mentioned before, generally, people from different expertise areas are included in FMEA process. Therefore, members of these cross-functional FMEA teams, usually view the system/product under investigation from various perspectives, responsibilities and concerns, as they have different levels of knowledge, skills, experiences and personalities (Liu et al. 2013c). For that reason, FMEA team members may use different linguistic term sets when evaluating and weighting the relevant risk factors (Herrera et al. 2000); in other words, they may give their judgments in different forms. In order to ensure that the aggregated assessment of the FMEA team reflects all members' viewpoints and priorities, Zhang and Chu (2011) claimed that, Fuzzy Weighted Least Squares Model (FWLSM) can be used. By using FWLSM, the total deviation degree between each individual assessment information and the aggregated assessment information can be easily determined. As being the only study in this category, Zhang and Chu (2011) proposed a new approach, integrating FWLSM, the method of imprecision (MOI) and the method of partial ranking for evaluating and ranking the FMs. In this approach, firstly, a FMEA team evaluates each FM by using linguistic term sets with different cardinalities (i.e., multi-granularity linguistic term sets). Then the individual assessments are aggregated by means of FWLSM. Following the aggregation step, nonlinear programming method incorporated with the MOI is used for calculating the fuzzy RPNs in order to address the compensation levels among risk factors. Finally, by using the Hamming distance between each two fuzzy RPNs, the partial ranking method that is based on fuzzy preference relations is applied to rank FMs. For illustrative purposes, Zhang and Chu (2011) have applied their approach to the case of a new product development application, and have concluded that their approach provides more precisely expressed individual assessments, more accurate fuzzy RPNs, and thus, more robust results.

Approaches Based on Possibility Theory
Fuzzy numerical technique for FMEA, as well as traditional FMEA technique and fuzzy rule-based technique, have some limitations. When defuzzified crisp risk values are used to obtain final ranking of FMs, as in fuzzy numerical technique, the entropy present in fuzzy sets is ignored (Mandal and Maiti 2014). Therefore, like other techniques, fuzzy numerical technique also suffers from the limitation of providing arbitrary final ranking of FMs. In response to this limitation, Mandal and Maiti (2014) developed a robust methodology that integrates the 'similarity value measure' of fuzzy numbers and 'possibility and necessity measures' of possibility theory. Similar to the fuzzy set theory, possibility theory is also an uncertainty theory devoted to the handling of incomplete, imprecise, and uncertain information. As it uses the possibility and necessity measures, it has the capability to capture partial ignorance (Dubois and Prade 2011). In their recent study, Mandal and Maiti (2014) firstly, used similarity measure approach to obtain FRPNs, and subsequently, relevant priority values are clustered by means of comparison with a standard linguistic scale. After partially ordering FRPNs, they used possibility theory for making comparison with conformance guidelines. To this end, after calculating the possibility and necessity measures, they combined these two dual measures to obtain 'credibility measure', and consequently, used this measure to compare FRPNs with compliance guidelines. Here, as the credibility measure gets closer to one, the possibility of the relevant risk being lower than or equal to the conformance guideline increases; on the other hand, as it gets closer to zero, then the possibility of the relevant risk being lower than or equal to the conformance guideline decreases.

From the review above, it can be concluded that although all approaches deal with uncertainty and subjectivity associated with risk assessment, each one addresses only a particular set of shortcomings of the conventional FMEA. In the following section, the most frequently studied and promising approaches are illustrated by a simple example.

10.3 Illustrative Example for Selected Approaches

In the previous section, many of the new approaches mentioned in the literature are reviewed, but here only six well-known ones (based on ordinary fuzzy sets, grey relational analysis, evidential reasoning, intuitionistic fuzzy sets, 2-tuple fuzzy linguistic representation, and rough set theory) will be discussed in detail. The main concern will not be the identification of risk factors, but their assessment and aggregation. First, a summary of their theoretical underpinnings will be presented. Then, to illustrate their basic steps and ability to deal with uncertainty, a simple example will be worked out for all six approaches. The example, adapted from Kutlu and Ekmekçioğlu (2012), involves the prioritization of risks in an assembly process at a manufacturing facility operating in the automotive industry. The potential failure modes (FMs) in the assembly process, identified by a group of experts, are: non-conforming material (FM$_1$), wrong die (FM$_2$), wrong program (FM$_3$), excessive cycle time (FM$_4$), wrong process (FM$_5$), damaged goods (FM$_6$), wrong part (FM$_7$), and incorrect forms (FM$_8$). For the rating of these FMs, with respect to three risk factors, experts use the linguistic terms given in Table 10.5, where each term corresponds to a triangular fuzzy number. If otherwise not stated, the relative importance of the risk factors S, O, and D are assessed by pairwise comparisons using the linguistic scale provided in Table 10.6. Note that, for the approaches using crisp ratings in the assessment of FMs or risk factors, only the midvalues of the triangular fuzzy numbers will be considered in the analyses.

The assessments of the FMs and risk factors using linguistic terms and numerical values were obtained from three experts as presented in Tables 10.7, 10.8, 10.9 and 10.10, respectively. For example, as shown in Table 10.7, the assessments of the three experts of FM$_1$ with respect to severity are "Medium", "Medium", and "Medium Low". These linguistic terms can be converted into the following crisp values 5, 5, and 3 as shown in Table 10.9. In another example, it can be seen from Table 10.8 that the comparison of the risk factors severity and occurrence is in favour of the former, as "Strongly Important", "Strongly Important", and "Very Important" (3/2, 3/2, and 2 in crisp values as given in Table 10.10).

Table 10.5 Linguistic scales used for rating FMs

Linguistic scale			Fuzzy scale
Severity	Occurrence	Detection	
Very low (VL)	Very low (VL)	Very high (VH)	(0, 0, 1)
Low (L)	Low (L)	High (H)	(0, 1, 3)
Medium low (ML)	Medium low (ML)	Medium high (MH)	(1, 3, 5)
Medium (M)	Medium (M)	Medium (M)	(3, 5, 7)
Medium high (MH)	Medium high (MH)	Medium low (ML)	(5, 7, 9)
High (H)	High (H)	Low (L)	(7, 9, 10)
Very high (VH)	Very high (VH)	Very low (VL)	(9, 10, 10)

Table 10.6 Linguistic scale used for pairwise comparisons

Linguistic scale	Fuzzy scale	Fuzzy reciprocal scale
Equally important (EI)	(1, 1, 1)	(1, 1, 1)
Weakly important (WI)	(1, 1, 3/2)	(2/3, 1, 1)
Strongly important (SI)	(1, 3/2, 2)	(1/2, 2/3, 1)
Very important (VI)	(3/2, 2, 5/2)	(2/5, 1/2, 2/3)
Absolutely important (AI)	(2, 5/2, 3)	(1/3, 2/5, 1/2)

Table 10.7 Linguistic scores of FMs with respect to each risk factor

Failure mode	S			O			D		
	E_1	E_2	E_3	E_1	E_2	E_3	E_1	E_2	E_3
FM_1	M	M	ML	M	MH	MH	L	ML	L
FM_2	L	ML	ML	VH	H	VH	MH	MH	H
FM_3	ML	L	ML	VH	H	H	VH	MH	H
FM_4	ML	M	ML	M	MH	MH	L	ML	L
FM_5	M	M	ML	MH	MH	H	L	VL	L
FM_6	MH	MH	M	MH	H	MH	MH	MH	M
FM_7	L	ML	VL	VH	VH	VH	VH	MH	H
FM_8	VL	VL	L	VL	VL	VL	VH	VH	VH

Table 10.8 Pairwise comparisons of risk factors using the linguistic scale (R: Reciprocal)

	S			O			D		
	E_1	E_2	E_3	E_1	E_2	E_3	E_1	E_2	E_3
Severity	EI	EI	EI	SI	SI	VI	WI	WI	WI
Occurrence	R	R	R	EI	EI	EI	WI	R	EI
Detection	R	R	R	R	SI	R	EI	EI	EI

Table 10.9 Crisp scores of FMs with respect to each risk factor

Failure mode	S			O			D		
	E_1	E_2	E_3	E_1	E_2	E_3	E_1	E_2	E_3
FM_1	5	5	3	5	7	7	9	7	9
FM_2	1	3	3	10	9	10	3	3	1
FM_3	3	1	3	10	9	9	0	3	1
FM_4	3	5	3	5	7	7	9	7	9
FM_5	5	5	3	7	7	9	9	10	9
FM_6	7	7	5	7	9	7	3	3	5
FM_7	1	3	0	10	10	10	0	3	1
FM_8	0	0	1	0	0	0	0	0	0

Table 10.10 Pairwise comparisons of risk factors using crisp values

	S			O			D		
	E_1	E_2	E_3	E_1	E_2	E_3	E_1	E_2	E_3
Severity	1	1	1	3/2	3/2	2	1	1	1
Occurrence	2/3	2/3	1/2	1	1	1	1	2/3	1
Detection	1	1	1	1	3/2	1	1	1	1

10.3.1 FMEA Using Fuzzy Evidential Reasoning

The illustrated fuzzy evidential reasoning based approach, suggested by Liu et al. (2011), offers a unique way for aggregating expert judgments and prioritizing FMs in FMEA. One of the main advantages of this approach is its ability to coherently model both accurate data and subjective judgments with various types of uncertainties (such as, incomplete and fuzzy information as well as complete ignorance) in a unified framework. A further strength of the approach is its ability to reflect the diversity in expert judgments. All these benefits are achieved by incorporating the experts' level of confidence in their assessments (i.e., belief degrees) into the analysis. Here, a subjective assessment is characterized by a belief structure that describes the intensity of the belief for each possible assessment value. The aggregation of such structures allows one to merge multiple sources of evidence (numerical or linguistic) for the same risk factor or FM (Yang et al. 2006). Unquestionably, the belief structures provide experts with an easy-to-use and flexible way to express their opinions and can better quantify risk factors than the traditional FMEA methods (Liu et al. 2011).

Below, it will be illustrated how the belief structures of each FM provided by each expert can be aggregated into a group belief structure and how the group belief structures of each FM with respect to the three risk factors can be synthesized into an overall belief structure.

Stage 1: Assessment of FMs Using Belief Structures

FMs are assessed using the linguistic terms provided in Table 10.5. For example, the set of evaluation grades for Severity are represented as $H_{FS} = \{$Very Low, Low, Medium Low, Medium, Medium High, High, Very High$\}$ For the sake of simplicity without losing generality, all seven individual assessment grades are approximated by triangular fuzzy numbers of which only two adjacent ones intersect. Then, let $\left\{ \left(H_{ij}, \beta_{ij}^{k}(\mathrm{FM}_n, \mathrm{RF}_l) \right), i = 1, \ldots, 7; j = 1, \ldots, 7 \right\}$ be the belief structure provided by expert E_k on the assessment of failure mode FM_n with respect to risk factor RF_l, where H_{ii} for $i = 1 - 7$ are the fuzzy assessment grades, H_{ij} for $i = 1 - 6$ and $j = i + 1$ to 7 are the interval fuzzy assessment grades between H_{ii} and H_{jj}, and $\beta_{ij}^{k}(\mathrm{FM}_n, \mathrm{RF}_l)$ are the belief degrees for the intervals H_{ij} (Liu et al. 2011). The interval fuzzy assessment grades, H_{ij}, define trapezoidal fuzzy sets that include the fuzzy assessment grades $H_{ii}, H_{(i+1)(i+1)}, \ldots, H_{jj}$ as shown in Fig. 10.3.

For Severity, the grades H_{ii} for $i = 1 - 7$ and the intervals H_{ij} for $i = 1 - 6$ and $j = i + 1$ to 7 all together can be expressed as $\hat{H}_{FS} = \{H_{ij}, i = 1, \ldots, 7; j = 1, \ldots, 7\}$ or equivalently as

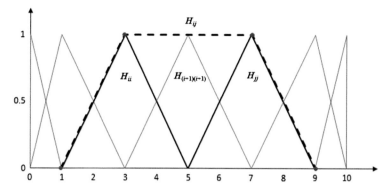

Fig. 10.3 Interval fuzzy grade H_{ij} (shown in *dashed line*)

$$\hat{H}_{FS} = \begin{cases} H_{11} & H_{12} & H_{13} & H_{14} & H_{15} & H_{16} & H_{17} \\ & H_{22} & H_{23} & H_{24} & H_{25} & H_{26} & H_{27} \\ & & H_{33} & H_{34} & H_{35} & H_{36} & H_{37} \\ & & & H_{44} & H_{45} & H_{46} & H_{47} \\ & & & & H_{55} & H_{56} & H_{57} \\ & & & & & H_{66} & H_{67} \\ & & & & & & H_{77} \end{cases}$$

This type of formulation allows experts to provide their subjective judgments in four possible ways (Liu et al. 2011):

- Certain. For example, "Medium" as a certain grade can be written as $\{(H_{33}, 1.0)\}$.
- Distribution. For example, a FM that is assessed as "High" with a confidence level of 0.3 and as "Very High" with a confidence level of 0.7, can be expressed as $\{(H_{66}, 0.3), (H_{77}, 0.7)\}$. This is a complete distribution. When all confidence levels do not sum to one, the distribution is known to be incomplete. The missing information in such cases is called local ignorance and it can be assigned any grade between "Very Low" and "Very High".
- Interval. For example, a FM that is assessed between "Medium" and "High", can be expressed as $\{(H_{46}, 1.0)\}$.
- Total Ignorance. The FM can be assigned any grade between "Very Low" and "Very High" and will be expressed as $\{(H_{17}, 1.0)\}$. In such cases the expert is whether unable or unwilling to provide an assessment.

Notice that FMs assessed to high values or intervals with high confidence levels are more risky than those assessed to low values or intervals with high confidence levels (Chin et al. 2009b). Table 10.11 presents the assessment information (in form of belief structures) on the eight FMs provided by three experts. The incomplete assessments and ignorance information are shaded and highlighted, respectively. For example, according to expert E_1, the Severity of 'using non-conforming

Table 10.11 Assessment information on eight FMs by three experts

Risk factors	Experts	Failure modes							
		FM_1	FM_2	FM_3	FM_4	FM_5	FM_6	FM_7	FM_8
Severity	E_1	$(H_{44}, 0.9)$	H_{22}	H_{33}	H_{33}	H_{44}	H_{55}	H_{22}	H_{11}
	E_2	H_{44}	H_{33}	H_{22}	H_{44}	H_{44}	H_{55}	H_{33}	H_{11}
	E_3	H_{33}	H_{33}	H_{33}	H_{33}	H_{33}	H_{44}	H_{11}	H_{22}
Occurrence	E_1	H_{44}	H_{77}	H_{77}	H_{44}	H_{55}	H_{55}	H_{77}	H_{11}
	E_2	H_{55}	H_{66}	H_{66}	H_{55}	H_{55}	H_{66}	H_{77}	H_{11}
	E_3	H_{55}	H_{77}	H_{66}	H_{55}	H_{66}	H_{55}	H_{77}	H_{11}
Detection	E_1	H_{66}	H_{33}	H_{11}	H_{66}	H_{66}	H_{33}	H_{11}	?*
	E_2	H_{55}	H_{33}	H_{33}	H_{55}	H_{77}	H_{33}	H_{33}	H_{11}
	E_3	H_{66}	$(H_{22}, 0.8)$	H_{22}	H_{66}	H_{66}	H_{44}	H_{22}	H_{11}

*:? Refers to total ignorance

material (FM_1)' is "Medium" with high confidence level (90 %). This assessment is incomplete with 10 % missing information.

Stage 2: Synthesis of Individual Belief Structures into Group Belief Structures
In this stage, the belief structures provided by the experts for each FM are synthesized into a group belief structure. Suppose that K experts, each given a weight $\lambda_k > 0$ ($k = 1, \ldots, K$) satisfying the condition $\sum_{k=1}^{K} \lambda_k = 1$ to reflect his/her relative importance, assess N failure modes with respect to L risk factors. Then, the aggregate assessment value, i.e., the fuzzy group belief structure, for each FM with respect to each risk factor is derived as follows (Liu et al. 2011, Chin et al. 2009b)

$$\tilde{X}_n(l) = \left\{ \left(H_{ij}, \beta_{ij}(FM_n, RF_l) \right), i = 1, \ldots, 7; j = 1, \ldots, 7 \right\}, n = 1, \ldots, N; l = 1, \ldots, L \tag{10.4}$$

where the group belief degree, $\beta_{ij}(FM_n, RF_l)$, is calculated as

$$\beta_{ij}(FM_n, RF_l) = \sum_{k=1}^{k} \lambda_k \beta_{ij}^k(FM_n, RF_l), i = 1, \ldots, 7; j = 1, \ldots, 7; \\ n = 1, \ldots, N; l = 1, \ldots, L \tag{10.5}$$

The relative importance of each expert should reflect the expert's experience and domain knowledge (Chin et al. 2009b). In the current example, the weights are supposed to be $\lambda_1 = 0.2$, $\lambda_2 = 0.5$, $\lambda_3 = 0.3$ and Table 10.12 presents the resulting group assessment values calculated using these weights. For example, given the individual belief structures $\{(H_{44}, 0.90)\}$, $\{(H_{44}, 1.00)\}, \{(H_{33}, 1.00)\}$ of three experts for FM_1 with respect to the risk factor Severity (see Table 10.11), the corresponding group belief structure is obtained as $\{(H_{33}, 0.30), (H_{44}, 0.68), (H_{17}, 0.02)\}$ where

Table 10.12 Fuzzy group belief structures

Failure mode	Severity	Occurrence	Detection
FM_1	$\{(H_{33}, 0.30), (H_{44}, 0.68), (H_{17}, 0.02)\}$	$\{(H_{44}, 0.20), (H_{55}, 0.80)\}$	$\{(H_{55}, 0.50), (H_{66}, 0.50)\}$
FM_2	$\{(H_{22}, 0.20), (H_{33}, 0.80)\}$	$\{(H_{66}, 0.50), (H_{77}, 0.50)\}$	$\{(H_{22}, 0.24), (H_{33}, 0.70), (H_{17}, 0.06)\}$
FM_3	$\{(H_{22}, 0.50), (H_{33}, 0.50)\}$	$\{(H_{66}, 0.80), (H_{77}, 0.20)\}$	$\{(H_{11}, 0.20), (H_{22}, 0.30), (H_{33}, 0.50)\}$
FM_4	$\{(H_{33}, 0.50), (H_{44}, 0.50)\}$	$\{(H_{44}, 0.20), (H_{55}, 0.80)\}$	$\{(H_{55}, 0.50), (H_{66}, 0.50)\}$
FM_5	$\{(H_{33}, 0.30), (H_{44}, 0.70)\}$	$\{(H_{55}, 0.70), (H_{66}, 0.30)\}$	$\{(H_{66}, 0.50), (H_{77}, 0.50)\}$
FM_6	$\{(H_{44}, 0.30), (H_{55}, 0.70)\}$	$\{(H_{55}, 0.50), (H_{66}, 0.50)\}$	$\{(H_{33}, 0.70), (H_{44}, 0.30)\}$
FM_7	$\{(H_{11}, 0.30), (H_{22}, 0.20), (H_{33}, 0.50)\}$	$\{(H_{77}, 1.00)\}$	$\{(H_{11}, 0.20), (H_{22}, 0.30), (H_{33}, 0.50)\}$
FM_8	$\{(H_{11}, 0.70), (H_{22}, 0.30)\}$	$\{(H_{11}, 1.00)\}$	$\{(H_{11}, 0.80), (H_{17}, 0.20)\}$

$$\beta_{33}(FM_1, RF_1) = \sum_{k=1}^{3} \lambda_k \beta_{33}^k(FM_1, RF_1) = 0.20 \cdot 0 + 0.50 \cdot 0 + 0.30 \cdot 1.00$$
$$= 0.30$$

$$\beta_{44}(FM_1, RF_1) = \sum_{k=1}^{3} \lambda_k \beta_{44}^k(FM_1, RF_1) = 0.20 \cdot 0.90 + 0.50 \cdot 1.00 + 0.30 \cdot 0$$
$$= 0.68$$

For the missing information in $\beta_{44}(FM_1, RF_1)$ any grade between "Very Low" and "Very High" can be assigned, thus

$$\beta_{17}(FM_1, RF_1) = \sum_{k=1}^{3} \lambda_k \beta_{17}^k(FM_1, RF_1) = 0.20 \cdot 0.10 + 0.50 \cdot 0 + 0.30 \cdot 0$$
$$= 0.02$$

Stage 3: Defuzzification

Before a group belief structure is aggregated into an overall belief structure, a defuzzification process is applied to convert fuzzy numbers into appropriate crisp values. Various defuzzification methods have been developed in the literature, including various forms of the centroid method, first (or last) of maxima, mean max membership, total integral value method, among others (see Ross 2009; Ramli and Mohamad 2009). Liu et al. (2011) suggests using the defuzzification method developed by Chen and Klein (1997) which is quite simple to carry out. The formula is as follows

$$h_{ij} = \frac{\sum_{i=0}^{n}(b_i - c)}{\sum_{i=0}^{n}(b_i - c) - \sum_{i=0}^{n}(a_i - d)}, i = 1, \ldots, 7; j = 1, \ldots, 7 \qquad (10.6)$$

where the values c and d denote the lower and upper limits of the linguistic scale, the values a_0 and b_0 (for a triangular membership function) represent the extreme limits of each linguistic term where the membership function is 0, and a_1 and b_1 are the values where the membership function is 1 (Pillay and Wang 2003). Here, h_{ij} is the defuzzified crisp value of H_{ij}. Accordingly, the crisp group belief structure can be represented as follows:

$$X_n(l) = \left\{ (h_{ij}, \beta_{ij}(FM_n, RF_l)), i = 1, \ldots, 7; j = 1, \ldots, 7 \right\}, n = 1, \ldots, N; l = 1, \ldots, L$$
$$(10.7)$$

For example, the linguistic term "Very Low–Very High", i.e., H_{17}, can be defuzzified as shown below (see Fig. 10.4):

$$h_{17} = \frac{[b_0 - c] + [b_1 - c]}{\{[b_0 - c] + [b_1 - c]\} - \{[a_0 - d] + [a_1 - d]\}} \qquad (10.8)$$

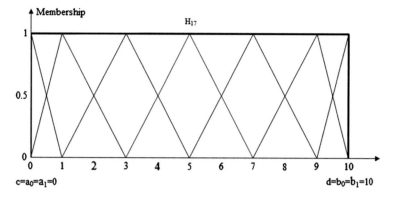

Fig. 10.4 Defuzzification of the linguistic term "Very Low–Very High"

$$h_{17} = \frac{[10-0]+[10-0]}{\{[10-0]+[10-0]\}-\{[0-10]+[0-10]\}} = 0.5$$

After calculating h_{33} and h_{44} in a similar way, the crisp group belief structure of FM_1 with respect to Severity can be stated as follows

$$X_1(1) = \{(0.33,0.30),(0.5,0.68),(0.5,0.02)\}$$

Table 10.13 presents all FMs' crisp group belief structures.

Stage 4: Aggregation of Defuzzified Group Belief Structures into Overall Belief Structure

Once the group belief structures of the FMs are defuzzified, they are aggregated into overall belief structures by using the following equation:

$$\bar{X}_n(l) = \sum_{i=1}^{7}\sum_{j=1}^{7} h_{ij}\beta_{ij}(FM_n,RF_l), n = 1,\ldots,N; l = 1,\ldots,L \qquad (10.9)$$

As an example, the overall belief structure of FM_1 with respect to the risk factor Severity is obtained as given below:

$$\bar{X}_1(1) = \sum_{i=1}^{7}\sum_{j=1}^{7} h_{ij}\beta_{ij}(FM_1,RF_1) = 0.333 \cdot 0.30 + 0.5 \cdot 0.68 + 0.5 \cdot 0.02$$
$$= 0.450$$

Note that, in the example above, values with zero belief degrees are omitted. The results for all FMs are presented in Table 10.14.

Stage 5: Introduction of Group Weights

In this stage, the relative importance of the risk factors, which will later be used to calculate the final priorities of the FMs, are determined. Instead of presenting the direct estimation method suggested by Liu et al. (2011), an alternative approach is introduced here. According to this method, the weight information can be elicited

Table 10.13 Crisp group belief structures

Failure mode	Severity	Occurrence	Detection
FM_1	{(0.333, 0.30), (0.5, 0.68), (0.5, 0.02)}	{(0.5, 0.20), (0.667, 0.80)}	{(0.667, 0.50), (0.826, 0.50)}
FM_2	{(0.174, 0.20), (0.333, 0.80)}	{(0.826, 0.50), (0.952, 0.50)}.	{(0.174, 0.24), (0.333, 0.70), (0.5, 0.06)}
FM_3	{(0.174, 0.50), (0.333, 0.50)}	{(0.826, 0.80), (0.952, 0.20)}	{(0.047, 0.20), (0.174, 0.30), (0.333, 0.50)}
FM_4	{(0.333, 0.50), (0.5, 0.50)}	{(0.5, 0.20), (0.667, 0.80)}	{(0.667, 0.50), (0.826, 0.50)}
FM_5	{(0.333, 0.30), (0.5, 0.70)}	{(0.667, 0.70), (0.826, 0.30)}	{(0.826, 0.50), (0.952, 0.50)}
FM_6	{(0.50, 0.30), (0.667, 0.70)}	{(0.667, 0.50), (0.826, 0.50)}	{(0.333, 0.70), (0.5, 0.30)}
FM_7	{(0.047, 0.30), (0.174, 0.20), (0.333, 0.50)}	{(0.952, 1.00)}	{(0.047, 0.20), (0.174, 0.30), (0.333, 0.50)}
FM_8	{(0.047, 0.70), (0.174, 0.30)}	{(0.047, 1.00)}	{(0.047, 0.80), (0.5, 0.20)}

Table 10.14 Aggregated values for the eight FMs

Failure mode	Severity	Occurrence	Detection
FM_1	0.450	0.633	0.746
FM_2	0.301	0.889	0.305
FM_3	0.254	0.851	0.228
FM_4	0.417	0.633	0.746
FM_5	0.450	0.714	0.889
FM_6	0.617	0.746	0.383
FM_7	0.216	0.952	0.228
FM_8	0.086	0.048	0.138

by means of pairwise comparisons using the linguistic scale introduced in Table 10.6. The typical question asked for the comparison of risk factors with respect to the overall prioritization goal is formulated as follows: "In assessing the risk level of failure modes, which of the two risk factors is more critical; and how much more?" (Asan et al. 2016). Table 10.8 presents the comparisons obtained from three experts. For example, when comparing the risk factors Severity and Occurrence, the responses of the three experts are "Strongly Important (SI)", "Strongly Important (SI)", and "Very Important (VI)", respectively. The assessments are then aggregated into group values. The group value of the comparison of the risk factors RF_l and RF_m for K experts is derived as:

$$\tilde{v}_{lm} = \sum_{k=1}^{K} \lambda_k \tilde{v}_{lm}^k = \left(\sum_{k=1}^{K} \lambda_k \tilde{v}_{lma}^k, \sum_{k=1}^{K} \lambda_k \tilde{v}_{lmb}^k, \sum_{k=1}^{K} \lambda_k \tilde{v}_{lmc}^k \right), l = 1, \ldots, L;$$
$$m = 1, \ldots, M$$

$$(10.10)$$

where $\tilde{v}_{lm}^k = \left(\tilde{v}_{lma}^k, \tilde{v}_{lmb}^k, \tilde{v}_{lmc}^k \right)$ denotes the fuzzy pairwise comparison of the risk factors RF_l and RF_m stated by expert k. For instance, the group value for the comparison pair Severity and Occurrence is calculated as follows:

$$\tilde{v}_{12} = (0.2 \cdot 1 + 0.5 \cdot 1 + 0.3 \cdot 1.5, 0.2 \cdot 1.5 + 0.5 \cdot 1.5 + 0.3 \cdot 2, 0.2 \cdot 2 + 0.5 \cdot 2 + 0.3 \cdot 2.5) = (1.15, 1.65, 2.15)$$

Before the weights can be derived from these group pairwise comparisons, shown in Table 10.15, the group comparison values need to be defuzzified. Using Eq. (10.6) (c = 1/3, d = 3), as explained above, the fuzzy comparison values are converted into crisp values (v_{lm}), see Table 10.16. Next, the additive normalization method is used to approximate the weights (w_l). It applies the following three-step

Table 10.15 Fuzzy group comparisons of the risk factors

	Severity	Occurrence	Detection
Severity	(1, 1, 1)	(1.15, 1.65, 2.15)	(1, 1, 1.5)
Occurrence	Reciprocal	(1, 1, 1)	(0.75, 0.83, 1.1)
Detection	Reciprocal	Reciprocal	(1, 1, 1)

Table 10.16 Defuzzified group comparisons of the risk factors

	Severity	Occurrence	Detection
Severity	0.250	0.495	0.314
Occurrence	2.021	0.250	0.223
Detection	3.182	4.488	0.250

Table 10.17 The normalized comparison matrix and estimated weights

	Severity	Occurrence	Detection	Weights
Severity	0.046	0.095	0.399	0.180
Occurrence	0.371	0.048	0.283	0.234
Detection	0.583	0.858	0.318	0.586

procedure (Saaty 1988, Asan et al. 2012): (1) the sum of the values in each column of the pairwise comparison matrix is calculated; (2) then, each column element is divided by the sum of its respective column; (3) finally, arithmetic mean of each row of the normalized comparison matrix is calculated. These final numbers provide an estimate of the weights for the risk factors being compared.

Table 10.17 provides the normalized pairwise comparison matrix and the resulting weights.

Stage 6: Ranking

In this final stage, the overall belief structures are synthesized over the three risk factors. Instead of presenting the weighted average method suggested by Liu et al. (2011), an alternative approach is introduced here where a process of weighting and multiplying is used as follows:

$$p_n = \prod_{l=1}^{L} \bar{X}_n^{w_l}(l), n = 1, \ldots, N \tag{10.11}$$

The resulting uni-dimensional values, p_n, represent the final priorities of the FMs that are used to produce a ranking. For example, the priority value of FM_1 is calculated as follows:

$$p_1 = \prod_{l=1}^{3} \bar{X}_1^{w_l}(l) = 0.450^{0.180} \cdot 0.633^{0.234} \cdot 0.746^{0.586} = 0.656$$

The results are shown in the last column of Table 10.18. The priority ranking of the eight failure modes is estimated as follows: FM_5, FM_1, FM_4, FM_6, FM_2, FM_3, FM_7, and FM_8.

10.3.2   FMEA Based on Ordinary Fuzzy Sets

Fuzzy set theory, as one of the most applied approaches to FMEA, comprises the following six main steps:

Table 10.18 Final priorities and their rank order

Failure mode	Severity	Occurrence	Detection	Priority	Rank
FM_1	0.450	0.633	0.746	0.656	2
FM_2	0.301	0.889	0.305	0.391	5
FM_3	0.254	0.851	0.228	0.317	6
FM_4	0.417	0.633	0.746	0.647	3
FM_5	0.450	0.714	0.889	0.747	1
FM_6	0.617	0.746	0.383	0.488	4
FM_7	0.216	0.952	0.228	0.316	7
FM_8	0.086	0.048	0.138	0.099	8
Weight	*0.180*	*0.234*	*0.586*		

Step 1: Listing FMs

The first step in applying fuzzy set theory to FMEA, as in all other approaches, is to list all possible FMs. The FMs, determined by a group of experts for further evaluation in the relevant assembly process, are given in the first part of this section.

Step 2: Determination of the Linguistic Scales

In the second step, firstly, the appropriate linguistic scales for the risk factors (S, O, and D), which will be used in the assessment of FMs, are selected. Both, the selected linguistic scales, expressed in positive triangular fuzzy numbers, and the membership function of the linguistic variables are given in Table 10.5 and Fig. 10.5, respectively. Note that, membership function values can be determined according to the historical data and judgment of domain experts (Liu et al. 2011). Secondarily, to obtain the importance weights of the risk factors, a linguistic scale for pairwise comparisons is provided in Table 10.6. However, it should be underlined here that, rather than selecting a linguistic scale for pairwise comparisons, a different linguistic scale for directly assessing the relative importance weights of the risk factors could be used. Here, in this study, the former one is preferred.

Step 3: Assessment of the Risk Factor Weights and FMs

Using the linguistic scale determined for pairwise comparisons in the prior step, experts are asked to pairwise compare the relative importance of the risk factors. Subsequently, they are also asked to assess the ratings of each FM. Risk assessment results for the FMs and pairwise comparison results for the risk factor weights are given in Tables 10.7 and 10.8, respectively.

Step 4: Aggregation of Individual Assessments

If it is supposed that there are K experts (E_1, ..., E_K) in a FMEA team who are responsible for the assessment of N failure modes (FM_1, ..., FM_N) with respect to L risk factors (RF_1, ..., RF_L), and $\tilde{r}_{nl}^k = \left(r_{nla}^k, r_{nlb}^k, r_{nld}^k \right)$ denotes the risk assessment rating of the failure mode FM_n given by the expert E_k with respect to the risk factor RF_l under consideration; the risk assessment rating of each FM with respect to all risk factors is aggregated into a group risk assessment rating as follows:

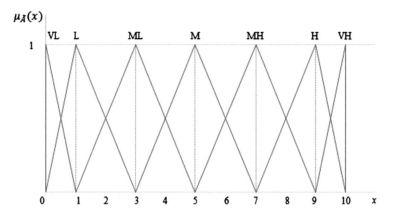

Fig. 10.5 Membership functions for rating the FMs

$$\tilde{r}_{nl} = \sum_{k=1}^{K} \lambda_k \tilde{r}_{nl}^k = \left(\sum_{k=1}^{K} \lambda_k \tilde{r}_{nla}^k, \quad \sum_{k=1}^{K} \lambda_k \tilde{r}_{nlb}^k, \quad \sum_{k=1}^{K} \lambda_k \tilde{r}_{nld}^k \right) \quad (10.12)$$

where $\lambda_k > 0$ $(k = 1,\ldots, K)$ reflects the relative importance weight of each expert E_k and satisfies the condition $\sum_{k=1}^{K} \lambda_k = 1$(Chin et al. 2009b, Liu et al. 2011). Here, λ_k should be determined according to the experts' level of domain knowledge, skills, and experience. In the current example, as indicated in Sect. 3.1, the expert weights are supposed to be $\lambda_1 = 0.2$, $\lambda_2 = 0.5$, and $\lambda_3 = 0.3$. Based on these weights, group risk assessment ratings of FMs are calculated and presented in Table 10.19. In order to demonstrate the aggregation step, an example calculation for FM$_1$ is shown below. As it can be seen from Table 10.7, risk assessment rating of FM$_1$, given by the experts E_1, E_2 and E_3 with respect to the Severity risk factor, are 'Medium (M)', 'Medium (M)' and 'Medium Low (ML)', respectively. Based on Table 10.5, it can be found that the corresponding fuzzy number of the linguistic term (M) is (3, 5, 7) and (ML) is (1, 3, 5). Then, the aggregated risk assessment rating of FM$_1$ with respect to the Severity risk factor $(\tilde{r}_{11})$ is calculated as follows:

Table 10.19 Fuzzy group risk assessment ratings of FMs

Failure mode	S	O	D
FM$_1$	(2.4, 4.4, 6.4)	(4.6, 6.6, 8.6)	(6.0, 8.0, 9.5)
FM$_2$	(0.8, 2.6, 4.6)	(8.0, 9.5, 10.0)	(0.7, 2.4, 4.4)
FM$_3$	(0.5, 2.0, 4.0)	(7.4, 9.2, 10.0)	(0.5, 1.8, 3.6)
FM$_4$	(2.0, 4.0, 6.0)	(4.6, 6.6, 8.6)	(6.0, 8.0, 9.5)
FM$_5$	(2.4, 4.4, 6.4)	(5.6, 7.6, 9.3)	(8.0, 9.5, 10.0)
FM$_6$	(4.4, 6.4, 8.4)	(6.0, 8.0, 9.5)	(1.6, 3.6, 5.6)
FM$_7$	(0.5, 1.7, 3.4)	(9.0, 10.0, 10.0)	(0.5, 1.8, 3.6)
FM$_8$	(0.0, 0.3, 1.6)	(0.0, 0.0, 1.0)	(0.0, 0.0, 1.0)

$$\tilde{r}_{11} = \sum_{k=1}^{3} \lambda_k \tilde{r}_{11}^{k} = \left(\sum_{k=1}^{3} \lambda_k r_{11a}^{k}, \quad \sum_{k=1}^{3} \lambda_k r_{11b}^{k}, \quad \sum_{k=1}^{3} \lambda_k r_{11d}^{k} \right) \quad (10.13)$$

$$\tilde{r}_{11} = ((0.20 \cdot 3 + 0.50 \cdot 3 + 0.30 \cdot 1), (0.20 \cdot 5 + 0.50 \cdot 5 + 0.30 \cdot 3), (0.20 \cdot 7 + 0.50 \cdot 7 + 0.30 \cdot 5))$$

$$\tilde{r}_{11} = (2.4, 4.4, 6.4)$$

Similarly, the relative importance weights of the risk factors are also aggregated as discussed above (for details, refer to the fifth step of Sect. 10.3.1).

Step 5: Defuzzification
Following the aggregation of individual assessments into a group assessment, the next step is the defuzzification of fuzzy risk assessment ratings into crisp values for ranking FMs according to their criticality levels. As indicated in the third step of Sect. 10.3.1, various methods of defuzzification are available in the literature. In this example, the method developed by Chen and Klein (1997), which can be expressed by the following equation, is selected due to its simplicity.

$$r_{nl}(x) = \frac{\sum_{i=0}^{n}(b_i - c)}{\sum_{i=0}^{n}(b_i - c) - \sum_{i=0}^{n}(a_i - d)} \quad (10.14)$$

where $r_{nl}(x)$ is the defuzzified (crisp) risk assessment rating of the failure mode FM_n with respect to the risk factor RF_l. Using this method, crisp risk assessment ratings of each FM are calculated and presented in Table 10.20. As an example, consider the defuzzification of the risk assessment rating of FM_1 with respect to the Severity risk factor as shown in Fig. 10.6. To obtain a crisp risk assessment rating, the relevant fuzzy risk assessment rating can be defuzzified as follows:

$$r_{11}(x) = \frac{[b_0 - c] + [b_1 - c]}{\{[b_0 - c] + [b_1 - c]\} - \{[a_0 - d] + [a_1 - d]\}} \quad (10.15)$$

$$r_{11}(x) = \frac{[6.4 - 0] + [4.4 - 0]}{\{[6.4 - 0] + [4.4 - 0]\} - \{[2.4 - 10] + [4.4 - 10]\}} = 0.450$$

Table 10.20 Crisp group risk assessment ratings of FMs

Failure mode	S	O	D
FM_1	0.450	0.633	0.745
FM_2	0.303	0.886	0.287
FM_3	0.255	0.850	0.234
FM_4	0.417	0.633	0.745
FM_5	0.450	0.713	0.886
FM_6	0.617	0.745	0.383
FM_7	0.223	0.952	0.234
FM_8	0.088	0.048	0.048

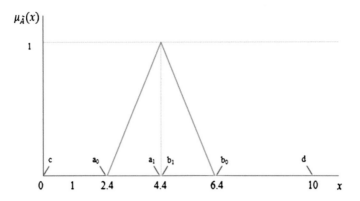

Fig. 10.6 Defuzzification

Following the defuzzification of group risk assessment rating of each FM, using the Eq. 10.14 as shown above, group pairwise comparisons of the relative importance of risk factors are defuzzified in a similar way. Then the additive normalization method is used to approximate the relative importance weights of risk factors (for details, refer to the fifth step of Sect. 10.3.1). The normalized pairwise comparison matrix, together with the resulting importance weights are given in Table 10.17.

Step 6: Ranking
In the final step, FMs are ordered according to their criticality levels. To this end, the RPN value of each FM is calculated by multiplying the crisp group risk assessment ratings of each FM (with respect to S, O, and D) with the relative importance weights of the risk factors as given in Eq. 10.16.

$$p_n = \sum_{l=1}^{L} w_l \cdot r_{nl}(x), n = 1, \ldots, N \qquad (10.16)$$

where p_n represents the RPN of a FM. As an example, RPN of FM_1 is calculated as follows:

$$p_1 = \sum_{l=1}^{3} w_l \cdot r_{1l}(x) = (0.450 \cdot 0.180 + 0.633 \cdot 0.234 + 0.745 \cdot 0.586) = 0.666$$

The final results are shown in the last column of Table 10.21. The priority rankings of eight failure modes are estimated as follows: FM_5, FM_1, FM_4, FM_6, FM_2, FM_7, FM_3, FM_8.

Table 10.21 Final RPNs of FMs and their rank orders

Failure mode	S	O	D	Priority	Rank
FM_1	0.450	0.633	0.745	0.666	2
FM_2	0.303	0.886	0.287	0.430	5
FM_3	0.255	0.850	0.234	0.382	7
FM_4	0.417	0.633	0.745	0.660	3
FM_5	0.450	0.713	0.886	0.767	1
FM_6	0.617	0.745	0.383	0.510	4
FM_7	0.223	0.952	0.234	0.400	6
FM_8	0.088	0.048	0.048	0.055	8
Weight	*0.180*	*0.234*	*0.586*		

10.3.3   FMEA Based on Grey Relational Analysis (GRA)

GRA is one of the most commonly used analytical methods used in FMEA literature, in terms of dealing with incomplete information and handling complicated inter-actions between multiple factors and variables. It combines the assessment ratings of all risk factors being considered into one single value, and thus, makes the comparison and ranking of FMs according to their criticality levels easier. Being firstly applied by Chang et al. in 1999, to FMEA with the aim of obtaining RPNs and prioritizing FMs, generally, GRA is composed of five major steps as detailed below, following the initial steps of (i) identification of all possible FMs (ii) determination of the scales, and (iii) assessment of the risk factor weights and FMs.

Step 1: Establishment of Comparative (Information) Series

As the first step of GRA, all assessment values for every FM are processed into a comparative series. In this context, if the measurement scales are different for each risk factor, then the risk assessment ratings should be normalized, in order to avoid any undervaluation of some factor's influence.

If it is supposed that there are N FMs (FM_1, ..., FM_N) and L risk factors (RF_1, ..., RF_L), the original vector of risk assessment ratings of FM_n can be expressed as $Y_n = (y_n(1), y_n(2), ..., y_n(L))$ where $y_n(l)$ is the assessment rating of FM_n with respect to the risk factor RF_l. Then, Y_n should be translated into the comparative series $X_n = (x_n(1), x_n(2), ..., x_n(L))$, by using one of the three equations given below.

$$x_n(l) = \frac{y_n(l) - Min\{y_n(l), n = 1, 2, ..., N\}}{Max\{y_n(l), n = 1, 2, ..., N\} - Min\{y_n(l), n = 1, 2, ..., N\}} \quad (10.17a)$$
$$for\ n = 1, 2, ..., N; l = 1, 2, ..., L$$

$$x_n(l) = \frac{Max\{y_n(l), n = 1, 2, ..., N\} - y_n(l)}{Max\{y_n(l), n = 1, 2, ..., N\} - Min\{y_n(l), n = 1, 2, ..., N\}} \quad (10.17b)$$
$$for\ n = 1, 2, ..., N; l = 1, 2, ..., L$$

$$x_n(l) = 1 - \frac{\left|y_n(l) - y^*(l)\right|}{Max\{y_n(l), n = 1,2,\ldots,N\} - Min\{y_n(l), n = 1,2,\ldots,N\}} \quad (10.17c)$$
$$for\ n = 1,2,\ldots,N; l = 1,2,\ldots,L$$

where Eq. 10.17a is used for the-larger-the-better factors, Eq. 10.17b is used for the-smaller-the-better factors and finally, the Eq. 10.17c is used for the-closer-to-the-desired-value-$y^*(l)$-the better factors. This translation process (translation of Y_n to X_n) is called grey relational generating in GRA (Geum et al. 2011). As all risk factors considered in FMEA are comparable to each other, the comparative series X can be defined as the following matrix:

$$X = \begin{bmatrix} x_1 \\ x_2 \\ \vdots \\ x_N \end{bmatrix} = \begin{bmatrix} x_1(1) & x_1(2) & \cdots & x_1(L) \\ x_2(1) & x_2(2) & \cdots & x_2(L) \\ \vdots & \vdots & \vdots & \vdots \\ x_N(1) & x_N(2) & \cdots & x_N(L) \end{bmatrix} \quad (10.18)$$

However, if the measurement scales used for each risk factor are all same, as in this example, then, the need for such translation (i.e., normalization) disappears; and the original vector of risk assessment ratings of the failure mode FM_n (Y_n), can be taken as the comparative series (X_n) of that FM. For this reason, in this example, the assessment ratings of the experts were directly considered in the analyses performed, without any further normalization calculations. In other words, the midvalues of the triangular fuzzy numbers representing each expert's opinion were directly taken into consideration without being normalized (See Table 10.9). To obtain the overall risk assessment ratings of the FMs with respect to each risk factor, the arithmetic mean of the corresponding assessment values of the experts were calculated. The resulting comparative series of each FM are given below:

$$X = \begin{bmatrix} x_1 \\ x_2 \\ x_3 \\ x_4 \\ x_5 \\ x_6 \\ x_7 \\ x_8 \end{bmatrix} = \begin{bmatrix} x_1(1) & x_1(2) & x_1(3) \\ x_2(1) & x_2(2) & x_2(3) \\ x_3(1) & x_3(2) & x_3(3) \\ x_4(1) & x_4(2) & x_4(3) \\ x_5(1) & x_5(2) & x_5(3) \\ x_6(1) & x_6(2) & x_6(3) \\ x_7(1) & x_7(2) & x_7(3) \\ x_8(1) & x_8(2) & x_8(3) \end{bmatrix} = \begin{bmatrix} 4.333 & 6.333 & 8.333 \\ 2.333 & 9.667 & 2.333 \\ 2.333 & 9.333 & 1.333 \\ 3.667 & 6.333 & 8.333 \\ 4.333 & 7.667 & 9.333 \\ 6.333 & 7.667 & 3.667 \\ 1.333 & 10.000 & 1.333 \\ 0.333 & 0.000 & 0.000 \end{bmatrix}$$

As an example, calculation of the overall risk assessment rating of FM_1 with respect to the Severity risk factor ($x_1(1)$) is given below. Risk assessment rating of FM_1, given by the experts E_1, E_2 and E_3 with respect to the Severity risk factor, are 'Medium (M)', 'Medium (M)' and 'Medium Low (ML)' (See Table 10.7), with the corresponding fuzzy numbers (3, 5, 7), (3, 5, 7), and (1, 3, 5), respectively (See Table 10.5). Therefore, by taking the midvalues of this triangular fuzzy numbers into consideration (i.e., 5, 5, and 3) to represent the opinion of each expert, the overall risk assessment rating of FM_1 (with respect to the Severity risk factor) is calculated as follows:

$$x_1(1) = \frac{(5+5+3)}{3} = 4.333$$

Step 2: Establishment of Standard (Reference) Series
In order to evaluate the comparative series calculated in Step 1, a standard series is obtained by determining the optimal levels of each risk factor. The standard series, which represents an original reference to be compared with the comparative series, is expressed as below:

$$X_0 = [x_0(l)] = [x_0(1) \quad x_0(2) \quad \dots \quad x_0(L)] \tag{10.19}$$

Since, the smaller the score for the risk factors, the less the risk in FMEA, the standard series includes the lowest level of all risk factors (Chang et al. 2001). In this example, the optimal level of the risk factors S, O, and D are VL, VL, and VH, respectively, with the corresponding fuzzy number (0, 0, 1). Hence, the standard series using the midvalues is defined as:

$$X_0 = [VL \quad VL \quad VH] = [0 \quad 0 \quad 0] \tag{10.20}$$

Step 3: Calculation of the Difference between Comparative Series and Standard Series
Following the establishment of comparative series and standard series, the difference between these series is calculated by using the Eq. 10.21 and reflected in a form of a matrix (D_0) as shown below:

$$\Delta_{0n}(l) = \|x_0(l) - x_n(l)\| \tag{10.21}$$

where $x_0(l)$ and $x_n(l)$ represent the standard series and comparative series of the relevant failure mode FM_n, respectively.

$$D_0 = \begin{bmatrix} \Delta_{01}(1) & \Delta_{01}(2) & \dots & \Delta_{01}(L) \\ \Delta_{02}(1) & \Delta_{02}(2) & \dots & \Delta_{02}(L) \\ \vdots & \vdots & \vdots & \vdots \\ \Delta_{0N}(1) & \Delta_{0N}(2) & \dots & \Delta_{0N}(L) \end{bmatrix} \tag{10.22}$$

As an example, the difference between comparative series and standard series of the failure mode FM_1 ($\Delta_{01}(1)$) for the Severity risk factor is calculated as follows:

$$\Delta_{01}(1) = \|0 - 4.333\| = 4.333$$

The rest of the calculated difference values are given in the matrix below.

$$D_0 = \begin{bmatrix} 4.333 & 6.333 & 8.333 \\ 2.333 & 9.667 & 2.333 \\ 2.333 & 9.333 & 1.333 \\ 3.667 & 6.333 & 8.333 \\ 4.333 & 7.667 & 9.333 \\ 6.333 & 7.667 & 3.667 \\ 1.333 & 10.000 & 1.333 \\ 0.333 & 0.000 & 0.000 \end{bmatrix}$$

Step 4: Calculation of Grey Relational Coefficient

In the fourth step of GRA, in order to determine how close $x_0(l)$ is to $x_n(l)$, the grey relational coefficient, $\gamma(x_0(l), x_n(l))$, is calculated using the equation given below:

$$\gamma(x_0(l), x_n(l)) = \frac{min_n min_l |x_0(l) - x_n(l)| + \zeta \cdot max_n max_l |x_0(l) - x_n(l)|}{\Delta_{0n}(l) + \zeta \cdot max_n max_l |x_0(l) - x_n(l)|}$$

$$\gamma(x_0(l), x_n(l)) = \frac{\Delta_{min} + \zeta \cdot \Delta_{max}}{\Delta_{0n}(l) + \zeta \cdot \Delta_{max}} \text{ for } n = 1, 2, \ldots, N; l = 1, 2, \ldots, L$$

(10.23)

where ζ is the distinguished coefficient which takes values in the range [0, 1] and which only affects the relative values of the risk factors without changing their priority (Chang et al. 2001). As it increases (decreases), it expands (compresses) the range of the grey relational coefficient and usually, is accepted as 0.5 (Geum et al. 2011). On the basis of the difference values given in the D_0 matrix above, the grey relational coefficient of the failure mode FM_1 for the severity risk factor is calculated as:

$$\gamma(x_0(1), x_1(1)) = \frac{\Delta_{min} + 0.5 \cdot \Delta_{max}}{\Delta_{01}(1) + 0.5 \cdot \Delta_{max}} = \frac{0.000 + 0.5 \cdot 10.000}{4.333 + 0.5 \cdot 10.000} = 0.536$$

The grey relational coefficients of each FM were calculated in a similar manner, and the results of these calculations are summarized in $\gamma(x_0(l), x_n(l))$ matrix below. Regarding the grey relational coefficient, it can be said that as it becomes larger, $x_0(l)$ and $x_n(l)$ values becomes closer.

$$\gamma(x_0(l), x_n(l)) = \begin{bmatrix} 0.536 & 0.441 & 0.375 \\ 0.682 & 0.341 & 0.682 \\ 0.682 & 0.349 & 0.789 \\ 0.577 & 0.441 & 0.375 \\ 0.536 & 0.395 & 0.349 \\ 0.441 & 0.395 & 0.577 \\ 0.789 & 0.333 & 0.789 \\ 0.938 & 1.000 & 1.000 \end{bmatrix}$$

Table 10.22 Grey relational grades of FMs and their rank orders

Failure mode	Grey relational coefficients			Grey relational grades	Rank
	S	O	D		
FM_1	0.536	0.441	0.375	0.419	2
FM_2	0.682	0.341	0.682	0.602	5
FM_3	0.682	0.349	0.789	0.667	6
FM_4	0.577	0.441	0.375	0.427	3
FM_5	0.536	0.395	0.349	0.393	1
FM_6	0.441	0.395	0.577	0.510	4
FM_7	0.789	0.333	0.789	0.683	7
FM_8	0.938	1.000	1.000	0.989	8
Weight	*0.180*	*0.234*	*0.586*		

Step 5: Calculation of Grey Relational Grade and Ranking
In the final step of GRA, the grey relational grades of each FM are calculated by multiplying the grey relational coefficients obtained in the previous step with the relative importance weights of the risk factors as given in Eq. 10.24. Since the calculation of the relative importance weights of risk factors is discussed in detail, in Sect. 10.3.1, here, the resulting importance weights of risk factors that were calculated previously were directly substituted into Eq. 10.24 (refer to the last column of Table 10.17 for the relative importance weights of the risk factors).

$$\Gamma(x_0, x_n) = \sum_{l=1}^{L} w_l \cdot \gamma(x_0(l), x_n(l)) \, for \, n = 1, 2, \ldots, N \qquad (10.24)$$

In order to demonstrate the calculation of grey relational grade, an example calculation for the degree of relation of the first failure mode FM_1 is seen below.

$$\Gamma(x_0, x_n) = \sum_{l=1}^{3} w_l \cdot \gamma(x_0(l), x_1(l))$$

$$\Gamma(x_0, x_n) = (0.180 \cdot 0.536) + (0.234 \cdot 0.441) + (0.586 \cdot 0.375) = 0.419$$

Grey relational grades of all FMs were calculated similarly, and presented in Table 10.22. The relevant grey relational grades estimates the priority ranking of the eight FMs as FM_5, FM_1, FM_4, FM_6, FM_2, FM_3, FM_7, and FM_8.

10.3.4 FMEA Based on Intuitionistic Fuzzy Sets (IFSs)

As a generalization of fuzzy sets, IFSs has a much better capability of coping with imperfect and/or imprecise information as it adds an extra degree of uncertainty for modelling the hesitation and uncertainty about the degree of membership (Da Costa

Table 10.23 Linguistic scales used for rating failure modes and IFNs

Linguistic scale			
Severity	Occurrence	Detection	IFNs
Very low (VL)	Very low (VL)	Very high (VH)	(0.10, 0.90)
Low (L)	Low (L)	High (H)	(0.25, 0.70)
Medium low (ML)	Medium low (ML)	Medium high (MH)	(0.35, 0.55)
Medium (M)	Medium (M)	Medium (M)	(0.50, 0.50)
Medium high (MH)	Medium high (MH)	Medium low (ML)	(0.60, 0.30)
High (H)	High (H)	Low (L)	(0.75, 0.20)
Very high (VH)	Very high (VH)	Very low (VL)	(0.90, 0.10)

et al. 2010). It consists of membership and non-membership functions and hence, provides a better representation of experts' assessments and more information than classic fuzzy sets. Therefore, it can be used in situations, where making assessments through linguistic variables based on only a membership function is considered to be inadequate. IFSs comprise the following four steps, performed just after all possible FMs are identified.

Step 1: Determination of the Linguistic Scales
In the first step, differently from the other approaches, the linguistic scales for the risk factors (S, O, and D) are defined as Intuitionistic Fuzzy Numbers (IFNs) and given in Table 10.23. Note here that, both membership and non-membership functions of the linguistic terms can be determined according to the suggestions of the domain experts.

Additionally, a linguistic scale for the pairwise comparison of the importance weights of risk factors is provided in Table 10.6.

Step 2: Assessment of the Risk Factor Weights and FMs
Using the linguistic scale defined by IFNs and set out in Table 10.23, experts are asked to assess the ratings of each FM, where the results are given in Table 10.7, and then are asked to pairwise compare the relative importance of the risk factors using the linguistic scale introduced in Table 10.6. The comparisons obtained from the experts are presented in Table 10.8.

Step 3: Aggregation of Individual Assessments
Following the assessment of FMs by each expert, the resulting individual expert assessments, which are quantified by the corresponding IFNs, are aggregated into group assessments. If it is supposed that there are K experts ($E_1, \ldots, E_K$) in a FMEA team who are responsible for the assessment of N failure modes ($FM_1, \ldots, FM_N$) with respect to L risk factors ($RF_1, \ldots, RF_L$); by using the Intuitionistic Fuzzy Weighted Averaging (IFWA) operator, the subjective risk assessment ratings of each FM are aggregated into a group risk assessment rating as follows (Liu et al. 2014):

$$\tilde{\alpha}_{nl} = IFWA\left(\tilde{\alpha}_{nl}^1, \tilde{\alpha}_{nl}^2, \ldots, \tilde{\alpha}_{nl}^K\right) = \sum\nolimits_{k=1}^{K} \lambda_k \cdot \tilde{\alpha}_{nl}^k$$

$$= \left[1 - \prod\nolimits_{k=1}^{K} \left(1 - \mu_{nl}^k\right)^{\lambda_k}, \quad \prod\nolimits_{k=1}^{K} \left(v_{nl}^k\right)^{\lambda_k}\right] for\, n = 1, 2, \ldots, N; l = 1, 2, \ldots, L$$

$$(10.25)$$

where $\tilde{\alpha}_{nl}^k = \left(\mu_{nl}^k, v_{nl}^k\right)$ denotes the IFN provided by E_k on the assessment of FM_n with respect to the risk factor RF_l and $\lambda_k > 0$ ($k = 1, \ldots, K$) reflects the relative importance weight of each expert E_k, satisfying the condition $\sum_{k=1}^{K} \lambda_k = 1$ (Liu et al. 2013c).

In the current example, the expert weights were supposed to be $\lambda_1 = 0.2$, $\lambda_2 = 0.5$, and $\lambda_3 = 0.3$ and Table 10.24 presents the resulting group assessment values calculated by using these weights. In fact, if needed, the expert weights can also be assessed in linguistic terms that will be defined as IFNs, instead of their crisp counterparts. As an example, given the risk assessment ratings "Medium ($\tilde{\alpha}_{11}^1 = (0.5, 0.5)$)", "Medium ($\tilde{\alpha}_{11}^2 = (0.5, 0.5)$)", and "Medium Low ($\tilde{\alpha}_{11}^3 = (0.35, 0.55)$)" of three experts, for FM_1 with respect to the Severity risk factor (RF_1), respectively (see Tables 10.7 and 10.23); the corresponding aggregated risk assessment rating, $\tilde{\alpha}_{11}$, is obtained as:

$$\tilde{\alpha}_{11} = \sum\nolimits_{k=1}^{3} \lambda_k \tilde{\alpha}_{11}^k = \left[1 - \prod\nolimits_{k=1}^{3} \left(1 - \mu_{11}^k\right)^{\lambda_k}, \quad \prod\nolimits_{k=1}^{3} \left(v_{11}^k\right)^{\lambda_k}\right] \quad (10.26)$$

$$\tilde{\alpha}_{11} = \left(1 - \left((1 - 0.5)^{0.2} \cdot (1 - 0.5)^{0.5} \cdot (1 - 0.35)^{0.3}\right), \quad \left((0.5)^{0.2} \cdot (0.5)^{0.5} \cdot (0.55)^{0.3}\right)\right)$$

$$\tilde{\alpha}_{11} = (0.459, 0.515)$$

Subsequently, the relative importance weights of the risk factors are then aggregated as shown in the fifth step of Sect. 10.3.1.

Table 10.24 Aggregated risk assessment ratings of FMs

Failure mode	S	O	D
FM_1	(0.459, 0.515)	(0.582, 0.332)	(0.684, 0.245)
FM_2	(0.331, 0.577)	(0.842, 0.141)	(0.321, 0.591)
FM_3	(0.302, 0.620)	(0.792, 0.174)	(0.276, 0.652)
FM_4	(0.43, 0.524)	(0.582, 0.332)	(0.684, 0.245)
FM_5	(0.459, 0.515)	(0.653, 0.266)	(0.842, 0.141)
FM_6	(0.572, 0.350)	(0.684, 0.245)	(0.399, 0.534)
FM_7	(0.263, 0.669)	(0.900, 0.100)	(0.276, 0.652)
FM_8	(0.148, 0.835)	(0.100, 0.900)	(0.100, 0.900)

Step 4: Calculation of FRPNs

In this step, aggregated risk assessment ratings of FMs with respect to each risk factor (S, O, and D), given in Table 10.24, are synthesized by using the Intuitionistic Fuzzy Weighted Geometric (IFWG) operator as follows (Wei 2010):

$$\tilde{\alpha}_n = IFWG(\tilde{\alpha}_{n1}, \tilde{\alpha}_{n2}, \ldots, \tilde{\alpha}_{nL}) = \prod_{l=1}^{L} (\tilde{\alpha}_{nl})^{w_l}$$

$$= \left[\prod_{l=1}^{L} (\mu_{nl})^{w_l} \quad , \quad 1 - \prod_{l=1}^{L} (1 - v_{nl})^{w_l} \right] for \, n = 1, 2, \ldots, N \qquad (10.27)$$

where $\tilde{\alpha}_n = (\mu_n, v_n)$ denotes the IFN representing the synthesized assessment of FM_n and $w_l > 0$ ($l = 1, \ldots, L$) reflects the subjective weight of risk factor RF_l, satisfying the condition $\sum_{l=1}^{L} w_l = 1$ (Wei 2010).

The relative importance weights of the risk factors calculated in Sect. 10.3.1, as $w_1 = 0.180$, $w_2 = 0.234$, and $w_3 = 0.586$, were directly substituted into Eq. 10.27, and the resulting synthesized assessments of FMs are presented in Table 10.25. As an example, given the aggregated risk assessment ratings of FM_1 as $\tilde{\alpha}_{11} = (0.459, 0.515)$, $\tilde{\alpha}_{12} = (0.582, 0.332)$, and $\tilde{\alpha}_{13} = (0.684, 0.245)$, regarding the risk factors S, O, and D, respectively, the corresponding synthesized assessment rating, $\tilde{\alpha}_1$, was obtained as:

$$\tilde{\alpha}_1 = \sum_{l=1}^{3} (\tilde{\alpha}_{1l})^{w_l} = \left[\prod_{l=1}^{3} (\mu_{1l})^{w_l} \quad , \quad 1 - \prod_{l=1}^{3} (1 - v_{1l})^{w_l} \right] \qquad (10.28)$$

$$\tilde{\alpha}_1 = \left(\left((0.459)^{0.2} \cdot (0.582)^{0.5} \cdot (0.684)^{0.3} \right), \quad 1 - \left((1 - 0515)^{0.2} \cdot (1 - 0.332)^{0.5} \cdot (1 - 0.245)^{0.3} \right) \right)$$

$$\tilde{\alpha}_1 = (0.613, 0.322)$$

Step 5: Ranking

In the final step, using the results obtained in the previous step as an input, the final rankings of the FMs are determined. To this end, firstly the score and accuracy degrees of each FM are calculated by Eqs. 10.29 and 10.30, and then, Xu and Yager

Table 10.25 Synthesized assessment ratings of FMs

Failure Mode	Synthesized assessment ratings
FM_1	(0.613, 0.322)
FM_2	(0.405, 0.511)
FM_3	(0.359, 0.568)
FM_4	(0.606, 0.325)
FM_5	(0.711, 0.253)
FM_6	(0.483, 0.446)
FM_7	(0.361, 0.570)
FM_8	(0.107, 0.891)

(2006)'s method given below, is used to compare the synthesized assessment ratings of FMs.

Let $\tilde{\alpha} = (\mu, v)$ be an IFN, the score function $S(\tilde{\alpha})$ and the accuracy function $H(\tilde{\alpha})$ of $\tilde{\alpha}$ can be represented as follows (Xu and Yager 2006):

$$S(\tilde{\alpha}) = \mu - v, \quad S(\tilde{\alpha}) \in [-1, 1] \tag{10.29}$$

$$H(\tilde{\alpha}) = \mu + v, \quad H(\tilde{\alpha}) \in [-1, 1] \tag{10.30}$$

Then the order relation between two IFNs, $\tilde{\alpha}_1 = (\mu_1, v_1)$ and $\tilde{\alpha}_2 = (\mu_2, v_2)$, can be defined as follows (Xu and Yager 2006):

$$
\begin{aligned}
&\rhd \quad \text{If } S(\tilde{\alpha}_1) < S(\tilde{\alpha}_2), \text{ then } \tilde{\alpha}_1 < \tilde{\alpha}_2 \\
&\rhd \quad \text{If } S(\tilde{\alpha}_1) = S(\tilde{\alpha}_2), \text{ and} \\
&\quad \bullet \ H(\tilde{\alpha}_1) < H(\tilde{\alpha}_2), \text{ then } \tilde{\alpha}_1 < \tilde{\alpha}_2 \\
&\quad \bullet \ H(\tilde{\alpha}_1) = H(\tilde{\alpha}_2), \text{ then } \tilde{\alpha}_1 = \tilde{\alpha}_2
\end{aligned}
\tag{10.31}
$$

where $S(\tilde{\alpha}_1)$ and $S(\tilde{\alpha}_2)$ are the score degrees, and $H(\alpha_1)$ and $H(\tilde{\alpha}_2)$ are the accuracy degrees of $\tilde{\alpha}_1$ and $\tilde{\alpha}_2$, respectively.

The score and accuracy degrees, and the resulting rankings of the FMs are presented in Table 10.26 The score and accuracy degrees estimate the priority ranking of the eight FMs as FM_5, FM_1, FM_4, FM_6, FM_2, FM_3, FM_7, and FM_8.

10.3.5 FMEA Using Rough Set Theory

As previously mentioned, rough set theory is a formal approximation of the classical set theory that can handle imprecise and subjective judgments without any assumption and additional information. In contrast to crisp and fuzzy values/intervals, which are all based on pre-defined membership functions and thus cannot truly accommodate inter-personal uncertainty, rough sets use flexible intervals

Table 10.26 Score and accuracy degrees of FMs and their rank orders

FM failure mode	Synthesized assessment ratings	$S(\tilde{\alpha})$	$H(\tilde{\alpha})$	Rank
FM_1	(0.613, 0.322)	0.290	0.935	2
FM_2	(0.405, 0.511)	−0.106	0.916	5
FM_3	(0.359, 0.568)	−0.209	0.926	6
FM_4	(0.606, 0.325)	0.281	0.931	3
FM_5	(0.711, 0.253)	0.458	0.964	1
FM_6	(0.483, 0.446)	0.037	0.929	4
FM_7	(0.361, 0.570)	−0.209	0.930	7
FM_8	(0.107, 0.891)	−0.783	0.998	8

(i.e., boundary regions) to reflect the diversity in expert judgments. A larger rough interval, then, indicates higher inconsistency among expert judgments (Song et al. 2014).

According to this theory, any vague concept (i.e., rough set) is represented with two precise concepts (i.e., crisp sets), called its lower and upper approximation (Pawlak 1997). The lower approximation consists of elements that certainly belong to the set, whereas the upper approximation contains all elements that possibly belong to the set. The difference between these two crisp sets constitutes the boundary region of the rough set and consists of the elements that cannot be classified uniquely to the set or its complement (Pawlak 1997). Below, more formal definitions are provided for the basic notions of rough set theory.

Let U be a non-empty set of finite elements (the universe), Y an arbitrary element of U, and $R = \{C_1, C_2, \ldots, C_n\}$ a set of n classes defined in the universe. For example, when assessing the severity of a FM, the distinct severity ratings provided by the experts can be viewed as classes associated with the FM (i.e., an element in the classification problem). If the classes are ordered in the following way $C_1 < C_2 < \ldots < C_n$, then for any class, $C_i \in R$, $1 \leq i \leq n$, the approximations of C_i and boundary region can be defined as follows (Zhai et al. 2007, 2009, see also Pawlak 1997):

Lower approximation of C_i:

$$\underline{Apr}(C_i) = \bigcup \{Y \in U/R : R(Y) \leq C_i\} \tag{10.32a}$$

Upper approximation of C_i:

$$\overline{Apr}(C_i) = \bigcup \{Y \in U/R : R(Y) \geq C_i\} \tag{10.32b}$$

Boundary Region of C_i:

$$\begin{aligned} Bnd(C_i) &= \bigcup \{Y \in U/R : R(Y) \neq C_i\} \\ &= \{Y \in U/R : R(Y) > C_i\} \cup \{Y \in U/R : R(Y) < C_i\} \end{aligned} \tag{10.32c}$$

Thus the class C_i can be represented by a rough number that is defined by its lower limit ($\underline{Lim}(C_i)$) and upper limit ($\overline{Lim}(C_i)$) as follows (Zhai et al. 2007, 2009):

$$\underline{Lim}(C_i) = \frac{1}{N_L} \sum R(Y) | Y \in \underline{Apr}(C_i) \tag{10.33a}$$

$$\overline{Lim}(C_i) = \frac{1}{N_U} \sum R(Y) | Y \in \overline{Apr}(C_i) \tag{10.33b}$$

where N_L and N_U are the number of elements contained in the lower and upper approximation of C_i, respectively. Then, a vague class C_i can be expressed in form of a rough number (interval) which is denoted as $RN(C_i)$ and defined as (Zhai et al. 2007, 2009):

$$RN(C_i) = \left[\underline{Lim}(C_i), \overline{Lim}(C_i)\right] \qquad (10.34)$$

The interval between the lower limit and the upper limit (i.e., the rough boundary interval of C_i) indicates the degree of preciseness.

Below, a slightly adapted version of the rough TOPSIS approach proposed by Song et al. (2014) is illustrated. The approach, which uses the definitions given above to represent the uncertainty in FM assessment, consists of two main stages. In the first stage, the weights of the risk factors S, O, and D are determined, while in the second stage the FMs are assessed and analyzed using the rough group TOPSIS approach.

Stage 1: Determination of Rough Interval Weights for Risk Factors
In this stage, the relative importance of the risk factors, which will later be used to calculate the weighted normalized rough matrix of FMs, are determined. Instead of using the direct estimation method suggested by Song et al. (2014), an alternative approach is presented here. According to this method, the weight information can be elicited by means of pairwise comparisons using the linguistic scale introduced in Table 10.6 (for the typical questions asked here see Sect. 10.3.1). Table 10.10 presents the crisp comparison values obtained from three experts. For example, when comparing the risk factors Severity and Occurrence, according to the first expert, Severity is strongly more important than Occurrence (i.e., $v_{SO}^1 = 3/2$). The additive normalization method is used here to approximate the crisp weights (v_l^k, where l denotes the risk factors and k the experts). It applies the following three-step procedure (Saaty 1988, Asan et al. 2012): (1) the sum of the values in each column of the pairwise comparison matrix is calculated; (2) then, each column element is divided by the sum of its respective column; (3) finally, arithmetic means of each row of the normalized comparison matrix are calculated. These final numbers provide an estimate of the weights for the risk factors being compared. Table 10.27 presents each expert's normalized pairwise comparison values and the corresponding crisp weights.

These crisp weights are then converted into rough numbers using Eqs. 10.32a, 10.32b, 10.32c–10.34. For example, the crisp weights (0.375, 0.379, and 0.411) obtained from three experts for "Severity" are converted into a rough number as follows:

Table 10.27 Normalized comparison values and crisp weights

	E_1				E_2				E_3			
	S	O	D	V^1	S	O	D	V^2	S	O	D	V^3
S	0.375	0.428	0.333	0.379	0.375	0.375	0.375	0.375	0.400	0.500	0.333	0.411
O	0.250	0.286	0.333	0.290	0.250	0.250	0.250	0.250	0.200	0.250	0.333	0.261
D	0.375	0.286	0.333	0.331	0.375	0.375	0.375	0.375	0.400	0.250	0.333	0.327

$$\underline{Lim}(0.375) = 0.375, \quad \overline{Lim}(0.375) = \frac{1}{3}(0.375 + 0.379 + 0.411) = 0.388$$

$$\underline{Lim}(0.379) = \frac{1}{2}(0.375 + 0.379) = 0.377, \quad \overline{Lim}(0.379) = \frac{1}{2}(0.379 + 0.411)$$
$$= 0.395$$

$$\underline{Lim}(0.411) = \frac{1}{3}(0.375 + 0.379 + 0.411) = 0.388, \quad \overline{Lim}(0.411) = 0.411$$

then

$$RN\left(w_S^1\right) = RN(0.375) = [0.375, 0.388]$$
$$RN\left(w_S^2\right) = RN(0.379) = [0.377, 0.395]$$
$$RN\left(w_S^3\right) = RN(0.411) = [0.388, 0.411]$$

The average rough weight of any risk factor $\left(\overline{RN(w_l)} = \left[\underline{Lim}(w_l), \overline{Lim}(w_l)\right]\right)$ can be calculated as follows (Song et al. 2014)

$$\underline{Lim}(w_l) = \left(\underline{Lim}\left(w_S^1\right) + \underline{Lim}\left(w_S^2\right) + \cdots + \underline{Lim}\left(w_S^K\right)\right)/K \qquad (10.35a)$$

$$\overline{Lim}(w_l) = \left(\overline{Lim}\left(w_S^1\right) + \overline{Lim}\left(w_S^2\right) + \cdots + \overline{Lim}\left(w_S^K\right)\right)/K \qquad (10.35b)$$

where $\underline{Lim}(w_l)$ and $\overline{Lim}(w_l)$ are the lower and upper limit of the average rough interval of risk factor l, respectively. According to Eqs. (10.35a and 10.35b), the average rough weight of Severity is $\overline{RN(w_S)} = [0.380, 0.398]$ where

$$\underline{Lim}(w_S) = \frac{(0.375 + 0.377 + 0.388)}{3} = 0.380$$

$$\overline{Lim}(w_S) = \frac{(0.388 + 0.395 + 0.411)}{3} = 0.398$$

The rough numbers and average rough weights of the other two risk factors can be obtained in a similar way (see Table 10.28).

Table 10.28 Rough numbers and average rough weights of risk factors

K	Severity		Occurrence		Detection	
	$\underline{Lim}(w_S^K)$	$\overline{Lim}(w_S^K)$	$\underline{Lim}(w_O^K)$	$\overline{Lim}(w_O^K)$	$\underline{Lim}(w_D^K)$	$\overline{Lim}(w_D^K)$
1	0.375	0.388	0.250	0.267	0.328	0.345
2	0.377	0.395	0.256	0.275	0.330	0.353
3	0.388	0.411	0.267	0.290	0.345	0.375
$\overline{RN(w_l)}$	[0.380, 0.398]		[0.257, 0.277]		[0.334, 0.358]	
Normalized	[0.955, 1.000]		[0.647, 0.697]		[0.839, 0.898]	

Stage 2: Failure Modes Assessment and Analysis

Experts use a crisp scale (i.e., midvalues of the fuzzy numbers in Table 10.5) to assess the FMs with respect to each risk factor. Table 10.9 presents the experts' assessments of the eight FMs. Since the assessment of FMs can be considered as a multi-criteria decision-making problem, the crisp ratings can be represented in the form of a decision matrix—which constitutes the main input of TOPSIS. Before the rough TOPSIS approach can be applied, the crisp ratings in the decision matrix need to be converted into rough numbers. The average rough assessments obtained by using Eqs. 10.32a, 10.32b, 10.32c–10.34 are given in Table 10.29.

Next, the average rough assessments are normalized in order to transform the various factor scales into comparable scales ranging from 0 to 1. The normalization method is conducted as follows (Song et al. 2014):

$$\underline{Lim}\left(x_{nl}'\right) = \frac{\underline{Lim}(x_{nl})}{\max_{n=1\text{to}N}\left\{\max\left[\underline{Lim}(x_{nl}), \overline{Lim}(x_{nl})\right]\right\}} \tag{10.36a}$$

$$\overline{Lim}\left(x_{nl}'\right) = \frac{\overline{Lim}(x_{nl})}{\max_{n=1\text{to}N}\left\{\max\left[\underline{Lim}(x_{nl}), \overline{Lim}(x_{nl})\right]\right\}} \tag{10.36b}$$

where $\left[\underline{Lim}\left(x_{nl}'\right), \overline{Lim}\left(x_{nl}'\right)\right]$ represents the lower and upper limits of the normalized average rough interval of failure mode FM_n with respect to risk factor RF_l. The normalized values are shown in Table 10.30.

Now, the normalized values are weighted by the relative importance of each risk factor using the following formulas.

$$\underline{Lim}(z_{nl}) = \underline{Lim}\left(w_l'\right) \cdot \underline{Lim}\left(x_{nl}'\right), n = 1, \ldots, N; l = 1, \ldots, L \tag{10.37a}$$

$$\overline{Lim}(z_{nl}) = \overline{Lim}\left(w_l'\right) \cdot \overline{Lim}\left(x_{nl}'\right), n = 1, \ldots, N; l = 1, \ldots, L \tag{10.37b}$$

Table 10.29 Average rough assessment matrix

Failure mode	Severity	Occurrence	Detection
FM$_1$	[3.889, 4.778]	[5.889, 6.778]	[7.889, 8.778]
FM$_2$	[1.889, 2.778]	[9.444, 9.889]	[1.889, 2.778]
FM$_3$	[1.889, 2.778]	[9.111, 9.556]	[0.611, 2.111]
FM$_4$	[3.222, 4.111]	[5.889, 6.778]	[7.889, 8.778]
FM$_5$	[3.889, 4.778]	[7.222, 8.111]	[9.111, 9.556]
FM$_6$	[5.889, 6.778]	[7.222, 8.111]	[3.222, 4.111]
FM$_7$	[0.611, 2.111]	[10, 10]	[0.611, 2.111]
FM$_8$	[0.111, 0.556]	[0, 0]	[0, 0]

Failure mode	Severity	Occurrence	Detection
FM$_1$	[0.574, 0.705]	[0.589, 0.678]	[0.826, 0.919]
FM$_2$	[0.279, 0.410]	[0.944, 0.989]	[0.198, 0.291]
FM$_3$	[0.279, 0.410]	[0.911, 0.956]	[0.064, 0.221]
FM$_4$	[0.475, 0.607]	[0.589, 0.678]	[0.826, 0.919]
FM$_5$	[0.574, 0.705]	[0.722, 0.811]	[0.953, 1.000]
FM$_6$	[0.869, 1.000]	[0.722, 0.811]	[0.337, 0.430]
FM$_7$	[0.090, 0.311]	[1.000, 1.000]	[0.064, 0.221]
FM$_8$	[0.016, 0.082]	[0.000, 0.000]	[0.000, 0.000]

Table 10.30 Normalized rough matrix

For example, $\underline{Lim}(z_{1S})$ and $\overline{Lim}(z_{1S})$ is calculated as follows:

$$\underline{Lim}(z_{1S}) = \underline{Lim}\left(w_S'\right) \cdot \underline{Lim}\left(x_{1S}'\right) = 0.955 \cdot 0.574 = 0.548$$

$$\overline{Lim}(z_{1S}) = \overline{Lim}\left(w_S'\right) \cdot \overline{Lim}\left(x_{1S}'\right) = 1.000 \cdot 0.705 = 0.705$$

The weighted normalized rough values of the eight FMs are given in Table 10.31.

Finally, the weighted normalized rough values are analyzed by means of TOPSIS. TOPSIS handles a multi-criteria problem as a geometric system with n points in an l-dimensional space. It is based on the concept that the chosen alternative should have the shortest distance from the positive-ideal solution (i.e., the solution composed of all the best attribute values achievable) and the longest distance from the negative-ideal solution (i.e., the solution composed of all the worst attribute values achievable) (Hwang and Yoon 1981). An index called relative closeness is used to rank the alternatives with respect to their similarity to the positive-ideal solution, as well as, the remoteness from the negative-ideal solution (Yoon and Hwang 1995).

Failure mode	Severity	Occurrence	Detection
FM$_1$	[0.548, 0.705]	[0.381 0.472]	[0.693 0.825]
FM$_2$	[0.266, 0.410]	[0.611 0.689]	[0.166 0.261]
FM$_3$	[0.266, 0.410]	[0.589 0.666]	[0.054 0.198]
FM$_4$	[0.454, 0.607]	[0.381 0.472]	[0.693 0.825]
FM$_5$	[0.548, 0.705]	[0.467 0.565]	[0.800 0.898]
FM$_6$	[0.829, 1.000]	[0.467 0.565]	[0.283 0.386]
FM$_7$	[0.086, 0.311]	[0.647 0.697]	[0.054 0.198]
FM$_8$	[0.016, 0.082]	[0.000 0.000]	[0.000 0.000]

Table 10.31 Weighted normalized rough matrix

In the context of rough sets, the Positive Ideal Solution (z_l^+) and Negative Ideal Solution (z_l^-) are identified as (Song et al. 2014):

$$z_l^+ = \left\{\max_{n=1\,to\,N}\left(\overline{Lim}(z_{nl})\right),\ \text{if}\ l \in B;\ \min_{n=1\,to\,N}\left(\underline{Lim}(z_{nl})\right),\ \text{if}\ l \in C\right\} \quad (10.38a)$$

$$z_l^- = \left\{\min_{n=1\,to\,N}\left(\underline{Lim}(z_{nl})\right),\ \text{if}\ l \in B;\ \max_{n=1\,to\,N}\left(\overline{Lim}(z_{nl})\right),\ \text{if}\ l \in C\right\} \quad (10.38b)$$

where B and C are associated with benefit and cost criterion, respectively. The Positive and Negative Ideal Solutions in our example are 0.016 and 1.000 for Severity, 0 and 0.697 for Occurrence, and 0 and 0.898 for Detection. The relative closeness (RC_n) of each FM is calculated using the l-dimensional Euclidean distances $(d_n^+$ and $d_n^-)$ as follows:

$$RC_n = \frac{d_n^-}{d_n^- + d_n^+},\ n = 1,\ldots,N \quad (10.39a)$$

where

$$d_n^+ = \left\{\sum_{l\in B}\left(\underline{Lim}(z_{nl}) - z_l^+\right)^2 + \sum_{l\in C}\left(\overline{Lim}(z_{nl}) - z_l^+\right)^2\right\}^{1/2},\ n = 1,\ldots,N$$
$$(10.39b)$$

$$d_n^- = \left\{\sum_{l\in B}\left(\overline{Lim}(z_{nl}) - z_l^-\right)^2 + \sum_{l\in C}\left(\underline{Lim}(z_{nl}) - z_l^-\right)^2\right\}^{1/2},\ n = 1,\ldots,N$$
$$(10.39c)$$

Using these equations, the distances and the relative closeness for FM_1 are calculated as follows

$$d_1^+ = \left\{(0.705 - 0.016)^2 + (0.472 - 0)^2 + (0.825 - 0)^2\right\}^{1/2} = 1.174$$

$$d_1^- = \left\{(0.548 - 1)^2 + (0.381 - 0.697)^2 + (0.693 - 0.898)^2\right\}^{1/2} = 0.589$$

then

$$RC_1 = \frac{d_1^-}{d_1^- + d_1^+} = \frac{0.589}{0.589 + 1.174} = 0.334$$

The relative closeness values and rank orders of the FMs are given in Table 10.32. The priority ranking of the eight FMs is estimated as follows: FM_5, FM_1, FM_6, FM_4, FM_2, FM_3, FM_7, and FM_8.

Table 10.32 Relative closeness values and their rank order

Failure mode	Positive ideal solution	Negative ideal solution	Relative closeness	Rank
FM_1	1.174	0.589	0.334	2
FM_2	0.835	1.040	0.555	5
FM_3	0.799	1.124	0.585	6
FM_4	1.119	0.663	0.372	4
FM_5	1.265	0.517	0.290	1
FM_6	1.199	0.678	0.361	3
FM_7	0.782	1.245	0.614	7
FM_8	0.066	1.504	0.958	8

10.3.6 FMEA Using 2-Tuple Fuzzy Linguistic Representation

A typical drawback in problems modelled with linguistic information is the lack of precision in the final results. This loss of information, which is inevitable when computing with words, can be avoided by using the fuzzy linguistic representation model proposed by Herrera and Martínez (2000). This model represents crisp or linguistic information in form of 2-tuples, and thereby, allows a continuous representation of the linguistic information on its domain. A 2-tuple, (s, α), is a pair of information composed by a linguistic term, s, and a numeric value, α, assessed in (-0.5, 0.5). Below, more formal definitions are provided for this symbolic translation.

Suppose that, the semantic element s_i is assessed by the linguistic variable s defined in the linguistic term set $S = \{s_0, s_1, \ldots, s_g\}$ where $i \in [0, g]$.

Definition 1 (Herrera and Martínez 2000): Let β be the result of an aggregation of the indices of a set of labels assessed in a linguistic term set S. $\beta \in [0, g]$, and $g + 1$ is the cardinality of S. Let $i = \text{round}(\beta)$ and $\alpha = \beta - i$ be two values such that $i \in [0, g]$ and $\alpha \in [-0.5, 0.5)$ then α is called a symbolic translation.

This can be expressed as follows, s_i represents the linguistic label center of the information; and α_i is a numerical value expressing the value of the translation from the original result β to the closest linguistic term i. Then, a 2-tuple that represents the equivalent information to β can be obtained as follows (Herrera and Martínez 2000):

$$\Delta : [0, g] \rightarrow S \times [-0.5, 0.5) \tag{10.40a}$$

$$\Delta(\beta) = (s_i, \alpha), \text{with} \begin{cases} s_i, & i = \text{round}(\beta) \\ \alpha = \beta - i, & \alpha \in [-0.5, 0.5) \end{cases} \tag{10.40b}$$

where round$(\cdot)$ is the usual round operation. The aggregation operator for 2-tuples equivalent to the arithmetic mean is formulated as given below. It calculates the mean of a set of linguistic values without any loss of information (Herrera and Martínez 2000). In our case, it will be used to aggregate the linguistic assessments of different experts.

Definition 2 (Herrera and Martínez 2000): Let $x = \{(r_1, \alpha_1), \ldots, (r_L, \alpha_L)\}$ be a set of 2-tuples. The 2-tuple arithmetic mean $\bar{x}$ is computed as:

$$\bar{x} = \Delta\left(\sum_{l=1}^{L} \frac{1}{L}\Delta^{-1}(r_l, \alpha_l)\right) = \Delta\left(\frac{1}{L}\sum_{l=1}^{L} \beta_l\right) \qquad (10.41)$$

In order to synthesize the linguistic assessment values (of FMs) with a weighted aggregation operator, in which the weights are not associated with a predetermined value but rather are associated to a determined position, an OWA operator for dealing with linguistic 2-tuples is developed.

Definition 3 (Herrera and Martínez 2000): Let $A = \{(r_1, \alpha_1), \ldots, (r_L, \alpha_L)\}$ be a set of 2-tuples and $W = (w_1, \ldots, w_L)$ be an associated weighting vector that satisfies (i) $w_l \in [0, 1]$ and (ii) $\sum w_l = 1$. The 2-tuple OWA operator F is calculated as follows:

$$F((r_1, \alpha_1), \ldots, (r_L, \alpha_L)) = \Delta\left(\sum_{l=1}^{L} w_l \cdot \beta_{\sigma(l)}\right) \qquad (10.42)$$

where $\sigma : \{1, \ldots, L\} \to \{1, \ldots, L\}$ is a permutation function such that $\left\{\beta_{\sigma(1)}, \ldots, \beta_{\sigma(L)}\right\}$ are in descending order. Finally, the comparison of these values is carried out according to an ordinary lexicographic order. Let (s_n, α_1) and (s_m, α_2) be two 2-tuples (Herrera and Martínez 2000):

- if $n < m$ then (s_n, α_1) is smaller than (s_m, α_2);
- if $n = m$ then

  i. if $\alpha_1 = \alpha_2$ then $(s_n, \alpha_1), (s_m, \alpha_2)$ represent the same information;
  ii. if $\alpha_1 < \alpha_2$ then (s_n, α_1) is smaller than (s_m, α_2);
  iii. if $\alpha_1 > \alpha_2$ then (s_n, α_1) is bigger than (s_m, α_2).

Below, the approach combining fuzzy linguistic representation and the OWA operator proposed by Chang and Wen (2010) is illustrated.

Stage 1: Assessment of Failure Modes
Suppose that experts use a linguistic scale $(S = \{s_0, s_1, s_2, s_3, s_4, s_5, s_6\})$ to assess the FMs with respect to each risk factor (see Table 10.5). The assessment values of the eight FMs obtained from three experts are presented in Table 10.7. For example,

the linguistic value provided by the first expert for FM_1 with respect to Severity can be represented by means of a 2-tuple as $\Delta(M) = (s_3, 0)$. Next, the assessment values provided by the experts for each FM are synthesized into group 2-tuples using Eq. 10.41. For example, given the individual assessment values "Medium", "Medium", "Medium Low" of three experts for FM_1 with respect to the Severity risk factor, the corresponding group 2-tuple $(\bar{x}_{S_1})$ is obtained as follows:

$$\bar{x}_{Severity_1} = \Delta\left(\frac{1}{3}\sum_{l=1}^{3}\beta_l\right) = \Delta\left(\frac{3+3+2}{3}\right) = \Delta(2.667) = (s_3, -0.333)$$

Table 10.33 presents the resulting group 2-tuples for all FMs and risk factors.

Stage 2: Calculation of the Weights
The weights used in Eq. 10.42 are derived by the method suggested by Fuller and Majlender (2001) who uses Lagrange multipliers to formulate a polynomial equation and then determine the optimal weighting vector by solving a constrained optimization problem. The weight vector is obtained as follows:

$$\ln w_j = \frac{j-1}{l-1}\ln w_l + \frac{l-j}{l-1}\ln w_1 \Rightarrow w_j = {}^{l-1}\sqrt{w_1^{l-j}w_l^{j-1}} \qquad (10.43)$$

$$w_l = \frac{((l-1)\gamma - l)w_1 + 1}{(l-1)\gamma + 1 - lw_1} \qquad (10.44)$$

$$w_1[(l-1)\gamma + 1 - lw_1]^l = ((l-1)\gamma)^{l-1}[((l-1)\gamma - l)w_1 + 1] \qquad (10.45)$$

where w is the weight vector, l is the number of factors, and γ is the situation parameter. For the weighting procedure, first, experts have to provide the situation parameter $(0 \leq \gamma \leq 1)$. Table 10.34 provides the optimal weight vectors (calculated using Eqs. 10.43–10.45) for different situation parameters.

Stage 3: Synthesis of Linguistic Assessment Values
In this stage, the group assessments are synthesized over the three risk factors into overall priorities. For the synthesis, the 2-tuple OWA operator (see Eq. 10.42) is used in which the derived weights are associated with particular ordered positions

Table 10.33 Group 2-tuples

Failure mode	Severity	Occurrence	Detection
FM_1	$(s_3, -0.333)$	$(s_4, -0.333)$	$(s_5, -0.333)$
FM_2	$(s_2, -0.333)$	$(s_6, -0.333)$	$(s_2, -0.333)$
FM_3	$(s_2, -0.333)$	$(s_5, 0.333)$	$(s_1, 0)$
FM_4	$(s_2, 0.333)$	$(s_4, -0.333)$	$(s_5, -0.333)$
FM_5	$(s_3, -0.333)$	$(s_4, 0.333)$	$(s_5, 0.333)$
FM_6	$(s_4, -0.333)$	$(s_4, 0.333)$	$(s_2, 0.333)$
FM_7	$(s_1, 0)$	$(s_6, 0)$	$(s_1, 0)$
FM_8	$(s_0, 0.333)$	$(s_0, 0)$	$(s_0, 0)$

Table 10.34 Optimal weight vectors for $l = 3$

Gamma	w_1	w_2	w_3
0.5	0.333	0.333	0.333
0.6	0.438	0.323	0.238
0.7	0.554	0.292	0.154
0.8	0.682	0.236	0.082
0.9	0.826	0.147	0.026
1.0	1	0	0

Table 10.35 Priority values and their rank order

Failure mode	$\Delta\left(\sum_{l=1}^{3} w_l \cdot \beta_{\sigma(l)}\right)$	(s_i, α)	Rank
FM_1	$\Delta(3.867)$	$(s_4, -0,133)$	2
FM_2	$\Delta(3.420)$	$(s_3, 0,420)$	5
FM_3	$\Delta(3.115)$	$(s_3, 0,115)$	7
FM_4	$\Delta(3.787)$	$(s_4, -0,213)$	3
FM_5	$\Delta(4.374)$	$(s_4, 0,374)$	1
FM_6	$\Delta(3.641)$	$(s_4, -0,359)$	4
FM_7	$\Delta(3.192)$	$(s_3, 0,192)$	6
FM_8	$\Delta(0.146)$	$(s_0, 0,146)$	8

of the aggregated values but have no connection with these values. For example, the priority value of FM_1 $(F_{FM_1}((s_3, -0.333), (s_4, -0.333), (s_5, -0.333)))$ is calculated as follows:

$$\Delta\left(\sum_{l=1}^{3} w_l \cdot \beta_{\sigma(l)}\right) = \Delta(0.438 \cdot 4.667 + 0.323 \cdot 3.667 + 0.238 \cdot 2.667)$$

$$= \Delta(3.867) = (s_4, -0.133)$$

Note that the situation parameter γ was chosen here as 0.6 and the corresponding optimal weight vector was calculated as (0.438 0.323 0.238). The results for all FMs are shown in Table 10.35. Eventually, the priority ranking of the eight failure modes is estimated as follows: FM_5, FM_1, FM_4, FM_6, FM_2, FM_7, FM_3, FM_8.

10.4 Comparison of the Approaches

The substantial concern here is whether the demonstrated approaches yield the same outcome or not. Table 10.36 summarizes the results of all six approaches plus the traditional FMEA method. Considering first the results of the traditional method, it is not surprising that the simple multiplication produces RPNs with exactly the same value from different combinations of S, O, and D (notice the RPNs

Table 10.36 Comparison of the rankings obtained by the approaches demonstrated

Failure mode	Traditional method		Evidential reasoning		Ordinary fuzzy sets		Grey relational analysis		Intuitionistic fuzzy sets			Rough set theory		2-Tuple fuzzy linguistic R.	
	Priority	Rank	Priority	Rank	Priority	Rank	Grade	Rank	Score	Accuracy	Rank	Closeness	Rank	2-tuple	Rank
FM$_1$	192	3	0.656	2	0.666	2	0.419	2	0.290	0.935	2	0.334	2	(s$_4$, -0,133)	2
FM$_2$	40	5	0.391	5	0.430	5	0.602	5	-0.106	0.916	5	0.555	5	(s$_3$, 0,420)	5
FM$_3$	18	6	0.317	6	0.382	7	0.667	6	-0.209	0.926	6	0.585	6	(s$_3$, 0,115)	7
FM$_4$	192	3	0.647	3	0.660	3	0.427	3	0.281	0.931	3	0.372	4	(s$_4$, -0,213)	3
FM$_5$	288	1	0.747	1	0.767	1	0.393	1	0.458	0.964	1	0.290	1	(s$_4$, 0,374)	1
FM$_6$	192	3	0.488	4	0.510	4	0.510	4	0.037	0.929	4	0.361	3	(s$_4$, -0,359)	4
FM$_7$	10	7	0.316	7	0.400	6	0.683	7	-0.209	0.930	7	0.614	7	(s$_3$, 0,192)	6
FM$_8$	0	8	0.099	8	0.055	8	0.989	8	-0.783	0.998	8	0.958	8	(s$_0$, 0,146)	8

Table 10.37 Comparison of the approaches with respect to their characteristic features

Approach	Factor weights	Uncertainty and subjectivity	Knowledge representation	Ranking (Distinctiveness)	Synthesis	Prior information
Traditional method	No	No (ignores uncertainty)	Is based on crisp assessment	Poorly distinguishing (RPNs are not continuous with many gaps)	Simple multiplication	Partly (equal weights assumptions)
Evidential reasoning	Yes	Partly (considers intra- and partly inter-personal uncertainty)	Uses different types of assessment information, confidence level, expert weights; but information loss is inevitable due to defuzzification	Well distinguishing (continuous overall group belief structures are used)	Weighted geometric mean of overall belief structures	Yes (belief structure, pre-determined functions)
Ordinary fuzzy sets	Yes	Partly (considers intra-personal uncertainty)	Uses expert weights, membership function; but information loss is inevitable due to defuzzification	Well distinguishing (continuous defuzzified risk assessment values are used)	Weighted average of defuzzified values	Yes (pre-determined membership function)
Grey relational analysis	Yes	Partly (considers incomplete information)	Uses grey relational coefficient; but restricted to crisp values	Well distinguishing (continuous grey relational grades are used)	Weighted average of grey relational grade	Yes (reference series)

(continued)

of

Table 10.37 (continued)

Approach	Factor weights	Uncertainty and subjectivity	Knowledge representation	Ranking (Distinctiveness)	Synthesis	Prior information
Intuitionistic fuzzy sets	Yes	Partly (considers intra-personal uncertainty)	Uses membership and non-membership functions, expert weights	Well distinguishing (continuous score and accuracy degrees are used)	Intuitionistic fuzzy weighted geometric aggregation of IFNs	Yes (pre-determined membership function, assumptions)
Rough set theory	Yes	Partly (considers inter-personal uncertainty)	Uses boundary region; but restricted to crisp values	Well distinguishing (rough numbers and continuous closeness coefficients are used)	Euclidean distance using rough values	No
2-Tuple fuzzy linguistic R.	Yes	Partly (considers inter-personal uncertainty)	Uses symbolic translation	Well distinguishing (continuous 2-tuples are used)	Ordered weighted averaging of group linguistic assessment values	Partly (label indices)

FM_1, FM_4, and FM_6 given in Table 10.36). This inevitably leads to confusion in prioritizing the FMs. On the other hand, it can be seen from the table that the scale and precision levels of the obtained values (i.e., priorities, scores, relative closeness, or linguistic representation values) differ significantly among the approaches, even though only slight differences can be noticed in the rankings. For example, the approaches based on 2-tuples and ordinary fuzzy sets suggest a ranking for the failure modes three and seven, which is exactly the opposite of the rankings suggested by the remaining five approaches. Such differences are obvious and should be expected since each approach suggests its own unique way to deal with uncertainty and subjectivity associated with risk assessment, as well as addresses only a particular set of conventional FMEA shortcomings. Thus, in order to examine the effectiveness of the approaches, it is not sufficient to merely compare the obtained rankings, but it is also necessary to examine the methodological differences. Below, the methodological differences of the demonstrated approaches are reviewed with respect to six characteristic features (see Table 10.37). Notice that the review considers only the six approaches as they are demonstrated here, and does not attempt to examine their underlying theories in detail. Certainly, the approaches can be extended or improved by integrating other techniques (for synthesis or weighting) and theories, but this option will be ignored here.

- **Factor Weights:** As discussed earlier, one of the most criticized shortcomings of the traditional FMEA is that all three factors are assumed to have the same importance. All the discussed approaches, on the other hand, suggests alternative mechanisms—such as weighted averaging, ordered weighted averaging, Euclidean distance—to incorporate the weight information into the analysis. In this respect, they do not really differ from each other and appear to be promising.
- **Uncertainty and Subjectivity:** This feature points out whether an approach can deal with subjectivity in both weight determination and failure assessment, as well as indicates the type of uncertainty it addresses. Except the traditional method, which completely ignores uncertainty, all the discussed approaches suggest alternative ways of dealing with uncertainty and subjectivity. For example, the FMEA approaches based on ordinary and intuitionistic fuzzy sets are mainly used to handle intra-personal uncertainty, which relates to assessments made under lack of knowledge and/or limited attention. The approaches based on rough set theory and 2-tuple fuzzy linguistic representation, on the other hand, basically, consider inter-personal uncertainty, which arises when a group of subjects delivers different judgments. Only evidential reasoning seems to allow dealing with both intra-personal and inter-personal uncertainty, however, with the latter only partly. The last approach based on grey relational analysis differs from the other approaches in that it handles partially known information. Consequently, although the alternative mechanisms suggested in these approaches provide the FMEA team a more realistic representation of subjective and uncertain information, there is still a need for an approach that properly captures both types of uncertainties (intra-personal and inter-personal).

- **Knowledge Representation**: How knowledge is extracted from a group of experts is another important feature that can be used to compare the approaches. With respect to this feature, evidential reasoning provides the most flexible and richest representation that allows experts using different possible ways to represent their assessments (e.g., certain, distribution, interval and total ignorance). Moreover, it considers experts' level of confidence in their assessments as well as their relative importance. Despite these benefits, an information loss to some degree is inevitable due to the defuzzification used in this approach. The approaches based on rough set theory and 2-tuple fuzzy linguistic representation also suggest unique ways to represent knowledge. The boundary region defined in rough set theory allows modelling the diversity and uncertainty of FMEA team members' assessment information; but using crisp values restricts this approach's potential. The symbolic translation used in 2-tuple linguistic representation avoids any loss of information produced by operations on linguistic values. Alternatively, knowledge can be represented by IFSs, which allows experts considering hesitation as part of their assessments by using both membership and non-membership functions. Thus, IFSs can represent the uncertainty in a more comprehensive manner than ordinary fuzzy sets. Finally, approaches based on grey relational analysis seem to be restricted to crisp values.
- **Ranking (Distinctiveness):** This feature indicates whether subtle differences in assessments are properly reflected in the rankings. All approaches, except the traditional one, suggest alternative mechanisms (e.g., 2-tuples, rough numbers, grey relational grades, belief structures, score and accuracy degrees) that allow ranking FMs on a continuous scale, and thereby, avoid unnecessary confusion. However, as mentioned earlier, the precision levels of the produced priority values differ significantly among the approaches.
- **Synthesis:** How the assessments are synthesized over the relevant risk factors into overall values is another valuable feature that needs to be considered. Not surprisingly, the method commonly preferred to synthesize assessment values is the weighted average (arithmetic or geometric) method. More advanced techniques, such as, the ordered weighted averaging and Euclidean distance from the ideal solution are also employed. In comparison to the simple weighted average, aggregation operators that weight assessment values according to their ordering, obtain more reasonable rankings of FMs. It is important to point out that the way the values are synthesized is largely independent of the method employed to represent and to analyze uncertain information.
- **Prior Information:** This feature describes whether an approach requires prior information, such as, assumptions or pre-defined functions (e.g., membership functions) to deal with uncertainty or not. Except the approach based on rough sets, all other approaches need a preliminary or additional information about data, such as, belief structures in evidential reasoning, membership functions in (intuitionistic) fuzzy sets, reference series in grey relational analysis and label indices in 2-tuples. With respect to this feature, rough set theory seems to be the most favorable approach. It focuses on the uncertainty caused by limited

distinction between assessments and extracts the facts hidden in the data by using only the available information. Thus, rough set theory maintains the objectivity of original information.

10.5 Conclusion and Suggestions for Future Research

The purpose of this study is to provide a comprehensive review of the approaches developed for failure mode and effects analysis under uncertainty and offer a brief tutorial for those who are interested in these approaches. According to the review above, we can safely conclude that the discussed approaches provide a better modeling of uncertainty, and thereby, a richer knowledge representation than the traditional FMEA. Although, the theories, on which the approaches are based, overlap to some extent in dealing with uncertainty and subjectivity, each can be viewed as an entity in its own right. These theories are neither competing nor the same; instead, they tend to complement each other. Especially, the recent shift towards hybrid/integrated methods (e.g., evidential reasoning combined with grey theory), seems to justify this argument (see also Liu et al. 2013b). Therefore, research studies are required that integrate different methods or theories to create synergies, and thereby, enhance the efficacy of risk assessment. A related issue that still remains unaddressed is the need for a prescriptive model that supports the FMEA team in deciding under which particular circumstances which FMEA approach to prefer. Hopefully, the present review may serve as a guidance for such attempts. Another main concern for future research is the lack of appropriate quality measures. These measures should check whether an approach is reliable and fulfills its intended purpose.

In the current study, two types of uncertainties (variation in one expert's understanding and variations in the understanding among experts) associated with the assessment information have been distinguished. The review reveals that none of the discussed approaches is able to sufficiently deal with both types. In order to address this issue, type-2 fuzzy sets can be used to develop an alternative approach, since they are capable of modelling both types of uncertainties. Finally, dealing with total ignorance, handling different types of assessment information simultaneously, and incorporating other factors, such as, costs are the other modelling issues that need further consideration in the FMEA literature.

References

Arabian-Hoseynabadi, H., Oraee, H., Tavner, P.: Failure modes and effects analysis (FMEA) for wind turbines. Int. J. Electr. Power Energy Syst. **32**, 817–824 (2010)
Asan, U., Soyer, A., Bozdag, C. E.: An interval type-2 fuzzy prioritization approach to project risk assessment. J. Multiple-Valued Logic Soft Comput. in press (2016)

Asan, U., Soyer, A., Bozdağ, C.E.: A fuzzy multi-criteria group decision making approach to project risk assessment. The 4th International Conference on Risk Analysis and Crisis Response. Istanbul, Turkey (2013)

Asan, U., Soyer, A., Serdarasan, S.: A Fuzzy Analytic Network Process Approach. In: Kahraman, C. (ed.) Computational Intelligence Systems in Industrial Engineering. Atlantis Press, Paris (2012)

Atanassov, K.T.: Intuitionistic fuzzy sets. Fuzzy Sets Syst. **20**, 87–96 (1986)

Bevilacqua, M., Braglia, M., Gabbrielli, R.: Monte Carlo simulation approach for a modified FMECA in a power plant. Quality Reliability Eng. Int. **16**, 313–324 (2000)

Bowles, J.B., Peláez, C.E.: Fuzzy logic prioritization of failures in a system failure mode, effects and criticality analysis. Reliability Eng. Syst. Saf. **50**, 203–213 (1995)

Braglia, M.: MAFMA: multi-attribute failure mode analysis. Int. J. Quality Reliability Manage. **17**, 1017–1033 (2000)

Braglia, M., Frosolini, M., Montanari, R.: Fuzzy TOPSIS approach for failure mode, effects and criticality analysis. Quality Reliability Eng. Int. **19**, 425–443 (2003)

Braglia, M., Fantoni, G., Frosolini, M.: The house of reliability. Int. J. Quality Reliability Manage. **24**, 420–440 (2007)

Bujna, M., Prístavka, M.: Risk Analysis of Production Process for Automotive and Electrical Engineering Industry Using FMEA. Acta Technologica Agriculturae **16**, 87–89 (2013)

Carlson, C.S.: Effective FMEAs: Achieving Safe, Reliable, and Economical Products and Processes Using Failure Mode and Effects Analysis. Wiley, Hoboken, New Jersey (2012)

Chang, K.-H.: Evaluate the orderings of risk for failure problems using a more general RPN methodology. Microelectron. Reliab. **49**, 1586–1596 (2009)

Chang, K.H., Cheng, C.H.: A risk assessment methodology using intuitionistic fuzzy set in FMEA. Int. J. Syst. Sci. **41**, 1457–1471 (2010)

Chang, K.-H., Cheng, C.-H.: Evaluating the risk of failure using the fuzzy OWA and DEMATEL method. J. Intell. Manuf. **22**, 113–129 (2011)

Chang, D.-S., Sun, K.-L.P.: Applying DEA to enhance assessment capability of FMEA. Int. J. Quality Reliability Manage. **26**, 629–643 (2009)

Chang, K.-H., Wen, T.-C.: A novel efficient approach for DFMEA combining 2-tuple and the OWA operator. Expert Syst. Appl. **37**, 2362–2370 (2010)

Chang, C.-L., Wei, C.-C., Lee, Y.-H.: Failure mode and effects analysis using fuzzy method and grey theory. Kybernetes **28**, 1072–1080 (1999)

Chang, C.-L., Liu, P.-H., Wei, C.-C.: Failure mode and effects analysis using grey theory. Integr. Manuf. Syst. **12**, 211–216 (2001)

Chang, K.-H., Cheng, C.-H., Chang, Y.-C.: Reprioritization of failures in a silane supply system using an intuitionistic fuzzy set ranking technique. Soft. Comput. **14**, 285–298 (2010)

Chang, K.-H., Chang, Y.-C., Wen, T.-C., Cheng, C.-H.: An innovative approach integrating 2-tuple and LOWGA operators in process failure mode and effects analysis. Int J Innov Comp Inf Control **8**, 747–761 (2012)

Chang, K.-H., Chang, Y.-C., Tsai, I.: Enhancing FMEA assessment by integrating grey relational analysis and the decision making trial and evaluation laboratory approach. Eng. Fail. Anal. **31**, 211–224 (2013)

Chen, C.-T.: Extensions of the TOPSIS for group decision-making under fuzzy environment. Fuzzy Sets Syst. **114**, 1–9 (2000)

Chen, J.K.: Utility priority number evaluation for FMEA. J. Fail. Anal. Prev. **7**, 321–328 (2007)

Chen, C.-B., Klein, C.M.: A simple approach to ranking a group of aggregated fuzzy utilities. Syst. Man Cybern. Part B: Cybernetics, IEEE Trans. **27**, 26–35 (1997)

Chen, L.-H., Ko, W.-C.: Fuzzy linear programming models for new product design using QFD with FMEA. Appl. Math. Model. **33**, 633–647 (2009)

Chin, K.-S., Wang, Y.-M., Poon, G.K.K., Yang, J.-B.: Failure mode and effects analysis by data envelopment analysis. Decis. Support Syst. **48**, 246–256 (2009a)

Chin, K.-S., Wang, Y.-M., Poon, G.K.K., Yang, J.-B.: Failure mode and effects analysis using a group-based evidential reasoning approach. Comput. Oper. Res. **36**, 1768–1779 (2009b)

Da costa, C.G., Bedregal, B.C., Neto, A.D.D: Intuitionistic fuzzy probability. In: Costa, A.C.D.R., Vicari, R.M., Tonidandel, F. (eds.) Advances in Artificial Intelligence–SBIA 2010. Springer-Verlag, Berlin Heidelberg (2010)

Deng, J.-L.: Control problems of grey systems. Syst. Control Lett. **1**, 288–294 (1982)

Deng, J.-L.: Introduction to grey system theory. J. Grey Syst. **1**, 1–24 (1989)

Dhillon, B.S.: Mining equipment safety: a review, analysis methods and improvement strategies. Int. J. Min. Reclam. Environ. **23**, 168–179 (2009)

Dubois, D., Prade, H.: Possibility Theory and its Applications: Where Do we Stand? Mathware Soft Comput. **18**, 18–31 (2011)

Franceschini, F., Galetto, M.: A new approach for evaluation of risk priorities of failure modes in FMEA. Int. J. Prod. Res. **39**, 2991–3002 (2001)

Fuller, R., Majlender, P.: An analytic approach for obtaining maximal entropy OWA operator weights. Fuzzy Sets Syst. **124**, 53–57 (2001)

Garcia, P.A.A., Schirru, R., Melo, P.: A fuzzy data envelopment analysis approach for fmea. Prog. Nucl. Energy **46**, 359–373 (2005)

Gargama, H., Chaturvedi, S.K.: Criticality Assessment Models for Failure Mode Effects and Criticality Analysis Using Fuzzy Logic. IEEE Trans. Reliab. **60**, 102–110 (2011)

Geum, Y., Cho, Y., Park, Y.: A systematic approach for diagnosing service failure: Service-specific FMEA and grey relational analysis approach. Math. Comput. Model. **54**, 3126–3142 (2011)

Guimarães, A.C.F., Lapa, C.M.F.: Fuzzy inference to risk assessment on nuclear engineering systems. Appl. Soft Comput. **7**, 17–28 (2007)

Hadi-Vencheh, A., Aghajani, M.: Failure mode and effects analysis: A fuzzy group MCDM approach. J. Soft Comput. Appl. **2013**, 1–14 (2013)

Herrera, F., Herrera-viedma, E., Martínez, L.: A fusion approach for managing multi-granularity linguistic term sets in decision making. Fuzzy Sets and Systems, **114**, 43–58 (2000)

Herrera, F., Martínez, L.: A 2-tuple fuzzy linguistic representation model for computing with words. Fuzzy Syst. IEEE Trans. **8**, 746–752 (2000)

Hu, A.H., Hsu, C.-W., Kuo, T.-C., Wu, W.-C.: Risk evaluation of green components to hazardous substance using FMEA and FAHP. Expert Syst. Appl. **36**, 7142–7147 (2009)

Hwang, C., Yoon, K.: Multiple Attribute Decision Making: Methods and Applications. Berlin (1981)

ISO/IEC-15026-1. Systems and software engineering-Systems and software assurance-Part 1: Concepts and vocabulary, ISO/IEC JTC 1/SC 7 Software and systems engineering subcommittee, Switzerland (2013)

Kerekes, L., Johanyák, Z.C.: Construction FMEA analysis of industrial sewing machines. In: 2nd International Conference on Innovation, Techniques and Education in the Textile, Garment, Shoe and Leather Industry, Budapest (1996)

Kostina, M., Karaulova, T., Sahno, J., Maleki, M.: Reliability estimation for manufacturing processes. J. Achiev. Mater. Manuf. Eng. **51**, 7–13 (2012)

Kuo, Y., Yang, T., Huang, G.-W.: The use of grey relational analysis in solving multiple attribute decision-making problems. Comput. Ind. Eng. **55**, 80–93 (2008)

Kutlu, A.C., Ekmekçioğlu, M.: Fuzzy failure modes and effects analysis by using fuzzy TOPSIS-based fuzzy AHP. Expert Syst. Appl. **39**, 61–67 (2012)

Lin, Q.-L., Wang, D.-J., Lin, W.-G., Liu, H.-C.: Human reliability assessment for medical devices based on failure mode and effects analysis and fuzzy linguistic theory. Saf. Sci. **62**, 248–256 (2014)

Liu, H.C., Liu, L., Li, P.: Failure mode and effects analysis using intuitionistic fuzzy hybrid weighted Euclidean distance operator. International Journal of Systems Science (2013c)

Liu, S., Lin, Y.: Grey Systems: Theory and Applications. Springer-Verlag, Berlin Heidelberg (2010)

Liu, H.-C., Liu, L., Li, P.: Failure mode and effects analysis using intuitionistic fuzzy hybrid weighted Euclidean distance operator. Int. J. Syst. Sci. 1–19 (2014)

Liu, H.-C., Liu, L., Bian, Q.-H., Lin, Q.-L., Dong, N., Xu, P.-C.: Failure mode and effects analysis using fuzzy evidential reasoning approach and grey theory. Expert Syst. Appl. **38**, 4403–4415 (2011)

Liu, H.-C., Liu, L., Liu, N., Mao, L.-X.: Risk evaluation in failure mode and effects analysis with extended VIKOR method under fuzzy environment. Expert Syst. Appl. **39**, 12926–12934 (2012)

Liu, H.-C., Liu, L., Lin, Q.-L.: Fuzzy failure mode and effects analysis using fuzzy evidential reasoning and belief rule-based methodology. IEEE Trans. Reliab. **62**, 23–36 (2013a)

Liu, H.-C., Liu, L., Liu, N.: Risk evaluation approaches in failure mode and effects analysis: A literature review. Expert Syst. Appl. **40**, 828–838 (2013b)

Mandal, S., Maiti, J.: Risk analysis using FMEA: Fuzzy similarity value and possibility theory based approach. Expert Syst. Appl. **41**, 3527–3537 (2014)

Mozaffari, F., Eidi, A., Mohammadi, L., Alavi, Z.: Implementation of FMEA to improve the reliability of GEO satellite payload. In: Reliability and Maintainability Symposium (RAMS), 2013 Proceedings-Annual, 2013. IEEE, 1–6

Pawlak, Z.: Rough set approach to knowledge-based decision support. Eur. J. Oper. Res. **99**, 48–57 (1997)

Pelaez, E.C., Bowles, J.B.: Using fuzzy cognitive maps as a system model for failure modes and effects analysis. Inf. Sci. **88**, 177–199 (1996)

Pillay, A., Wang, J.: Modified failure mode and effects analysis using approximate reasoning. Reliab. Eng. Syst. Saf. **79**, 69–85 (2003)

Ramli, N., Mohamad, D.: A comparative analysis of centroid methods in ranking fuzzy numbers. Eur. J. Sci. Res. **28**, 492–501 (2009)

Ross, T.J.: Fuzzy Logic With Engineering Applications. Wiley, England (2009)

Saaty, T.L.: Multicriteria Decision Making: The Analytic Hierarchy Process. Pittsburgh, RWS Pub (1988)

Sant'Anna, A.P.: Probabilistic priority numbers for failure modes and effects analysis. Int. J. Qual. Reliab. Manage. **29**, 349–362 (2012)

Seyed-Hosseini, S.M., Safaei, N., Asgharpour, M.J.: Reprioritization of failures in a system failure mode and effects analysis by decision making trial and evaluation laboratory technique. Reliab. Eng. Syst. Saf. **91**, 872–881 (2006)

Sharma, R.K., Kumar, D., Kumar, P.: Application of fuzzy methodology to build process reliability: a practical case. Int. J. Prod. Dev. **5**, 125–152 (2008a)

Sharma, R.K., Kumar, D., Kumar, P.: Fuzzy modeling of system behavior for risk and reliability analysis. Int. J. Syst. Sci. **39**, 563–581 (2008b)

Song, W., Ming, X., Wu, Z., Zhu, B.: Failure modes and effects analysis using integrated weight-based fuzzy TOPSIS. Int. J. Comput. Integr. Manuf. **26**, 1172–1186 (2013)

Song, W., Ming, X., Wu, Z., Zhu, B.: A rough TOPSIS approach for failure mode and effects analysis in uncertain environments. Qual. Reliab. Eng. Int. **30**, 473–486 (2014)

Stamatis, D.H.: Failure mode and effect analysis: FMEA from theory to execution. ASQ Quality Press Pub, Milwaukee, Wisconsin (2003)

Wang, Y.-M., Yang, J.-B., Xu, D.-L., Chin, K.-S.: The evidential reasoning approach for multiple attribute decision analysis using interval belief degrees. Eur. J. Oper. Res. **175**, 35–66 (2006)

Wang, Y.-M., Chin, K.-S., Poon, G.K.K., Yang, J.-B.: Risk evaluation in failure mode and effects analysis using fuzzy weighted geometric mean. Expert Syst. Appl. **36**, 1195–1207 (2009)

Wei, G.: Some induced geometric aggregation operators with intuitionistic fuzzy information and their application to group decision making. Appl. Soft Comput. **10**, 423–431 (2010)

Welborn, C.: Applying failure mode and effects analysis to supplier selection. IUP J Supply Chain Manage. **7**, 7–14 (2010)

Wu, D., Mendel, J.M.: A comparative study of ranking methods, similarity measures and uncertainty measures for interval type-2 fuzzy sets. Inf. Sci. **179**, 1169–1192 (2009)

Xiao, N., Huang, H.-Z., Li, Y., He, L., Jin, T.: Multiple failure modes analysis and weighted risk priority number evaluation in FMEA. Eng. Fail. Anal. **18**, 1162–1170 (2011)

Xu, Z.: Approaches to multiple attribute group decision making based on intuitionistic fuzzy power aggregation operators. Knowl.-Based Syst. **24**, 749–760 (2011)

Xu, Z., Yager, R.R.: Some geometric aggregation operators based on intuitionistic fuzzy sets. Int. J. Gen Syst **35**, 417–433 (2006)

Xu, K., Tang, L.C., Xie, M., Ho, S., Zhu, M.: Fuzzy assessment of FMEA for engine systems. Reliab. Eng. Syst. Saf. **75**, 17–29 (2002)

Yang, J.-B., Singh, M.G.: An evidential reasoning approach for multiple-attribute decision making with uncertainty. Syst. Man Cybern. IEEE Trans. **24**, 1–18 (1994)

Yang, J.-B., Wang, Y.-M., Xu, D.-L., Chin, K.-S.: The evidential reasoning approach for MADA under both probabilistic and fuzzy uncertainties. Eur. J. Oper. Res. **171**, 309–343 (2006)

Yang, Z., Bonsall, S., Wang, J.: Fuzzy rule-based Bayesian reasoning approach for prioritization of failures in FMEA. Reliab. IEEE Trans. **57**, 517–528 (2008)

Yang, J., Huang, H.-Z., He, L.-P., Zhu, S.-P., Wen, D.: Risk evaluation in failure mode and effects analysis of aircraft turbine rotor blades using Dempster-Shafer evidence theory under uncertainty. Eng. Fail. Anal. **18**, 2084–2092 (2011)

Yoon, K.P., Hwang, C.-L.: Multiple attribute decision making: an introduction. Sage Publications, Thousand Oaks (1995)

Zammori, F., Gabbrielli, R.: ANP/RPN: a multi criteria evaluation of the Risk Priority Number. Qual. Reliab. Eng. Int. **28**, 85–104 (2012)

Zhai, L.Y., Khoo, L.P., Zhong, Z.W.: Integrating rough numbers with interval arithmetic: a novel approach to QFD analysis. HKIE Trans. **14**, 74–81 (2007)

Zhai, L.-Y., Khoo, L.-P., Zhong, Z.-W.: A rough set based QFD approach to the management of imprecise design information in product development. Adv. Eng. Inform. **23**, 222–228 (2009)

Zhang, Z., Chu, X.: Risk prioritization in failure mode and effects analysis under uncertainty. Expert Syst. Appl. **38**, 206–214 (2011)

Chapter 11
Intelligent Quality Function Deployment

Huimin Jiang, C.K. Kwong and X.G. Luo

Abstract Quality function deployment (QFD) is commonly used in the product planning stage to define the engineering characteristics and target value settings of new products. However, some QFD processes substantially involve human subjective judgment, thus adversely affecting the usefulness of QFD. In recent years, a few studies have been conducted to introduce various intelligent techniques into QFD to address the problems associated with subjective judgment. These studies contribute to the development of intelligent QFD. This chapter presents our recent research on introducing intelligent techniques into QFD with regard to four aspects, namely, determination of importance weights of customer requirements, modeling of functional relationships in QFD, determination of importance weights of engineering characteristics and target value setting of engineering characteristics. In our research, a fuzzy analytic hierarchy process with an extent analysis approach is proposed to determine the importance weights for customer requirements to capture the vagueness of human judgment and a chaos-based fuzzy regression approach is proposed to model the relationships between customer satisfaction and engineering characteristics by which fuzziness and nonlinearity of the modeling can be addressed. To determine importance weights of engineering characteristics, we propose a novel fuzzy group decision-making method to address two types of uncertainties which integrates a fuzzy weighted average method with a consensus ordinal ranking technique. Regarding the target value setting of engineering characteristics, an inexact genetic algorithm is proposed to generate a family of inexact optimal solutions instead of determining one set of exact optimal target values. Possible future research on the development of intelligent QFD is provided in the conclusion section.

Keywords Quality function deployment · Intelligent techniques

H. Jiang · C.K. Kwong (✉)
Department of Industrial and Systems Engineering, The Hong Kong Polytechnic University,
Hong Kong, China
e-mail: c.k.kwong@polyu.edu.hk

X.G. Luo
Department of Systems Engineering, State Key Laboratory of Synthetical Automation for
Process Industries, Northeastern University, Shenyang, China

© Springer International Publishing Switzerland 2016   327
C. Kahraman and S. Yanık (eds.), *Intelligent Decision Making
in Quality Management*, Intelligent Systems Reference Library 97,
DOI 10.1007/978-3-319-24499-0_11

11.1 Introduction

Quality function deployment (QFD) is a systematic method of translating customer requirements (CRs) into engineering characteristics (ECs) in the product planning stage (Terninko 1997). QFD provides a visual relationship to help engineers focus on design requirements instead of design function in the whole development process. QFD uses the voices of the customers from the beginning of product development and deploys it throughout the whole product design process. Customer requirements (CRs) in a new product are collected. Then, product development teams map the CRs to ECs based on their knowledge, experience, and judgement. A QFD system comprises four inter-linked phases: product planning phase, part deployment phase, process planning phase, and production/operation planning phase (Karsak 2004). Figure 11.1 shows the four phases of QFD.

The implementation of QFD is called house of quality (HoQ), which offers a global view of information on a new product and on how CRs can be met at different stages of new product development. A HoQ typically contains information on "what to do" (CRs or voice of customers), importance weights of CRs, "how to do" (ECs), importance weights of ECs, the relationship matrix (relationships between CRs and ECs), technical correlation matrix, benchmarking data, and the target values settings of the ECs (Govers 1996). QFD was proposed to develop products with higher quality to meet or surpass customer's needs through collecting and analysing the voice of the customer (Chan and Wu 2002). It has been applied successfully in many industries. New product designs with QFD can enhance

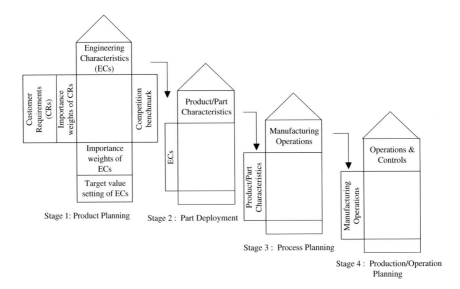

Fig. 11.1 The four phases of QFD

organizational learning and improve customer satisfaction. QFD can also decrease product costs, simplify manufacturing processes, and shorten the development time of new products (Vonderembse and Raghunathan 1997).

11.2 Literature Review

Determining the importance weights of CRs is a crucial QFD step because these weights can largely affect the target value setting of ECs in the later stage. Conducting surveys, such as lead user and focus group surveys, is common to determine the importance weights. The weights are then determined by analyzing the survey data. Respondents in surveys are always asked to rate various CRs, such as good quality and user-friendliness (Mochimaru et al. 2012). The product ratings of respondents involve subjective judgments. CRs always contain ambiguity and multiplicity of meaning. The description of CRs is usually linguistic and vague. Thus, conventional methods, which determine the importance weights of CRs on the basis of crisp numerical data, are inadequate. Some intelligent techniques have been introduced in previous studies to determine the importance weights of CRs, such as artificial neural network (Che et al. 1999), fuzzy and entropy methods (Chan et al. 1999), fuzzy AHP (Kwong and Bai 2002), supervised learning with a radial basis function (RBF) neural network (Chen 2003), fuzzy group decision-making approach (Zhang 2009), and fuzzy decision making trial and evaluation laboratory (DEMATAL) method (Shahraki and Paghaleh 2011).

Another important step of QFD is to prioritize ECs to facilitate resource planning. However, the inherent vagueness or impreciseness of QFD makes the prioritization of ECs ineffective. Two types of uncertainties in QFD exist. The first type is human assessment and judgment on qualitative attributes, which are always subjective and imprecise; thus, the input information of human perception can be ambiguous. The second type is the involvement of various stakeholders and/or the number of customers in the assessment of the importance of CRs, as well as the degree of relationships between CRs and ECs in QFD. Uncertainty that is associated with group assessment will exist because of individual heterogeneity. Previous studies have employed intelligent techniques to prioritize ECs, such as fuzzy outranking approach (Wang 1999), fuzzy set theory of axiomatic design review (Huang and Jiang 2002), a combined analytic network process (ANP) and goal programming approach (Karsak et al. 2002), fuzzy ANP (Buyukozkan et al. 2004), an integrated fuzzy weighted average method and fuzzy expected value operator method (Chen et al. 2006), and a methodology of determining aggregated importance of ECs with the considerations of fuzzy relation measures between CRs and ECs as well as fuzzy correlation measures among ECs (Kwong et al. 2007). However, most previous studies only address one of the two types of uncertainties that can affect the robustness of prioritizing ECs.

The success of products is heavily dependent on the associated customer satisfaction level. If customers are satisfied with a new product, the chance of that product being successful in marketplaces would be higher. A product usually is associated with a number of ECs, such as size, weight, and power consumption that could affect customer satisfaction. In this regard, modeling the functional relationships between CRs and ECs for product design is crucial. The developed models can be employed to formulate an optimization model to determine an optimal EC setting that leads to maximum customer satisfaction. Some techniques such as statistical regression (Han et al. 2000) and fuzzy rule-based method (Fung et al. 1998) were adopted to model the functional relationships. However, only a small number of data sets are usually available from the HoQ for modeling. On the other hand, the CRs obtained from market surveys are commonly ambiguous and qualitative in nature, and the relationships between CRs and ECs can be highly nonlinear. The customer satisfaction models developed based on QFD should be able to address nonlinearity and fuzziness. Intelligent techniques such as fuzzy regression (FR) (Chan et al. 2012), fuzzy least squares regression (Kwong et al. 2010) and genetic programming (Chan et al. 2011) have been recently adopted to model nonlinear and fuzzy relationships. However, the modeling methods of nonlinear and fuzzy relationships and the development of explicit customer satisfaction models based on a small number of data sets have not been addressed well in previous studies.

One of the key issues in QFD is how the EC settings of new products can be determined such that a high degree or even maximum customer satisfaction can be obtained. This involves a complex decision-making process with multiple variables. Currently, the setting of target EC values relies heavily on the professional experience and intuition of engineers; thus, the setting of such values is accomplished in a subjective manner. However, determining the optimal setting for target EC values based on this approach is very difficult. Some previous research has attempted to develop systematic procedures and methods for setting optimal target values of ECs in QFD, such as linear programming (Wasserman 1993), integer programming (Kim and Park 1998), mixed integer linear programming (Zhou 1998), fuzzy inference system (Fung et al. 1999), nonlinear mathematical program (Dawson and Askin 1999), and prescriptive fuzzy optimization (Kim et al. 2000).

Few studies have been conducted to introduce intelligence techniques in QFD. The results of previous studies on the introduction of intelligent techniques into QFD undoubtedly contribute to the development of intelligent QFD. In the following sections, our research on the development of intelligent QFD with regard to four aspects, namely, determination of importance weights of CRs, modeling of functional relationships in QFD, determination of importance weights of ECs and target value setting of ECs, is described.

11.3 Determination of Importance Weights of CRs by Using Fuzzy AHP with Extent Analysis

Quite a number of techniques were introduced to determine the importance weights of CRs in QFD. One of them is AHP which is very popular to be used in determining importance weights. AHP was used to determine the importance weights for product planning (Lu et al. 1994; Armacost et al. 1994; Hsiao, 2002) but has been mainly applied in crisp (non-fuzzy) decision applications. However, human judgment on the importance of CRs is always imprecise and vague. To address this deficiency in AHP, a fuzzy AHP with an extent analysis approach is proposed to determine the importance weights for CRs. In this method, the linguistic assessment of CRs is converted to triangular fuzzy numbers. These triangular fuzzy numbers are used to build a pairwise comparison matrix for the AHP. By applying the fuzzy AHP with extent analysis, the importance weights for the CRs can be obtained. Extent analysis refers to the "extent" in which an object can be satisfied for the goal. In this approach, the "satisfied extent" is defined by triangular fuzzy numbers. The use of the extent analysis technique and principles for the comparison of fuzzy numbers allows the calculation of weight vectors for fuzzy AHP. The new approach can improve the imprecise ranking of CRs inherited from studies based on conventional AHP. The fuzzy AHP with extent analysis is simple and easy to implement for prioritizing CRs in the QFD process compared with conventional AHP. The details of the fuzzy AHP are described as follows.

11.3.1 Development of a Hierarchical Structure for CRs

All CRs are initially structured into different hierarchical levels to apply AHP in prioritizing CRs. An affinity diagram, a tree diagram, or a cluster analysis can be used for this purpose. The voices of the customers can be gathered by a variety of methods. All of these approaches aim to ask customers to express their needs of a particular product. Such needs are usually expressed in words that are too general to use as CRs directly. However, by sorting, classifying, and structuring the voices of customers, useful CRs can be obtained.

11.3.2 Construction of Fuzzy Judgment Matrixes for AHP

The hierarchy of attributes is the subject of a pairwise comparison of AHP. After constructing a hierarchy, decision makers are asked to compare the elements at a given level on a pairwise basis to estimate their relative importance in relation to the element at the immediately proceeding level. Figure 11.2 shows an example of a hierarchy of attributes. The total importance weights of CRs can be calculated based

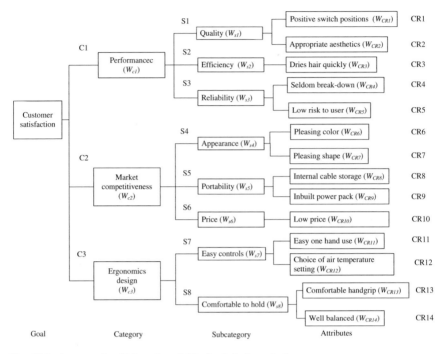

S1		Positive switch positions (W_{CR1})	CR1
	Quality (W_{s1})		
S2		Appropriate aesthetics (W_{CR2})	CR2
	Efficiency (W_{s2})	Dries hair quickly (W_{CR3})	CR3
S3		Seldom break-down (W_{CR4})	CR4
	Reliability (W_{s3})		
		Low risk to user (W_{CR5})	CR5

Fig. 11.2 An example of hierarchy of CRs for hair dryer design

on the expression, $W^*_{CRi} = W_{cj} \cdot W_{sk} \cdot W_{CRi}$, where W^*_{CRi} is the total importance weight of the ith CR, W_{cj}, W_{sk} and W_{CRi} are the importance weights of the jth, kth and ith element in the category level, subcategory level and attributes level, respectively. In conventional AHP, the pairwise comparison is made by using a ratio scale. A nine-point scale is commonly used to show the judgment or preference of participants between options as equally, moderately, strongly, very strongly, or extremely preferred. Even though a discrete scale of one to nine has the advantages of simplicity and ease of use, such a scale does not consider the uncertainty associated with the mapping of one's perception (or judgment) to a number. The linguistic terms that people use to express their feelings or judgments are vague. The use of this objective is unfeasible for defining the precise numbers to present linguistic assessment. Fuzzy set theory was first advocated to manage ambiguity in a system. The widely adopted triangular fuzzy number technique is employed to represent a pairwise comparison of CRs.

A fuzzy number is a special fuzzy set $F = \{(x, \mu_F(x)), x \in R\}$, where x takes its values on the real line R_1: $-\infty < x < \infty$, and $\mu_F(x)$ is a continuous mapping from R_1 to the close interval $[0, 1]$. A triangular fuzzy number can be denoted as $M = (l, m, u)$. The membership function $\mu_M(x) : R \rightarrow [0, 1]$ of a triangular fuzzy number is equal to the following:

$$\mu_M(x) = \begin{cases} \frac{x}{m-l} - \frac{l}{m-l}, & x \in [l, m] \\ \frac{x}{m-u} - \frac{u}{m-u}, & x \in [m, u] \\ 0, & otherwise, \end{cases} \tag{11.1}$$

where $l \le m \le u$; l and u stand for the lower and upper values of the support of M, respectively; m is the mid-value of M. When $l = m = u$, it is a non-fuzzy number by convention.

The main operational laws for two triangular fuzzy numbers, M_1 and M_2, are as follows:

$$\begin{aligned} M_1 + M_2 &= (l_1 + l_2, m_1 + m_2, u_1 + u_2) \\ M_1 \otimes M_2 &\approx (l_1 l_2, m_1 m_2, u_1 u_2) \\ \lambda \otimes M_1 &= (\lambda l_1, \lambda m_1, \lambda u_1), \lambda > 0, \lambda \in R \\ M_1^{-1} &\approx (\frac{1}{u_1}, \frac{1}{m_1}, \frac{1}{l_1}). \end{aligned} \tag{11.2}$$

To consider vagueness in an assessment during the pairwise comparison of CRs, triangular numbers M_1, M_3, M_5, M_7, M_9 are used to represent the assessment from "equal to extremely preferred"; M_2, M_4, M_6, M_8 are middle values. Figure 11.3 shows the triangular fuzzy numbers $M_t = (l_t, m_t, u_t)$ and $t = 1, 2 \ldots 9$, where l_t and u_t are the lower and upper values of fuzzy number M_t, respectively; m_t is the middle value of fuzzy number M_t. δ represents a fuzzy degree of judgment where $u_t - l_t = l_t - u_t = \delta$, which should be larger than or equal to 0.5. A larger δ value implies a higher fuzzy degree of judgment. When $\delta = 0$, the judgment is a non-fuzzy number.

Participants of the survey use triangular numbers $(M_1 - M_9)$ to express their preferences between options. For example, a participant may consider that element i is very important compared with element j under certain criteria; he/she may set

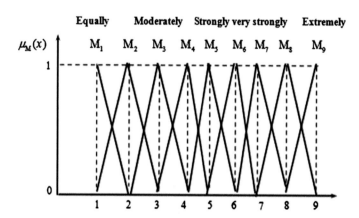

Fig. 11.3 The membership functions of the triangular numbers

$a_{ij} = (4,5,6)$. If element j is thought to be less important than element i, the pairwise comparison between j and i can be presented by using the fuzzy number, $a_{ij} = (\frac{1}{6},\frac{1}{5},\frac{1}{4})$. On the basis of the results of pairwise comparisons for the CRs obtained from participants, Eq. (11.2) is applied to obtain the fuzzy judgment matrixes, FCM_n, for the CRs.

The AHP methodology provides a consistency index to measure any inconsistency within the judgments in each comparison matrix and for the entire hierarchy. The index can be used to indicate whether the targets can be arranged in an appropriate order of ranking and the consistency of pairwise comparison matrixes. The defuzzification method of triangular fuzzy numbers was employed to convert fuzzy comparison matrixes into crisp matrixes, which are then used to investigate consistency. The consistency index, CI, and the consistency ratio, CR, for a comparison matrix can be computed by the following equations:

$$CI = {(\lambda_{max} - n)}/{n - 1}, \tag{11.3}$$

$$CR = \left({(CI)}/{RI(n)}\right)100\% \tag{11.4}$$

where λ_{max} is the largest eigenvalue of the comparison matrix, n is the dimension of the matrix, and $RI(n)$ is a random index that depends on n.

If the calculated CR of a comparison matrix is less than 10 %, the consistency of the pairwise judgment can be considered acceptable; otherwise, the judgments expressed by the experts are considered inconsistent and the decision maker has to repeat the pairwise comparison matrix.

A triangular fuzzy number, denoted as $M = (l,m,u)$ can be defuzzified to a crisp number:

$$M\_crisp = {(4m+l+u)}/{6}. \tag{11.5}$$

11.3.3 Calculation of Weight Vectors for Individual Levels of a Hierarchy of the CRs

The extent analysis method and the principles for the comparison of fuzzy numbers were employed to obtain estimates for the weight vectors for individual levels of a hierarchy of CRs. The extent analysis method is used to consider the extent of an object to be satisfied for the goal, that is, the satisfied extent. In this method, the "extent" is quantified by a fuzzy number. On the basis of the fuzzy values for the extent analysis of each object, a fuzzy synthetic degree value can be obtained.

Let $X = \{x_1,x_2,\ldots,x_n\}$ be an object set and $U = \{u_1,u_2,\ldots,u_m\}$ be a goal set. According to the extent analysis method, each object can be used to perform extent analysis for each goal. Therefore, m extent analysis values for each object can be obtained as follows:

$$M_{g_i}^1, M_{g_i}^2, \ldots, M_{g_i}^m, i = 1, 2 \ldots n, \tag{11.6}$$

where all $M_{g_i}^j$ ($j = 1, 2, \ldots, m$) are triangular fuzzy numbers. The value of fuzzy synthetic degree with respect to the ith object is thus defined as follows:

$$D_i = \sum_{j}^{m} M_{g_i}^j \otimes \left(\sum_{i}^{n} \sum_{j}^{m} M_{g_i}^j \right)^{-1}. \tag{11.7}$$

According to the above definition, the fuzzy synthetic degree values of all elements in the kth level can be calculated by using Eq. (11.7) based on the fuzzy judgment matrix of the kth level:

$$D_i^k = \sum_{j=1}^{n} a_{ij}^k \otimes \left(\sum_{i=1}^{n} \sum_{j=1}^{n} a_{ij}^k \right)^{-1}, i = 1, 2, \ldots, n, \tag{11.8}$$

where D_i^k is the fuzzy synthetic degree values of element i in the kth level and $A^k = (a_{ij}^k)_{nn}$ is the fuzzy judgment matrix of the kth level.

The principles for the comparison of fuzzy numbers were introduced to derive the weight vectors of all elements for each level of the hierarchy with the use of fuzzy synthetic values. The principles that allow the comparison of fuzzy numbers are as follows:

Definition 1 M_1 and M_2 are two triangular fuzzy numbers. The degree of possibility of $M_1 \geq M_2$ is defined as $V(M_1 \geq M_2) = \sup_{x \geq y}[\min(\mu_{M_1}(x), \mu_{M_2}(y))]$.

Theorem 1 *If M_1 and M_2 are triangular fuzzy numbers denoted by (l_1, m_1, u_1) and (l_2, m_2, u_2), respectively, then the following holds:*

(1) The necessary and sufficient condition of $V(M_1 \geq M_2) = 1$ is $m_1 \geq m_2$.
(2) If $m_1 \leq m_2$, let $V(M_1 \geq M_2) = \text{hgt}(M_1 \cap M_2)$. $V(M_1 \geq M_2) = \mu(d) =$

$$\begin{cases} \dfrac{l2 - u1}{(m1 - u1) - (m2 - l2)}, & l2 \leq u1 \\ 0 & otherwise \end{cases}$$, where d is the crossover point's abscissa

for M_1 and M_2.

Definition 2 The degree of possibility for a fuzzy number to be greater than k fuzzy numbers $M_i (i = 1, 2, \ldots, k)$ can be defined by $V(M \geq M_1, M_2, \ldots, M_k) = \min V(M \geq M_i), i = 1, 2, \ldots, k$.

Table 11.1 Pairwise comparison for category level

		C_1	C_2	C_3
	C_1	(1,1,1)	(1,2,3)	(1,1,2)
			(2,3,4)	(1,1,2)
			(1,1,2)	(1,2,3)
	C_2	(1/3,1/2,1/1)	(1,1,1)	(1,1,2)
		(1/4,1/3,1/2)		(1,2,3)
		(1/2,1/1,1/1)		(1,1,2)
	C_3	(1/2,1/1,1/1)	(1/2,1/1,1/1)	(1,1,1)
		(1/2,1/1,1/1)	(1/3,1/2,1/1)	
		(1/3,1/2,1/1)	(1/2,1/1,1/1)	

Let $d(p_i^k) = \min V(S_i^k \geq S_j^k)$, where p_i^k is the ith element of the kth level, $j = 1, 2, \ldots, n; \; j \neq i$. The number of elements in the kth level is n. The weight vector of the kth level is then obtained as follows:

$$W_k' = (d(p_1^k), d(p_2^k), \ldots, d(p_n^k))^T. \tag{11.9}$$

After normalization, the normalized weight vector, W_k, is expressed as follows:

$$W_k = (w(p_1^k), w(p_2^k), \ldots, w(p_n^k))^T. \tag{11.10}$$

Taking the hierarchy of CRs for the hair dryer design (Fig. 11.2) as an example, the pairwise comparisons of C_1, C_2 and C_3 are shown in Table 11.1.

Equation (11.2) is applied to obtain the fuzzy judgment matrixes, FCM_1, for the category level.

$$FCM_1 = \begin{array}{c} \\ C1 \\ C2 \\ C3 \end{array} \begin{array}{ccc} C1 & C2 & C3 \\ \begin{bmatrix} (1,1,1) & (1.33, 2, 3) & (1.00, 1.33, 2.33) \\ (0.33, 0.50, 0.75) & (1, 1, 1) & (1.00, 1.33, 2.33) \\ (0.43, 0.75, 1.00) & (0.43, 0.75, 1.00) & (1, 1, 1) \end{bmatrix} \end{array}$$

The fuzzy synthetic degree values of the element C_1, D_{C_1}, can be calculated based on (11.8) as follows:

$$D_{C_1} = \sum_{j=1}^{3} a_{1j} \otimes \left(\sum_{i=1}^{3} \sum_{j=1}^{3} a_{ij} \right)^{-1}$$
$$= ((1, 1, 1) + (1.33, 2, 3) + (1.00, 1.33, 2.33)) \otimes ((1, 1, 1) + (1.33, 2, 3) + \cdots (1, 1, 1))^{-1}$$
$$= (0.25, 0.45, 0.84)$$

Following the similar calculation, $D_{C_2} = (0.17, 0.29, 0.54)$ and $D_{C3} = (0.14, 0.26, 0.40)$ can be obtained. The following comparison results are then derived based on Theorem 1.

$$V(D_{C_1} \geq D_{C_2}) = 1; \ V(D_{C_1} \geq D_{C_3}) = 1;$$

$$V(D_{C_2} \geq D_{C_1}) = \frac{(0.25 - 0.54)}{(0.29 - 0.54) - (0.45 - 0.25)} = 0.65;$$

$$V(D_{C_2} \geq D_{C_3}) = 1; \ V(D_{C_3} \geq D_{C_1}) = \frac{(0.25 - 0.40)}{(0.26 - 0.40) - (0.45 - 0.25)} = 0.44;$$

$$V(D_{C_3} \geq D_{C_2}) = \frac{(0.17 - 0.40)}{(0.26 - 0.40) - (0.29 - 0.17)} = 0.88.$$

Based on Definitions 1, 2 and Eq. (11.9), the weight vector W'_C of the category level can be calculated by using the following formula:

$$d(C_1) = \min(V(D_{C_1} C_2, D_{C3}) = \min\{1, 1\} = 1.00;$$
$$d(C_2) = \min(V(D_{C_2} \geq D_{C_1}, D_{C3}) = \min\{0.65, 1\} = 0.65;$$
$$d(C_3) = \min(V(D_{C_3} \geq D_{C_1}, D_{C_2}) = \min\{0.44, 0.88\} = 0.44;$$
$$W'_C = (d(C_1), d(C_2), d(C_3))^T = (1.00, 0.65, 0.44).$$

Based on (11.10), the normalized weight vectors of the category level are obtained as follows:

$$(W_{C_1}, W_{C_2}, W_{C_3}) = (0.48, 0.31, 0.21).$$

Following the similar calculation, the weight vectors of the subcategory level, W_{sk}, and attributes level, W_{CRi}, can be calculated. Hence, the total importance weights of CRs can be calculated based on the expression, $W^*_{CRi} = W_{cj} \cdot W_{sk} \cdot W_{CRi}$. For example, referring to Fig. 11.2, $W^*_{CR4} = W_{c1} \cdot W_{s3} \cdot W_{CR4} = 0.078$.

11.4 Modeling of Functional Relationships in QFD by Using Chaos-Based FR

As mentioned in Sect. 11.1, the methods of modeling nonlinear and fuzzy relationships in QFD and the development of explicit customer satisfaction models based on a small number of data sets have not been addressed in previous studies. In this section, a novel FR approach, namely chaos-based FR, is described to model the relationships between customer satisfaction and ECs in order to address the limitations of previous studies. This approach employs a chaos optimization algorithm (COA) to generate the polynomial structures of customer satisfaction models with second- and/or higher-order terms and interaction terms. COA employs chaotic dynamics to solve an optimization problem. COA does not rely on learning factors, has fast convergence, and can search for accurate solutions compared with the conventional optimization methods. COA also has better capacity in searching for the global optimal solution of an optimization problem and can escape from a local minimum easier than conventional optimization algorithms. However, COA cannot address the fuzziness of survey data. The FR method of Tanaka et al. (1982) was introduced to determine the fuzzy coefficients of models. FR is effective for modeling problems wherein the degree of fuzziness of the data sets for modeling is high and only a small amount of data sets is available for modeling. However, the FR method can yield only linear type models. The chaos-based FR approach combines the advantages of COA and FR to generate customer satisfaction models wherein the modeling fuzziness can be explicitly addressed and nonlinear models can be developed. The proposed approach to the modeling of functional relationships in QFD mainly involves four processes: development of HoQ, generation of polynomial structures of customer satisfaction models by COA, determination of fuzzy coefficients of customer satisfaction models by Tanaka's FR, and generation of fuzzy polynomial customer satisfaction models.

11.4.1 Fuzzy Polynomial Models with Second- and/or Higher-Order Terms and Interaction Terms

Kolmogorov–Gabor polynomial has been widely used to evolve general nonlinear models but is incapable of addressing the fuzziness of modeling data. In fuzzy polynomial models developed based on the proposed approach, nonlinear terms and interaction terms between independent variables are represented in a form of a higher-order Kolmogorov–Gabor polynomial. The fuzzy coefficients of the models are determined by using Tanaka's FR method. The proposed models can overcome the limitation of conventional FR models where only first-order terms are generated. A fuzzy polynomial model based on the chaos-based FR approach can be expressed as follows:

$$\tilde{y} = \tilde{f}_{NR}(x)$$
$$= \tilde{A}_0 + \sum_{i_1=1}^{N} \tilde{A}_{i_1} x_{i_1} + \sum_{i_1=1}^{N}\sum_{i_2=1}^{N} \tilde{A}_{i_1 i_2} x_{i_1} x_{i_2} + \sum_{i_1=1}^{N}\sum_{i_2=1}^{N}\sum_{i_3=1}^{N} \tilde{A}_{i_1 i_2 i_3} x_{i_1} x_{i_2} x_{i_3} + \cdots \sum_{i_1=1}^{N} \cdots \sum_{i_d=1}^{N} \tilde{A}_{i_1 \ldots i_d} \prod_{j=1}^{d} x_{i_j} ,$$
$$(11.11)$$

where $\tilde{y}$ is the dependent variable, x_{i_j} is the i_jth independent variable, $i_j = 1, \ldots, N$ and $j = 1, 2, \ldots d$. N and d denote the number of design variables. $\tilde{A}$ is the fuzzy coefficients of the linear, second order, and/or higher-order terms and the interaction terms of the model.

$$\tilde{A}_0 = \left(a_0^c, a_0^s\right), \tilde{A}_1 = \left(a_1^c, a_1^s\right), \tilde{A}_2 = \left(a_2^c, a_2^s\right), \ldots, \tilde{A}_N = \left(a_N^c, a_N^s\right),$$

$$\tilde{A}_{11} = \left(a_{11}^c, a_{11}^s\right), \tilde{A}_{12} = \left(a_{12}^c, a_{12}^s\right), \tilde{A}_{13} = \left(a_{13}^c, a_{13}^s\right), \ldots, \tilde{A}_{NN} = \left(a_{NN}^c, a_{NN}^s\right),$$

$$\tilde{A}_{111} = \left(a_{111}^c, a_{111}^s\right), \tilde{A}_{112} = \left(a_{112}^c, a_{112}^s\right), \tilde{A}_{113} = \left(a_{113}^c, a_{113}^s\right), \ldots,$$

$$\tilde{A}_{NNN} = \left(a_{NNN}^c, a_{NNN}^s\right),$$

$$\ldots \ldots \tilde{A}_{N\ldots N} = \left(a_{N\ldots N}^c, a_{N\ldots N}^s\right),$$

where a^c and a^s are the central value and spread of fuzzy numbers, respectively.
The fuzzy polynomial model (11.11) can be rewritten as follows:

$$\tilde{y} = \tilde{A}_0' + \tilde{A}_1' * x_1' + \tilde{A}_2' * x_2' \ldots + \tilde{A}_{N_{NR}}' * x_{N_{NR}}', \qquad (11.12)$$

or

$$\tilde{y} = \left(a_0^{c\prime}, a_0^{s\prime}\right) + \left(a_1^{c\prime}, a_0^{s\prime}\right) * x_1' + \left(a_2^{c\prime}, a_2^{s\prime}\right) * x_2' + \ldots + \left(a_{N_{NR}}^{c}{}', a_{N_{NR}}^{s}{}'\right) * x_{N_{NR}}',$$

$$(11.13)$$

where $\tilde{A}_0' = \tilde{A}_0, \quad \tilde{A}_1' = \tilde{A}_1, \quad \tilde{A}_2' = \tilde{A}_2, \quad \ldots, \quad \tilde{A}_{N_{NR}}' = \tilde{A}_{N\ldots N}; \quad \tilde{A}_0' = \left(a_0^{c\prime}, a_0^{s\prime}\right),$
$\tilde{A}_1' = \left(a_1^{c\prime}, a_1^{s\prime}\right), \quad \ldots, \quad \tilde{A}_{N_{NR}}' = \left(a_{N_{NR}}^{c}{}', a_{N_{NR}}^{s}{}'\right),$ and $x_0' = 1, \ x_1' = x_1, \ x_2' = x_1 x_2, \ \ldots,$
$x_{N_{NR}}' = x_1 x_2 \ldots x_d.$ x_j' and $\tilde{A}_j'$ $(j = 0, 1, 2 \ldots, N_{NR})$ denote the transformed variables
and fuzzy coefficients, respectively.
The vector of the fuzzy coefficients can be defined as follows:

$$\tilde{A}' = \left(\tilde{A}_0', \tilde{A}_1', \ldots \tilde{A}_{N_{NR}}'\right) = \left(\left(a_0^{c\prime}, a_0^{s\prime}\right), \left(a_1^{c\prime}, a_1^{s\prime}\right), \ldots, \left(a_{N_{NR}}^{c}{}', a_{N_{NR}}^{s}{}'\right)\right), \quad (11.14)$$

$$a^{c\prime} = (a_0^{c\prime}, a_1^{c\prime}, \ldots, a_{N_{NR}}^{c}{}'), \qquad (11.15)$$

$$a^{s\prime} = (a_0^{s\prime}, a_1^{s\prime}, \ldots, a_{N_{NR}}^{s}{}') \qquad (11.16)$$

11.4.2 Determination of Model Structures Utilizing COA

In this approach, COA was introduced to determine the polynomial structure of
fuzzy models. COA is a stochastic search algorithm wherein chaos is introduced
into the optimization strategy to accelerate the optimum seeking operation and
determine the global optimal solution. Two phases exist in searching for an optimal
solution in the chaos optimization process. The first phase is called wide search and
involves the whole solution space according to an ergodic track. When the end
condition is satisfied, the current optimal state becomes close to the optimal solution
and the second phase starts. The second phase is based on the results of the first

phase and involves a narrow search focused on a local region. The second phase adds a small disturbance term until the final requirement is met. The added disturbance can be a chaos variable, a random variable based on Gaussian distribution/uniform distribution, or a bias generated by the mechanism of gradient descent. Current COAs use the carrier wave method to map chosen chaos variables linearly to the space of optimization variables and then search for optimal solutions on the basis of the ergodicity of chaos variables. COAs focus on chaos variable-based optimization rather than on introducing chaos variables as a small disturbance in search optimization.

The logistic model used in chaos optimization is shown in Eq. (11.17). Logistic mapping can generate chaos variables by iteration:

$$c_{n+1} = f(c_n) = \mu c_n (1 - c_n), \tag{11.17}$$

where $\mu = 4$ and $c_n \in [0, 1]$ is the nth iteration value of the chaos variable c.

The optimization process uses the chaos variables generated from logistic mapping to search through its own locomotion law. Chaos has dynamic properties including ergodicity, intrinsic stochastic property, and sensitive dependence on initial conditions. The characteristic of randomness ensures the possibility of large-scale search. Ergodicity allows COA to traverse all possible states without repetition and to overcome the limitations caused by ergodic search in general random methods.

The linear mapping for converting chaos variables into optimization variables is formulated as follows:

$$q_n = a + (b - a) \cdot c_n, \tag{11.18}$$

where q_n is the optimization variable; a and b are the lower and upper limits of the optimization variable q, respectively.

According to the iteration, the chaos variables traverse in $[0, 1]$ and the corresponding optimization variables traverse in the corresponding range $[a, b]$. Therefore, the optimal solution can be found in the area of feasible solution.

Each optimization variable represents the polynomial structure of a fuzzy model and is described by the input variables $[x_1, x_2, \ldots and \ x_m]$ and arithmetic operations. The two arithmetic operations of addition ("+") and multiplication ("*") are employed in the fuzzy polynomial model (Eq. 11.12). The optimization variable at the n th generation is defined as follows:

$$q_n = \left[q_n^1, \ q_n^2, \ \ldots, \ q_n^{N_c} \right], \tag{11.19}$$

where N_c is an odd number representing the number of elements in a chaos variable.

For example, if four variables exist in the model, the value of N_c is first set to seven with four elements representing four design variables. Another three elements in the middle of every two adjacent design variables represent the arithmetic operations to guarantee that the optimization variable, q_n, can include all four variables. If the error requirement is not satisfied, the N_c value is adjusted until a satisfactory modeling error that is close to zero and is smaller than the modeling errors based on the previous studies is achieved.

The elements in odd numbers $(q_n^1, q_n^3, \ldots, q_n^{N_c})$ are used to represent variables in a nonlinear structure. For the odd number k, if $(l-1)/m < q_n^k \leq l/m$ (m is the number of variables in a nonlinear fuzzy model, $1 \leq l \leq m$), the position of q_n^k is represented by the lth variable x_l. The elements in even numbers $(q_n^2, q_n^4, \ldots, q_n^{N_c-1})$ are used to determine the arithmetic operations. For even number k, if $0 < q_n^k \leq 1/2$ and $1/2 < q_n^k \leq 1$, the arithmetic operations are addition ("+") and multiplication ("*"), respectively. For example, an optimization variable with nine elements is used to represent the structure of a fuzzy polynomial model with four dependent variables. If the optimal variable is obtained as $q = [x_2, +, x_3, *, x_4, *, x_4, +, x_1]$, the polynomial structure can be expressed as $x_2 + x_3 x_4^2 + x_1$. The transformed variables are $x_0' = 1, x_1' = x_2, x_2' = x_3 x_4^2, x_3' = x_1$. Therefore, the fuzzy polynomial model with fuzzy coefficients can be represented as

$$\tilde{y} = \tilde{A}_0' + \tilde{A}_1' x_2 + \tilde{A}_2' x_3 x_4^2 + \tilde{A}_3' x_1$$

where $\tilde{A}_0' = (a_0^{c'}, a_0^{s'})$, $\tilde{A}_1' = (a_1^{c'}, a_1^{s'})$, $\tilde{A}_2' = (a_2^{c'}, a_2^{s'})$ and $\tilde{A}_3' = (a_3^{c'}, a_3^{s'})$. The central values $a_j^{c'}$ and spread values $a_j^{s'}$ ($j = 0, 1, \ldots, 4$) of the fuzzy coefficients are determined by using Tanaka's FR analysis.

11.4.3   Determination of Fuzzy Coefficients by Using FR Analysis

Tanaka's FR aims to use fuzzy functions to describe an imprecise and vague phenomenon. All input and output variables, as well as the coefficients of the relationships, are considered as fuzzy numbers. Two different criteria (i.e., the least absolute deviation and the minimum spread) are used to evaluate the fuzziness of the output. Deviations between observed and estimated values depend on the indefiniteness of system structures and are considered as the fuzziness of system parameters. The fuzzy parameters of FR models indicate the possibility distribution and are obtained as fuzzy sets that represent the fuzziness of models. The objective of FR analysis is to minimize the fuzziness of the model by minimizing the total spread of the fuzzy coefficients, thus leading to the minimum uncertainty of the output.

On the basis of chaos optimization, a fuzzy model containing second- and/or higher-order terms and interaction terms is represented in a polynomial structure. Tanaka's FR analysis is employed to determine the fuzzy coefficients for each term of the fuzzy polynomial model. Fuzzy coefficients with the central point $a^{c'}$ and spread value $a^{s\prime}$ are determined by solving the following linear programming (LP) problem:

$$\text{Min } J = \sum_{j=0}^{N_{NR}} \left(a_j^{s\prime} \sum_{i=1}^{M} \left| x_j^{\prime}(i) \right| \right), \tag{11.20}$$

where J is the objective function that represents the total fuzziness of the system, $1 + N_{NR}$ is the number of terms of the fuzzy polynomial model, M is the number of data sets, $x_j^{\prime}(i)$ is the jth transformed variable of the ith data set in the fuzzy polynomial model, and $|.|$ refers to absolute value of the transformed variable. The constraints can be formulated as follows:

$$\sum_{j=0}^{N_{NR}} a_j^{c\prime} x_j^{\prime}(i) + (1 - h) \sum_{j=0}^{N_{NR}} a_j^{s\prime} \left| x_j^{\prime}(i) \right| \geq y_i \quad i = 1, 2, \ldots, M, \tag{11.21}$$

$$\sum_{j=0}^{N_{NR}} a_j^{c\prime} x_j^{\prime}(i) - (1 - h) \sum_{j=0}^{N_{NR}} a_j^{s\prime} \left| x_j^{\prime}(i) \right| \leq y_i \quad i = 1, 2, \ldots, M, \tag{11.22}$$

$$a_j^{s\prime} \geq 0, \ a_j^{c\prime} \in R, j = 0, 1, 2, \ldots N_{NR},$$

$$x_0^{\prime}(i) = 1 \text{ for all } i \text{ and } 0 \leq h \leq 1.$$

where h refers to the degree of fit of the fuzzy model in a given data (between zero and one), and y_i is the value of the ith dependent variable. Constraints (11.21) and (11.22) ensure that each objective y_i has at least h degree of belonging to $\tilde{y}_i$ as $\mu_{\tilde{y}_i}(y_i) \geq h$, $i = 1, 2, \ldots, M$. The last constraints for the variables ensure that $a_j^{s\prime}$ and $a_j^{c\prime}$ are non-negative. Therefore, the fuzzy parameters $\tilde{A}_j^{\prime}(j = 0, 1, 2, \ldots N_{NR})$ can be determined by solving the LP problem subject to $\mu_{\tilde{y}_i}(y_i) \geq h$.

11.4.4 Algorithms of Chaos-Based FR

The algorithms of the proposed chaos-based FR method are described as follows:

(1) The number of iterations and the number of elements N_c in a chaos variable are initialized. N_c is an odd number, and N_c values are chosen randomly in the range of $[0, 1]$ to decide the value of the initialized chaos variable.

(2) The iteration starts from $n = 1$. The chaos variables c_n are generated based on the logistic model in Eq. (11.17) and are transformed into the optimization variables, q_n, by using Eq. (11.18). The polynomial structure of the fuzzy

model is determined based on the value of the optimization variable q_n. According to the rules described in Sect. 11.4.2, the elements in odd numbers and even numbers are substituted by input variable x_k ($k = 1, \ldots, N$ and N is the number of variables) and arithmetic operations "+" and "*," respectively. Subsequently, the transformed variable x'_j with linear, second order, and/or higher-order terms and interaction terms are generated based on arithmetic operations. If the operation is "*," the second- and/or higher-order terms and interaction terms are determined by multiplying the variables on both sides of "*". The arithmetic operation "+" is used to add all terms, including linear terms, to generate the final polynomial structure of the fuzzy model.

(3) According to the generated polynomial structure, the fuzzy coefficients $\tilde{A}'_j = \left(a^{c\prime}_j, a^{s\prime}_j\right)$ are assigned to all transformed variables of the developed structure. The values of the fuzzy coefficients are calculated by solving the LP problem as shown in Eqs. (11.20–11.22).

(4) Based on the developed fuzzy polynomial model, the predicted variable $\tilde{y}_i$ can be calculated. The fitness value RE can then be obtained by calculating the relative error between $\tilde{y}_i$ and the actual data y_i.

(5) Step (2) is repeated. The algorithms are again executed by another training data set until all training data sets are employed. The mean fitness value $MRE(n)$ for all training data sets is calculated.

(6) The iteration continues by $n + 1 \rightarrow n$ and stops after the number of iterations reaches the predefined value. The values of $MRE(n)$ are recorded for each iteration and the solution with the smallest mean fitness value is selected. The fuzzy polynomial model with the smallest training error is then found. Finally, the chaos-based FR model is generated.

We have applied the proposed approach on modeling the functional relationships in QFD for mobile phone products. The following shows an example of a customer

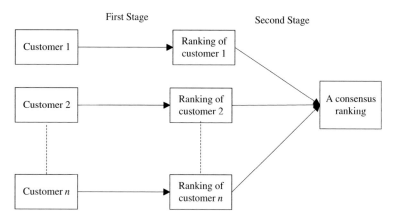

Fig. 11.4 Flowchart of the fuzzy group decision-making method

satisfaction model for the CR "comfortable to hold" based on the chaos-based FR approach.

$$
\begin{aligned}
\tilde{y} = & \left(14.9991, 2.8422 * 10^{-14}\right) + \left(-1.5036, 2.8422 * 10^{-14}\right)x_3 \\
& + \left(-0.2890, 0\right)x_4 + \left(-0.3634, 5.6843 * 10^{-14}\right)x_1 \\
& + \left(0.0045, 0.0077\right)x_1 x_2,
\end{aligned} \tag{11.23}
$$

where $\tilde{y}$ is the predicted value of "comfortable to hold"; x_1, x_2, x_3, and x_4 are four ECs that denote weight, height, width, and thickness, respectively.

11.5   Determination of Importance Weights of ECs by Using Fuzzy Group Decision-Making Method

Regarding the determination of importance weights of ECs in QFD, most previous studies only address one type of uncertainties as described in Sect. 11.1 that would adversely affect the robustness of prioritizing ECs. Thus, it is necessary to consider the two types of uncertainties simultaneously while determining the importance weights of ECs. In this section, a novel fuzzy group decision-making method that integrates a fuzzy weighted average method with a consensus ordinal ranking technique is described to address the two types of uncertainties. The approach consists of two stages. The first stage involves the determination of the importance weights of ECs with respect to each customer by using fuzzy weighted average method with the fuzzy expected operator. The second stage determines a consensus ranking by synthesizing all customer preferences on the ranking of ECs. The flowchart for the proposed methodology is presented in Fig. 11.4.

11.5.1   Determination of the Importance Weights of ECs Based on the Fuzzy Weighted Average Method with Fuzzy Expected Operator

For the first type of uncertainty in QFD, fuzzy set theory can be an effective tool to deal with uncertain inputs. On the basis of fuzzy set theory, the inputs required for QFD are represented with linguistic terms characterized by fuzzy sets. The two sets of input data should be expressed as fuzzy numbers, namely, the relative weights of CRs and the relationship measures between CRs and ECs, to determine the importance weights of ECs.

Some notations that are used in this section are shown as follows:

CR_i The ith customer requirement where $i = 1, 2, \ldots, m$

EC_j The jth engineering characteristic where $j = 1, 2, \ldots, n$

C_l The lth customer surveyed in a target market where $l = 1, 2, \ldots, K$
m The number of CRs
n The number of ECs
K The number of customers surveyed in a target market
$\tilde{W}_i^l$ The lth customer's individual preference on the ith customer need, which is a triangular fuzzy number belonging to certain predefined linguistic terms, such as "very important," "important," and "moderately important."
$\tilde{R}_{ij}$ The relationship measure between the ith CR and jth EC, which is a triangular fuzzy number belonging to some predefined linguistic terms, such as "strong," "moderate," and "weak."

The relative importance of the ith CR with respect to the lth customer and relationship measures between CRs and ECs are expressed as triangular fuzzy numbers. Thus, the determination of the importance of ECs falls under the category of fuzzy weighted average. The fuzzy importance of EC_j with respect to the lth customer, which is denoted by $\tilde{Z}_j^l$, can be expressed as follows:

$$\tilde{Z}_j^l = \sum_{i=1}^{m} \tilde{W}_i^l \tilde{R}_{ij} \Big/ \sum_{i=1}^{m} \tilde{W}_i^l. \tag{11.24}$$

Several methods can be devised to calculate the fuzzy weighted average. A common method for calculating the fuzzy ECs of Eq. (11.24) was proposed by Kao and Liu (2001):

Let $(W_i^l)_\alpha$, $(R_{ij})_\alpha$ denote the α-level sets of $\tilde{W}_i^l, \tilde{R}_{ij}$, respectively, which can be defined as follows:

$$(W_i^l)_\alpha = \left[(W_i^l)_\alpha^L, (W_i^l)_\alpha^U \right] = \left[\min_{w_i^l} \left\{ w_i^l \in W_i^l \Big/ \mu_{\tilde{W}_i^l}(w_i^l) \geq \alpha \right\}, \max_{w_i^l} \left\{ w_i^l \in W_i^l \Big/ \mu_{\tilde{W}_i^l}(w_i^l) \geq \alpha \right\} \right]$$

$$(R_{ij})_\alpha = \left[(R_{ij})_\alpha^L, (R_{ij})_\alpha^U \right] = \left[\min_{r_{ij}} \left\{ r_{ij} \in R_{ij} \Big/ \mu_{\tilde{R}_{ij}}(r_{ij}) \geq \alpha \right\}, \max_{e_{ik}} \left\{ r_{ij} \in R_{ij} \Big/ \mu_{\tilde{R}_{ij}}(r_{ij}) \geq \alpha \right\} \right]. \tag{11.25}$$

These intervals indicate where the relative weight of customer attributes and the relationship between CRs and ECs are located at possibility level α. According to the extension principle of Zadeh (1978), the membership function, $\mu_{\tilde{Z}_j^l}$, can be derived from the following equation:

$$\mu_{\tilde{Z}_j^l}\left(z_j^l\right) = \sup_{r,w} \quad \min\left\{ \mu_{\tilde{W}_i^l}(w_i^l), \mu_{\tilde{R}_{ij}}(r_{ij}) \quad \forall i,j \Big/ z_j^l = \sum_{i=1}^{m} w_i^l r_{ij} \Big/ \sum_{i=1}^{m} w_i^l \right\}. \tag{11.26}$$

Therefore, the upper and lower bounds of the α-level of $\tilde{Z}_j^l$ can be obtained. The solution of the upper and lower bounds can be attained by solving the following LP model:

$$\left(Z_j^l\right)_\alpha^U = \max \ \sum_{i=1}^m w_i^l r_{ij} \bigg/ \sum_{i=1}^m w_i^l$$

s.t.

$$\left(W_i^l\right)_\alpha^L \le w_i^l \le \left(W_i^l\right)_\alpha^U \qquad\qquad (11.27)$$
$$\left(R_{ij}\right)_\alpha^L \le r_{ij} \le \left(R_{ij}\right)_\alpha^U$$
$$i = 1,\ldots,m \qquad\qquad\qquad ,$$

and

$$\left(Z_j^l\right)_\alpha^L = \min \ \sum_{i=1}^m w_i^l r_{ij} \bigg/ \sum_{i=1}^m w_i^l$$

s.t.

$$\left(W_i^l\right)_\alpha^L \le w_i^l \le \left(W_i^l\right)_\alpha^U \qquad\qquad (11.28)$$
$$\left(R_{ij}\right)_\alpha^L \le r_{ij} \le \left(R_{ij}\right)_\alpha^U$$
$$i = 1,\ldots,m \qquad\qquad\qquad .$$

Assume that $q = 1\bigg/ \sum_{i=1}^m w_i^l$ and $t_i = qw_i^l \ \ i = 1,2,\ldots,m$, Eqs. (11.27) and (11.28) can be transformed into the following LP model:

$$\left(Z_j^l\right)_\alpha^U = \max \ \sum_{i=1}^m t_i \left(R_{ij}\right)_\alpha^U$$

s.t.

$$q(W_i)_\alpha^L \le t_i \le q(W_i)_\alpha^U \qquad\qquad (11.29)$$
$$\sum_{i=1}^m t_i = 1$$
$$i = 1,\ldots,m;\ q,\ t_i \ge 0 \qquad ,$$

and

$$\left(Z_j^l\right)_\alpha^L = \min \ \sum_{i=1}^m t_i \left(R_{ij}\right)_\alpha^L$$

$$s.t.$$

$$q(W_i)_\alpha^L \le t_i \le q(W_i)_\alpha^U \tag{11.30}$$

$$\sum_{i=1}^m t_i = 1$$

$$i = 1,\ldots,m; \ \ q, t_i \ge 0.$$

According to the method of Chen et al. (2006), the expected value of $\tilde{Z}_j^l$ can be calculated by the following:

$$E\left[\tilde{Z}_j^l\right] = \frac{1}{2L} \sum_{f=1}^L \left(\left(\tilde{Z}_{ji}^l\right)_{\alpha_f}^U + \left(\tilde{Z}_{Ji}^l\right)_{\alpha_f}^L\right), j = 1, 2, \ldots, m. \tag{11.31}$$

The amount of information is best reflected by a single value derived by using the fuzzy expected value operator. This condition is caused by the fuzzy importance weights of CRs lying in a range, as well as the different α-cuts providing different intervals and uncertainly levels of importance weights of CRs. When the importance weights of ECs are calculated, their ordinal rankings can also be derived. Details of the methods to determine importance weights of ECs are given by Chen et al. (2006).

11.5.2   Synthesis of Individual Preferences on the Ranking of ECs

In synthesizing individual preferences on ECs in terms of ordinal rankings to address the second type of uncertainty, a form of consensus can be derived by simply adding up the member preferences and taking their average. However, such an approach does not necessarily lead to a consensus ranking because the ranking derived by taking a simple arithmetic average may not be robust. The ranking of ECs should be viewed in terms of a "distance" measure. Such a measure relative to a pair of rankings will be an indicator of the degree of correlation between rankings. In this research, a method is proposed to deal with the problem of synthesizing ordinal preferences expressed as priority vectors to form a consensus. This method suggests the problem of determining a compromise or consensus ranking that best agrees with all individual rankings through an assignment problem.

In the proposed approach, a metric or distance function for a set of rankings is introduced. The consensus ranking approach can minimize the total absolute distance (disagreement). We begin by examining some conditions where such a distance, d, should be satisfied. First, the following axioms are required:

Axiom 1 $d(A,B) \succ 0$ with equality if $A = B$

Axiom 2 $d(A,B) = d(B,A)$

Axiom 3 $d(A,C) \prec d(A,B) + d(B,C)$ with equality if the ranking B is between A and C.

Axiom 4 (Invariance)

(1) If A' results from A by a permutation of the objects, and B' results from B by the same permutation, then $d(A',B') = d(A,B)$

(2) If $\bar{A}$ and $\bar{B}$ result from A and B by reversing the order of the objects, then $d(\bar{A},\bar{B}) = d(A,B)$

Axiom 5 (Lifting-moving from n to $n+1$ dimensional space)
If A and B are two rankings of n objects and $A*$ and $B*$ are the rankings that result from A and B, then $d(A^*,B^*) = d(A,B)$ by listing the same $(n+1)^{st}$ object last.

Axiom 6 (Scaling)
The minimum positive distance is one.

The axioms are consistent and can characterize a unique distance function. We can consider the problem wherein K customers provide the ordinal rankings $\{A^l\}_{l=1}^{K}$ of n ECs. Let $A^l = (a_1^l, \ldots, a_n^l)$ and $b_j \in \{1, 2, \ldots, n\}$, where b_i represents the ordinal ranking of the ECs.

Definition 1 The median or consensus ranking $\hat{B}$ refers to the ranking that minimizes the total absolute distance.

$$M(B) = \sum_{l=1}^{K} d(A^l, B) = \sum_{l=1}^{K} \sum_{j=1}^{n} |a_j^l - b_j|. \qquad (11.32)$$

The median ranking is in the best agreement with the set of selected rankings, thus providing an objective criterion to arrive at a consensus.

Let B_0 be the set of all rankings of n objects. Thereafter, the following is obtained:

$$\min_{B \in B_0} \sum_{l=1}^{K} d(A^l, B) \geq \min_{B \in R^n} \sum_{l=1}^{K} d(A^l, B) = \min_{B \in R^n} \sum_{l=1}^{K} \sum_{i=1}^{n} |a_j^l - b_j|. \qquad (11.33)$$

Equation (11.33) attains its minimum when the following is satisfied:

$$b_j = median\{a_j^l\}_{l=1}^{m} = b_j'. \qquad (11.34)$$

Let $B' = (b_i', \ldots, b_n')$. Thereafter, we obtain the following:

$$M(B') \leq M(B) \text{ For all } B \in B_0. \tag{11.35}$$

Thus, we have:

Theorem 1 *Let* $\{A^l\}_{l=1}^K$ *be a set of rankings and* B' *be given by* (11.34). *If* $B' \in B_0$, *then* B' *is the median ranking.*

The determination of the median ranking requires a specialized algorithm. However, if consideration is restricted to the set of complete rankings, the problem can be represented by an LP formulation.

The problem can be solved effectively by representing the problem as an assignment problem. d_{jq} is defined as follows:

$$d_{jq} = \sum_{l=1}^K \left| a_j^l - q \right|. \tag{11.36}$$

Considering the following expression:

$$\sum_{l=1}^K d\left(A^l, B\right) = \sum_{l=1}^K \sum_{j=1}^n \left| a_j^l - b_j \right|. \tag{11.37}$$

Equation (11.36) is the value of the j th sum in Eq. (11.37) if b_j is set equal to $q(q \in \{1, 2, \ldots, n\})$.

If we define the following expression

$$x_{jq} = \begin{cases} 1 & \text{if } b_j = q \\ 0 & \text{otherwise}, \end{cases} \tag{11.38}$$

then the restricted problem is equivalent to the following assignment problem:

$$\begin{aligned} \min_{x_{jq}} & \quad \sum_{j=1}^n \sum_{q=1}^n d_{jq} x_{jq} \\ \text{s.t.} & \quad \sum_{j=1}^n x_{jq} = 1 \quad \text{for all} \quad q \\ & \quad \sum_{q=1}^n x_{jq} = 1 \quad \text{for all} \quad j \\ & \quad x_{jq} \geq 0 \quad \text{for all} \quad j, q. \end{aligned} \tag{11.39}$$

The above integer programming model is capable of handling large problems. By solving the model, the consensus rankings of ECs can be obtained.

11.5.3 Evaluation of Robustness

The measurement of robustness depends on the perspective of robustness. Kim and Kim (2009) proposed an index to evaluate the robustness of ordinal ranking, namely, priority relationships among ECs. This approach indicates the relative priority order among two or more ECs. The robustness on the priority relationships among ECs can be measured as the degree in which the relative priority relationships among ECs are maintained despite the presence uncertainty. In this viewpoint, the robustness index on the priority relationships among ECs are expressed as the likelihood that a priority relationship in V is retained. For instance, if V is $[j_1 j_2]$, V represents a priority relationship wherein EC_{j_1} has a higher priority than EC_{j_2}. The robustness index on the priority relationship, denoted as $RI(V)$, can be calculated as follows:

$$RI(V) = \frac{\sum_l^L y(l)}{L}$$
$$where \begin{cases} y(l) = 1, if \ ranking\left(EC_{V(v)}, l\right) \leq ranking\left(EC_{V(v+1)}, l\right) \forall v = 1, \ldots, N(V) - 1 \\ y(l) = 0, \quad otherwise \end{cases},$$

$$(11.40)$$

where $y(l)$ is an indicator variable, $EC_{V(v)}$ denotes the vth EC in V, and $N(V)$ denotes the array size of V.

Considering that $RI(V)$ is expressed as a type of likelihood measured by an empirical probability, it has a value between zero and one. The larger value of $RI(V)$ implies that higher robustness on the absolute ranking of ECs or the priority relationship in V can be obtained. If $RI(V)$ is equal to one, the priority relationship in V is always retained despite the variability. By using the robustness index, the robustness of the prioritization decision, EC or V, can be evaluated.

The design of a flexible manufacturing system (FMS) (Liu 2005; Chen et al. 2006) is used to illustrate the proposed method. Assume that ten customers, denoted as $C_l, l = 1, 2, \ldots, 10$, are surveyed in a target market. Seven fuzzy numbers are used to express their individual assessments on the eight CRs, as shown in Table 11.2. $W_1 \sim W_7$ are the importance weights of CRs, which represent very unimportant, quite unimportant, unimportant, slightly important, moderately important, important and very important, respectively. The relationship measures between CRs and ECs are shown in Table 11.3. $R_1 \sim R_4$ denote four relationship linguistic terms, which are very weak, weak, moderate, and strong, respectively.

The proposed approach was applied to compute the ranking of ECs and the final ordinal rankings of ECs can be obtained.

$$EC_3 \succ EC_7 \succ EC_1 \succ EC_2 \succ EC_4 \succ EC_8 \succ EC_9 \succ EC_{10} \succ EC_6 \succ EC_5$$

Based on the method of Chen et al. (2006), the ordinal ranking of ECs is shown as follows:

Table 11.2 The fuzzy importance of eight CRs assessed by ten customers using fuzzy numbers

	CR_1	CR_2	CR_3	CR_4	CR_5	CR_6	CR_7	CR_8
C_1	W_7	W_2	W_7	W_2	W_7	W_3	W_4	W_1
C_2	W_3	W_3	W_6	W_7	W_1	W_2	W_6	W_7
C_3	W_1	W_7	W_2	W_5	W_1	W_1	W_5	W_7
C_4	W_7	W_4	W_7	W_1	W_3	W_6	W_7	W_2
C_5	W_5	W_2	W_2	W_6	W_5	W_1	W_7	W_6
C_6	W_7	W_4	W_1	W_5	W_3	W_2	W_7	W_2
C_7	W_7	W_2	W_6	W_1	W_4	W_5	W_1	W_6
C_8	W_6	W_7	W_5	W_7	W_7	W_6	W_2	W_2
C_9	W_7	W_1	W_6	W_3	W_7	W_5	W_7	W_7
C_{10}	W_2	W_4	W_7	W_2	W_7	W_1	W_6	W_2

Table 11.3 The relationship matrix between CRs and ECs

R_{ij}	EC_1	EC_2	EC_3	EC_4	EC_5	EC_6	EC_7	EC_8	EC_9	EC_{10}
CR_1	R_3	R_2	R_3	R_3	R_1	R_1	R_3	R_2	R_2	R_2
CR_2	R_2	R_3	R_4	R_2	R_1	R_3	R_3	R_3	R_3	R_2
CR_3	R_4	R_3	R_4	R_2	R_1	R_1	R_4	R_2	R_2	R_2
CR_4	R_3	R_3	R_3	R_3	R_2	R_2	R_2	R_3	R_3	R_2
CR_5	R_4	R_4	R_3	R_4	R_2	R_2	R_3	R_3	R_3	R_3
CR_6	R_2	R_2	R_3	R_3	R_2	R_2	R_2	R_2	R_2	R_2
CR_7	R_2	R_3	R_3	R_2	R_3	R_3	R_4	R_3	R_2	R_2
CR_8	R_3	R_2	R_3	R_3	R_2	R_2	R_4	R_3	R_2	R_2

$$EC_3 \succ EC_7 \succ EC_1 \succ EC_2 \succ EC_4 \succ EC_8 \succ EC_9 \succ EC_6 \succ EC_{10} \succ EC_5$$

From the ranking results of ECs of the two methods, we can find the difference in the order of EC_{10} and EC_6. Based on the method of Chen et al. (2006), the ordinal ranking of EC_{10} and EC_6 is $EC_6 > EC_{10}$. Based on the proposed approach, the result is opposite. Therefore, the robustness index is used to evaluate the ordinal ranking of EC_{10} and EC_6 between the method of Chen et al. (2006) and the proposed approach. In the prior method, the prioritization relationship V is [6 10]. For the first customer, $ECI_6 > ECI_{10}$ is consistent with V. Hence, $y(1)$ is equal to 1. In the proposed approach, V is [10 6]. For the first customer, $ECI_{10} < ECI_6$ is not consistent to V. Hence, $y(1)$ is equal to 0. Similarly, the value of $y(l)$ for the ten customers can be derived, as shown in Table 11.4. Then, $RI(V)$ value based on the Chen's method can be calculated as follows:

$$RI(V) = \frac{1+0+0+1+0+0+1+0+1+0}{10} = 0.4$$

Table 11.4 An illustration of the robustness index between the two methods

Customer	$ECI_6(l)$	$ECI_{10}(l)$	Method of Chen et al. (2006) with traditional average arithmetic EC = {EC$_6$, EC$_{10}$}, V = [6 10]	The proposed approach with an 0–1 integer programming EC = {EC$_{10}$, EC$_6$}, V = [10 6]
C_1	0.4561 (8)[a]	0.4164 (9)	$y(1) = 1$	$y(1) = 0$
C_2	0.3276 (9)	0.3669 (8)	$y(2) = 0$	$y(2) = 1$
C_3	0.2499 (9)	0.3411 (8)	$y(3) = 0$	$y(3) = 1$
C_4	0.4432 (8)	0.3951 (9)	$y(4) = 1$	$y(4) = 0$
C_5	0.3540 (9)	0.3725 (8)	$y(5) = 0$	$y(5) = 1$
C_6	0.4219 (9)	0.4280 (8)	$y(6) = 0$	$y(6) = 1$
C_7	0.3969 (8)	0.3327 (9)	$y(7) = 1$	$y(7) = 0$
C_8	0.4146 (9)	0.4301 (8)	$y(8) = 0$	$y(8) = 1$
C_9	0.3916 (8)	0.3526 (9)	$y(9) = 1$	$y(9) = 0$
C_{10}	0.3719 (9)	0.4023 (8)	$y(10) = 0$	$y(10) = 1$
Robustness index value			RI(V) = 0.4	RI(V) = 0.6

Note [a]Denotes the ranking of the corresponding ECI for the *l*th customer

and $RI(V)$ value based on the proposed approach can be calculated as shown below.

$$RI(V) = \frac{0+1+1+0+1+1+0+1+0+1}{10} = 0.6$$

Finally, the index values for the ranking based on the prior method and the proposed approach are calculated as 0.4 and 0.6 respectively. Therefore, the proposed approach outperforms the method of Chen et al. (2006) in prioritizing the ECs in terms of robustness.

11.6 Target Value Setting of ECs by Using Fuzzy Optimization and Inexact Genetic Algorithm

In the product design stage, product development teams may need to consider various design scenarios while determining design specifications. However, in previous studies of target value setting of ECs in QFD, only a single solution is obtained and target value setting for different design scenarios cannot be considered. In this section, a fuzzy optimization model is presented to determine the target value setting for ECs in QFD. An inexact genetic algorithm approach is described to solve the model that takes the mutation along the weighted gradient direction as a genetic operator. Instead of obtaining one set of exact optimal target values, the approach can generate a family of inexact optimal target values setting within an acceptable satisfaction degree. Through an interactive approach, a product development team can determine a combination of preferred solution sets from which a set of preferred target values of ECs based on a specific design scenario can be obtained.

11.6.1 Formulation of Fuzzy Optimization Model for Target Values Setting in QFD

The processes of determining the target values for ECs in QFD can be formulated as shown below:

Determine target values $x_1, x_2, \ldots, x_n$ by maximizing the overall customer satisfaction:

$$y_i = f_i(X), i = 1, \ldots, m,$$

$$X_j = g_j(X^j), j = 1, \ldots, n,$$

where

$$X = (x_1, x_2, \ldots, x_n)^T,$$

$$X^j = (x_1, \ldots, x_{i-1}, x_{j+1}, \ldots, x_n)^T,$$

y_i is the customer perception of the degree of satisfaction of the ith CR $i = 1, \ldots, m,$

x_j is the target value of the jth EC, $j = 1, \ldots, n,$

f_i is the functional relationship between the ith CR and ECs, i.e. $y_i = f_i(x_1, \ldots, x_n)$, and

g_j is the functional relationship between the jth EC and other ECs, i.e., $x_j = g_j(x_1, \ldots, x_{j-1}, x_{j+1}, \ldots x_n)$

The above equation is a general model to determine the target values of ECs. Additional constraints can be added to the above formulation as appropriate. In reality, many design tasks are performed in an environment wherein system parameters, objectives, and constraints are not known precisely. Therefore, developing a crisp optimization model to set the target values of ECs in QFD is difficult. For the establishment of an objective function, customers usually cannot provide a precise satisfaction value and instead express their satisfaction in linguistic terms such as "quite satisfied" and "very satisfied." The relationships between CRs and ECs, as well as among ECs, can be very complicated. Engineers usually do not have full knowledge of the impact of an EC on CRs or on other ECs. Thus, setting the relationship values between a CR and an EC is also imprecise. Regarding the constraints, vagueness also exists. For example, the cost is usually described as a function of CR and should not exceed a predetermined upper limit. The cost constraint can then be formulated as follows:

$$c_1x_1 + c_2x_2 + \cdots + c_nx_n \leq \tilde{c},$$

where, $\tilde{c}$ is the upper limit of cost and $c_1, c_2, \ldots c_n$ are the coefficients. Owing to the imprecise and incomplete design information available in the early design stage, the values of $\tilde{c}$ may be imprecise.

The fuzziness presents a special challenge to model effectively the process of target values setting by using traditional mathematical programming technique. One way to deal with such vagueness quantitatively is to employ fuzzy set theory, which can be used to develop a fuzzy optimization model for the target value setting of ECs in QFD. On the basis of the work of Kim et al. (2000), a general fuzzy optimization model to set target values of ECs in QFD is proposed as follows:

$$\begin{aligned}
&\tilde{M}ax \ \mu(y_1, y_2, \ldots, y_n) \\
&s.t. \\
&C1: \ y_i \cong f_i(X), \quad i = 1, \ldots, m \\
&C2: X_j \cong g_j(X^j), \quad j = 1, \ldots, n \\
&C3: C(X) \leq \tilde{c},
\end{aligned} \qquad (11.41)$$

where μ is the satisfaction degree of customers to CRs; $C1$ and $C2$ are the fuzzy relationship constraints; $C3$ is the cost constraint.

11.6.2 Tolerance Approach to the Determination of an Exact Optimal Solution from the Fuzzy Optimization Model

The symmetric models, which are based on the definition of fuzzy decision, were frequently adopted in fuzzy LP models. They assumed that the objective and constraints in an imprecise situation could be represented by fuzzy sets. A decision could be stated as the union of the fuzzy objective and constraints and defined by a max–min operator. A fuzzy objective set G and a fuzzy constraint set C are assumed to be given in a space X. G and C are then combined to form a decision D, which is a fuzzy set resulting from the intersection of G and C, and the corresponding μ_D is equal $\mu_G \cap \mu_c$. Lai and Hwang (1992) mentioned that the approaches of Zimmermann, Werner, Chanas and Verdegay are the most practical approaches among various techniques in fuzzy LP. Transforming a fuzzy optimization model into a crisp model is the common idea of these approaches. In this research, Zimmerman's tolerance approach (Zimmerman 1996) is adopted to solve the fuzzy optimization model. First, the membership function of fuzzy constraints and the fuzzy objective have to be determined. Customers usually cannot provide a satisfaction value precisely. Customers express satisfaction in fuzzy terms such as "quite satisfactory" and "not very satisfactory." Let y_i^{min} and y_i^{max} represent the lower and upper bounds of aspirations, respectively, with respect to y_i. A customer would then express complete dissatisfaction of a design (X) in which $y_i(X) \leq y_i^{min}$; but would express complete satisfaction if $y_i(X) \geq y_i^{max}$. A membership function $\mu_{y_i}(X)$ can be introduced to measure the satisfaction degree of customers to the ith CR at various ECs for design (X). The membership function $\mu_{y_i}(X)$ can be represented as follows:

$$\mu_{y_i}(X) = \begin{cases} 0 & if & y_i(X) \leq y_i^{min} \\ \tau(X) & if & y_i^{min} \leq y_i(X) \leq y_i^{max} \\ 1 & if & y_i(X) \geq y_i^{max}, \end{cases} \tag{11.42}$$

where $\tau(X)$ could be linear or nonlinear.

The membership functions for all CRs $\mu_{y_i}(X)$, where $i=1,2,\ldots,m$, form the membership function of a fuzzy objective function. Each fuzzy constraint in the fuzzy optimization model can be represented by a fuzzy set. The membership function of the fuzzy relationship constraints are $\mu_{f_i}(X,Y)$ and $\mu_{g_j}(X,Y)$, where $i=1,2\ldots,m$ and $j=1,2\ldots,n$, respectively. The membership functions of a fuzzy constraint "$AX \overset{\sim}{=} b$" can be represented as follows (Zimmerman 1996):

$$\mu(X) = \begin{cases} 0 & if & AX \leq b - d \ or \ AX \geq b + d \\ 1 - \frac{|AX-b|}{d} & if & b - d < AX < b + d \\ 1 & if & AX = b \end{cases} \tag{11.43}$$

where A is a row vector, b is a constant, and d is a subjectively chosen constant of the admissible violations of the constraint.

The membership of the cost constraint, $\mu_c(X)$, can be represented in the following form:

$$\mu_c(X) = \begin{cases} 1 & if \ CX < c \\ 1 - \frac{(CX - c)}{t} & if \ c \leq CX \leq c + t \\ 0 & if \ CX > c + t \end{cases},$$

where t is a pre-specified non-negative tolerance level to cost c. The above membership function denotes the following:

$\mu(X)$ is zero if the constraints are strongly violated,
$\mu(X)$ is one if the constraints are very satisfied, and
$\mu(X)$ increases monotonously from zero to one

With the use of Zimmerman's tolerance approach, the crisp form of the fuzzy optimization model in Eq. (11.41) can be formulated as follows:

$$Maximize\ \lambda$$

$$subject\ to$$

$$\lambda \le \mu_{y_i}(X), \qquad i = 1, 2, \ldots, m$$
$$\lambda \le \mu_{f_i}(X, Y), \quad i = 1, 2, \ldots .m \qquad\qquad (11.44)$$
$$\lambda \le \mu_{g_j}(X, Y), \quad j = 1, 2, \ldots, n$$
$$\lambda \le \mu_c(X)$$

where $\lambda\ (0 \le \lambda \le 1)$ represents the overall value of membership functions.

A unique optimal solution for the above model can be obtained by using an LP technique, which is a set of optimal target values for ECs$x_1, x_2, \ldots, x_n$. The exact setting of optimal target values for ECs obtained by using the LP technique may not be acceptable to a product development team in a new product design. This condition is caused by the inherent or permitted possibility and flexibility in the target values setting of ECs existing in QFD and the obtained solutions allowing a product development team to reconcile tradeoffs among the CRs and ECs under various design scenarios. The provision of a family of inexact satisfactory target values setting for ECs would then be very useful to the product development team. In this chapter, an inexact genetic algorithm is employed to generate a family of inexact satisfactory target values setting for ECs from the fuzzy optimization model. Detailed descriptions are shown below.

11.6.3 Inexact Genetic Algorithm Approach to the Generation of a Family of Inexact Solutions from the Fuzzy Optimization Model

During the last 30 years, interest in problem solving systems based on principles of evolution and hereditary has grown. Even though many different classes of the systems exist, such as genetic algorithms, evolutionary programming, and evolution strategies, they all rely on the same concept of mimicking mechanisms of biological evolution. Admittedly, the gap among them is getting smaller and smaller. They have also been called as some common terms such as evolutionary algorithms and evolution programs. The inexact genetic algorithm is a specially designed one to solve these problems with fuzziness.

The basic idea of the inexact genetic algorithm (Wang 1997) is that the mutation operator moves along a weighted gradient direction. An individual is led by this mutation operator to arrive at inexact solutions within an acceptable range of the fuzzy optimal solution sets. By means of an interactive approach, a set of preferred solutions can be sought by a convex combination of the solutions selected from the family. The basic idea of the method is applied to solve the fuzzy optimization model Eq. (11.44) to obtain a set of preferred solutions corresponding to a particular design scenario.

Generally, two types of genetic operators exist for the genetic algorithm: crossover and mutation. For the linear problem, only the mutation operator is utilized. In the inexact genetic algorithm, the mutation operator moves along a weighted gradient direction. For model Eq. (11.44), the mutation operator can be induced as follows.

Based on the tolerance method, the fuzzy optimal solution set of the fuzzy optimization model Eq. (11.44) is a fuzzy set defined by the following:

$$\tilde{S} = \{(X, \mu_{\tilde{S}}(X, Y)) | X, Y \in R^n_+ \}, \tag{11.45}$$

where

$$\mu_{\tilde{S}}(X, Y) = \min\{\mu_{y_i}(X), \mu_{f_i}(X, Y), \mu_{g_j}(X, Y), \mu_c(X)\} \ i = 1, 2, \ldots, n; j = 1, 2, \ldots, m, \tag{11.46}$$

and R^n_+ is the non-negative n-dimensional space. Based on the equivalent unconstrained max-min optimization problem, a weighted gradient of $\mu_{\tilde{S}}(X, Y)$ can be defined as follows:

$$\begin{aligned} G(X, Y) &\equiv \nabla \mu_{\tilde{S}}(X, Y) \\ &= -\Gamma(\mu_{yi}) \cdot \sum_i \mu_{yi}(X) + \Gamma(\mu_{fi}) \cdot \sum_i \mu_{fi}(X, Y) + \Gamma(\mu_{gj}) \cdot \sum_j \mu_{gj}(X, Y) + \Gamma(\mu_c) \cdot \mu_c(X), \\ &i = 1, 2, \ldots, n; j = 1, 2, \ldots, m \end{aligned} \tag{11.47}$$

where $\Gamma(\mu)$ represents the weight of the corresponding membership, which can be designed as follows:

Let $\mu_{\min} = \min\{\mu_{yi}(X), \mu_{f_i}(X, Y), \mu_{g_j}(X, Y), \mu_c(X), i = 1, 2, \ldots, n; j = 1, 2, \ldots, m\}$; then

$$\Gamma(\mu) = \begin{cases} 1, & \mu = \mu_{\min} \\ \dfrac{\sigma}{\mu}, & \mu_{\min} < \mu < 1 \\ 0, & \mu = 1 \end{cases} \tag{11.48}$$

where σ is a sufficiently small positive number.

If $(x, y)^{k+1}$ is the child of the individual, $(x, y)^k$, the mutation along the weighted gradient direction can be described as follows:

$$(x, y)^{k+1} = (x, y)^k + \theta G(x, y), \tag{11.49}$$

where θ is a random step length of the Erlang distribution, which is generated by a random number generator.

By employing the inexact genetic algorithm to solve the fuzzy optimization model, a family of inexact optimal solutions can be obtained. The interactive approach allows a product development team to select a preferred solution from the fuzzy optimal solutions. First, the team provides an acceptable membership degree level of the fuzzy optimization. They then choose their preference structure utilizing the human-computer interaction. The product development team needs to point out which criteria are of utmost concern to them. The criteria could be the objective, constraints, or decision variables. The solutions in the α-cut of the fuzzy solutions set, $\tilde{S}_\alpha$, with the highest and lowest values of these criteria are stored in memory and updated in each interaction. Given the large number of the visited points, the solutions preferred by the product development team can be found with high probability. Considering that in general, more than one criterion of concerns from the product development team exists, more than one solution would be derived. When the iteration terminates, the solutions with their criteria values will be displayed to the product development team. The product development team can then select a couple of preferred solutions each time. By repeating the above procedures, the preferred final solution can be generated.

The proposed approaches have been applied to car door design. Table 11.5 shows the 5 sets of optimal solutions, which correspond to maximum values of the decision variables. In this table, y_1–y_5 are the satisfaction values of five CRs, which are "easy to close from outside," "stays open on a hill," "rain leakage," "road noise," and "cost," respectively; x_1 to x_6 are target values setting of six ECs, which are "energy to close the door," "check force on level ground," "check force on 10 % slope," "door seal resistance," "road noise reduction," and "water resistance," respectively; $\mu(z)$ is the membership function of the fuzzy optimal solution set, which can be calculated based on Eq. (11.46); The last column shows the maximum values of y_1 to y_5. For example, the maximum of $y_1 = max\{4.9273, 3.3547, 4.9273, 4.0517, 3.1515\} = 4.9273$.

From the results, various sets of preferred target values setting ECs can be obtained for different design scenarios rather than the only one exact optimal target values setting. For example, the design team would like to have the maximum satisfaction values of all the CRs. In this case, the design team should select the solutions, $Z_{max(1)}, Z_{max(2)}, Z_{max(3)}, Z_{max(4)}$ and $Z_{max(5)}$. The satisfaction values of the individual CRs, Y^{final}, and the target values setting of the ECs, X^{final} can be calculated using the following linear combination.

$$Z = Z_{max(1)} * \omega_1 + Z_{max(2)} * \omega_2 + Z_{max(3)} * \omega_3 + Z_{max(4)} * \omega_4 + Z_{max(5)} * \omega_5$$

where $Z = \left(Y^{final}, X^{final}\right)$, and $\omega_1 = 0.3, \omega_2 = 0.2, \omega_3 = 0.1, \omega_4 = 0.1$ and $\omega_5 = 0.3$ are the importance weights of the corresponding CRs.

Table 11.5 A family of satisfactory solutions corresponding to the maximum values of decision variables

Solution	y_1-y_5	x_1-x_6	$\mu(z)$	Meaning
$Z_{max(1)}$	4.9273 3.2594 5.9671 4.9027 3.8029	7.91431 4.5618 7.8602 5.0459 6.1948 74.7782	0.3365	Maximum of y_1 $y_1 = 4.9273$
$Z_{max(2)}$	3.3547 3.2897 5.7525 4.8893 3.7896	8.0849 15.1257 7.8171 4.7470 6.2084 74.7672	0.3443	Maximum of y_2 $y_2 = 3.2897$
$Z_{max(3)}$	4.9273 3.2594 5.9671 4.9027 3.8029	7.9143 14.5618 7.8602 5.0459 6.1948 74.7782	0.3365	Maximum of y_3 $y_3 = 5.9671$
$Z_{max(4)}$	4.0517 3.2626 5.1167 4.9183 3.6646	8.7023 14.0130 7.5656 4.4619 6.6273 74.6778	0.5414	Maximum of y_4 $y_4 = 4.9183$
$Z_{max(5)}$	3.1515 2.3806 4.8255 3.9799 5.2530	9.5063 12.8237 7.0370 4.2907 8.0396 74.2166	0.3580	Maximum of y_5 $y_5 = 5.2530$

Based on the above linear combination, Y^{final} and X^{final} can be determined as follows:

$$Y^{final} = \left(y_1^{final}, y_2^{final}, y_3^{final}, y_4^{final}, y_5^{final}\right) = (3.9925, 3.0021, 5.4967, 4.6247, 4.2214)$$

$$X^{final} = \left(x_1^{final}, x_2^{final}, x_3^{final}, x_4^{final}, x_5^{final}, x_6^{final}\right) = (8.5048, 14.0983, 7.5752, 4.7012, 6.7942, 74.5975)$$

11.7 Conclusion

In this chapter, our recent research on the development of intelligent QFD is described. First, a fuzzy AHP with extent analysis approach is described to determine the importance weights for CRs. The approach is effective to determine the importance weights as it is capable of capturing the vagueness of human judgment in assessing the importance of CRs. The fuzzy AHP with extent analysis is also easy to implement because the tedious calculation of eigenvectors required by the conventional AHP is no longer necessary. Second, a chaos-based FR approach is described to model the functional relationships between CRs and ECs. This approach can address the issues pertaining to the modeling of the functional relationships based on QFD: (1) only a small number of data sets are available for modeling, (2) relationships between CRs and ECs are nonlinear in nature, (3) data sets for modeling contain a high degree of fuzziness, and (4) explicit customer satisfaction models are preferred. Third, a fuzzy group decision-making method is described to address the two uncertainties simultaneously in prioritizing ECs and to determine the importance weights of ECs. It mainly involves an ordinal ranking of

ECs based on a fuzzy weighted average method with fuzzy expected operator and a consensus ranking method. Finally, a fuzzy optimization model is presented and an inexact genetic algorithm approach is described to solve the model to determine the target value setting of ECs. Considering that product development teams may consider product design under various design scenarios, a unique optimal solution obtained from the fuzzy model may not be acceptable to them. The proposed approaches are capable of generating a family of optimal target values setting for ECs, from which various sets of preferred target values setting for ECs can be obtained for different design scenarios rather than the only one exact optimal target values setting.

Future research on developing intelligent QFD can involve the detection and elimination of outliers to improve survey data quality. Outliers may exist in the survey data and can affect the predictive accuracy of the models. Regarding the chaos-based FR approach for modeling functional relationships, minimizing the complexity of the generated fuzzy polynomial models could be considered in future work as an objective together with minimizing errors in the formulation of the fitness function. This approach would help develop fuzzy polynomial models with simpler structures and good modeling accuracy. Future works can also consider cost minimization as an objective in the fuzzy optimization model apart from maximizing customer satisfaction. The optimization problem thus becomes a multi-objective one. Other solving techniques such as multi-objective genetic algorithms and particle swarm optimization need to be introduced to solve the optimization problem.

Acknowledgments The work described in this chapter was fully supported by a grant from the Research Grants Council of the Hong Kong Special Administrative Region, China (Project No. PolyU 517113).

References

Armacost, R.T., Componation, P.J., Mullens, M.A., Swart, W.W.: An AHP framework for prioritizing custom requirements in QFD: an industrialized housing application. IIE Trans. **26** (4), 72–79 (1994)

Buyukozkan, G., Ertay, T., Kahraman, C., Ruan, D.: Determining the Importance Weights for the Design Requirement in the House of Quality Using the Fuzzy Analytic Network Approach. Int. J. Intell. Syst. **19**(5), 443–461 (2004)

Chan, K.Y., Kwong, C.K., Dillon, T.S.: Development of product design models using fuzzy regression based genetic programming. Stud. Comput. Intell. **403**, 111–128 (2012)

Chan, K.Y., Kwong, C.K., Wong, T.C.: Modelling customer satisfaction for product development using genetic programming. J. Eng. Des. **22**(1), 55–68 (2011)

Chan, L.K., Kao, H.P., Ng, A., Wu, M.L.: Rating the importance of customer needs in quality function deployment by fuzzy and entropy methods. Int. J. Prod. Res. **37**(11), 2499–2518 (1999)

Chan, L.K., Wu, M.L.: Quality function deployment: a literature review. Eur. J. Oper. Res. **143**(3), 463–497 (2002)

Che, A., Lin, Z.H., Chen, K.N.: Capturing weight of voice of the customer using artificial neural network in quality function deployment. J. Xi'an Jiaotong Univ. **33**(5), 75–78 (1999)

Chen, C.H., Khoo, L.P., Yan, W.: Evaluation of multicultural factors from elicited customer requirements for new product development. Res. Eng. Des. **14**(3), 119–130 (2003)

Chen, Y., Fung, R.Y.K., Tang, J.F.: Rating technical attributes in fuzzy QFD by integrating fuzzy weighted average method and fuzzy expected value operator. Eur. J. Oper. Res. **174**(3), 1553–1556 (2006)

Dawson, D., Askin, R.G.: Optimal new product design using quality function deployment with empirical value functions. Qual. Reliab. Eng. Int. **15**(1), 17–32 (1999)

Fung, R., Law, D., Ip, W.: Design targets determination for inter-dependent product attributes in QFD using fuzzy inference. Integr. Manufact. Syst. **10**(6), 376–383 (1999)

Fung, R.Y.K., Popplewell, K., Xie, J.: An intelligent hybrid system for customer requirements analysis and product attribute targets determination. Int. J. Prod. Res. **36**(1), 13–34 (1998)

Govers, C.P.M.: What and how about quality function deployment (QFD). Int. J. Prod. Econ. **46–47**, 575–585 (1996)

Han, S.H., Yun, M.H., Kim, K.J., Kwahk, J.: Evaluation of product usability: development and validation of usability dimensions and design elements based on empirical models. Int. J. Ind. Ergon. **26**(4), 477–488 (2000)

Hsiao, S.W.: Concurrent design method for developing a new product. Int. J. Ind. Ergon. **29**(1), 41–55 (2002)

Huang, G.Q., Jiang, Z.: FuzzySTAR: Fuzzy set theory of axiomatic design review. Artif. Intell. Eng. Des. Anal. Manuf. **16**(4), 291–302 (2002)

Kao, C., Liu, S.T.: Fractional programming approach to fuzzy weighted average. Fuzzy Sets Syst. **120**(3), 435–444 (2001)

Karsak, E.E., Sozer, S., Alptekin, S.E.: Product planning in quality function deployment using a combined analytical network process and goal programming approach. Comput. Ind. Eng. **44**(1), 171–190 (2002)

Karsak, K.: Fuzzy multiple objective decision making approach to prioritize design requirments in quality function deployment. Int. J. Prod. Res. **42**(18), 3957–3974 (2004)

Kim, D.H., Kim, K.J.: Robustness indices and robust prioritization in QFD. Expert Syst. Appl. **36**(2), 2651–2658 (2009)

Kim, K., Moskowitz, H., Dhingra, A., Evans, G.: Fuzzy multicriteria models for quality function deployment. Eur. J. Oper. Res. **121**(3), 504–518 (2000)

Kim, K., Park, T.: Determination of an optimal set of design requirements using house of quality. J. Oper. Manage. **16**(5), 569–581 (1998)

Kwong, C.K., Bai, H.: A fuzzy AHP approach to the determination of importance weights of customer requirements in quality function deployment. J. Intell. Manuf. **13**(5), 367–377 (2002)

Kwong, C.K., Chen, Y., Chan, K.Y., Luo, X.: A generalised fuzzy least-squares regression approach to modelling relationships in QFD. J. Eng. Des. **21**(5), 601–613 (2010)

Kwong, C.K., Chen, Y.Z., Bai, H., Chan, D.S.K.: A methodology of determining aggregated importance of engineering characteristics in QFD. Comput. Ind. Eng. **53**(4), 667–679 (2007)

Lai, J.Y., Hwang, C.L.: Fuzzy mathematical programming: method and application. Springer, New York (1992)

Liu, S.T.: Rating design requirements in fuzzy quality function deployment via a mathematical programming approach. Int. J. Prod. Res. **43**(3), 497–513 (2005)

Lu, M.H., Madu, C.N., Kuei, C., Winokur, D.: Integrating QFD, AHP and benchmarking in strategic marketing. J. Bus. Indust. Mark. **9**(1), 41–50 (1994)

Mochimaru, M., Takahashi, M., Hatakenaka, N., Horiuchi, H.: Questionnaire survey of customer satisfaction for product categories towards certification of ergonomic quality in design. Work: J. Prev. Assess. Rehabil. **41**(1), 956–959 (2012)

Shahraki, A.R., Paghaleh, M.J.: Ranking the voice of customer with fuzzy DEMATEL and fuzzy AHP. Indian J. Science and Technol. **4**(12), 1763–1772 (2011)

Tanaka, H., Uejima, S., Asai, K.: Linear regression analysis with fuzzy model. IEEE Trans. Syst. Man Cybern. **12**(6), 903–907 (1982)

Terninko, J.: Step-by-step QFD: customer-driven product design. St. Lucie Press, Boca Raton (1997)

Vonderembse, M.A., Raghunathan, T.S.: Quality function deployment's impact on product development. Int. J. Qual. Sci. 2(4), 253–271 (1997)

Wang, D.: An inexact approach for linear programming problems with fuzzy objective and resources. Fuzzy Sets Syst. 89(1), 61–68 (1997)

Wang, J.: Fuzzy outranking approach to prioritize design requirements in quality function deployment. Int. J. Prod. Res. 37(4), 899–916 (1999)

Wasserman, G.S.: On how to prioritize design requirements during the QFD planning process. IIE Trans. 25(3), 59–65 (1993)

Zadeh, L.A.: Fuzzy sets as a basis for a theory of possibility. Fuzzy Sets Syst. 1, 3–28 (1978)

Zhang, Z.F., Chu, X.N.: Fuzzy group decision-making for multi-format and multi-granularity linguistic judgments in quality function deployment. Expert Syst. Appl. 36(5), 9150–9158 (2009)

Zhou, M.: Fuzzy logic and optimization models for implementing QFD. Comput. Ind. Eng. 35(1–2), 237–240 (1998)

Zimmermann, H.J.: Fuzzy Set Theory and Its Applications. Kluwer, Dordrecht (1996)

Chapter 12
Process Improvement Using Intelligent Six Sigma

James Fogal

Abstract Six Sigma is a well-regarded and proven methodology for improving the quality of products and services by removing inconsistencies in processes. Insomuch of the early Six Sigma initiatives was focused on process effectiveness in meeting quality expectations and process efficiency for achieving maximum producer value; the future trends is moving towards utilizing feedback loops to create intelligent processes that enhances the adaptability to changing conditions. The purpose of this chapter is to extend understanding of what performance measures can be applied to processes in order to gain useful information and the emerging application of artificial neural networks to handle concurrent multiple feedback loops.

Keywords Six Sigma · Process improvement · Performance measures · Artificial intelligence · Neural networks

12.1 Introduction

Quality is a key concern for all businesses seeking profitability and long term success and managing quality—at all levels and across all company components- is crucial in enabling organizations to make decisions that will achieve those goals. Whether deciding the best way to increase brand reputation, minimize liabilities, or become distinguished against the competition, quality is a critical factor in how a company is viewed in the marketplace. In fact, quality is so important, it is fairly common to see success in this area not only expressed in measures of product availability, price, and customer satisfaction, but also in terms of process management and efficacy.

J. Fogal (✉)
School of Business and Management, Notre Dame de Namur University,
Belmont, CA, USA
e-mail: jfogal@ndnu.edu

© Springer International Publishing Switzerland 2016   363
C. Kahraman and S. Yanık (eds.), *Intelligent Decision Making
in Quality Management*, Intelligent Systems Reference Library 97,
DOI 10.1007/978-3-319-24499-0_12

The far reaching value of quality as an important differentiator is why managing it effectively is essential to the success of an organization and why vast amounts of time, attention, and resources are committed annually by organizations to achieve levels of quality necessary to ensure their prosperity. To that end, there are a number of approaches put forth by various industries as the most effective way to improve quality in products or services. These approaches range from Quality Control, a focus on ensuring results are as expected and Quality Assurance, which advocates doing the right things the right way, to the pursuit of a philosophy of Total Quality Management, which emphasizes continuous quality improvements. However, the framework that has garnered the most worldwide interest in recent years is Six Sigma. Six Sigma is a disciplined, data-driven approach and strategy for process improvement and is seen as an amalgamation of the best elements of all other quality approaches. It differentiates itself by promoting a stronger emphasis on monitoring yields and costs associated with a quality improvement effort. In other words, Six Sigma is not just about seeking quality based on the output of product and services, but also as a means for achieving efficacy in how things are accomplished.

The origins of the Six Sigma movement can be traced back to the mid-1980s with Motorola's ground breaking work to seek ways to significantly improve quality in order to obtain bottom-line production results in their organization. The simplistic view of Six Sigma often describes it as a methodology for taking quality measurements in order to achieve reductions in process variation. However, the Six Sigma strategy and the results it can achieve is much more than just the simple application of a set of statistical tools and techniques. At its most fundamental level, the objective of Six Sigma is the implementation of a measurement-based strategy focused on achieving perfection in quality, but the framework is also being used widely today on process improvements to drive organizational effec-tiveness and gain greater value from processes. The drive to improve process efficacy is being fueled by ever-changing requirements for products and services in areas such as technology, consumption demands, and legal standards and these changes require organizations to adapt and instill processes with capabilities to meet increased expectations. These continuously rising expectations necessitate that organizations understand quality beyond the basic level of what is working and what isn't in individual components of the company. Six Sigma is a strategy that can be the means for managing the whole organization as well as its indi-vidual components to achieve success. A key benefit of Six Sigma is it can be useful for tackling issues that are often embedded in complexities of processes (Zu et al. 2010). The thorough, end-to-end strategy that Six Sigma brings to quality improvement is why organizations around the world have adopted it as their primary approach for doing business.

12.2 Processes

Processes are an inescapable part of doing business and for this reason are integral components within the fabric of all organizations. By definition, a process is any ordered collection of activities that takes one or more kinds of input and transforms it into an output that is valuable for internal and/or external use or consumption. These outputs can be tangible or even intangible goods or services. Similarly, the output from one process can be the inputs (sub-process) to another downstream process. Whether the outputs are intended for internal use or delivery to a customer or end user, the point is with every process there is a value, causal dependency, and expectation associated with the output from a process. For this reason, processes performing to expectation are maximizing the value of everything as well ensuring that the results can be leveraged elsewhere. In addition organizations are wholly dependent upon processes because they assist with knowledge formalization by translating narrative guidelines into structured knowledge that can be readily communicated and whereby consistency is achieved with outputs derived from a process. The seminal message is that processes are inescapable because they are the glue that binds together assets and rules which enable organizations to operate within a framework of roles and responsibilities needed to accomplish all of necessary steps in a correct order. As a result, organizations that get it right tend to have high efficiency, lower operating costs, reduced waste, and proper utilization of human resources. By their very nature, processes are a good thing, however, for reasons good or bad, over time each process leaves an organizational legacy in its wake. It is not uncommon to find once simple processes that have morphed to become unduly complex and multi-layered with sometimes conflicting purposes because they failed to adapt to changing conditions. The two most common reasons for this are that the process was never required to be overly efficient from the beginning and that over time, the effect of short term measures overlaid on existing processes, making them inefficient and overly complex.

Clearly the improvements of processes are the intrinsic focus of Six Sigma. However the management of Six Sigma initiatives are commonly administered with a project-wide focus. Typically Six Sigma projects undertaken are broadly classified according to the use of one of two Six Sigma sub-methodologies: DMAIC and DMADV. Determining which one will be implemented is often dependent upon the state of processes currently in place. If the objective is to improve an existing process, the Six Sigma roadmap selected for quality improvement will utilize the methodology steps of: Define, Measure, Analyze, Improve, and Control (DMAIC). However in those situations absent any existing process, Six Sigma provides an alternate methodology using the steps of: Define, Measure, Analyze, Design, and Verify (DMADV). While DMAIC and DMADV approaches have a general similarity in that that the outcome will be a process that is more efficient and effective; the difference one approach over another often comes downs to whether there is an existing process for which changes can be measured accurately (DMAIC) or if it is a case requiring a new start to meet requirements (DMADV). Regardless of either framework, project initiatives implemented in organizations will always be focused at a process level.

It is not uncommon for an organization to endeavor to move directly from the perception of a problem to an attempt to solve it, which is fundamentally the wrong approach to take. Before undertaking any attempt to find a solution, there needs not only to be a perceived problem, but it must be properly defined. The difficulty of properly defining the problem can sometimes be underestimated because often a finalized description of a problem is often very different from initial perception of it. Without the essential step of consensus on the definition of the problem, the implemented solution may turn out to treat the symptoms rather than cure the underlying problem. Consider that within any given organization there can be a myriad of processes comprising any number of ordered or sequenced activities (with rules and logic dictating behavior). These processes are in place to ultimately transform inputs by which (hopefully) value is added by the actions, methods, and operations into useful and meaningful outputs. Recognize though, that any processes also most likely interact with other processes throughout an organization; as outputs from one process form the inputs to another. It is crucial to see how any one process is often part of a larger process; where the organizations as the amalgamation of complex networks of interconnecting processes can be seen. The simple truth is that processes are the fundamental building blocks of all organizations. Since with Six Sigma the primary objective of improvement is to reduce variation in process output, it is a necessary step to first identify causes of variation. One technique which aids in considering processes in the overall realm of an organization is to map the Suppliers, Inputs, Process, Output, Customers (SIPOC). This enables the studying of complex interrelationships likely to exist within and externally to a process and facilitates identify sources of variation within processes. Conceptually, a SIPOC map shown Fig. 12.1 can be used to identify and categorize the parts of a process as relating to either suppliers, inputs, process, outputs, or customers. In particular it becomes a useful starting point for not only identifying sources of variation, but is invaluable way to describe the relationship between the variables and help in ascertaining what metrics are capable of being captured at each point.

Viewing it at this level, it is easy to see how each process consists of inputs and outputs. Process inputs may be raw materials, human resources, or the result of some upstream process. Particularly important is the feedback from downstream process measurements that can also be used to improve an upstream process.

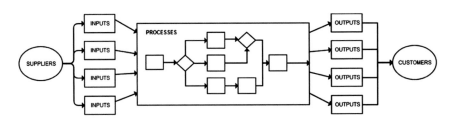

Fig. 12.1 Mapping of process Suppliers, Inputs, Process, Output, Customers (SIPOC)

All inputs have some quantifiable measurement, including human effort and skill level. Therefore it is imperative that process input requirements should be stated so that key measures of input quality can be controlled. Once process capabilities are known, output measures can then be used to monitor whether or not a process has remained in control. Finally once monitoring strategies are implemented, the measurements derived are capable changing the behavioral characteristics or can even effectively control processes. Ultimately the starting point is gaining understanding of processes and installing useful measurement strategies as the underlying aim with Six Sigma is to identify, measure, and reduce sources of variation within processes.

12.3 Variation ≈ Opportunity

Variation represents the dispersion between an ideal standard of perfection and what actually occurs. Quite simply variation is a fact of life in that it is all around us and present in everything we do. No matter how hard we try, there can never be two identical actions that generate exactly the same result. Therefore you can safely assume that processes themselves are subject to being affected by variation. However instead of regarding variation as the enemy, it is more productive to see it as an opportunity since it is through the reduction of variation from requirements and the centering of performance on targets that improvement is achieved with Six Sigma.

Operationally variation reflects any change or difference from the desired ideal state or central tendency (in statistics also called distribution of dispersion denoting relatively spread of values). The two statistical measures of dispersion that are the most widely used to provide indication of this are variance (σ^2) and standard deviation (σ). In addition Statistical Process Control charts (a.k.a., SPC or Shewhart charts) are often employed to distinguish between these two types of variation and uses what are known as control charts to analyze variation that is occurring within processes (as shown in Fig. 12.2). Since variation is an inevitable part of all processes, it needs to be measured over time to determine whether changes occurring are the result of *common* and/or *special-cause* conditions.

The first sources of variation to look for in processes are of the category *special-cause*. These occur because of the presence of assignable (yet unpredictable) factors external to the process. These types of variation can be traced to a specific reason. For example defective materials, operator error, incorrect training, and faulty machinery are sources of assignable variation. Unfortunately special-cause variations are not readily predictable. From a management perspective, once this situation occurs, the task is to uncover the source and take corrective action to prevent it from reoccurring. Unlike special cause variation, *common cause* variations are due to factors internal to the process. Examples of common cause variations are: normal wear and tear, temperature and humidity, inappropriate procedures, poor design, and even measurement errors. The point is any and all of these are predictable causes and as such are ultimately controllable. Common cause variation is

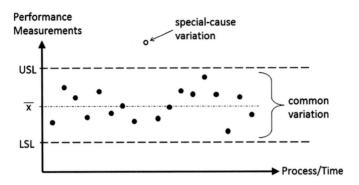

Fig. 12.2 Control chart mapping of process variations

readily observed in the random distribution of output about the average of measurements ($\overline{x}$). Again variation is not evil, rather there is a need for a change of
perspective to view common cause variation as a measure of the process's potential.
In other words it shows how well the process could perform when special cause
variation is removed. Anecdotally Edward Deming estimated that the way processes where constructed was responsible for more than 85 % of the causes of
variation. This statistic alone is evidence of an abundance of opportunities to make
sustained improvements.

Six Sigma seeks to reduce quality dispersions by minimizing the occurrences of
variation at the earliest stage in a processes life cycle; whereby preventing failure in
the downstream stages. Since variation happens as a result of how we directly or
indirectly construct processes, variation can be treated as a causal factor. As such it
can be quantified, predicted, and even reduced if the investment is worth the gains
realized. However be aware that individual sources of variation within a process are
also additive. What this means is often it is any number of small sources of variations which are underlying cause disrupting output quality rather just one a solitary
significant source. Identification and classification of the inputs and outputs of
every step in a process as either critical, noise, standard operating procedure, or
controllable, provides a knowledge basis necessary that along with monitoring
outputs enables the establishment of key decision points within a process on how
best to affect sustained changes. For this reason having a working foundation of
where all possible sources of variation can occur in a process is a requisite in
designing an effective quality management measurement system.

12.4 Measures for Performance

A **metric** can be a measurement, but a measurement is not necessarily a metric.
A measurement is simply some dimension, quantity, or determination of capacity.
Examples of metrics capture might be cycle times, defect rates, wait times,

headcount, inventory levels, or service times. A metric can also be defined as a standard of measurement. Thus, it is more appropriate to view Six Sigma as the set of standards of measurement in an organization that help to it achieve quality goals for delivering value to improve bottom line performances.

Measurements in and of themselves are not capable of delivering process improvements in efficiency or effectiveness. In fact, there are many methodologies for achieving improved performances at the process level. Business Process Management (BPM) and Eliyahu Goldratt's Theory of Constraints (TOC) share similarities emphasizing process improvements by way of focusing on streamlining processes with the elimination of bottlenecks, waste, and any other non-valued added activities. However, Six Sigma differs from these other methodologies in that it promotes the use of statistical measures to drive quality improvements and sustain improvements realized. The philosophy underlying Six Sigma is that *variation* within any given process is likely the main driver for defects occurring. Specifically the need to measure variation over time and in order is a requisite for determining whether changes occurring are the result of common and/or special causes. Thereby when an organization is able to minimize process variation it will have the effect of reducing defects causing unnecessary waste and customer dissatisfaction.

Six Sigma, while borne out of manufacturing, has actually proven to be a well-suited strategy and can be found implemented across all forms of industries; both manufacturing and service. However there tends to still be a great deal of misunderstanding this important quality improvement strategy. In part this can be attributed to the term itself; sigma (σ) which is actually the mathematical symbol related to the standard deviation. Central then to Six Sigma is calculating the "sigma" level of a process where variation is measured in terms of sigma values or thresholds. To the uninitiated, Six Sigma's emphasis on statistics can be seen as unduly burdensome. However, unless there is a willingness to measure something, you can't manage for improvement because there has to be a baseline to see what is getting better and what isn't. The underlying techniques of Six Sigma are by no means new since seminal works like Walter Shewhart's concept of Statistical Process Control (SPC) and Edwards Deming's Total Quality Management (TQM) had been popularized early on in the 20th century (Deming, 1982; Pyzdek & Keller, 2009). However the Six Sigma methodology has evolved in recent years out of a need to spur greater reductions of manufacturing defects; measuring such not in the thousands of opportunities as was the norm, but to achieve a granularity whereby Defects Per Million Opportunities (DPMO) is the aim. Therefore a Six Sigma (6σ) process will statistically be expected to have a 99.99966 % yield free from defects (or inversely, 3.4 DPMO). In fact the term *Six Sigma* is prominently being used these days to denote comparisons of performance in nonmanufacturing settings. Therefore Six Sigma should not be viewed as applicable only to the manufacture of products or just to improve the consistency of services provided, rather Six Sigma has evolved into an organizational philosophy connoting the application of improvement methods using systematic quality tools to achieve quality goals to both deliver value and improve bottom line performance in any industry.

12.5 Quality and Value

Value is the overall subjective and objective assessment of the utility of a product or service based on perceptions of what is received and what is given. It is for this reason why processes are seen as playing such an important role in organization as they serve as one of the primary mechanisms for value creation. From the perspective of a process, value is the net result of reducing the cost (e.g., procurement, transformation, waste) to as small amount as possible while maximizing qualities that makes the outcome desirable. It appears that value happens to be one of those traits which seem to have the greatest diversity in how it can be defined or assessed. However, what distinguishes successful organizations is when they are able to quantify it relative to its ultimate use and consumption. Quality on the other hand has a pragmatic interpretation as an indicator non-inferiority or superiority related to some dimension of a product or service. What Joseph Juran related to as *fitness for purpose* implied that quality was really nothing more than a judgment on how well the output of a process met, exceeded, or alternatively did not measure up to expectations. Consider the case of a product or service deemed high quality. If there is not a demand for it because it is too costly for consumption, its value will not be perceived as being as high as that of a nearly comparable product seen as more affordable even though it has lower quality. Thus, quality and value, while two distinct attributes, have differences which are inextricably linked as tradeoffs which can have profound implications for how creation or transformations of products and/or services are performed. Whereas there are numerous techniques and strategies that help lower cost and others that are focused on achieving highest degrees of quality, Six Sigma has been so widely adopted because of the desire to streamline cost expenditures while increasing quality in order to maximize value creation achieved through processes. For this reason, successful organizations are those who are best able to best quantify dimensions of both relative to the ultimate use and consumption of any product or service. And it is in this manner by which organizations tend to differentiate in their views on quality. Often a cost strategy is employed where the focus is on reducing costs in order to capture as much value as possible. The counter-approach to this would be a quality strategy where creating higher quality justified any costs so long as sufficient demand for the perceived high-costing value exists. Neither approach is general is better; just different. What dictates which approach is ultimately preferred is the premium placed on the final value perceived from the products or services. However, Six Sigma sidesteps this conundrum by emphasizing both approaches. For purposes of Six Sigma, value is treated as the ratio of *quality to costs*; where the aim is to increase quality and simultaneously reduce cost. Therefore quality costs (i.e., cost of quality) is but a means to quantify the total cost of quality related efforts and deficiencies resulting from processes. In this way, the total cost of quality is comprised to two types of costs: prevention and noncompliance. The first are those costs incurred to inspect for defects, prevent defects, and even those steps to correct defects in order to make the outcome acceptable. The costs associated with noncompliance would be such as those as waste, returns, and even loss of sales. Obviously as more costs are attributed to prevention quality will be expected to improve. Conversely, if prevention costs are reduced, it is reasonable to assume that noncompliance costs will rise as well. Ideally then, it is finding the right balance of these two costs that is where optimal ratio of quality to costs (as shown in Fig. 12.3) will be derived.

In fact this strategy has characterized most approaches to managing quality where quality is viewed in terms of conforming within a range of Lower Specification Limit (LSL) and some Upper Specification Limit (USL) which are used to dictate whether it is either good/acceptable or it isn't. However, an underlying principle of Six Sigma is based on continuous improvement and not being satisfied just with an optimal ratio of quality to costs. Genichi Taguchi in the 1980s pioneered the notion that even if a product or service attributes lies within these limits of acceptability, there will still some degradation with the perceived value. What he described was that even when an attribute (product or service) is within compliance (i.e., acceptable), the closer the outcome is to its ideal state will have higher value than one nearer one of the outlier specification limits (Fig. 12.4). This is reflected in Taguchi's *loss function*:

$$L(x) = k(x - m)^2 \qquad (12.1)$$

where deviation from the utmost ideal value is monetized as Loss (L), m is the point at representing the ideally desired characteristic should be set, x denoting what the characteristic's value is actually (measured or perceived), and k is a constant corresponding to the magnitude of the characteristic typically the same as the monetary unit involved. Recognize this is a dramatic contrast from the traditional understanding of cost of poor quality, where Taguchi's loss function facilitates the

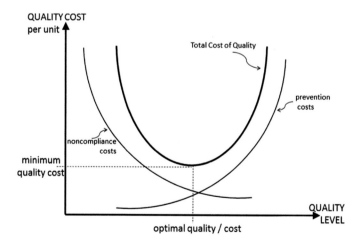

Fig. 12.3 Achieving optimized total cost of quality

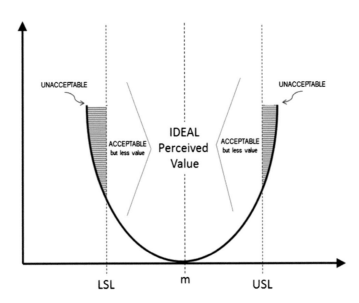

Fig. 12.4 Taguchi's total loss function

measuring of financial impact of any process deviation from its target. Even further, Taguchi's philosophy is clearly aimed at the continuous reduction of common cause variations within processes. It is because of this ability to quantify the value associated with process improvement as part of a financial assessment and benefit evaluation that it has become a technique use inclusive of most Six Sigma initiatives.

Traditionally organizations encounter not one performance issue related to processes but two: effectiveness and efficiency. Effectiveness relates to processes ability to conform to quality expectations (i.e., the initial focus of Six Sigma). Efficiency, on the other hand, relates to processes operating with the least amount of associated costs to provide the maximum amount of value (i.e., focus of Lean methodologies). There are many examples illustrating where processes can operate effectively (high quality) but not efficiently; or where they are efficient (high value) but lacking in effectiveness. The goal with Six Sigma projects is therefore to seek both quality and value as it is needed. To do so and achieve effectiveness and efficiency within a process often necessitates mapping the linkages of activities within and between processes. The aim is to identify and measure each of the inputs and outputs of every step in a process in order to see what variation is occurring and the effect. Once that is known, there is a basis to ascertain what is causing sub-optimal processes and thereby differentiating the total cost of quality; even when everything is operating within specification limits.

12.6   Process Stability

Process stability refers to when all of the response parameters being measured about the process are characterized by stable and consistent patterns of variation over time and where the average process value show a constant distribution as well. This of course does not mean there will not be any variation; rather, it means that what dispersion does occur will be random in nature stemming from common-cause sources of variation. Remember that variation is an integral part of any process. When variation is found to occur from common causes, these are considered controllable since they can be reduced or even eliminated by those responsible for the process. A stable process is one that is predictable and consistent over time. In other words the process HAS performed in a certain way in the past and WILL continue to do so in the future. On the other hand, process instability occurs when there exist abnormal or *special causes* of variation acting upon a process. Unstable processes by definition will not have predictable outcomes. Note that achieving stability reflects efficiency, but not necessarily effectiveness.

Process stability reflects process performance over time and therefore any assessment of stability should precede analyzing process capability. The easiest way to ascertain stability is with the use of control charts. A control chart is a scatter or line graph of the measurements collected from response measurements plotted against a center line representing the mean value for the in-control process and two sigma lines called the upper control limit (UCL) and the lower control limit (LCL) to show the boundaries of what is acceptable behavior for stability (Fig. 12.5).

Specification limits are specified by the customer whereas chart control limits are a calculated as 3 standard deviations above and below the grand average of the response measurement data ($\bar{x} \pm 3\sigma$). Take care not to use interchangeably specification limits and control limits. A stable process will present all data 99.7 % of all response measurement occurring between the UCL (+3 sigma line) and LCL (−3 sigma line). An instable or process out-of-control is detected on a control chart

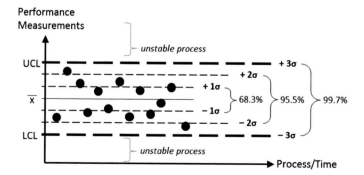

Fig. 12.5 Process stability relation of process deviations to established control limits

either by having any points outside the control limits or by unnatural patterns of
variability.

12.7 Process Capability

Variation is a perfectly normal part of any process. This is because all the activities
involved in creating the product or service outputs have naturally occurring fluctuations in how they perform. The problem arises when there is so much variation
that the outcome is a process generating unpredictable results. Clearly as variation
is reduced, quality will be seen as improving. Therefore, the question really
becomes one of just how much variation is then acceptable? This is where it
becomes necessary to measure the process capability or extent to which a process is
able to provide output within an allowable deviation from what is desired of it. The
output goal or target becomes important as well as any specification limits which
are imposed to define what is an acceptable distribution of occurring output values.
In this way a process capability study can be used to determine whether a process is
and will likely continue to be capable of producing results within tolerance
requirements. Consider the two distributions of output measures taken for the
output of a process shown in Fig. 12.6. In the first distribution all output measures
lay wholly between the Lower Specification Limit (LSL) and the Upper
Specification Limit (LSL). In this case the process is considered to be capable of
meeting the long term performance needs with regard to tolerances or specifications
range set. However in the second distribution it is evinced that some of the measurements are occurring outside permissible limits. This is an example where the
process is unable to consistently produce results within specification.

Specification limits can be product specific (e.g., length = 4.25 m ± 0.02 cm) or
just as easily related to services generated by processes (e.g., average time to
respond = 6.00 ± 0.50 min). Again, variation does not necessarily indicate that there
are any problems. If the process is producing desirable outputs and is stable, then it
is safe to expect outputs are meeting requirements as well future variations from it

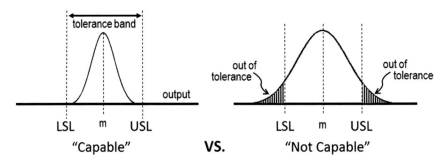

Fig. 12.6 Process capability established by specification limits

are able to be predicted. This is why in order to achieve process stability where all special-causes of variation have been eliminated is needed; otherwise, any measurements for process capability will be meaningless. Also where process stability makes use of control limits based on actual process output; process capability charts use specification limits (LSL and USL) which are independently set in order to narrow the distribution of a product's properties. The performance and capability of a process in this way can simply be defined as the variability inherent within a process; absent any special-cause influences.

When evaluating the quality associated with a process, a Six Sigma level of quality is defined as achieving 3.4 or less defects per million opportunities. Therefore, when discussing the relative measure of quality for a process, in Six Sigma terminology this is referred to as the *sigma of a process* or simply the *process score*. This reflects the number of standard deviations from the mean or target goal. For example one standard deviation in either direction accounts for 68 % of the metrics captured. The process score is based on the statistical modified Z score which takes into account that over the long term, disturbances to the process cause it to expand and ultimately create defects beyond the specification limit. For this reason the process score is calculated as the statistical Z score adjusted by 1.5 standard deviations. Similar to the statistical Z-scores, it too is applicable for only normal distributions; which if special-cause variations have been removed, should be the observed behavior for processes. The usefulness of this is in knowing how far a particular score is away from the mean as it can be used as an indicator for the likelihood of process defects occurring. For example a process score of 0.67 means there is a 31 % probability of encountering a defect; a process score of 1.00 means there is a 6.7 % probability of a defect; and for a process that is 6σ the probability of a defect is 0.00034 % or 3.4 Defects Per Million Opportunities (DPMO). Shown in the table is a summary of expected defects and yields corresponding to associated process scores. These scores are important in providing a benchmark when calculating the capability and performance of processes (Table 12.1).

Assessing the capability of a process first begins with establishing whether it stable and free of external sources is causing special-cause variations to occur. In all cases where a process is deemed to be stable and normal, *what it will do* can be described by its standard deviation (σ) and process mean (μ). However for new

Table 12.1 Process scores for corresponding level of statistical variation

Sigma	% defective (%)	% yield (%)	DPMO	Process score
1	69	31	691,462	0.33
2	31	69	308,538	0.67
3	6.7	93.3	66,807	1.00
4	0.62	99.38	6,210	1.33
5	0.023	99.977	233	1.67
6	**0.00034**	**99.99966**	**3.4**	**2.00**
7	0.0000019	99.9999981	0.019	2.33

processes lacking sufficient data, a pilot run can be used to generate control chart data for estimating the process capability where the standard deviation can be estimated from:

$$\sigma \approx \frac{\overline{R}}{d_2} \tag{12.2}$$

Similarly in lieu of the mean, the sample mean ($\overline{x}$) is all that is needed to begin calculating the performance capabilities of any process.

12.7.1 Process Capability Ratio: C_p

Since quality goals are geared towards the minimization of variation around appropriate targets, understanding how well processes are performing becomes a necessary evaluation. This first begins with gaining an understanding of the total variability in a process. For this the process capability ratio (C_p) is used to provide a measure of potential or maximum process capability which relates how well the process is capable of adhering to specification limits (USL, LSL).

$$C_p = \frac{(USL - LSL)}{6\sigma} \tag{12.3}$$

A process is then defined as being capable if C_p values fall with ±3 standard deviations (a.k.a., having a tolerance of 6σ). By simply by knowing the dispersion of the process output, the Lower Specification Limit (LSL), and the Upper specification Limit (USL); then any variation identified beyond the specified limits are considered a defect.

The greater the C_p values are, the higher levels of quality are achieved. For example a C_p value of 1.0 indicates a process consistently yielding 3-sigma quality with only 6.7 % defects. A 4σ process would have a critical value of 1.33; a 5σ process a C_p value of 1.67; and in order to achieve a 6σ process it would need to achieve a C_p value of 2.0.

12.7.2 Capability Index: C_{pk}

While C_p provides a good measure of variation present in a process (higher the value, less variation), this does not provide any information about how close it is to meeting the mean capacity of the process. For this the capability index (C_{pk}) is used to relate the centeredness of performance expected from the process. For this, C_{pk} is defined as:

$$C_{pk} = \min\left[\frac{(\mu - LSL)}{3\sigma}, \frac{(USL - \mu)}{3\sigma}\right] \qquad (12.4)$$

With this metric, the minimum value of the two ratios is needed as it represents the worst-case performance for the process. In effect, what this indicates is that if C_{pk} is less than the critical value, then either it is because the process average is too close to one of the tolerance limits or because the process variability measured is too large. Here the process is capable only when the capability ratio is greater than the critical value and the process distribution is centered about the mean output value. Simply C_{pk} is an index which indicates how close a process is running to its specification limits. Consider a process with minimal variability (high C_p score) but producing results skewed closer to one of the specification limits would be characteristic of lower C_{pk} index. Alternatively a process with lots of variation (low C_p score) but with a distribution centered about the mean would have a high C_{pk} index. Ideally the processes will consistently meet the mean (high C_{pk}) with minimal variation (high C_p).

12.7.3 Taguchi Capability Index: C_{pm}

While the process capability indices C_p and C_{pk} are widely used to provide unitless measures of process potential and performance, there is a drawback in that both operate under the assumption that the process mean coincides with the target goal. In fact, the process index C_p is considered location independent as it only provides indication of whether a process is operating within limits. The process index C_{pk} does consider location, but only to that of the mean of a process. Simply put, it is not enough to know that processes are conforming to tolerances, it is equally important to know just how good they are. This is where work based on Genichi Taguchi's Loss Function has led to relating process performance to consistency in meeting the target goal. The Taguchi Capability Index combines variability and distance from the target into one measure; where T represents the optimized goal or ideal target value. For this, C_{pm} is defined as:

$$C_{pm} = \frac{USL - LSL}{6\sqrt{\sigma^2 + (\mu - T)^2}} \qquad (12.5)$$

This unique index is then able to take into account the degree of relationship that exists between the goal target and the specification limits. While C_p and C_{pk} are useful measures for estimating the impact on the fraction nonconforming of a shift in the process mean, variance, or specification limits; it is only with C_{pm} that you are able to generate a sensitivity measure of the impact any process changes might have on conforming outputs. This becomes yet another component in establishing a quality measurement system. Commonly C_p and C_{pk} are classified as

Critical-To-Quality (CTQ) evaluators as they relate process output to the wants and needs of the customer; whereas C_{pm} is considered an indication of Critical-To-Customer (CTC). It is not uncommon to hear confusion between CTQ and CTC, however recognize the perspectives provided a very different. While CTQs are what is important to ensure that the output of the process is conforming to defined quality standards, it is ultimately CTCs which are what is most valued. In this fashion knowing the C_{pm} of a process becomes important in order to make adjustments in optimizing a process to consistently produce as close as possible the ideal product or service desired.

12.8   Feedback Gives Rise to Intelligent Processes

An axiom of organizations is that as they change over time, so too will their processes inevitably evolve in both complexity and scope. The challenge therefore becomes one of deciding how best to design a process that exhibits inherent self-control that is able to quickly respond and adapt to changing conditions. In large order this reflects the underlying objectives of any Six Sigma initiative which involves the translation of needs into the quantifiable and measurable outcomes, evaluating process inputs for an understanding of root causes of variation, and then using that knowledge to optimize processes based on uncertainties and opportunities discovered (Elshennawy, 2004). This is why such a heavy reliance is placed on incorporating the use of feedback with processes for relating measurement data about outputs to better enable an analysis of performance behaviors stemming from a process. The simple fact is that to reach a state of continuous improvement; outputs must be taken into consideration where corrections are based on feedback. The progression of quality systems has used numerous analytical approaches to gain richer insights and "intelligence" into process performance in order to drive process effectiveness for optimum compliance. While there is little question that effectiveness drives better process outcomes, organizations have come to realize that efficiency from processes is also just a necessary and vital condition to realize optimum value from outputs (Nold, 2011). This was in fact the impetus for Six Sigma, to optimize process Effectiveness (quality) and Efficiency (cost). However just as Six Sigma is used as a means achieving continuous improvement, so too has it evolved over the years. Most notably this was where Joseph Juran began to extol **Adaptability** as the third dimension for which processes should be designed and evaluated. This implies that in addition to using performance feedback (CTQ and CTC) metrics about the process to achieve optimization, there should also be feedback control loops embedded within the process that lend themselves to analyzing the Critical-To-Adaptability (CTA) or capability in behaviors to accommodate changing circumstances that invariably do occur over time. What this further necessitates is capturing internal metrics about outputs from any activity or sub-stage within a process and relating them to another interdependent activity or sub-stage to understand the impact they might have on the process and in turn influence when and what actions are necessary. CTA evaluates the number of non-compliance events recorded (δ = count of measurements exceeding LSL or USL) by the reciprocal of the square root of the product of interactions between distinct activities or sub-stages occurring within a process (a_{ij} since activities and sub-stages can have multiple interactions) and count of process outputs (O) over an interval of time.

$$\text{CTA} = \frac{\sum \delta}{\sqrt{\sum a_{ij}\, O}} \qquad (12.6)$$

When CTA approaches 0 the process is demonstrating adaptability for handling changes. Alternatively as it approaches or exceeds 1.0 the process is demonstrating an inability to adapt to changing conditions. Insomuch as the environments and conditions that organizations operate in are constantly changing, so too must it have the ability to remain effective and efficient in the face of change. This was a major extension of how Six Sigma has evolved, looking not only at how processes are behaving today, but ensuring that they have the necessary feedback mechanisms to be to recognize and adapt as needed.

For this very reason the concept of feedback within any given processes is important in case future changes are proposed to any complex processes (Savolainen & Haikonen, 2007). This begins by instilling feedback loops on any causal relationships that can possibly be used to affect change upon a process (*anything that affects an effect is a factor of that effect*). In fact much of the art of Six Sigma lies in discovering and representing the feedback processes and other elements of complexity for a given process. The aim therefore is to extract the essence of pertinent information while being able to ignore any irrelevant context.

12.9 Emergent Quality Assurance Practices: Neural Networks

Since the underlying theme with any Six Sigma initiative is the analyses of data gathered from processes to detect a changing process mean (μ), this is accomplished by continuously sampling attributes about a process or its outputs (Conklin, 2006). Whereby for each new sample collected, a new sample mean ($\bar{x}$) is and compared with previous data points to ascertain whether the process mean is shifting; an indication of the presence of deviation within the process. While there are many associated strengths with Six Sigma methodology; it is however not without its share of weaknesses. As a process improvement method, Six Sigma is dependent upon users involved having an understanding of statistical sampling methods and then having the confidence in making changes to processes. As noted, if $\bar{x}$ is detected as changing, then the likely course of action is to stop the process and attempt and conduct an investigation. The outcome of such investigation will be either because the process mean is changing due to occurrences of common-cause variation (where the process does need to be corrected), special-cause variation (where the process does not need to be corrected), or a false-positive measurement (type I error) has been recorded. This in fact lends to one of the complaints around Six Sigma where analyses performed tend to be focused on a given single parameters and ultimately disruptive while processes are interrupted in the investigation of the problems identified. Further the effects from changes implemented are not immediately known since past performance is no longer a viable indicator of future results. In fact, one of the chief limitations of Six Sigma methods is it only looks only at the last point as opposed to any pattern as a predictor of future performance. As a result this has led to efforts which are less reliant on traditional description statistics and instead incorporate using inferential statistics combined with heuristic learning. Following is an overview/synopsis where research has made substantive strides in adapting artificial intelligence to Six Sigma methods.

12.9.1 Adapting Neural Network Approach to Six Sigma

One of the viable trends emerging is in the application of artificial neural networks for ascertaining whether a process is under statistical in-control or one which is about to become an out of control process. This is accomplished in its ability to extract patterns and detect trends from dynamic and nonlinear inputs and then adaptively learn from these inputs to better predict when a process might be out of control. First, neural networks are ideally suited for modeling process and collecting in real-time any number of inputs of performance measurement data. Secondly it associated weights to each of the inputs in order to produce a linear output. However what truly makes a neural network so remarkable is by its ability to not only compare the outputs to some expected result, but in its ability to modify the weighting associated with each of the inputs in order to further refine the input parameters. Consider in Fig. 12.7 a process comprised of any number of activities which quality response measurement data (X_n) is collected. Each input is given an initial weighting (W_n) to each of the inputs to produce a linear output for deciding conformity or not.

In this example the output is dependent upon the weighted sum of all its inputs where the weighting (W_n) represents the difference between the actual and the desired output.

$$\sum_{i=1}^{n} x_i \cdot w_i$$

This is in fact where the learning function of a neural network happens as the weights are adjusted with each subsequent data point collected so that the error between the desired output and the actual output is reduced. Simply with each pass

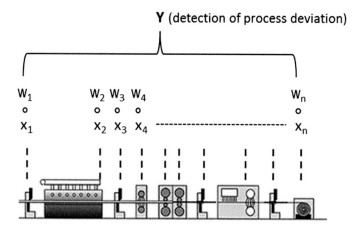

Fig. 12.7 Example process producing multitude of performance data

of data being collected, corresponding weights are increased or decreased slightly with each recalculation by an amount which is proportional to the rate at which the error changes as the weight is changed.

12.9.2   Integrating Multiple Quality of Service Measurements

The architecture of a neural network is not limited by the number of inputs nor by the amount of abstraction where weighted functions are processing this information; rather it can easily encompass many input sources linked together with any number of distinct functional processing QoS objectives that together are able to transform the inputs into meaningful outputs. What is presented in Fig. 12.8 is an example of a simple network consisting of one input layer ($\sum x_n$), one hidden layer where processing and sigmoidal activation functions ($\sum a_j$) reside that culminates in an output layer (Y) where a decision result is presented.

Fig. 12.8 Multilayered network

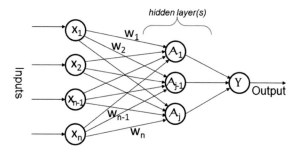

The topology for neural networks can then be constructed in layers of inter-connected nodes with the first representing the input layer where data inputs are received (e.g., x_1, x_2, .., x_{n-1}, x_n). The next layer of interconnected nodes is then referred to as a hidden layer. The nodes in the hidden layer represent the activation function whereby the actual processing of the data is done via weighted connections (w_1, w_2, ..., w_{n-1}, w_n). Each of these nodes in the hidden layer are then capable of processing like-data. For example one node could be responsible for processing data from similar activities the C_p; while another node process the C_{pk}, the C_{pm}, and so on. However there are no limitations on the number of hidden layers within a neural network which can then be used as additional layers of processing. In this way a network can process dissimilar data from a multitude of sources with their own distinct specification limits and target values. For example in the first hidden layer there could be a node associated to one activity processing its C_{p1}; and another node processing the C_{p2} value for another activity. However with the addition of another hidden layer, it is possible to then look at the grouping of like activation functions and process that information to globally look at the state response of the entire process; and not just one activity.

Neural networks are able learn by applying training algorithms to recode weights based on the structure of underlying data inputs. In fact there are not one but several variations neural network architectures that employ learning rules based on modi-fying the weights of the input connections. Some of these which have been adapted to measuring accuracy include Radial Basis Function (RBF) networks, Hopfield Recurrent Neural Networks (RNN), Kohonen Self-Organizing Mapping (SOM) networks, and the Associative Neural Networks (ASNN). However the architecture which has shown great promise inquality process control analysis are Back-Propagation (BP) neural networks. For purposes of discussion in this paper, the focus will be on the adaption of BP networks to creating Six Sigma measure-ment systems.

12.9.3   Back-Propagational Learning

Back-Propagation (BP) neural networks are an adaptation of the *delta rule* whereby error in the activation functions are minimalized with the input gradient recalculate and then recoding within each node each of the hidden layers.

Learning is thereby achieved through a two-step process. The first step is con-sidered a *forward pass* where inputs measurements received are compared within the nodes of the hidden layer to some baseline threshold value that results a gradient descent minimization of the output error being determined. The net effect is whereby the activation function becomes is a simple linear function of the inputs. The second step then involves a back-propagation of the information where newly derived output error is used to revise the associated connection weights so with the adjustment made produces a better comparative baseline for the next round or pass of data inputs. This two-step process is then repeated over and over again with receipt of new data inputs until the iterative error becomes so small that the degree of learning occurring is no longer significant. One of the more popular activation functions suited for statistical process control with BP networks is the sigmoid function since slope calculated always yield a descent direction that is either increasing or decreasing. This particular activation function is defined by the expression:

$$A(x) = \frac{1}{1 + e^{-x}} \tag{12.7}$$

In order for a BP network to learn first necessitates iterative processing of historical runs utilizing training data. The training data set consists of input signals assigned with corresponding desire target output. It is crucial that training data be representative of the problem space for the processes which are modeled and where output is evaluated against a separate and independent data set. By using historical measurement data used to adjust the weights and thresholds, the network is able to eventually make gains in minimizing the error and learn to detect whether a process shift has occurred for runs other than those used for the training process. What occurs then is a form of supervised or inductive learning of which combinations of parameters are adequate to produce meaningful results derived from the training data. If the error of the output is after one a training run is deemed too large, this data set then further acts to retrain the learning algorithms. This is repeated with additional separate and independent data sets until the error signal is longer statistically significant in regards to the underlying input data. In this way a multilayer BP network can be ultimately be trained to yield a maximum-likelihood estimation of the weight values.

12.9.4 Survey of Neural Network Applications in the Design of Quality Measurement Systems

The driving force for applying neural networks to the area of quality measurement and control is because of the unique nature or processes which in themselves can behave as systems comprised of interconnected elements. In such situation this can result where data independence makes it extremely difficult for traditional Six Sigma charting to be used as a tool for identifying shifts in correlated processes. For this very reason neural networks are especially useful in error approximation, classification, and as work well in parallel structures whereby quality parameters can be self-tuned within closed-loop systems. Now with the introduction of the first neural model by McCulloch and Pitts (1943), there has continued to be numerous advancements in the field of artificial neural networks with a particular focus on quality diagnostics and evaluation of process. One of the most notable advancements was with the back-propagation algorithm formulated by Rumelhart and McClelland (1986) which apportioned error responsibility throughout a multi-layer network. In fact this work has provided much of the basis for the overwhelming application of neural networks to the correlation of diagnostics from inputs and outputs with multistage processes. The following is a review of the state of Back-Propagation (BP) neural networks in recent literature where they have demonstrated an ability to self-learn and adapt to dynamic quality control problems.

What differentiates back-propagation handling of data (i.e., BP networks) from other neural network architectures and which makes it ideally suited for statistical process control applications was first given notice by Alwan and Roberts' (1988) work on how correlated data could be monitored in lieu of traditional Six Sigma control charts. Just as with traditional Six Sigma measurement systems which seek to detect whether there is a causal shift in the process means, a BP type measurement architecture also similarly samples process metrics. In both approaches for each new sample collected, only one of two binary decisions can be made: the process mean did not shift or ascertain that the process is no longer operating normally. However with Six Sigma if it is ascertained there has been shift in the mean, then the process is stopped and an attempt is made to discover the reason for the change in the process mean. What Alwan and Roberts (1988) demonstrated with a BP network is how it is ideal for statistical process control is that where multiple means must be evaluated in context of the entire process. One of the first to experiment in comparing back-propagation algorithms for replacing traditional statistical control charts for arriving at decision criteria governing a multistage quality control system was Puch (1991) where he proved that such a network is capable of detecting both in-control and out-of-control signals. A shortcoming with traditional Six Sigma quality control charts it the inability to differentiate between existences of data dependency within processes. This is further compounded when underlying processes changes and previously developed control charts can become misleading if control limits are not readjusted. This is where BP network models have a distinct advantage in that they require only failure history as input and not assumptions about either the development environment or external parameters. For this reason in order for a BP network to be considered successful, it must correctly identify the types of the underlying disturbances and of the out-of-control data vectors as well. Building upon this, work by Guo and Dooley (1992) introduced an activation function to detect whether a process change was due to a shift in the mean or because of special-cause variability. This work was particularly seminal in that found where BP network could successfully identify of mean shifts whereas conventional Shewart or cumulative sum control charts effectiveness would degrade as sampling rates increased. One of the earliest to demonstrate BP networks could be used in lieu of traditional Six Sigma charting for identifying patterns that signified mean shift in processes was Hwarng and Hubele (1993). Hwarng (2004) applied this later to an industrial application of where a BP network was compared to statistics-based control charts of an associated process for producing paper. What was evinced was that the quality monitoring with the BP network did not require any readjustment (unlike SPC charts) about the time series data inputs or underlying data distributions as it correctly identified mean shift and correlation magnitudes as underlying processes where changed.

While much of the previous work was aimed at identifying patterns of systematic variation, the research performed by Perry et al. (2001) developed a BP network that generated results similar to that of traditional X-bar and R-charts which uniquely address the identification of process instability. Whereas with previous research the focus was on detecting mean-shift, here their work focused on the early detection of a change before a pattern would be recognized by use of traditional charting alone. After all it is not unnatural that multiple patterns may be present in a given process. This was demonstrated by Perry et al. (2001) with in a two-layer BP construct for 64 processing inputs where the first layer tested for early indications of change and the second layer identify a specific pattern of distribution. The outcome of their experiment showed significant reduction in Type I and Type II errors over using a single network alone. The seminal promise shown by their work was that BP networks can do more than just recognize patterns; they can be adapted to a closed-loop whereby adjustments to the process in near real-time in order that quality problems are averted before noncompliance patterns fully develop. The application of BP networks is however not limited to just manufacturing processes. A novel approach adapted the BP algorithm to sampling quality indicators of teachers to predict whether a teacher was developing good or bad traits which ultimately would impact the overall quality of teaching (Lianxin et al. 2014). In this case the process was itself an evaluation system comprised of 16 quality metrics. The training set consisted of subjective historical data for 20 teachers which then utilized two hidden layers; the first to normalize the data and the second to rate overall teaching effectiveness. The takeaway from their research showed was that a BP network can be an effective means to model the relationship between the prediction and analysis of pre-normalized data when sufficient training data is available.

12.10 Quality is Itself Decision

Decision making is an inescapable part of life. Organizations are no different in that they are also being driven by the demand for information in order to make better decisions and deliver solutions more widely and faster. Therefore, achieving sustained and effective well-informed choices requires use of an organized and logical approach for understanding what you have and what you need. For this reason is why so many organizations have adopted Six Sigma as a means to provide a structured approach to baseline performance for any products, services, and processes in order to achieve higher effectiveness and efficiency from processes. The benefit of Six Sigma is that it ensures gains made do not have to be redone at a later time. It facilitates understanding of how processes compare to one another and assists in focusing resources effectively toward the organizations largest problems. At a minimum, it helps puts in place the right feedback measurements on processes to answer how current performance has changed from past performance. This however also becomes its *Achilles heel* since to ever fully optimize process quality, the sum total sources of disparate feedback measurements need to be evaluated in parallel; something that traditional Six Sigma control charts were not envisioned for. For this reason the use of neural networks is being seen as an emergent means by which to improve performance within process by way of its ability to differentiate concurrent patterns from individual patterns. In addition when presented with a pattern the neural network is able to make a guess as to what it might be, determine how far the outcome was off, and is then able to make an adjustment to its connection weights. It is this ability to both exhibit capability for generalization beyond its initial training data (i.e., self-learn) and in focusing on the cumulative set of patterns (as opposed to just the last data point) which is giving promise to using neural networks for real-time quality measurement systems in improving processes. However use of artificial neural networks should not viewed simply as a replacement to Six Sigma, but rather should be viewed as the next extension of quality management systems wherever data is incomplete or noisy and functional relationships amongst disparate data sources from processes are needed to be considered in whole to enable improvements. While neural networks are seen as promising in providing a self-adaptive approach to handling non-parametric data, it does not eliminate the need understanding cause-and-effect relationships inherent in the process. When implemented as part of a Six Sigma architecture, a neural network goes beyond just being able to predict how often a process will meet specifications and hold tolerances. It provides guidance on where improvements in the process can increase value. Most importantly, it facilitates understanding whether processes are able to adapt to changes which invariably will occur in the future. The true determinant of how intelligent processes are in an organization, as well as its ability to prosper over time, is the degree to which a structured quality measurement system is put in place to promotes process effectiveness, efficiency, and adaptability in responding to changes.

References

Alwan, L., Roberts, H.: Time series modeling for statistical process control. J. Bus. Econ. Stat. **6**(1), 87–95 (1988)

Conklin, J.: Measurement system analysis for attribute measuring processes. Qual. Prog. **39**(3), 50–53 (2006)

Deming, E.: Quality, Productivity, and Competitive Position. MIT, Cambridge (1982)

Elshennawy, A.: Quality in the new age and the body of knowledge for quality engineers. Total Qual. Manag. Bus. Excell. **15**(5/6), 603–614 (2004)

Goh, T.N.: Six sigma at a crossroads. Curr. Issues Bus. Law **7**(1), 17–26 (2012)

Guo, Y., Dooley, K.: Identification of change structure in statistical process control. Int. J. Prod. Res. **30**(7), 1655–1669 (1992)

Hwarng, B.: Detecting process mean shift in the presence of autocorrelation: a neural-network based monitoring scheme. Int. J. Prod. Res. **42**(3), 573–595 (2004)

Hwarng, H.B., Hubele, N.F.: X-bar control chart pattern identification through efficient off-line neural network training. IIE Trans. **25**(3), 27–40 (1993)

Lianxin, L., Yu, L., Guangxia, S.: Evaluation of quality of teaching based on BP neural network. J. Chem. Pharm. Res. **6**(2), 83–88 (2014)

McCulloch, W., Pitts, W.: A logical calculus of the ideas immanent in nervous activity. Bull. Math. Biophys. **5**(4), 115–133 (1943)

Nold, H.: Merging knowledge creation theory with the six-sigma model for improving organizations: the continuous loop model. Int. J. Manag. **28**(2), 469–477 (2011)

Perry, M.B., Spoerre, J.K., Velasco, T.: Control chart pattern recognition using back propagation artificial neural networks. Int. J. Prod. Res. **39**(15), 3399–3418 (2001)

Pugh, A.: A comparison of neural networks to SPC charts. Comput. Ind. Eng. **21**(1), 253–255 (1991)

Pyzdek, T., Keller, P.: The Six Sigma Handbook, 3rd edn. McGraw-Hill, New York (2009)

Rumelhart, D., McClelland, J.: Parallel Distributed Processing. MIT, Cambridge (1986)

Savolainen, T., Haikonen, A.: Dynamics of organizational learning and continuous improvement in six sigma implementation. TQM Mag. **19**(1), 6–17 (2007)

Zu, X., Robbins, T., Fredendall, L.: Mapping the critical links between organizational culture and TQM/Six Sigma practices. Int. J. Prod. Econ. **123**(1), 86–106 (2010)

Chapter 13
Taguchi Method Using Intelligent Techniques

Kok-Zuea Tang, Kok-Kiong Tan and Tong-Heng Lee

Abstract The Taguchi method has been widely applied in quality management applications to identify and fix key factors contributing to the variations of product quality in manufacturing processes. This method combines engineering and statistical methods to achieve improvements in cost and quality by optimizing product designs and manufacturing processes. There are several advantages of the Taguchi method over other decision making methods in quality management. Being a well-defined and systematic approach, the Taguchi method is an effective tuning method that is amenable to practical implementations in many platforms. To build on this, there are also merits, in terms of overall system performance and ease of implementation, by utilizing the Taguchi method with some of the artificial intelligent techniques which require more technically involved and mathematically complicated processes. To highlight the strengths of these approaches, the Taguchi method coupled with intelligent techniques will be employed on the fleet control of automated guided vehicles in a flexible manufacturing setting.

Keywords Taguchi method · Artificial intelligent techniques · Automated guided vehicles

13.1 Introduction

Taguchi method is an experimental design method that has been developed by Dr. Genichi Taguchi (1993). It is called quality engineering (or Taguchi methods in the United States). The Taguchi method combines engineering and statistical methods

K.-Z. Tang (✉)
Engineering Design and Innovation Centre, Faculty of Engineering,
National University of Singapore, Singapore, Singapore
e-mail: engtkz@nus.edu.sg

K.-K. Tan · T.-H. Lee
Department of Electrical and Computer Engineering, Faculty of Engineering,
National University of Singapore, Singapore, Singapore

© Springer International Publishing Switzerland 2016   389
C. Kahraman and S. Yanık (eds.), *Intelligent Decision Making
in Quality Management*, Intelligent Systems Reference Library 97,
DOI 10.1007/978-3-319-24499-0_13

(Mori 1993; Ealey 1994) to achieve improvements in cost and quality by optimizing product design and manufacturing processes. The main advantage of Taguchi method over other search and tuning method is the twofold benefit of both efficiency and simplicity. Efficiency provides an affordable avenue for problem solving. Simplicity results in a set of tools more easily adopted and embraced by the non-statistical expert.

Conventionally, the Taguchi method has been widely applied in quality control applications to identify and fix key factors contributing to the variation of product quality in manufacturing process (Peace 1993; Ross 1988; Hong 2012; Rao et al. 2008; Taguchi et al. 2004; Chen et al. 1996). These successful quality management applications can be found in a diverse range of industries, ranging from chemical plants, electrical and electronics manufacturing, mechanical productions, software testing, and biotechnology-related fields like fermentation, food processing, molecular biology, wastewater treatment and bioremediation biological sciences.

There are several strengths of the Taguchi method that is worth mentioning here (Ealey 1994; Peace 1993). The Taguchi method is an effective tuning method that has well-defined, systematic and simple steps. The optimum values of the factors to be tuned are determined in relatively shorter and limited steps. All the above mentioned points make the Taguchi method amenable to practical implementation. Some of the artificial intelligent methods, like the genetic algorithm, the neural networks and the evolutionary algorithm, propose more involved and complicated search methods for optimization (Tortum et al. 2007). Furthermore, the Taguchi method selects a combination of the factors (i.e., that needs to be tuned) that are robust against the changes in the environment. The Taguchi method has the additional advantage of being amenable to analyzing the sensitivity of the individual factor that is to be tuned on the final objective performance. The above mentioned artificial intelligent methods, on the other hand, are not able to provide this analysis. In this perspective, there are some merits in utilizing the Taguchi method together with some of these artificial intelligent methods.

In the current literature, researchers and engineers have also built on the successful track records of the Taguchi method. Being an optimization tool that is amenable to practical implementation, the Taguchi method is able to complement the strengths of various tools in artificial intelligence (Tortum et al. 2007; Chang 2011; Ho et al. 2007; Yu et al. 2009; Khaw et al. 1995; Hissel et al. 1998; Tsai 2011; Chou et al. 2000; Hoa et al. 2009; Woodall et al. 2003; Hwang et al. 2013; Huang et al. 2004; Tan and Tang 2001). For example, Hissel et al. have evaluated the robustness of a fuzzy logic controller using the Taguchi quality methodology and experimental product-plans (Hissel et al. 1998). The Taguchi method provides an effective means to enhance the performance of the neural network in terms of the speed for learning and the accuracy for recall (Tortum et al. 2007; Khaw et al. 1995). The determination of the parameters in engineering systems can be improved using hybrid methods that employ evolutionary algorithms, genetic algorithms and the Taguchi method (Chang 2011; Yu et al. 2009).

It is to be noted that the Taguchi method has some weak areas in comparison to some of the above mentioned artificial intelligent methods, as a search and tuning method. Due to the simplicity of the Taguchi method, it may not be as effective in situations where the search region is complicated. Also, Taguchi method may not be as precise a tuning method, as compared to the more elaborate artificial intelligent methods (Chang 2011).

In this chapter, a brief overview of the Taguchi method will first be provided, focusing on the practical aspects of applications. Leveraging on the quality methodology of the Taguchi method, hybrid approaches of combining the strengths of the Taguchi method with intelligent techniques like the fuzzy logic and the neural network. The case study at the end of the chapter will provide simulation results of applying the Taguchi method and the hybrid approaches (i.e., combining the Taguchi method with artificial intelligence) on the fleet control of automated guided vehicles (AGVs) in a flexible manufacturing setting.

13.2   Taguchi Method

The Taguchi method is a statistical search for a set of optimum factors that could affect the quality outcome of a process or final product. Given a problem, the objective value is expressed as a quality characteristic, i.e., productivity, durability, number of products completed in a given duration of time, etc. The scaling factors are referred to as the control factors in the Taguchi method. Determination of the optimum values of the control factors occurs during the experimental design phase. In this phase, the different combinations of the control factors are used in the runs to investigate their effects on the final quality characteristic. The results of the experiments are then used to adjust the control factors for iteratively improved performance.

13.2.1   Selection of Quality Characteristics

Quality characteristics may be classified into a few general categories such as smaller-the-better (STB), larger-the-better (LTB) or nominal-the-best (NTB). For a given problem, a number of measures of the quality characteristics are possible and associated with the nature of the problem itself.

13.2.2   Control Factors and Levels

The control factors are to be tuned by the Taguchi method for optimal performance. The number of levels of each control factor is the number of different values that are to be assigned to the control factor. The number of levels for each factor is assigned to be two, three or four (Mori 1993), depending on the nature of the overall problem. Using a larger number of levels reduces the number of control factors that are can be effectively analyzed.

13.2.3   Selection of Orthogonal Array and Linear Graph

For the control factors considered, it will not be practically possible to analyze the effects of all the different combinations on the quality characteristic, as in a full factorial experimental design (Mori 1993) in a practical sense. Given the constraints, the optimum levels of the control factors must be determined using a practical and limited number of experiments on a rich sample of the possible combinations. While this set may be generated by a random parameter picking procedure, there is little assurance that the thus generated set will offer a good variation of possibilities within the finite set.

To this end, a special table called the orthogonal array due to Fisher (Ealey 1994) is used to generate an efficient set of parameters for experimentation. By drawing a relatively small amount of data which are statistically balanced and independent, meaningful and verifiable conclusions can be derived from the orthogonal array. Complete compilation of orthogonal arrays can be easily available (Mori 1993; Ealey 1994).

One of the main steps in the experimental design is to choose the most appropriate orthogonal array for the particular problem. There are many issues to be considered in the final selection, i.e., the number of control factors, the number of levels of the control factors, number of interactions of interest, etc. After the orthogonal array is selected, the linear graph (Peace 1993) may be used to assign the control factors and interactions to the appropriate columns of the orthogonal array. Incorrect analysis and faulty conclusions may result if the control factors and interactions are assigned to the any columns of the orthogonal array. By using the linear graph, a systematic way of assigning the control factors and interactions to the respective columns of the orthogonal array is developed by Taguchi. The linear graph is made up of nodes and connecting arcs. The nodes represent the control factors and the arcs represent the interactions or the relationship between the control factors.

The systematic experimentation procedure of the Taguchi method can be summarized as follows:

1. Determine the number of factors and interactions to be considered in the experiment and the number of levels (i.e., or values) of the factors. Typically, the number of levels of the factors is two or three.
2. Select the appropriate orthogonal array.

    (a) Determine the required degrees of freedom (DOF) (Peace 1993) from the factors and interactions. The DOF of a factor is one less than the number of levels of the factor. The DOF of a particular orthogonal array is obtained by the sum of the individual DOF for each column in the array.

(b) The appropriate orthogonal array is the one with has the DOF that is equal to or more than the required DOF of the factors. The smallest array satisfying this requirement is normally chosen for efficiency.

3. With the appropriate orthogonal array chosen, choose the linear graph that fits the relationships of the factors of interest. The factors can then be assigned to the columns of the orthogonal array according to the linear graph.
4. Conduct the experiments and analyze the results.
5. Finally, run a confirmation experiment using the results obtained.

13.2.4 Interpretation and Validation of Experimental Results

After the experimental data from the set of weighting parameters is obtained, there are various approaches to verify the results. One of the common approaches adopted by the industry is to use past data published in the literature. To provide a more systemic approach, statistical tools like the F-test and the T-test could be used to interpret the experimental data mathematically and to provide indications of whether the decision related to the management of the quality is adequate and if further tuning is necessary to improve the quality of the final output (Mori 1993).

13.2.5 Literature Review on Taguchi Method

In the literature, there are many applications using the Taguchi method. These applications include a wide range of fields, from manufacturing processes to design and development of systems and devices. The publication frequencies of the Taguchi method for the past decade (i.e., from 2004 to 2013) are shown in Fig. 13.1. There has been increasing interests on the Taguchi method, since Dr. Genichi Taguchi introduced this practical statistical tool in the 1980s to improve the quality of manufactured goods. These publications are in the form of journal articles, books, technical notes and theses. The different publication categories on the Taguchi method for the past decade are shown in Fig. 13.2.

It is to be noted that these publications focus heavily on the application and verification of the Taguchi method. For example, Sreenivasulu (2013) employed the Taguchi method to investigate the machining characteristics on a glass fiber reinforced polymeric composite material (GFRP) during end milling. He then used the ANOVA method to verify significant parameters and confirm the optimum values obtained by the Taguchi method. These results are even compared with the neural networks tuning method. The results show that the Taguchi method and the neural networks produce very similar results. The Taguchi method is also employed to

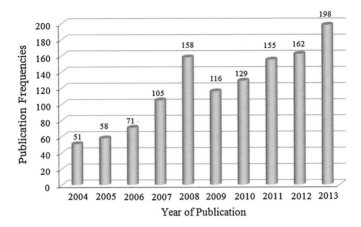

Fig. 13.1 Publication frequencies of the Taguchi method for the past decade (i.e., from 2004 to 2013)

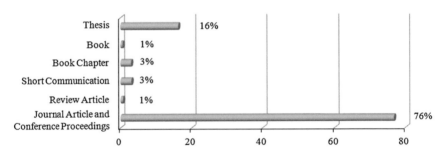

Fig. 13.2 Publication categories on the Taguchi method for the past decade (i.e., from 2004 to 2013)

understand the fermentative hydrogen production process (Wang and Wan 2009). The effects of the various factors in this complex process could be investigated and analyzed using the Taguchi method. The results of this work highlights that the simplicity of the Taguchi method is an advantage over other more involved optimization tools. Asafa and Said (2013) integrates the Taguchi method and the neural networks together to model and control the stresses induced during the plasma enhanced chemical vapor deposition process on hydrogenated amorphous silicon thin films. The Taguchi method with ANOVA is able to obtain the significance of the various network parameters on the overall model. Using these significances, the authors are able to verify the trend between the deposition parameters and the resulting intrinsic stresses during the process. The obtain results concur with other published data in the literature on plasma enhanced chemical vapor deposition process. In a similar way, there are other works in the literature that integrate the Taguchi method with other tools in artificial intelligence (i.e., genetic algorithm, neural networks, fuzzy logic and regression analysis) to optimize other processes and systems (Lin et al. 2012; Sun et al. 2012; Mandal et al. 2011; Chang 2011; Tsai 2011; Tansel et al. 2011; Tzeng et al. 2009). In Lin et al. (2012), the Taguchi method is integrated with the neural networks and the genetic algorithm to improve a particular manufacturing process of the solar energy selective absorption film. For this problem, the Taguchi method is able to perform well to optimize a given search space; whereas the genetic algorithm and other evolutionary numerical methods can be used to control a poorly defined search space. The results in the literature show that the integrated approach of the Taguchi method with other artificial intelligence tools produces better results than just utilizing a single optimization tool alone.

Besides modeling and optimizing manufacturing processes, the Taguchi method has also been widely applied in other fields, like the supply chain management (Yang et al. 2011) and clinical diagnostics (De Souza et al. 2011). Yang et al. (2011) used the Taguchi method to study the robustness of different supply chain strategies under various uncertain environments. The complexity of the problem is accentuated by the variations in the business environments. The performance of the Taguchi method is shown to compare well with other multiple criteria decision-making techniques, like the simple multiple attribute rating technology (SMART), the technique for order performance by similarity to ideal solution (TOPSIS), and grey relational analysis (GRA). The Taguchi method is utilized to optimize the Molecular assay for venous thrombo-embolism investigation (De Souza et al. 2011). There are various risk factors that are patient-dependent and render the investigation process uncertain and difficult. The application of the Taguchi method can lessen the time and cost necessary to achieve the best operation condition for a required performance. The results is proven in practice and confirmed that the Taguchi method can really offer a good approach for clinical assay efficiency and effectiveness improvement even though the clinical diagnostics can be based on the use of other qualitative techniques.

13.3 Tuning Fuzzy Systems Using the Taguchi Method

The statistical potential of the Taguchi method can be harnessed to tune the performance of many intelligent systems. As highlighted briefly in Sect. 13.2.5 above, there are other works in the literature that integrate the Taguchi method with other tools in artificial intelligence (i.e., genetic algorithm, neural networks, fuzzy logic and regression analysis) to optimize other processes and systems (Lin et al. 2012; Sun et al. 2012; Mandal et al. 2011; Chang 2011; Tsai 2011; Tansel et al. 2011; Tzeng et al. 2009). The results show that the integrated approach of the Taguchi method with other artificial intelligence tools produces better results than just utilizing a single optimization tool alone.

In this section, the integration of the Taguchi method with fuzzy systems will be described in details to show how these two methodologies can be utilized to complement each other. The work of Zadeh (1973) provides a comprehensive review on fuzzy logic, as an alternative branch of mathematics.

In a fuzzy system, the objects are associated with attributes. These attributes can be computed from fuzzy operations on a combination of variables which are used to describe mathematically the real-time conditions of the given problem. Decisions that affect the system's performance will be driven primarily by these attributes.

To incorporate the Taguchi method into the fuzzy system, the fuzzy system has to be posed as a quality control scenario, with appropriate performance measures as the quality characteristics. The Taguchi method then can be utilized to tune the parameters in the fuzzy rules in the inference engine.

13.3.1 Takagi and Sugeno's Fuzzy Rules

To illustrate how the Taguchi method could be used to tune fuzzy systems, the K attributes will be inferred from a Takagi and Sugeno type of fuzzy inference (Takagi and Sugeno 1985). Consider the following p rules governing the attribute a_k:

$$IF\ x_{k1}^i\ is\ F_1^i\ \otimes \cdots \otimes\ IF\ x_{km_i}^i\ is\ F_{km_i}^i,\ THEN\ u_k^i\ is\ \alpha^i,\quad i = 1 \ldots p. \tag{13.1}$$

with $\sum_{i=1}^p \alpha^i$ and $k = \{k \in \mathbb{Z}\,|\,0 \le k \le K\}$, where F_j^i are fuzzy sets, $x^i = \left[x_{k1}^i, \ldots, x_{km_i}^i\right]^T \in \mathbb{R}$ are the input linguistic variables identified to affect the attribute for rule i, $\otimes$ is a fuzzy operator which combine the antecedents into premises, and u_k^i is the crisp output for rule i. α^i is the scaling factor for rule i reflecting the weight of the rule in determining the final outcome.

The value of each attribute a_k is then evaluated as a weighted average of the u^i's.

$$a_k = \frac{\sum_{i=1}^P w_k^i u_k^i}{\sum_{i=1}^P w_k^i} \tag{13.2}$$

where the weight w_k^i implies the overall truth value of the premise of rule i for the input and is calculated as:

$$w_k^i = \prod_{j=1}^m \mu_{F_{kj}^i} x_{kj}^i \tag{13.3}$$

13.3.2 Incorporating Fuzzy Logic with Taguchi Method

In a fuzzy system, the approach to the decision making is based on a fuzzy inference engine which is able to specialize in a multiple criteria satisfaction. The overall effectiveness of this fuzzy logic approach in a given scenario is critically dependent on the appropriateness of the fuzzy rules, in particular, the weight or the scaling factors of each of the fuzzy rules on the final decision. One of the main issues is to address the selection of the scaling factors, i.e., α's in Eq. (13.1). By a manual trial-and-error adjustment method, the search for the optimal set is by far too tedious and time consuming. On the other hand, an exhaustive search for the optimal set of scaling factors would be unrealistic and impractical. To this end, a systematic search procedure within a balanced set of possible weighting parameters will be both desirable and practical. The Taguchi method would be an ideal candidate that could be deployed to obtain the optimal set of the scaling factors for the fuzzy rules.

The main strength of the Taguchi method over other search and tuning methods is the twofold benefit of both efficiency and simplicity. Efficiency provides an affordable avenue for problem solving. Simplicity results in a set of tools more easily adopted and embraced by the non-statistical expert. The quality characteristics used in the Taguchi methodology may be classified into a few general categories such as smaller-the-better (STB), larger-the-better (LTB) or nominal-the-best (NTB). For a given problem, usually a number of measures of the quality characteristics are possible and associated with the nature of the problem itself.

To determine the control factors and the assigned levels, the scaling factors associated with the attributes in the fuzzy rules are the control factors to be tuned by the Taguchi method for optimal performance. The number of levels of each control factor is the number of different values that are to be assigned to the control factor. The values are dependent upon the constraints (i.e., which could be related to the hardware and other constraints of the system) of the given problem. In the next section, the case studies will demonstrate the potential of the Taguchi method combined with the fuzzy system.

13.4   Using Taguchi Method with Neural Networks

To explore further the potential of the Taguchi method as a search and tuning method, the Taguchi method is applied together with another intelligent method, i.e., neural network system, for quality management. Neural networks are inherently useful for approximating non-linear and complex functions. This is especially true for functions when only the input/output pairs are available and the explicit relationships are unknown. Considering such strengths, neural networks could be good candidates for quality management problems, like monitoring and controlling quality-critical processes with complex system dynamics in the manufacturing environment. This is because such processes display non-linear behavior and it is very difficult to obtain closed-form models of such processes to describe the overall system characteristics and dynamics perfectly. By gathering input and output data pairs of such processes, neural networks could be employed to model the process and provide corrective control action in various manufacturing environments, requiring quality management.

Artificial neural networks are an alternative computing technology that have proven useful in a variety of pattern recognition, signal processing, estimation, and control problems. Indeed, many real-world processes are multidimensional, highly non-linear and complex. It is extremely difficult and time-consuming to develop an accurate analytical model based on known mathematical and scientific principles. Moreover, it is often found that the simplified analytical model is not accurate enough to model these complex processes, resulting in poor performance or results. Artificial neural networks allows one to consider them as a black box can be taught using actual data to act as the accurate model for the processes (Haykin 1994).

There are two common configurations which we could utilize neural networks for quality management (Figs. 13.3 and 13.4). In Fig. 13.3, the neural network 'learns' the process dynamics using input/output data pairs. This type of learning is done using past batches of input/output data pairs to tune the neural network's parameters. This neural network can then be used to monitor the performance of the process during the manufacturing cycle. Corrective actions in the form of user alerts will be invoked whenever, the output of the process deviates from a known pattern or when other abnormal conditions are detected by the neural network controller.

Besides the configuration shown in Fig. 13.3, the neural network could also be incorporated into the process control loop as shown in Fig. 13.4. Here, the neural network provides control actions that adapt to changes in the system output during the process run. The learning phase of the neural network could be online or offline. Online learning of the neural network refers to tuning of the internal parameters while the process is given new inputs; whereas offline learning refers to tuning of the neural network's parameters using past batches of input/output data pairs.

In both configurations shown in Figs. 13.3 and 13.4, the effectiveness of the neural network relies critically on how well it is trained. Training of the neural network can be seen as allowing its parameters to learn the patterns of the process using input/output data pairs. The neural network is then able to model the

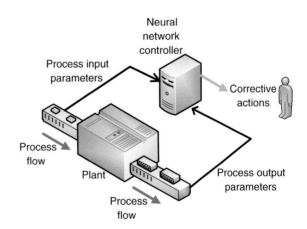

Fig. 13.3 Utilizing neural networks for quality management with the controller outside the control loop

Fig. 13.4 Utilizing neural networks for quality management with the controller integrated within the control loop

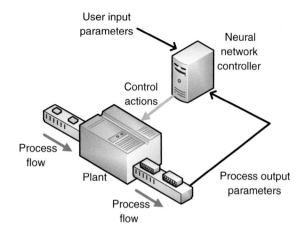

non-linear and complex function that represents the input/output data pairs. This training process may entail an exhaustive search for optimum weights, using a conventional Newton-Raphson type of gradient search (Haykin 1994), or adopting a backpropagation approach (Zurada 1992). The effectiveness of these approaches is highly dependent on the proximity of the initial set to the optimum set, and typically a localized optimum point can be located at best.

As highlighted briefly in Sect. 13.2.5 above, there are other works in the literature that integrate the Taguchi method with other tools in artificial intelligence (i.e., genetic algorithm, neural networks, fuzzy logic and regression analysis) to optimize other processes and systems (Lin et al. 2012; Sun et al. 2012; Mandal et al. 2011; Chang 2011; Tsai 2011; Tansel et al. 2011; Tzeng et al. 2009). The results show that the integrated approach of the Taguchi method with other artificial intelligence tools produces better results than just utilizing a single optimization tool alone. It would be interesting to compare the results of integrating the Taguchi method with the neural networks. For illustration purpose, a radial basis function neural network (RBFNN) is used in this section to show how the Taguchi method can be used effectively with neural networks for process control applications. RBFNNs are one popular and commonly used configuration of neural networks (Haykin 1994; Zurada 1992). RBFNNs use a set of basic functions in the hidden units. Other types of neural networks could use different types of functions in the hidden units. The training process of the different types of neural networks may differ in the methodology. But the objectives are the same, i.e., that is to improve the performance of the neural network when the network is deployed for its purpose. More details of the different types of neural networks could be found in Haykin (1994).

As mentioned earlier, the effective modeling of the given functions entails training the RBFNNs which, in turn, filter down to proper selection and tuning of the parameters (i.e., weighting factors) in the RBFNNs. This training process may entail an exhaustive search for optimum parameters. The effectiveness of the tuning approach is highly dependent on the proximity of the initial set to the optimum set, and typically a localized optimum point can be located at best. The effectiveness and success of the Taguchi statistical method in experiment design to yield an optimum result is well demonstrated and proven in many cases. Considering complex processes in quality management, Taguchi-tuned neural networks could provide good solutions for control and monitoring applications. In the following, the application of the Taguchi method to RBFNN tuning is briefly discussed.

13.4.1 Taguchi Method Applied to RBFNN Tuning

In this section, the Taguchi method is used to obtain optimum weights associated with a RBFNN which is used for the purpose of modeling uncertain nonlinear functions which are subsequently applied for process control. Conventionally, these weights are obtained via an exhaustive search or a localized search using an iterative gradient search algorithm.

To state the problem for the RBFNN, $f(x)$ is a nonlinear smooth function (i.e., which is unknown) which can be represented by

$$f(x) = \sum_{i=0}^{m} w_i \emptyset_i(x) \qquad (13.4)$$

where $\emptyset_i(x)$ denotes the RBF, i.e.,

$$\emptyset_i(x) = \exp\left(-\frac{|x - c_i|^2}{2\sigma_i^2}\right) \bigg/ \sum_{j=0}^{m} \exp\left(-\frac{|x - c_j|^2}{2\sigma_j^2}\right) \qquad (13.5)$$

where the vector x represents the states of the system. The ideal weights are bounded by known positive values such that

$$|w_i| \leq w_M \qquad (13.6)$$

where $i = \{i \in \mathbb{Z} \, | \, 0 \leq i \leq m\}$. Let the RBF functional estimates for $f(x)$ be given as:

$$\widehat{f}(x) = \sum_{i=0}^{m} \widehat{w}_i \emptyset_i(x) \qquad (13.7)$$

where $\widehat{w}_i$ are the estimates of the ideal RBF weights.

Therefore, there are $(m + 1)$ parameters (i.e., or weights) to be tuned. It is a difficult *NP*-complete problem to determine the optimum weights especially for large m. A gradient search method is sometimes used, which is sensitive to the initial set selected and is faced with a convergence problem. Even if the search is convergent, usually only a localized optimum point is obtained. The Taguchi method can thus be applied to systematically search for the optimum set.

As mentioned earlier in this chapter, quality characteristics in a Taguchi experiment refer to the assessment factors which are used for measuring how good the objectives of the experiment are met. For machine learning, the quality characteristic can be based on an appropriate measure of the deviation of the RBFNN from the actual nonlinear function, i.e., the residue. Since the objective here is for the RBFNN to approximate the actual function closely, this measure is desired to be as small as possible (i.e., smaller-the-better, STB). Even then, the quality characteristic can also be formulated in various ways, e.g., maximum error, sum of absolute error, sum of squares of error, etc.

The control factors are the parameters to be tuned using the Taguchi method. Clearly, the weights of the RBFNN are the control factors of the Taguchi experiment. Depending on the number of control factors and assigned levels in the RBFNN, the appropriate orthogonal array and linear graph can be used to generate an efficient set of parameters for experimentation. The systematic experimentation procedure of the Taguchi method can then be employed for tuning the neural networks. The testing and validation process of the neural network (which employed the Taguchi method for training purpose) is the same as that when the other conventional training methods, like the back propagation and other gradient-based methods, are employed. The main advantage of employing the Taguchi method for training the neural network is to provide a systemic view for the search space. Problems encountered by neural networks employing the traditional methods, such as over-training and local minima, could be avoided.

13.5 Case Studies

In this case study, we would like to study the performance of integrating Taguchi method with another artificial intelligence tool on the overall system performance. A particular fuzzy system for a vehicle dispatching platform involving an automated guided vehicle system (AGVS) will be used for this purpose. With just fuzzy logic alone, the approach to the dispatching of AGVs is based on a self-adapting fuzzy method. This method is provided for a unit load transport system in a manufacturing environment. In this method, it allows for the possibility to formulate more versatile and flexible rules. Thus, the AGVS no longer have to operate in a single criterion satisfaction. The AGVS is able to specialize in a multiple criteria satisfaction. Also, the balance between rules can be precisely adapted to a production environment through real-time parameterization from available information.

The overall effectiveness of this fuzzy logic approach to the vehicle dispatching system for the AGVS is critically dependent on the appropriateness of the fuzzy rules, in particular, the weight or the scaling factors of each of the fuzzy rules on the final dispatch decision. One of the main issues in the fleet control of the AGVS is to address the selection of the scaling factors, i.e., α's as mentioned in Eq. (13.1). The scaling factors could be set based on the experience of the operator and fixed throughout the production process. By this manual adjustment method, the search for the optimal set is by far too tedious and time consuming. Also, an exhaustive search for the optimal set of scaling factors would be unrealistic and impractical. To this end, a systematic search procedure within a balanced set of possible weighting parameters will be both desirable and practical. In this section, the results of using the Taguchi method to tune the fuzzy rules in the fuzzy vehicle dispatching system will be elaborated.

The key idea in the proposed approach is to associate all the work centers in the system with two attributes for each respective vehicle. The two attributes are *PARTS_IN* and *PARTS_OUT*, which are associated with the extent of demand-driven and source-driven needs of the work center with respect to the vehicle (Egbelu and Tanchoco 1984). These attributes are fuzzy variables (i.e., *PARTS_IN*, *PARTS_OUT* $\in [0, 1]$) computed from a fuzzy operation on a combination of variables which are expected to influence the extent of the demand and source-driven needs of the work center. Decisions for material movement will be driven primarily by these attributes.

The two attributes, *PARTS_IN* and *PARTS_OUT* can be inferred from a Takagi and Sugeno type of fuzzy inference (Haykin 1994). The fuzzy rules will then govern the *PARTS_IN* and *PARTS_OUT* attributes of the work centers. For example from Eq. (13.1), x_{kj}^i may be the linguistic variable *CYCLE_TIME* and F_j^i may be the fuzzy set *SHORT* for the *PARTS_IN* attribute; similarly, x_{kj}^i may be the linguistic variable *WAITING_TIME* and F_j^i may be the fuzzy set *LONG* for the *PARTS_OUT* attribute. The value of the *PARTS_IN$_k$* and *PARTS_OUT$_k$* attributes are then evaluated separately as a weighted average of the u^i's in Eq. (13.2).

With these attributes, the work centers may be sorted in the order of their demand and source-driven needs. In a pull-based situation, an idle vehicle searches for the highest inflow demand station from the *PARTS_IN* attribute. This station may then be paired off with a station having the highest *PARTS_OUT* attribute, identified from a set of K stations supplying the parts to the kth station in demand. The converse is true for a push-based system.

Thus far, dispatching rules are rigidly based on either a demand or a source-driven procedure. There are attractive features associated with a demand-driven rule as they provide the load movement flexibility for just-in-time manufacturing concepts. However, under certain circumstances, it might be more advantageous to revert to a source-driven rule. For example, by reverting to a source-driven rule when there is a low level of demand but a large number of parts to be cleared, machine blockage may be reduced. Furthermore, reverting to the source-driven rule is also a hedge against an unanticipated surge in vehicle demand at a future time.

Instead of rigidly commissioning a push or a pull-based concept, it is viable to view each of these concepts as being suited to different operating conditions, and switch between them when crossing these different operating regions. Clearly, some mechanism to trigger this switch between a pull and a push-based environment is needed. To this end, a methodology may be formulated as follows:

Denote $SCE(k)$ as the set of work centers supplying to the input buffer of work center k, and $DES(k)$ as the set of work centers to which work center k supplies parts. The work centers, k_u^* and k_v^* are identified where

$$PARTS\_IN_{k_u^*} = max\ (PARTS\_IN_k),\ where\ SCE(k_u^*) \not\subset \emptyset$$
$$PARTS\_OUT_{k_v^*} = max\ (PARTS\_OUT_k),\ where\ DES(k_v^*) \not\subset \emptyset$$

Based on these attributes, the current state system towards a push or a pull operation may be determined. For example, a simple formulation may be to compute the following ratio,

$$STRATEGY = \frac{PARTS\_OUT_{k_v^*}}{PARTS\_IN_{k_u^*} + PARTS\_OUT_{k_v^*}} \qquad (13.8)$$

If $STRATEGY > \gamma$ (where suitable values for γ may be in the range $0.6 < \gamma < 0.7$, depending on the desired level of $PULL$ dominance), a $PUSH$ operation may be initiated, otherwise a $PULL$ operation will be initiated. The $PULL$ dominance is necessary for reasonably busy facilities where it has been shown (Egbelu and Tanchoco 1984) that a PULL strategy is more efficient in moving parts through the facility.

A graphical representation of the self-adapting fuzzy dispatching algorithm is illustrated in Fig. 13.5. The vehicles are all indexed. When the vehicle in consideration becomes idle, it is available for reassigned to pick up the load from the source work center and then deliver the load to the destination work center. With the idle vehicle in consideration, the fuzzy dispatching algorithm block (as shown in Fig. 13.5) is then invoked. The fuzzy algorithm is represented as an operation block in Fig. 13.5. In Fig. 13.6, the fuzzy algorithm is expanded into further details. Here, the needs of all the work centers are prioritized according to their demand and source driven needs (i.e., the demand and source driven needs are computed using the fuzzy $PARTS\_IN$ and $PARTS\_OUT$ attributes). At the end of the execution of the fuzzy dispatching algorithm, the work center (i.e., this work center could either be a source work center or a destination work center; if it is a source work center most in need of service, push-based operation or a source-driven procedure is then selected. If it is a destination work center most in need of service, pull-based operation or a demand-driven procedure is then selected) most in need of service is selected. The vehicle is then invoked to pick up the load from the assigned source work center and then deliver the load to the assigned destination work center. The whole procedure is then repeated for the vehicle, next in the index sequence. If the vehicle next in the sequence is not idle, the fuzzy dispatching algorithm would not be invoked. Instead, it would only continue in its travel to its assigned pick-up work center and then the destination work center.

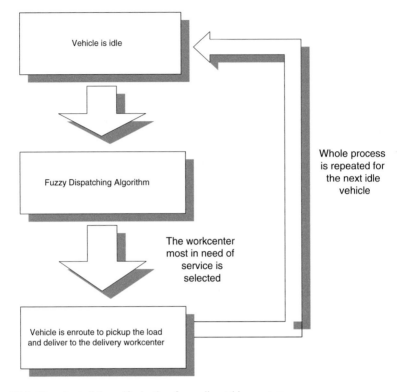

Fig. 13.5 Flowchart of the self-adapting fuzzy dispatching system

While the algorithm is essentially based on vehicle-initiated rules, there are situations such as during start-up, when more than one vehicle is available for dispatching. Since these are insignificant for reasonably busy facilities, a simple random procedure is used to select the vehicle for dispatch under these situations.

13.5.1 Test Facility

The simulation analysis is based on a hypothetical facility as given in Fig. 13.7. The facility operating data is provided in Table 13.1. There are 9 work centers (WC1 to WC9) or departments, and a warehouse (WH) for the raw materials and finished products.

- Job routing = WH, WC1, (WC2, WC3), WC4, WC5, WC6, (WC7, WC8), WC9, WH
- Load pickup/delivery time = 10 s
- Vehicle length = 3 ft.

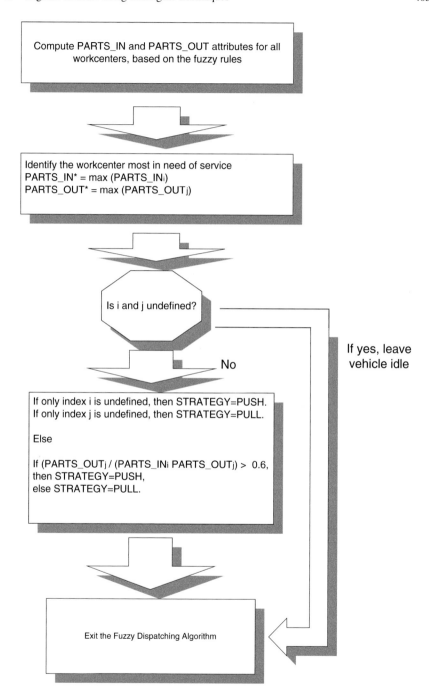

Compute PARTS_IN and PARTS_OUT attributes for all workcenters, based on the fuzzy rules

Identify the workcenter most in need of service
PARTS_IN* = max (PARTS_IN$_i$)
PARTS_OUT* = max (PARTS_OUT$_j$)

Is i and j undefined?

No

If yes, leave vehicle idle

If only index i is undefined, then STRATEGY=PUSH.
If only index j is undefined, then STRATEGY=PULL.

Else

If (PARTS_OUT$_j$ / (PARTS_IN$_i$ PARTS_OUT$_j$) > 0.6,
then STRATEGY=PUSH,
else STRATEGY=PULL.

Exit the Fuzzy Dispatching Algorithm

Fig. 13.6 Fuzzy algorithm block (more detailed version)

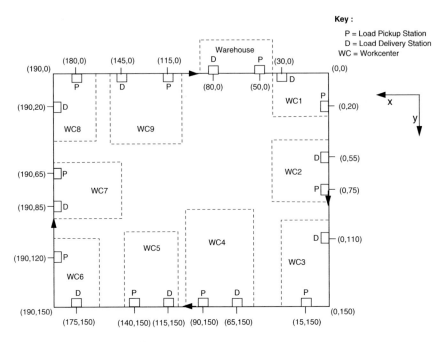

Fig. 13.7 Layout of the test facility

Table 13.1 The test facility operating data

Work center	Processing time/unit load (min)	Input queue size	Output queue size
1	1	3	5
2	3	2	3
3	3	2	3
4	2	3	2
5	1	1	4
6	3	2	3
7	3	2	3
8	2	3	2
9	3	4	4

- Vehicle speed = 200 fpm
- Pickup and delivery spur capacity = 1 vehicle

13.5.2   *Simulation Language*

The control simulation language, MATLAB is used for implementing the study. The language may be used for simulation of both continuous and discrete-time systems.

In this case, it is applied to discrete event investigation, where the AGV guide path is modeled as a directed network consisting of nodes and arcs. Point locations in the network are uniquely identified by their Cartesian coordinates. Traffic conflicts at the load pickup/delivery points are explicitly modeled.

13.5.3   Computation of Attributes

In this simulation, the input variables chosen for the computation of the *PARTS_IN* attributes are:

- Length of time before incoming queue is empty, **LT_IN**
- Shortest travel distance of vehicle to source work centers, and to the work center concerned, **STD_IN**.
- Shortest length of time before the outgoing queue of source work centers is full, **SLT_IN**
- Number of parts completed already by the workstation, **PC_IN**.

The time taken for the processing of the load at each work center is assumed to be fixed and shown as in Table 13.1 so that these variables can be directly computed. The 4 rules formulated for the computation of the *PARTS_IN* attribute for the *k*th work center are:

IF **LT_IN$_k$** is SHORT, THEN $u_k = \mu_1$
IF **STD_IN$_k$** is SHORT, THEN $u_k = \mu_2$
IF **SLT_IN$_k$** is SHORT, THEN $u_k = \mu_3$
IF **PC_IN$_k$** is LOW, THEN $u_k = \mu_4$

PARTS_IN is then computed as in (13.2). The input variables chosen for the computation of the *PARTS_OUT* attributes are:

- Shortest length of time before outgoing queue of work center is full, **SLT_OUT**.
- Shortest travel distance of vehicle to work center concerned, and to target work centers, **STD_OUT**.
- Length of time before the incoming queue of destination work center is empty, **LT_OUT**.
- Number of parts completed already by the workstation, **PC_OUT**.

Similarly, the 4 rules formulated for the computation of the *PARTS_OUT* attribute for the *k*th work center are:

IF **SLT_OUT$_k$** is SHORT, THEN $v_k = \mu_5$
IF **STD_OUT$_k$** is SHORT, THEN $v_k = \mu_6$
IF **LT_OUT$_k$** is SHORT, THEN $v_k = \mu_7$
IF **PC_OUT$_k$** is LOW, THEN $v_k = \mu_8$

PARTS_OUT is then computed as in (13.2). The membership functions are made time varying according to the set of assigned tasks at any point in time. In this

approach, a linear interpolation between the maximum and minimum values of the
variables serves as the membership function. As an example, consider the following
STD_IN variable

$$\mu_{SHORT}(STD\_IN_k) = \frac{STD\_IN_k - \min(STD\_IN)}{\max(STD_{IN}) - \min(STD\_IN)} \tag{13.9}$$

13.5.4 Rule Comparison

The performances of the dispatching system with the deployment of three different
deployments will be illustrated. The three different combinations are, just the fuzzy
logic alone, the Taguchi-tuned fuzzy system and the demand driven (DEMD) rules
of Egbelu (Mandal et al. 2011). The comparisons were carried out under the following three different cases.

Case I Given an equal number of vehicles, the same facility scenario, and the
same length of time or shift duration, how does the facility throughput compares
between the two sets of rules? Throughput is defined as the total number of parts
completed and removed from the facility shop floor during the shift. The following
parameters are used for Case I analysis:

- The facility operates a 2-h shift.
- Three vehicles are in use.
- Infinite number of loads were available for processing at time, $t = 0$.

Due to the nature of the dispatching problem, there are a number of quality
characteristics that could be considered in the Taguchi-tuned fuzzy system. This
quality characteristic is a LTB characteristic.

Case II Given the same conditions as in Case I, how long does it take for the
facility to produce a known number of parts under the three different sets of rules?
The analysis was done with 3 vehicles and 30 parts or unit loads to be produced and
centers on the determination of the length of time it will take the facility to produce
the 30 parts under each of the dispatching methodology. The facility operating
duration is a measure of the rule's ability to accelerate the unit loads through the
facility. In the Taguchi-tuned fuzzy system, this quality characteristic is a STB
characteristic.

Case III Given the same conditions as in Case I and a production target over a
fixed time period, how many vehicles are required to meet the production target
under the three different sets of rules? The conditions for the analysis are the
following:

- There are a fixed number (30) of unit loads to be produced.
- The production of the fixed number of unit loads must be satisfied within the
  time interval specified (2 h).

If all other factors remain the same, it seems that the number of vehicles required will be a function of the dispatching rule (i.e., or control factors in the Taguchi method sense) in force, since the rules act differently with the vehicles. This is a LTB quality characteristic.

13.5.5 Deployment of Taguchi-Tuned Fuzzy System

In the deployment of the Taguchi method to the AGVS, there are some further considerations to modify the problem as a quality optimization problem.

13.5.5.1 Control Factors and Number of Levels

As 4 fuzzy rules are chosen for the *PARTS_IN* fuzzy attribute, there are thus 4 scaling factors (denoted as α_1, α_2, α_3 and α_4) for the *PARTS_IN* fuzzy attribute. Similarly, for the *PARTS_OUT* fuzzy attribute, 4 scaling factors (denoted as β_1, β_2, β_3 and β_4) are selected. These 8 scaling factors for the fuzzy rules are the control factors to be tuned by the Taguchi method for optimal performance. These scaling factors are the parameters as mentioned in Eq. (13.1). The sum of these scaling factors is fixed, i.e., $\alpha_1 + \alpha_2 + \alpha_3 + \alpha_4 = 1$ and $\beta_1 + \beta_2 + \beta_3 + \beta_4 = 1$. This implies that within the set, the factors cannot be varied arbitrarily. But rather the choice of one factor limits the possibilities for the other factors. Due to this constraint, the nested-factor design (Peace 1993) is used for the assignment of the levels of the control factors (shown in Table 13.2). The layout of the first four control factors for the *PARTS_IN* attribute, i.e., α_1, α_2, α_3 and α_4 denoted as A, B, C and D respectively, is shown in Table 13.2. The other four scaling parameters for the *PARTS_OUT* attribute, i.e., β_1, β_2, β_3, and β_4 denoted as E, F, G and H respectively, have a similar layout as shown in Table 13.2. The factors are scaled up by a factor of 100 for ease of calculations.

Each of the control factors is assigned 3 levels. A 3-layered structure for the control factors is used. Factors C and D are nested within Factor B which is in turn nested within Factor A. For example, consider Factor A at level 1, i.e., A_1, Factor B can be at level 1, level 2 or level 3, i.e., B'_1, B'_2 and B'_3 respectively. For Factor A at level 2, i.e., A_2, Factor B can be at another set of level 1, level 2 or level 3, i.e., B''_1, B''_2 and B''_3 respectively. The set of levels of Factor B at A_1 is different from that at A_2. The same layout is also applied to Factor B and Factors C, D.

As shown in Table 13.3 for Factor A (traversing the second column), there are 3 levels, i.e., 1, 50 and 90. Factor A at level 1, level 2 and level 3 is denoted as A_1, A_2 and A_3 respectively. For A_1, Factor B has three levels: 1, 50 and 90 (denoted as B'_1, B'_2 and B'_3 respectively). For A_2, Factor B has another 3 levels: 1, 25 and 40 (denoted as B''_1, B''_2 and B''_3 respectively). For A_3, Factor B has another 3 levels: 1, 5 and 9 (denoted as B'''_1, B'''_2 and B'''_3 respectively).

Table 13.2 Layout for the layered structure of the control factors Control factors

Control factors	α_1	α_2	α_3	α_4
Symbol used	A	B	C	D
Actual level value	1	1	97	1
			49	49
			1	97
		50	48	1
			24.5	24.5
			1	48
		90	8	1
			4.5	4.5
			1	8
Actual level value	50	1	48	1
			24.5	24.5
			1	48
		25	24	1
			12.5	12.5
			1	24
		40	9	1
			5	5
			1	9
Actual level value	90	1	8	1
			4.5	4.5
			1	8
		5	4	1
			2.5	2.5
			1	4
		9	0.9	0.1
			0.5	0.5
			0.1	0.9

Table 13.3 Optimum levels of the various control factors Control factors

Control factor	Optimum level	Value
A	A_1	0.01
B	B'_1	0.01
C	$(CD)'^1_1$	0.49
D	$(CD)'^1_2$	0.49
E	E_1	0.01
F	F'_3	0.90
G	$(GH)'^3_1$	0.08
H	$(GH)'^3_1$	0.01

13.5.5.2   Orthogonal Array and Linear Graph

For this vehicle dispatching problem considered here, the required DOF is 36, obtained as follows.

Axial factor (A)	$1 \times (3 - 1) = 2$ DOF
Nested factor (B)	$1 \times (3 - 1) = 2$ DOF
Nested factor $(C \& D)$	$1 \times (3 - 1) = 2$ DOF
Axial nesting for $A \& B$	$1 \times (3 - 1)(3 - 1) = 4$ DOF
Axial nesting for $A \& B \& C, D$	$1 \times (3 - 1)(3 - 1)(3 - 1) = 8$ DOF
Total DOF needed for $A, B, C, D = 18$	
Total DOF needed for $A, B, C, D \& E, F, G, H = 36$	

The L_{81} (3^{40}) orthogonal array (Peace 1993) is used, being the smallest array with a degree of freedom larger than 36. The subscript refers to the number of experiments or rows in the array whereas the superscript refers to the number of factors or columns in the array. For this array, there are many different linear graphs with different structures (Peace 1993). The linear graph chosen is as shown in Fig. 13.8. The nodes represent the control factors. The joining arcs represent the interactions or the relationship between the control factors. The numbers in brackets refer to the column number in which the particular control factor is assigned to in the orthogonal array.

13.5.5.3   Results of the Experiments

After the experiments are run, the optimum condition is determined by selecting the best levels of each factor. Note that the choice of one factor limits the choice of subsequent ones due to nesting. The average level of a factor is obtained by summing the experimental data for each particular level of a factor and averaging the sum by the number of experiments. The best level of a factor is the level with the highest average level (for a LTB quality characteristic) or the level with the

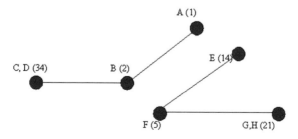

Fig. 13.8 Linear graph for the control factors in the L_{81} (3^{40}) orthogonal array

lowest average level (for a STB quality characteristic). The best levels for the factors are thus obtained as illustrated from Fig. 13.9 to 13.14.

In this experiment, the quality characteristic of concern is a LTB quality characteristic. From the factors' effects plot above (Figs. 13.9, 13.10, 13.11, 13.12, 13.13 and 13.14), the optimum condition is thus obtained by choosing the factors combination A_1, B'_1 and $(CD)'^1_1, E_1, F'_3$ and GH'^3_1. The next best factors combination is A_1, B'_1 and $(CD)'^1_2, E_1, F'_3$ and GH'^3_1. There is little difference between Factor $(CD)'^1_1$ and $(CD)'^1_2$ on the quality characteristic. Using the two optimum sets of factors, a final test run is done. The results of the final test run confirm that the optimum condition obtained for the control factors at the given levels as shown in Table 13.3.

Here, Factor A at Level 1, Factor B at Level 1 and Factors C and D at Level 2 are chosen, whereas Factor E at Level 1, Factor F at Level 3, and Factors G and H at Level 1 are chosen.

13.5.5.4 Validation with Statistical Tools

As mentioned at the start of this chapter, after the experimental data from the set of weighting parameters is obtained, the statistical tools could be used to interpret the experimental data mathematically and to provide indications of whether the choice is adequate and if further experimentation is necessary. For this purpose in this chapter, the analysis of variance (ANOVA) is chosen. This is because when results of laboratories or methods are compared where more than one factor can be of influence and must be distinguished from random effects, then the ANOVA is a powerful statistical tool to be used. Furthermore, the ANOVA also allows one to assess the degree to which a factor affects variation and the quality characteristic. ANOVA is used as a supplement to the experimental design (Sreenivasulu 2013; Asafa and Said 2013; Mandal et al. 2011; Tzeng et al. 2009; Howanitz and Howanitz 1987; International Organization for Standardization 1981; Bauer 1971).

Analysis results from the ANOVA method are summarized in Table 13.4. The results obtained by the ANOVA method confirm the results obtained earlier. Referring to Table 13.4, the second column (F) refers to the degrees of freedom of each factor. The third column (S) refers to the factor variation and error variation. The fourth column (V) refers to the factor variances and the error variance. The fifth

Fig. 13.9 Factor effects of Factor A

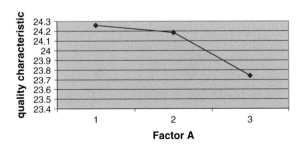

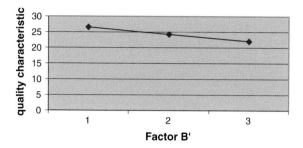

Fig. 13.10 Factor effects of Factor B'

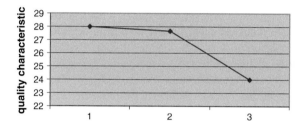

Fig. 13.11 Factor effects of Factor CD'^1

Fig. 13.12 Factor effects of Factor E

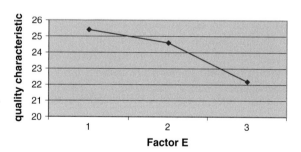

Fig. 13.13 Factor effects of Factor F'

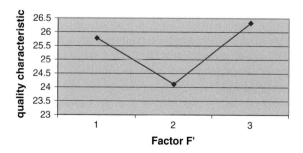

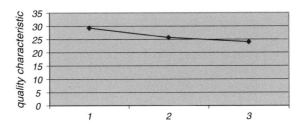

Fig. 13.14 Factor effects of Factor GH'^3

column (F) refers to the variance ratio of each the factor variance to the error variance. The last column (ρ) refers to the contribution ratio, i.e., significance or importance with regards to the particular quality characteristic, of each factor.

There are two conclusions that can be drawn from the ANOVA table to support the optimum combination of factors obtained earlier. First, there is about 11 % of significance to the quality characteristic due to uncontrolled or unwanted factors. This implies positively that the factors chosen earlier to study their effects on the quality characteristic has a major influence of about 89 % on the quality characteristic. Secondly, the factors chosen earlier in Table 13.3 are indeed those with most significant contribution to the quality characteristic, as shown in the last column of Table 13.4 by their contribution ratio. Referring to the contribution ratio (last column of Table 13.4), Factor A has a significance of about 0.259 % on the quality characteristic. From the factor effects graph of Factor A (Fig. 13.9), Level 1 is chosen as the optimum level. For Factors B', B'' and B''', B' with Level 1 is chosen in the optimum factor combination from the factor effects' graph as A_1 is chosen. This is due to the nesting arrangement in Table 13.2. Factor B' has the highest contribution ratio among B', B'' and B'''. For Factor CD, the optimum factor chosen from the factor effects' graph is $(CD)'^1$. This is again due to the nesting arrangement in Table 13.2 as we have chosen A_1 and B'_1. Factors $(CD)'^1$, $(CD)'^2$ and $(CD)'^3$ have quite similar contribution ratios.

13.5.6 Simulation Results

The values of the manually adjusted (in a trial-and-error manner) scaling factors (i.e., α's as mentioned in Eq. (13.1)) obtained manually by trial and error method, are shown in Table 13.5. For this approach, the scaling parameters are changed on a rather ad hoc basis, relying solely on the experience and the judgment of the operator. For the Taguchi method, the obtained weighting parameters at the end of the tuning procedure are given in the third row of Table 13.5.

Based on the scaling factors in Table 13.5, the performances of the various dispatching methods are compared and summarized in Table 13.6. It can be seen that the Taguchi-tuned fuzzy dispatching methodology has outperformed the

Table 13.4 Table showing the ANOVA results

Factor	F	S	V	F	ρ %
A	2	4.247	2.123	2.279	*0.259*
B'	2	93.407	46.704	100.298	*9.950*
B''	2	23.407	11.704	25.134	2.340
B'''	2	3.185	–	–	–
CD'^1	2	29.556	14.778	31.736	*3.008*
CD'^2	2	29.556	14.778	31.736	3.008
CD'^3	2	52.667	26.334	56.552	5.519
CD''^1	2	24.889	12.445	26.725	2.501
CD''^2	2	4.222	2.111	4.553	0.256
CD''^3	2	10.889	5.445	11.692	0.980
CD'''^1	2	2.889	–	–	–
CD'''^2	2	2.889	–	–	–
CD'''^3	2	3.556	–	–	–
E	2	151.580	75.790	162.762	*16.261*
F'	2	24.074	12.037	25.850	2.412
F''	2	18.074	9.037	19.407	1.761
F'''	2	10.296	5.148	11.056	*0.916*
GH'^1	2	94.889	47.445	101.056	*10.104*
GH'^2	2	81.556	40.778	87.573	8.656
GH'^3	2	44.667	22.334	47.962	4.649
GH''^1	2	82.889	41.445	89.004	8.801
GH''^2	2	28.667	14.334	30.782	2.911
GH''^3	2	0.222	–	–	–
GH'''^1	2	24.667	12.334	26.487	2.477
GH'''^2	2	0.222	–	–	–
GH'''^3	2	26.889	13.445	28.873	2.718
Total	80	920.691	–	–	100
Error	64	59.603	0.932	1	10.513

Table 13.5 Values of the scaling factors for the different approaches

Scaling factors	α_1	α_2	α_3	α_4	β_1	β_2	β_3	β_4
Trial and error setting	0.1	0.5	0.2	0.2	0.2	0.5	0.1	0.2
Taguchi method	0.01	0.01	0.49	0.49	0.01	0.50	0.48	0.01

manually tuned fuzzy dispatching rules of and DEMD dispatching rules of Egbelu for all the three cases studied. It is also interesting to note the improvement of the performance of the system with the introduction of fuzzy logic to this vehicle dispatching platform.

Table 13.6 Rule comparisons of the various methods

	DEMD (Egbelu)	Fuzzy	% Improvement (Fuzzy vs. DEMD) (%)	Taguchi-tuned Fuzzy	% Improvement (Taguchi-Fuzzy vs. Fuzzy) (%)
Case I: Throughput	16	27	69	31	14.81
Case II: Production time/h	2.99	2.14	28	1.95	19.74
Case III: No of vehicles	9	5	44	3	40

The results in this study concur with the other works in the literature that integrate the Taguchi method with other tools in artificial intelligence (i.e., genetic algorithm, neural networks, fuzzy logic and regression analysis) to optimize processes and systems (Lin et al. 2012; Sun et al. 2012; Mandal et al. 2011; Chang 2011; Tsai 2011; Tansel et al. 2011; Tzeng et al. 2009). The integrated approach of the Taguchi method with other artificial intelligence tools produces better results than just utilizing a single optimization tool alone. For this study, it is also interesting to note that there is a marked improvement in the fuzzy system for the third case. It is to be noted that the Taguchi method performs well when the boundaries of the search space is well-defined. With the production time and other manufacturing constraints fixed, the Taguchi method seeks to only optimize the vehicles deployed.

13.6 Conclusions

Since Dr. Genichi Taguchi introduced the Taguchi method as a practical statistical tool in the 1980s to improve the quality of manufactured goods, there has been increasing interests on the Taguchi method in various industries and the academia. It is to be noted that in the literature, there is a strong focus on the application and verification of the Taguchi method. Recently, there are several works in the literature that integrate the Taguchi method with other tools in artificial intelligence (i.e., genetic algorithm, neural networks, fuzzy logic and regression analysis) to optimize processes and systems. This is interesting in that the main strengths of the Taguchi methods are accentuated with a well-defined boundary for a given search space. The results in the literature show that the integrated approach of the Taguchi method with other artificial intelligence tools produce better results than just utilizing a single optimization tool alone. Besides modeling and optimizing manufacturing processes, the Taguchi method has also been widely applied in other non-conventional fields.

In this chapter, a few possible approaches of integrating the Taguchi method combined with some artificial intelligent techniques, to create hybrid approaches with improved overall system performance, are elaborated. Particularly, details and illustrations of combining the Taguchi method with a fuzzy system and a radial basis neural network are provided. In the simulation study, the adaptive fuzzy rules are formulated to base the decision making process and the Taguchi method is applied to fine tune the rules for optimal performance. The experimental design is performed and simulations are conducted to compare the Taguchi-tuned fuzzy method to other earlier reported methods. The results in this study concur with the other works in the literature that integrate the Taguchi method with other tools in artificial intelligence.

Acknowledgments Special thanks to Ms. Chua Xiaoping Shona and Mr. Lee Tat Wai David for their efforts in the initial drafting of this chapter.

References

Asafa, T.B., Said, S.A.M.: Taguchi method–ANN integration for predictive model of intrinsic stress in hydrogenated amorphous silicon film deposited by plasma enhanced chemical vapour deposition. Neurocomputing **106**, 86–94 (2013)

Bauer, E.L.: A Statistical Manual for Chemists. Academic Press, New York (1971)

Chang, K.Y.: The optimal design for PEMFC modeling based on Taguchi method and genetic algorithm neural networks. Int. J. Hydrogen Energy **36**, 13683–13694 (2011)

Chen, Y.H., Tam, S.C., Chen, W.L., Zheng, H.Y.: Application of Taguchi method in the optimization of laser micro-engraving of photomasks. Int. J. Mater. Prod. Technol. **11**, 333–344 (1996)

Chou, J.H., Chen, S.H., Li, J.J.: Application of the Taguchi-genetic method to design an optimal grey-fuzzy controller of a constant turning force system. J. Mater. Process. Technol. **105**, 333–343 (2000)

De Souza, H.J.C., Moyses, C.B., Pontes, F.J., Duarte, R.N., Da Silva, C.E.S., Alberto, F.L., Ferreira, U.R., Silva, M.B.: Molecular assay optimized by Taguchi experimental design method for venous thrombo-embolism investigation. Mol. Cell. Probes **25**(5), 231–237 (2011)

Ealey, L.A.: Quality by Design. Irwin Professional Publishing, Illinois (1994)

Egbelu, P.J.: Pull versus push strategy for automated guided vehicle load movement in a batch manufacturing system. J. Manuf. Syst. **6**, 209–221 (1987)

Egbelu, P.J., Tanchoco, J.M.A.: Characterisation of automated guided vehicle dispatching rules. Int. J. Prod. Res. **22**, 359–374 (1984)

Haykin, S.: Neural Networks—A Comprehensive Foundation. MacMillan Publishing Company, New York (1994)

Hissel, D., Maussion, P., Faucher, J.: On evaluating robustness of fuzzy logic controllers through Taguchi methodology. In: Proceedings of the IEEE Industrial Electronics Society 24th Annual Conference, IECON'98, pp. 17–22 (1998)

Ho, W.H., Tsai, J.T., Chou, J.H.: Robust-stable and quadratic-optimal control for TS-fuzzy-model-based control systems with elemental parametric uncertainties. IET Control Theory Appl. **1**, 731–742 (2007)

Hoa, W.H., Tsai, J.T., Lin, B.T., Chou, J.H.: Adaptive network-based fuzzy inference system for prediction of surface roughness in end milling process using hybrid Taguchi-genetic learning algorithm. Expert Syst. Appl. **36**, 3216–3222 (2009)

Hong, C.W.: Using the Taguchi method for effective market segmentation. Expert Syst. Appl. **39**, 5451–5459 (2012)

Howanitz, P.J., Howanitz, J.H.: Laboratory quality assurance. McGraw-Hill, New York (1987)

Huang, S., Tan, K.K., Tang, K.Z.: Neural Network Control—Theory and Applications. Research Studies Press, London (2004)

Hwang, C.C., Chang, C.M., Liu, C.T.: A fuzzy-based Taguchi method for multiobjective design of PM motors. IEEE Trans. Magn. **49**, 2153–2156 (2013)

International Organization for Standardization: Statistical methods. ISO Standards Handbook 3, 2nd edn. ISO Central Seer., Genève (1981)

Khaw, F.C., Lim, B.S., Lim, E.N.: Optimal design of neural networks using the Taguchi method. Neurocomputing **7**, 225–245 (1995)

Lin, H.C., Su, C.T., Wang, C.C., Chang, B.H., Juang, R.C.: Parameter optimization of continuous sputtering process based on Taguchi methods, neural networks, desirability function, and genetic algorithms. Expert Syst. Appl. **39**(17), 12918–12925 (2012)

Mandal, N., Doloi, B., Mondal, B., Das, R.: Optimization of flank wear using Zirconia Toughened Alumina (ZTA) cutting tool: Taguchi method and regression analysis. Measurement **44**(10), 2149–2155 (2011)

Mori, T.: The New Experimental Design. American Supplier Institute, Michigan (1993)

Peace, G.S.: Taguchi Methods. Addison-Wesley Publishing Company, New York (1993)

Rao, R.S., Kumar, C.G., Prakasham, R.S., Hobbs, P.J.: The Taguchi methodology as a statistical tool for biotechnological applications—a critical appraisal. Biotechnol. J. **3**, 510–523 (2008)

Ross, J.R.: Taguchi Techniques for Quality Engineering. McGraw-Hill, Columbus (1988)

Sreenivasulu, R.: Optimization of surface roughness and delamination damage of GFRP composite material in end milling using Taguchi design method and artificial neural network. Procedia Eng. **64**, 785–794 (2013)

Sun, J.H., Fang, Y.C., Hsueh, B.R.: Combining Taguchi with fuzzy method on extended optimal design of miniature zoom optics with liquid lens. Optik—Int. J. Light Electr. Opt. **123**(19), 1768–1774 (2012)

Taguchi, G., Yokoyama, T.: Taguchi Methods—Design of Experiments. Dearborn, ASI Press, Tokyo (1993)

Taguchi, G., Chowdhury, S., Wu, Y.: Taguchi's Quality Engineering Handbook. Wiley, Hoboken (2004)

Takagi, T., Sugeno, M.: Fuzzy identification of systems and its applications to modelling and control. IEEE Trans. Syst. Man Cybern. **15**, 116–132 (1985)

Tan, K.K., Tang, K.Z.: Taguchi-tuned radial basis function with application to high precision motion control. Artif. Intell. Eng. **15**, 25–36 (2001)

Tansel, I.N., Gülmez, S., Aykut, S.: Taguchi Method–GONNS integration: complete procedure covering from experimental design to complex optimization. Expert Syst. Appl. **38**(5), 4780–4789 (2011)

Tortum, A., Yaylab, N., Celikc, C., Gökdag, M.: The investigation of model selection criteria in artificial neural networks by the Taguchi method. Phys. A **386**, 446–468 (2007)

Tsai, T.N.: Improving the Fine-Pitch Stencil printing capability using the Taguchi method and Taguchi fuzzy-based model. Robot. Comput.-Integr. Manuf. **27**, 808–817 (2011)

Tzeng, C.J., Lin, Y.H., Yang, Y.K., Jeng, M.C.: Optimization of turning operations with multiple performance characteristics using the Taguchi method and grey relational analysis. J. Mater. Process. Technol. **209**(6), 2753–2759 (2009)

Wang, J.L., Wan, W.: Experimental design methods for fermentative hydrogen production. Int. J. Hydrogen Energy **34**(1), 235–244 (2009)

Woodall, W.H., Koudelik, R., Tsui, K.L., Kim, S.B., Stoumbos, G., Carvounis, C.P., Jugulum, R., Taguchi, G., Taguchi, S., Wilkins, J.O., Abraham, B., Variyath, A.M., Hawkins, D.M.: Review and analysis of the Mahalanobis-Taguchi system. Technometrics **45**, 1–30 (2003)

Yang, T., Wen, Y.F., Wang, F.F.: Evaluation of robustness of supply chain information-sharing strategies using a hybrid Taguchi and multiple criteria decision-making method. Int. J. Prod. Econ. **134**(2), 458–466 (2011)

Yu, G.R., Huang, J.W. Chen, Y.H.: Optimal fuzzy control of piezoelectric systems based on hybird Taguchi method and particle swarm optimization. In: Proceedings of the 2009 IEEE International Conference on Systems, Man, and Cybernetics, pp. 2794–2799. IEEE Press (2009)

Zadeh, L.A.: Outline of a new approach to the analysis of complex systems and decision process. IEEE Trans. Syst. Man Cybern. **3**, 28–44 (1973)

Zurada, J.M.: Introduction to Artificial Neural Systems. West Publishing Company, New York (1992)

Chapter 14
Software Architecture Quality of Service Analysis Based on Optimization Models

Pasqualina Potena, Ivica Crnkovic, Fabrizio Marinelli and Vittorio Cortellessa

Abstract The ability to predict Quality of Service (QoS) of a software architecture supports a large set of decisions across multiple lifecycle phases that span from design through implementation-integration to adaptation phase. However, due to the different amount and type of information available, different prediction approaches can be introduced in each phase. A major issue in this direction is that QoS attribute cannot be analyzed separately, because they (sometime adversely) affect each other. Therefore, approaches aimed at the tradeoff analysis of different attributes have been recently introduced (e.g., reliability versus cost, security versus performance). In this chapter we focus on modeling and analysis of QoS tradeoffs of a software architecture based on optimization models. A particular emphasis will be given to two aspects of this problem: (i) the mathematical foundations of QoS tradeoffs and their dependencies on the static and dynamic aspects of a software architecture, and (ii) the automation of architectural decisions driven by optimization models for QoS tradeoffs.

Keywords Quality of service · Software architecture · Optimization · Medical informatics system

P. Potena
Computer Science Department, University of Alcalà, 28871 Alcalà de Henares,
Madrid, Spain
e-mail: p.potena@uah.es

I. Crnkovic
School of Innovation, Design and Engineering, Mälardalen University,
72123 Västerås, Sweden

F. Marinelli (✉)
Dipartimento di Ingegneria dell'Informazione, Università Politecnica delle Marche,
60131 Ancona, Italy

V. Cortellessa
Dipartimento di Ingegneria e Scienze dell'Informazione, e Matematica,
Università dell'Aquila, 67010 Coppito, AQ, Italy

© Springer International Publishing Switzerland 2016
C. Kahraman and S. Yanık (eds.), *Intelligent Decision Making in Quality Management*, Intelligent Systems Reference Library 97,
DOI 10.1007/978-3-319-24499-0_14

421

14.1 Introduction

The presence in the market of standard off-the-shelf components/services has drastically changed in the last decade the development process of component-based and service-based systems (as claimed, e.g., in Szyperski 2002).

A software system today is no more conceived as a product to be built "from scratch"; rather software engineers aim at building a system where several software units–components/services, each satisfying a certain number of requirements— interact each other and with users to accomplish the tasks required.

Requirements can be partitioned in functional and non-functional. The former concerns "what" the software has to do, while the latter concern "how" the software works. In a service-oriented architecture, in practice, functional requirements determine the services the system should provide, whereas non-functional requirements (that determine the Quality of Service of the system), are constraints on the services offered by the system, such as timing constraints or constraints on the development process (Sommerville 2004).

The properties (functional and non-functional) of the final software product therefore heavily depend on (i) the properties of the reused software units and those of newly built software units, as well as on (ii) the way these software units are assembled (i.e. the software architecture).

In the last years several research efforts have been devoted to the definition of models representing dependencies between non-functional properties of single elements and the properties of the whole system.

External properties (i.e., system attributes) are functions of both internal properties (i.e. attributes of elementary components or services) and other factors, such as system architecture or usage profile. Developers must therefore address how the integrated system inherits attributes of elementary parts. For example, if you integrate several high performance or high-reliability components, what can you say about the performance or reliability of the system as a whole? Similarly, if you integrate a combination of low and high-quality components, how can you assess and improve the resulting system's quality? (Brereton and Budgen 2000). The formulation of such models is not easy due to complex relationships between components that may be hard to express in a closed form.

Component-Based Software Engineering (CBSE) and Service-Oriented Software Engineering (SOSE) are the most dominant disciplines that deal with problems of building software systems based on (reused and newly built) components/services (Breivold and Larsson 2007). The ability to predict QoS of a software architecture has to be supported by a large set of decisions arising from several phases of the software lifecycle that span from design, through implementation-integration, to adaptation phase. However, due to the different amount and type of information available, different prediction approaches should be introduced in each phase.

In the design and implementation phase, elementary software units are typically selected and verified/tested alone or in combination with other (selected) software units at the aim of choosing the combination that best fits the goals. In fact, it is well known (Wallnau and Stafford 2002) that even if isolated components correctly work, an assembly of them may fail due to not immediately apparent dependencies and relationships, such as shared data and resources. Besides, since the software units always have to be deployed on an hardware platform, the best mapping of software onto hardware with respect to certain criteria (e.g. performance of the whole system) has to be considered as well. Finally, an existing software unit (or a set of units) would be replaced and/or new units would be adopted in the maintenance phase, e.g., when the requirements of the system evolve or when the vendor of the component releases an updated version,[1] while keeping other units unchanged.

On the basis of the above considerations, it is evident that the architectural decisions must be carefully carried on taking into account non-functional properties (besides functional ones). In fact, functionally equivalent software units (to be used for replacing existing software units or to be added to the system) may heavily differ in their non-functional properties, affecting in this way the QoS at various extents. We hence forth refer to functionally equivalent software units that differ for their non-functional properties as to *instances*.

One of the most prominent characteristic of a software unit is its cost. In general, the cost of an in-house developed component depends (among others) on the development and testing effort required to deliver the component. On the other hand, the cost of a purchased component depends (among others) on its buying price and on the effort needed to adapt it to the working context.

The non-functional properties and the cost of a software unit are typically tied. Indeed, components and services with high quality value in general result to be the more expensive ones. Hence there is an intrinsic trade-off between the cost of a software product, which results from the costs of its elementary elements plus, e.g., the cost for component/service adaptation, and its quality that result from both the non-functional properties of its elementary elements and other characteristics such as the architecture of the system.

In general, the definition of architectural decision criteria based on non-functional properties is not easy. In fact, an elementary unit could be the best one with respect to a certain property, but at the same time it could be either too expensive or not compliant with possibly constraints on other non-functional properties.

Due to the complexity of addressing non-functional criteria in the architectural decision-making process, and given the extremely high number of parameters to consider in order to achieve a (near-) optimal decision, the introduction of quantitative methods and automatic tools would help the software engineers to raise their focus from a human-based search to a machine-based search.

Quantitative methods find their natural definition in the field of optimization. An optimization model allows, for example, to find a solution that minimizes the cost

---

[1]A deeper discussion on the peculiarities of the component selection activity within each phase of a development process can be found in Cortellessa et al. (2008).

of a software system while satisfying requirements that can be expressed as a set of mathematical constraints. Optimization techniques have been already proposed and used for the analysis of QoS tradeoffs of a software architecture (Aleti et al. 2013). In Sect. 14.2 we discuss this aspect.

In this chapter we focus on the optimization-based modeling and analysis of software architecture QoS tradeoffs. A particular emphasis will be given to two aspects of these tasks: (i) the mathematical foundations of QoS tradeoffs and their dependencies on the static and dynamic aspects of a software architecture, and (ii) the automation of the architectural decision-making process driven by QoS tradeoff optimization models.

In particular, we present a general optimization model that minimizes the total costs subject to constraints on the level quality of the software architecture. The model can be adopted in (specialized for) one of the lifecycle phases by leveraging available information and parameters, the level of detail of which obviously increases as the development progresses. Then, each specialized form of the general model can be either separately used and solved, if required in a certain lifecycle phase, or used in pipeline feeding with each other, as we will show in our example.

In the context of a waterfall development process, we implement three models: one for the architectural design (i.e. the software architecture driven model applicable before the release of a system), one for the implementation/deployment phase (we show how the QoS of a software architecture depends on the hardware architecture), and one for the maintenance phase (i.e. the software architecture driven model applicable after the release of a system).

In order to show the usefulness of our approach, we run these models on an example coming from the domain of medical information systems. We also study the sensitivity of the solutions to changes of parameters; we analyze, in particular, the behavior of the system costs at varying of non-functional requirements, see Potena et al. (2016).

Although here we describe the phases and interactions that fit well in a waterfall approach and show how our models can be employed in such a context, our approach is not limited to the waterfall design process only. Different paradigms can be considered, provided that the interactions between phases are properly taken into account. Indeed, the interactions between phases may change among different design approaches. For example, the interaction between the requirements and design phases will repeat when performed within agile, iterative or incremental development frameworks. In such cases the decision-making process would converge faster, e.g., due the know-how acquired and/or the activities performed in the previous iterations of the process. Also, in case of selecting new software units for new requirements, potential compatibility problems with existing units can be already recognized in the early phase of the process.

Our major contribution is to show how effectively optimization modeling techniques can capture relevant aspects of the architectural decision-making process in different lifecycle phases, thus representing a very relevant support for the software engineer's tasks.

All the proposed models belong to the class of mixed-integer nonlinear programming problems and therefore can be solved by means of exact and heuristic optimization techniques such as spatial branch-and-bound (Belotti et al. 2009) and tabu search. Although such problems are generally hard to solve due to non-linearities and integrality, they can be handled by common solvers (we used LINGO http://www.lindo.com in our computational assessment) since usually they are small for most of the common software domains. For large scale problems, however, search-based techniques, e.g., tabu search or genetic algorithms (Blum and Roli 2003), can be successfully adopted. Indeed, such techniques have been applied for obtaining solutions for several problems in the software engineering domain, from requirements and project planning to maintenance and re-engineering (Harman et al. 2012).

The chapter is organized as follows. In Sect. 14.2 we present related works and discuss the novelty of our contribution. In Sect. 14.3 the most common problems encountered for QoS tradeoffs analysis are discussed, and in Sect. 14.4 we introduce the general formulation of an optimization model for such kind of analysis. Section 14.5 describes the distributed medical informatics system adopted as example. Sections 14.6, 14.7 and 14.8 detail the optimization models and their application to the example for the architectural design, maintenance, and implementation/deployment phases, respectively. Finally, conclusions are delineated in Sect. 14.9. In Potena et al. (2016) we have collected all the further details that are not strictly necessary for this chapter understanding.

14.2 Related Work

A quite extensive collection of papers on decision-making processes across life-cycle phases and on methods/tools able to predict and evaluate the QoS of a software architecture can be found in literature. Decision-making frameworks have been introduced to facilitate the reasoning process for different goals and from different perspectives. For example, software architecture has been used for documenting and communicating design decisions and architectural solutions (Clements et al. 2011). However, being the focus of this chapter on the QoS tradeoffs' analysis of a software architecture, we report only papers that present similar criteria for this task. This helps us to clearly describe, at the end of this section, the novelty of this chapter with respect to the existing related work.

Several qualitative methods have been proposed in order to explicitly analyze the impact of architectural decisions on system quality, among which the well-known Architecture Tradeoff Analysis Method (Kazman et al. 1998) and Cost Benefit Analysis Method (CBAM) (see, for example, the survey in Breivold et al. 2012). These evaluation techniques suffer of some weaknesses that mainly are the subjective point of view of the analysts and the heavyweight process, which requires many steps and intense participation of stakeholders (Kim et al. 2007).

In order to overcome these limitations, qualitative attributes are transformed into quantitative figures, e.g., see the Multi-Criteria Decision Analysis (MCDA) technique that combines Analytic Hierarchy Process (AHP) and CBAM (Lee et al. 2009; Kim et al. 2007). Other common techniques such as AHP and Weighted Scoring Method (WSM) are used, for example, by component selection approaches (Kontio 1996). In particular, WSM estimates how to modify a software architecture, e.g. by introducing a different COTS component, with respect to a set of weighted criteria. The score of the change is calculated by the weighted sum of the criteria values. Alternatively, AHP suggests to define a hierarchy of criteria. Modification choices are compared in pairs and finally ranked on the basis of a score that combines the results of the comparison. Both the above methods come with serious drawbacks: the combinatorial explosion of the number of pair-wise comparisons, the need of extensive a priori preference information, and the highly problematic assumption of linear utility functions. Optimization techniques may solve some of these drawbacks because, in general, they do not need any weighting and/or ranking of the evaluation criteria (Neubauer and Stummer 2007).

Several research efforts have also been devoted in the last years to the designing of optimization methods for the analysis of software architectures (a quite extensive list of these approaches can be found in Aleti et al. 2013). Mostly depending on the lifecycle phase, different types of decisions and quality analysis methods are considered. Typically the decisions span from the service/component selection (e.g., Cardellini et al. 2012; Yang et al. 2009) through the deployment of components/services (e.g., Malek et al. 2012; Vinek et al. 2011) to the application of recurring software designs solutions[2] (e.g., Mirandola and Potena 2011). All these approaches basically provide guidelines to automate the search for an optimal architecture design based on the QoS tradeoffs.

The QoS tradeoffs analysis of such approaches basically is based on simple optimization models (see, e.g., Cortellessa et al. 2010) or multi-objective optimization models that, for example, maximize both reliability and performance (see, e.g., Cardellini et al. 2012). Different techniques are used to solve such optimization models, such as metaheuristic techniques, integer programming, or a combination of both (see, for example, surveys Harman et al. 2012; Aleti et al. 2013). For example, the work in Grunske (2006) shows how evolutionary algorithms and multi-objective optimization strategies, based on architecture refactorings, can be implemented to identify architecture designs, which can be used as an input for architecture tradeoffs analysis techniques.

Usually the goal of the existing approaches is to predict and/or analyze QoS attribute, like performance or reliability, starting from the architectural description of the system, or to select the architecture of the system, among a finite set of candidates, that better fulfill the required quality. In our previous works (Cortellessa et al. 2010; Potena 2013), we have addressed the problem of system quality from a

---

[2]They provide a generic solution to address issues pertaining to quality attributes, like the architectural tactics (Vinek et al. 2011).

different point of view: starting from the description of the system and from a set of new requirements, we devise the set of actions to be accomplished to obtain a new architecture. This is able to fulfill the new requirements with the minimum cost based QoS tradeoffs (i.e., reliability vs. availability, and vs. performance).

Other challenges related to the quality analysis are represented by the lots of different type of uncertainties that can be faced during the decision-making process. The specification of the effect of architectural decisions on goals (e.g., functional or non-functional requirements) is a difficult task. As a consequence, the process of making early architectural choices is a risky proposition mired with uncertainty (Esfahani et al. 2012). Several interesting approaches have been introduced in order to make the uncertainty explicit and using it to drive the production process itself (see, for example, Esfahani et al. 2012; Autili et al. 2011; Ghezzi et al. 2013) some of which are detailed below. In particular, for the design time (early phases of the software development process), the GuideArch framework (Esfahani et al. 2012) guides the exploration of alternative architectures under uncertainty by exploiting fuzzy mathematical methods. GuideArch allows to compare alternative architectures with respect to system's properties (like cost and battery usage). The ADAM (Adaptive Model-driven execution) framework (Ghezzi et al. 2013), based on probability theory and probabilistic model checking, supports the development and execution of software that tolerates manifestations of uncertainty by self-adapting to changes in the environment, trying to do its best to satisfy certain non-functional requirements (i.e., response time and the faulty behavior of components integrated in a composite application).

Research efforts have also been spent in order to deal with parameters' uncertainty (Doran et al. 2011; Meedeniya et al. 2012; Wang et al. 2012; Wiesemann et al. 2008). In particular, in Meedeniya et al. (2012), a robust optimization approach allows to deal with the impact of inaccurate design-time estimates of parameters. A Bayesian approach has been introduced in Doran et al. (2011), in order to systematically consider parametric uncertainties in architecture-based analysis. In Wang et al. (2012), the propagation of a single parameter's uncertainty on the overall system reliability estimation is analyzed. Finally, in Wiesemann (2008), the stochastic programming is exploited to support the service composition under quality attributes tradeoffs. In particular, the service composition problem is formulated as a multi-objective stochastic program which simultaneously optimizes some quality-of-service parameters (i.e., workflow duration, service invocation costs, availability, and reliability).

The originality of this chapter mainly consists in showing how effectively optimization modeling techniques can capture relevant aspects of the architectural decision making process in different lifecycle phases, thus representing a very relevant support for the software engineers' decisions. Our overall approach of embedding optimization models for different lifecycle phases is, at the best of our knowledge, the first example of an integrated framework for supporting developers' decisions based on cost/QoS tradeoffs during the whole software development process. Moreover, our optimization models are not tied to any particular development process as well as they do not depend on the specific application domain.

14.3 Typical Problems of QoS Tradeoffs Modeling

There are some limitations in the analytical formulation of non-functional aspects of components/services-based software systems mostly due to the intrinsic complexity of the component/service inter-relationships. Here below we summarize the major points.

In general, the quality attributes (such as response time and availability) depend on many observable parameters (such as size of messages exchanged, number of function points, etc.) that might be tightly correlated to each other. Some assumptions are typically made in order to keep as simple as possible the model formulation. For example, most reliability models for systems composed by basic elements (e.g. objects, components or services Becha and Amyot 2012; Goseva-Popstojanova and Trivedi 2001; Immonen and Niemelä 2008; Krka et al. 2009) assume that the elements are independent, namely the models do not take into account the dependencies that may exist between elements. They assume that the failure of a certain element provokes the failure of the whole system. What is basically neglected under this assumption is the error propagation probability, which in several real domains (such as control systems) is not an issue, because component/service errors are straightforwardly exposed as system failures. In order to relax such an assumption, an error propagation model must be introduced (see, for example, the reliability model for service-based systems introduced in our previous work Cortellessa and Potena 2007).

Also the non-functional properties are tightly correlated, and often depend on each other. In fact, some conflicts could exist among quality attributes (Boehm and In 1996), e.g., suitable tradeoffs between modifiability and performance have to be provided while building a software architecture, as remarked in Lundberg et al. (1999).

Sometimes the providers of pre-existing components/services are not able to come up with the exact values of some non-functional properties, and simply get a set of ranges over which the values may lie. For example, the component reliability for a given component cost is usually specified over a range based on prior experience (Gokhale 2007). If only ranges are available, then optimization can be performed on a parametric model, i.e., a model with some parameters ranging within provided limits, in order to observe the trend and sensibility of solutions.

In other cases, the information provided by vendors are not enough to estimate the non-functional properties of a given component/service since some of its parameters (e.g. cost or reliability) may be characterized by a not negligible uncertainty. In the case of component reliability, the propagation of such uncertainty is analyzed by Goseva-Popstojanova and Kamavaram (2004), and Dai et al. (2007). However, it was out of the scope of this chapter to deal with this kind of sensitivity analysis.

The reliability estimation methods typically deal with the operational profile (Musa 1993; Chandran et al. 2010) which is another factor that brings uncertainty in QoS analysis. In fact, the operational profile of the system is in general different from the one adopted to estimate the non-functional properties of elementary components/services. As remarked in Becker and Koziolek (2005), no standard model are available for describing the operational profile and hence it is necessary to take into account the transformations that the components may provide on it. "Inputs on the provided interfaces of a component are transformed along the control flow down to the required interfaces. Thus, the provided interfaces of subsequent components connected with the required interfaces receive a different operational profile than the first component. The transformations form a chain through the complete architecture of components until the required interfaces of components only execute functions of the operating system or middleware" (Becker and Koziolek 2005). However, if the operational profile of the system is not (fully) available at the design phase, the domain knowledge and the information provided by the software architecture in general are sufficient for estimating it, as suggested in Roshandel and Medvidovic (2007) or in Musa (1993).

The integration of components/services often entails mismatches whose handling cost should be included into the QoS tradeoffs modeling. Several approaches have been introduced to deal with the mismatches problems (e.g., see Park 2006; Younas et al. 2005 for the integration of web services in distributed system). For solving a mismatch between a requirement and a pre-existing software unit, different actions are possible, and different existing works could be exploited, such as the approach presented in Mohamed et al. (2007), which supports the resolution of mismatches during and after a COTS selection process by using an optimization model.

As far as concerns the non-functional requirements, the task of handling mismatches between the properties of single components/services and the quality required for the whole system is even harder than one for the functional mismatches, e.g., sometimes the improvement of a single software unit could not affect the quality of the whole system. Clearly, closed formulas for estimating the quality of the system as a function of the properties of components/services would be very helpful, but many problems have to be faced for defining them.

14.4 A General Formulation for Architectural Decisions Versus Quality

In this section we propose a general optimization model that helps developers to make the QoS tradeoffs analysis of a software architecture.

Let $S = \{u_1, \ldots, u_n\}$ be a software architecture made of n software units $u_i \, (1 \le i \le n)$ the composition of which results in services that the system offers to users.

Since the proposed model may support different lifecycle phases, we adopt a general definition of software unit: it is a self-contained deployable software module containing data and operations, which provides/requires services to/from other elementary elements. A unit instance is a specific implementation of a unit.[3] For each unit u_i, let J_i be the set of instances available by vendors and $\bar{J}_i$ the set of possible options for developing the instance in-house. Let u_{ij} be the jth instance of $J_i \cup \bar{J}_i$.

The analysis of the QoS tradeoffs is a broad decision-making process that consists of a set of actions aiming to modify the static and dynamic structure of the software architecture. The decisions within the different life-cycle phases are basically related to the following software actions:

1. *Introducing new software units*: One or more new software units may be embedded into the system.[4] We call *NewS* the set of new available software units that can provide different functionalities.
2. *Replacing existing unit instances with functionally equivalent ones available on the market*: The employed instance u_{ik} of a software unit u_i may be replaced with an element of the set J_i, i.e., with of the instances available for it on the market (e.g. a Commercial-Off-The-Shelf (COTS) component/web service). We assume that all the instances in J_i are functionally compliant with u_{ik}, i.e., each of them provides at least all services provided by u_{ik} and requires at most all services required by u_{ik}.[5] The instances in J_i may differ from u_{ik} for cost and quality attribute (e.g. reliability and response time).
3. *Replacing existing unit instances with functionally equivalent ones developed in-house*: An existing instance of a software unit u_i may be replaced with one developed in-house. Developers could opt for different building strategies resulting in different in-house instances, i.e., the elements of the set $\bar{J}_i$. The values of quality attributes of such optional instances (e.g., reliability, response time) could vary due to the values of the development process parameters (e.g. experience and skills of the developing team).
4. *Modifying the interactions among software units in a certain functionality*: The system dynamics may be modified by introducing/removing interactions among software units within a certain functionality.

Clearly, the system quality heavily depends on the hardware features, e.g., response time decreases as the processing capacity improves, and therefore decisions on software architecture must also take into account the decisions on the hardware characteristics of the system. Hardware decisions typically span from the deployment of software units on hardware nodes through to modify the characteristics of the underlying hardware resources (e.g., CPU, disk, memory, network

---

[3]The optimization model can work for any semantics given to software units under the condition that the parameters are associated to the correct units. The only difference, of course, is in the techniques needed to estimate the model parameters, but this is out of the scope of this chapter.

[4]Notice that such type of action has to be associated to another action that indicates how this unit interacts with existing units, therefore it modifies the interactions within certain functionalities (see last type of software action).

[5]As remarked in Cortellessa et al. (2010), such an assumption could be relaxed by introducing integration/adaptation costs.

throughput, etc.) to introducing/removing connection links among hardware nodes.[6] Indeed, depending on the adopted engineering paradigm (e.g., CBSE or SOSE), different types of hardware changes may be performed. For example, as explained in Mirandola and Potena (2011), in the SOA domain, due to the fact that the services are not acquired in terms of their binaries and/or source code, but they are simply used while they run within their own execution environment (that is not necessarily under the control of the system using them), hardware changes can be suggested by the service providers.

Optimization Model Formulation

All the above actions can be modeled by decision variables that describe the software architecture instances selection process. In particular, let x_{ij} $(1 \leq i \leq n, j \in J_i \cup \bar{J}_i)$ be the binary variable that is equal to 1 if the instance j is chosen for the software unit i, and 0 otherwise. Moreover, let z_h $(1 \leq h \leq |NewS|)$ be the binary variable that is equal to 1 if the new software units h is chosen and 0 otherwise.

Let us suppose to analyze the system on the base of p quality attributes (such as cost, response time, availability, etc.). Suppose moreover that each attribute of any software unit depends on the value of parameters α_i^k's, β_i^k's, and γ_{ij}^k's, where (i) the vector α_i^k describes the (at most) u software architecture observable parameters, e.g., the average number of invocations of a software unit within the execution scenarios considered for the software architecture, (ii) the vector β_i^k contains the (at most) v hardware observable parameters, e.g., the processing capacity of the node hosting the software unit, that is measured, for example, as the average number of instructions per second that the source can execute, and (iii) the vector γ_{ij}^k represents the (at most) w features of the implementation of u_i, e.g., the reliability of the instance used for replacing the existing unit. For the k quality attributes of a provided instance, the value of the features γ_{ij}^k's is assumed to be either given from the software unit provider or estimated from the customer. On the contrary, for an in-house developed instance the γ_{ij}^k's can be predicted by considering variables of the decision planning. For example, in Sect. 14.6, we express the reliability of an in-house instance as a function of a variable representing the amount of testing N_i^{tot} to be performed on that instance.

Let $\Gamma_k : \mathbb{R}^u \times \mathbb{R}^v \times \mathbb{R}^w \to \mathbb{R}$ $(\bar{\Gamma}_k : \mathbb{R}^u \times \mathbb{R}^v \times \mathbb{R}^w \to \mathbb{R})$ be the function that, on the base of the above parameters, returns the value of the kth quality attribute $(1 \leq k \leq p)$ of an existing (new) software unit. In particular, let $\Lambda_{ij}^k = \Gamma_k \left(\alpha_i^k, \beta_i^k, \gamma_{ij}^k \right)$ the value of the kth attribute of the provided/in-house instance u_{ij}.

For sake of readability, we introduce here a formulation without correlations among Γ_k's, where each quality attribute does not affect other attributes and a self-contained analytical expression can be formulated for it. Obviously this is not always true, as it depends on the considered quality attributes and the model complexity. If

---

[6]A deeper discussion on the hardware changes can be found in Mirandola (2011).

quality attributes have to be correlated (Bass et al. 2002) (e.g., when performability is considered) then additional constraints may be needed, which can be expressed as *contingent decisions* (Jung and Choi 1999).

We can represent the value of the *k*th quality attribute of the *i*th existing software unit as a function of the decisional strategy **x**:

$$\theta_i^k = \sum_{j \in \bar{J}_i \cup J_i} \Lambda_{ij}^k x_{ij} \tag{14.1}$$

Similarly, we can represent the value of the *k*th quality attribute of the *h*th new software unit as a function of the decisional strategy **z**:

$$\bar{\theta}_h^k = z_h \bar{\Gamma}_k \left(\alpha_i^k, \beta_i^k, \gamma_{ij}^k \right) \tag{14.2}$$

Let $G_k : \mathbb{R}^n \times \mathbb{R}^{|News|} \to \mathbb{R}$, with $1 \leq k \leq p$, be the function that returns the *k*th quality attribute of the whole system on the base of the same attributes of each existing/new software unit. And let us assume (without loss of generality) that the values of each quality attribute *k* are constrained to be above a lower threshold value Θ^k. Assume, moreover, that the cost is the first quality attribute, i.e., θ_i^0 ($\bar{\theta}_i^0$) express the cost of the existing (new) software units. Finally, let $Cost : \mathbb{R}^n \times \mathbb{R}^{|NewS|} \to \mathbb{R}$ be the cost function of the whole system that clearly depends on the costs of all the existing (new) software units. Different cost models could be used to define *Cost*, e.g., it may also include the potential costs of software unit adaption (i.e. the glue ware). For the sake of readability, we introduce here a formulation without correlation between the software unit costs and the other software/hardware quality attributes.

The general formulation of the optimization model for the QoS tradeoffs analysis is given by:

$$\min_{\mathbf{x}, \mathbf{z}} Cost(\theta^0, \bar{\theta}^0) \tag{14.3}$$

s.t.

$$G_k(\theta^0, \bar{\theta}^0) \geq \Theta^k \quad \forall k = 1 \dots p$$

$$\sum_{j \in \bar{J}_i \cup J_i} \Lambda_{ij}^k x_{ij} = \theta_i^k \quad \forall k = 1 \dots p, \quad \forall i = 1 \dots n$$

$$z_h \bar{\Gamma}_k \left(\alpha_h^k, \beta_h^k, \gamma_h^k \right) = \bar{\theta}_i^k \quad \forall k = 1 \dots p, \quad \forall h = 1 \dots |NewS|$$

$$\sum_{j \in \bar{J}_i \cup J_i} x_{ij} = 1 \quad \forall i = 1 \dots n$$

$$x_{ij} \in \{0,1\} \quad \forall i = 1\ldots n, \quad \forall j = 1\ldots p$$

$$z_h \in \{0,1\} \quad \forall h = 1\ldots|NewS|$$

Other constraints (e.g., equations to predict α_i^k's and β_i^k's).

14.5   An Example: A Distributed Medical Informatics System

In this section we describe the main features of an example that we will use for illustrating the application of our approach (see Sects. 14.6, 14.7 and 14.8). For sake of readability, a description of the high-level structure of the system, together with all the details on the models, i.e., the meaning of additional parameters and constraints and on the computational results, is available in Potena et al. (2016).

We have considered the distributed medical informatics system described by Yacoub et al. (1999) mainly because its features allow us to show how effectively optimization modeling techniques can capture relevant aspects of the architectural decision making process in different lifecycle phases. Shortly, medical institutions need in general to exchange information, e.g., medical images, between each other. Actually, they form a client/server system where the *AE Client* subsystem is connected to the *AE Server* subsystem by the *Network* subsystem. The communication between the entities of the system is performed using Digital Imaging and Communication in Medicine (DICOM) standard,[7] which is typically used for producing, processing and exchanging medical images: "The DICOM specifies the transport and presentation layer for a network protocol as DICOM Upper Layer (*DICOM UL Client and Server* subsystems)" (Yacoub et al. 1999).

In the following sections, we will analyze the three scenarios identified by Yacoub et al.: We will consider *AE Client, Network, AE Server, DICOM UL Client* and *Server* subsystems as architectural elementary elements of the system. Moreover, we will suppose that *Network* subsystem does not identify all the network, but a component which is deployed along the network.

14.6   Architectural Design Phase

14.6.1   *Before Release (Platform Independent)*

For the design phase, the general optimization model (14.3) is instantiated with a mathematical formulation that stems from our previous work in the context of component based software (Cortellessa et al. 2006). Specifically, we consider the

---

[7]http://medical.nema.org/.

following architectural decisions: (i) replacing existing unit instances with functionally equivalent ones available on the market, and (ii) replacing existing unit instances with functionally equivalent ones developed in-house.

We report the model formulation by plugging the problem in a general application domain, where the build-or-buy decisions refer to general software unit rather than components. Additional constraints on delivery time and reliability of the system are considered, and decision planning variables associated to the amount of testing to be performed on each in-house instance are introduced.

Our model definition makes the following significant assumptions. (i) We assume that the pattern of interactions within each scenario does not change by changing the software unit instance. (ii) We only consider the sequential execution of the software units, and we assume that the units communicate by exchanging synchronous messages. (iii) From a reliability viewpoint, we suppose that the software units are independent, namely we assume that the failure of a unit provokes the failure of the whole system. We only consider crash failures that are failures that (immediately and irreversibly) compromise the behavior of the whole system. Besides, we suppose that a unit shows the same reliability across different invocations. (iv) We assume that the operational profile of the system is the same one used for certifying the component. (v) Finally, we assume that sufficient manpower is available to independently develop in-house unit instances. Note that the above assumptions are shared with most of the models in this domain, as discussed in Sect. 14.3.

Let us suppose to be committed to assemble the system by the time T while ensuring a minimum reliability level R and spending the minimum amount of money. Let N_{ij}^{tot} be the integer variable representing the total number of tests performed on the in-house developed instance j of the ith unit.[8] Figure 14.1 summarizes the parameters and the expressions used in the model formulation. Specifically, (i) the development cost and the delivery time of an in-house instance are computed by considering the development time, the testing time and the number of tests. (ii) The reliability of the whole system can be obtained as a function of the probability of failure on demand of its elementary elements. In particular, the expression of the system reliability reported in Fig. 14.1 is the probability of a failure-free execution of the system, and hence the reliability constraint is $RelSyS \geq R$. (iii) The delivery time constraints can be expressed as $DT_1 \leq T \ldots DT_n \leq T$.

Experimenting the model on an example

In order to show the practical usefulness of the model, we apply it to the example presented in Sect. 14.5.

Figure 14.2 reports a synthesis of the results obtained by solving the optimization model with different values of T and R. The former spans from 4 to 30 whereas the latter from 0.89 to 0.99.

---

[8]The effect of testing on cost, reliability and delivery time of provided units is instead assumed to be accounted in the parameters.

---

1. Model Parameters

s_i average number of invocations
μ_{ij} probability of failure on demand of the provided instance j
c_{ij} cost of the provided instance j
d_{ij} delivery time of the provided instance j
τ_{ij} average time to perform a test case on the in-house instance j
p_{ij} probability that the in-house instance j is faulty
π_{ij} testability of the in-house instance j
$\bar{c}_{ij}$ unitary development cost of the in-house instance j
t_{ij} estimated development testing time of the in-house intance j

N_{ij}^{suc} number of successful (i.e. failure-free) tests performed on the in-house j

$N_{ij}^{suc} = (1 - \pi_{ij})N_{ij}^{tot}$

2. Cost Objective Function:

$$COF = \sum_{i=1}^{n}\left(\sum_{j \in J_i} \bar{c}_{ij}(t_{ij} + \tau_{ij}N_{ij}^{tot})x_{ij} + \sum_{j \in J_i} c_{ij}x_{ij} \right)$$

3. System Reliability :

$$RelSys = \prod_{i=1}^{n} e^{-\left(\sum_{j \in \bar{J}_i} \theta_{ij}s_i x_{ij} + \sum_{j \in J_i} \mu_{ij}s_i x_{ij} \right)}$$

4. Probability of failure on demand of the jth in-house developed instance:

$$\theta_{ij} = \frac{\pi_{ij} \cdot p_{ij}(1 - \pi_{ij})^{N_{ij}^{suc}}}{(1 - p_{ij}) + p_{ij}(1 - \pi_{ij})^{N_{ij}^{suc}}}$$

5. Delivery time of the software unit i :

$$DT_i = \sum_{j \in \bar{J}_i}(t_{ij} + \tau_{ij}N_{ij}^{tot})x_{ij} + \sum_{j \in J_i} d_{ij}x_{ij}$$

Fig. 14.1 Design phase: parameters and cost, reliability and delivery time expressions

As expected, the total cost of the application decreases for the same value of the reliability bound R and increasing values of the delivery time limit T. On the other hand, for the same value of T the total cost decreases while decreasing the reliability bound R (i.e. less reliable application required).

As shown in Potena et al. (2016), the model tends to select in-house instances for increasing values of T because they become cheaper than the available provided instances. The total cost decreases while T increases because it is possible to

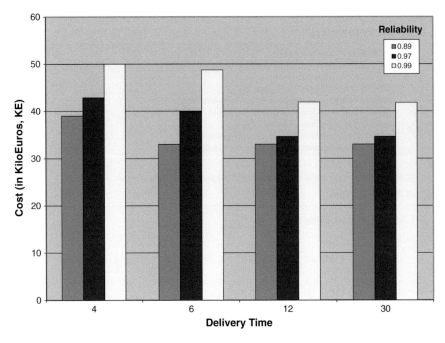

Fig. 14.2 Solutions for the design phase

increase the amount of testing to perform. The in-house instances remain cheaper than the corresponding provided instances even in cases where a non negligible amount of testing is necessary to make them more reliable with respect to the available provided instances.

In this example, the in-house instances result cheaper than the provided instances, but real situations may be different. In fact, an in-house unit could be built by adopting different strategies of development. Therefore, its values of cost, reliability and delivery time could vary due to the values of the development process parameters (e.g. experience and skills of the developing team). In Potena et al. (2016) we also study the sensitivity of the model to changes in its parameters (we analyze, in particular, the behavior of the system costs at varying of non-functional requirements).

14.7 Maintenance Phase

14.7.1 After Release (Platform Independent)

In this section, we instantiate the general optimization model (14.3) for supporting the maintenance phase. Specifically, we show how an optimization model can support the software unit replacement maintenance activity for overcoming an *unexpected* system failure. *Unexpected* means that, on the basis of the certified reliability of the elementary software units, a failure shall not occur so early. Under the assumption that exactly one faulty software unit is present in the system, the proposed optimization model aims to maintain the system by suggesting how to reconfigure it. After a software failure occurs, our approach searches for a different system configuration (e.g. by replacing a (some) unit(s)) that minimizes the costs while raising the system reliability by a fair amount that (hopefully) allows in future to avoid unexpected failures. Indeed, the model solution may suggest either to replace a faulty software unit by a provided instance or to perform on the faulty software unit an additional number of test cases if it has been developed in-house.

The mathematical formulation, similar to that described in Sect. 14.6, has been presented in Cortellessa and Potena (2009) in the context of component-based software. In this chapter we plug the model in a general application domain, where the decisions refer to general software units rather than components.

Let S be the software architecture of a deployed system that has been assembled following the architectural approach presented in Sect. 14.6. In particular, let $(\bar{x}, N^{tot})$ be the description of the instances chosen to build S at minimum cost while assuring (among others) a system reliability greater than the threshold R. For sake of readability, suppose that the possibly in-house built instance for the software unit i is included in the set J_i (and therefore $\bar{x}_{i0} = 1$ means that the ith software unit has been developed inhouse). Moreover, assume that an *unexpected* system failure occurs and that no specific monitoring action is devised to identifying the faulty unit originating the failure.

Let R' be the new reliability threshold required for the whole system (i.e. $R' > R$) and T' be the time limit for this maintenance action to be completed.

Given the current solution $(\bar{x}, N^{tot})$, let $NTest_i$ $(\forall i = 1, \ldots n)$ be the number of test cases required for the unit i in order to satisfy the new reliability threshold R'. The number ΔN_i of possible additional test cases to be performed on the ith unit is given by $\Delta N_i = max\{0, Ntest_i - N_i^{tot}\}$. Since the system has been already assembled, new costs incur only if additional tests are performed on in-house instances, i.e., $\Delta N_i > 0$, and/or existing instances are replaced by new instances bought by vendors, i.e., $\bar{x}_{ij} = 1$ and $x_{ij} = 0$. The latter case can be modeled by introducing a new binary variable $y_{ij} \geq x_{ij} - \bar{x}_{ij}$. Differently from the model presented in Sect. 14.6, the objective function and the constraints of the maintenance model take into account only such kind of costs, see Fig. 14.3.

Experimenting the model on an example

In order to show the practical usefulness of the model, we apply it to the example presented in Sect. 14.5. In particular, among the results of the architectural design phase (see Sect. 14.6), we picked the system configuration $[u_{11}, u_{21}, u_{32}, (u_{40}, 128), u_{51}]$ corresponding to the case $(T = 4, R = 0.97)$. Here, $(u_{40}, 128)$ means that the fourth software unit has been built in-house and 128 test cases has been performed on it.

Figure 14.4 reports the results obtained from solving the optimization model for different values of T' and R'. Each bar represents the minimum cost for a given

1. Model Parameters

s_i	average number of invocations
μ_{ij}	probability of failure on demand of the provided instance j
c_{ij}	cost of the provided instance j
d_{ij}	delivery time of the provided instance j
τ_i	average time to perform a test case on the in-house instance of the unit i
$\bar{c}_i$	unitary development cost of the in-house instance of the unit i
t_i	estimated development testing time of the in-house instance of the unit i
$(\bar{\mathbf{x}}, \mathbf{N}^{tot})$	description of the assembled system
	$\Delta N_i = \max\{0, Ntest_i - N_i^{tot}\}$

2. Cost Objective Function:
$$COF = \sum_{i=1}^{n}\left(\bar{c}_i t_i y_{i0} + \bar{c}_i \tau_i \Delta N_i + \sum_{j=1}^{|J_i|} c_{ij} y_{ij}\right)$$

3. System Reliability:
$$RelSys = \prod_{i=1}^{n} e^{-\left(\theta_i s_i x_{i0} + \prod_{j=1}^{|J_i|} \mu_{ij} s_i x_{ij}\right)}$$

4. Delivery time of the software unit i:
$$DT_i = t_i y_{i0} + \tau_i \Delta N_i + \sum_{j=1}^{|J_i|} d_{ij} y_{ij}$$

Fig. 14.3 Maintenance phase: cost, reliability and delivery time expressions

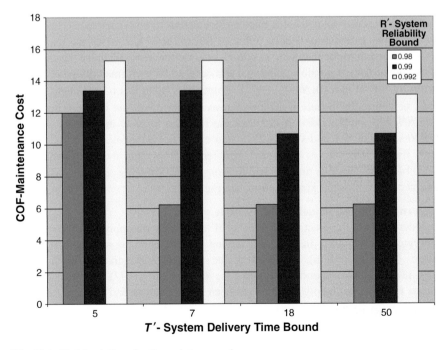

Fig. 14.4 Model solutions for the maintenance phase

value of the delivery time bound T' and a given value of the reliability bound R'. The former spans from 5 to 50 whereas the latter from 0.98 to 0.992.

As expected, the maintenance cost of the system increases for given T' and increasing R'. However, for the same value of R' the cost decreases while increasing T' which means that a larger availability of time helps to reduce maintenance cost.

The model suggests restructuring the system by working on the second and fourth software units: in some cases it suggests to perform additional testing on the fourth unit, while in all cases it argues to replace the second software unit with either its in-house instance or with its second or third provided instance available.

If we increase the value of R' to 0.995 and set T' = 18, then the model provides the solution $[u_{11}, u_{23}, u_{32}, (u_{40}, 452), u_{52}]$, with a maintenance cost equal to 26.268 KE and a system reliability equal to 0.996227. In this case the model suggests replacing also the fifth unit. If it would keep the first provided instance for the fifth unit (i.e. if the fifth software unit would not be replaced), the reliability constraint would be not satisfied. In fact, the system reliability would be equal to 0.992548.

In Potena et al. (2016) we study the sensitivity of the model to changes in its parameters. We also show how, under no-monitoring assumptions and in case a monitoring action allows identifying the faulty software unit, the model can leverage the approach to overcome an *unexpected* failure of a system, see Cortellessa and Potena (2009).

14.8 Implementation/Deployment Phase

In this section, we instantiate the general optimization model (14.3) in order to support the activities of the implementation/development phase. In particular, we show how changes in the hardware features may affect the system quality and therefore the software decisions. As in the previous phases, the model's solution describes the instances to choose for build up a minimum cost software architecture that satisfies reliability and performance constraints. In addition, the model of the deployment phase also suggests the hardware nodes on which the software unit shall be deployed.

The mathematical formulation makes the following significant assumptions. (i) We assume that an UML Sequence Diagram (SD) describes the dynamic of each available functionality in terms of interactions that take place between software units (however, multiple Sequence Diagrams could be lumped by using the methodology suggested in Uchitel et al. 2003). (ii) The communication between two components co-located in the same node is assumed totally reliable, because it does not use any hardware links. (iii) Finally, we make all the assumptions of the model that we have introduced for the architectural design phase (see Sect. 14.6).

Let H be the set of hardware nodes on which the software units can be deployed, and L the set of (uni-directional) network links between hardware nodes. A link implements a connector between components deployed on different hardware nodes.

Additional binary variables d_{ik} $(i \in S, k \in H)$ and $h_{ii'}^l$ $(l \in L, i, i' \in S)$ are needed to describe how to deploy software units on hardware nodes and how connect-software units to each other. In particular, (i) d_{ik} is equal to 1 if the node k is chosen for software unit i, and 0 otherwise, and (ii) $h_{ii'}^l$ is equal to 1 if the link l is chosen to connect the software units i and i', and 0 otherwise. Each software unit i must be deployed on exactly one node k, i.e., $\sum_{k \in H} d_{ik} = 1, \forall i \in S$, and a path must exist between the components i and i' if a call exists between them. The latter condition can be easily expressed as network flow constraints. Also constraints on the capacity of the nodes and the bandwidth of the network links have to be considered, see Potena et al. (2016) for details.

Assume that the performance of the system is measured in terms of calls' response time, and that a maximum threshold $ResT$ has been given. The response time RT_f of the functionality f can be obtained as a function of the processing time and the network time, see Fig. 14.5. In a worst-case scenario, all the functionalities should satisfy the performance threshold, hence the constraints $RT_1 \leq ResT \ldots RT_{|F|} \leq ResT$ have to be included in the formulation. Alternatively, in an average-case scenario, the response time RT of the whole system can be computed in terms of arrival rate λ_f of the calls for the fth functionality as $RT = \sum_{f \in F} \frac{\lambda_f}{\sum_{i \in F} \lambda_i} RT_f$, and therefore the performance constraint can be simply expressed as $RT \leq ResT$.

The evaluation of the reliability of each functionality, see Fig. 14.5, takes into account that two software units may be connect from a path of more than one link. Note that the communication between two software units co-located in the same node is assumed totally reliable, because it does not use any hardware link. Again, in a worst-case scenario, the constraints $REL_f \geq R$ $(f \in F)$ must be considered, whereas in an average-case scenario, the reliability of the system is

$$REL = \sum_{f \in F} \frac{\lambda_f}{\sum_{i \in F} \lambda_i} REL_f \text{ and the reliability constraint is } REL \geq R.$$

Experimenting the model on an example

In this section we conclude the example presented in Sect. 14.5.

Since in general the implementation phase takes place between the architectural design and the deployment, at the deployment time no real distinction, for sake of modeling, needs to be made between in-house and provided instances. This is why the in-house instances indicated by the model solution of the architectural design phase in the scenario $(T = 30, R = 0.99)$ are now simply considered as possible provided instances. The hardware architecture consists of three hardware nodes, see Potena et al. (2016) for details on the model parameters.

Figure 14.6 reports the results provided by the optimization model by setting the probability of failure of the links to a value between 0.00001 and 0.0004, the processing speed of the links to 200 bits/s (measured as the average number of bits per second), the arrival rate for a service provided by the system to 1, and the reliability required to 0.97 and 0.99. Two configurations of the processing capacities of the nodes (measured as the average number of instructions per second, ips)

1. Model Parameters

c_{ij}	cost of the provided instance j
I_f	set of software unit involved in the f-th scenario
θ_{ij}	probability of failure on demand of the provided instance j
φ_l	probability of failure on demand of the l-th link
bp_{if}	number of busy periods that the unit i shows in the SD f
$\|Interact(i,i',f)\|$	number of interactions that the units i and i' exchange in the SD f
TS_{ij}	task size of the provided instance j
PC_k	processing capacity of the hardware node k
$MS_{ii'}$	average size of an exchanged message between unit i and i'
PS_l	processing speed of the link l

2. Cost Objective Function:

$$COF = \sum_{i=1}^{n}\left(\sum_{j\in J_i} c_{ij}x_{ij}\right)$$

3. Reliability of the f-th system functionality :

$$REL_f = \prod_{i\in I_f}\left(\sum_{j\in J_i}x_{ij}(1-\theta_{ij})^{bp_{if}}\cdot\prod_{l\in L}\left(\prod_{i'\in I_f}(1-\varphi_l)^{|Interact(i,i',f)|h^l_{ii'}}\right)\right)$$

4. Response time of the f-th system functionality:

$$RT_f = \sum_{k\in H}\sum_{i\in I_f}bp_{if}d_{ik}\left(\sum_{j\in J_i}\frac{TS_{ij}}{PC_k}x_{ij}\right)+\sum_{l\in L}\left(\sum_{i,i'\in I_f}\left(|Interact(i,i',f)|h^l_{ii'}\frac{MS_{ii'}}{PS_l}\right)\right)$$

Fig. 14.5 Implementation/deployment phase: cost, reliability and performance expressions

have been considered: the first with 60, 80, and 90 ips for the first, second and third node, respectively; the second with 50, 80, and 50 ips.

As expected, for a given configuration of processing capacities of the hardware nodes and for the same value of the probability of failure of the links, the cost decreases while decreasing the reliability required for the system. On the other hand, for the same values of reliability and probability of failure of the links, the second configuration of processing capacities requires a more expensive solution.

The deployment of the software unit could change as the probability of the failure of the links varies, even when the total cost of the system remains unchanged. Indeed, in some cases it is not possible to deploy the software unit in the same way, because this does not guarantee the reliability threshold of the

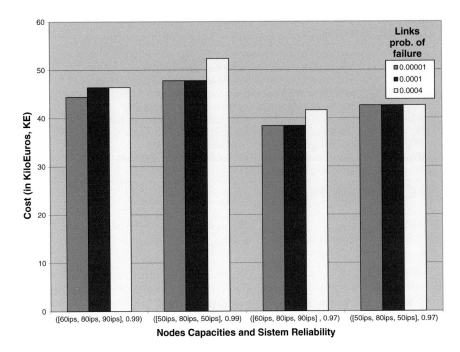

Fig. 14.6 Model solutions for the implementation/deployment phase

system. For example, for the scenario (([60, 80, 90], 0.99), 0.0001) the model suggests a configuration of nodes that is different from the one suggested for the scenario (([60, 80, 90], 0.99), 0.0004). In fact the reliability achieved with the former configuration of nodes would be equal to 0.98853 with the probability of failure of the links fixed to 0.0004. In other cases, it is possible to deploy the software units on the hardware nodes in the same way. For example, the configuration of nodes that the model returns for the scenario (([50, 80, 50], 0.99), 0.00001) is optimal also for the scenario (([50, 80, 50], 0.99), 0.0001). In fact, the reliability achieved with the former configuration would be equal to 0.99099 with the probability of failure of the links fixed to 0.0001.

Therefore, the probability of failure of the links, that would have not emerged during the architectural design phase where the information of the links (i.e. the hardware architecture) is not taken into account (see Sect. 14.6), may sensibly affect the reliability of the system. As we have remarked in Sect. 14.1, the QoS prediction gets more accurate while progressing in the development process because more knowledge is available about the features of the system. In Potena et al. (2016) we also study the sensitivity of the model to changes in its parameters.

14.9 Conclusions

In this chapter, we have showed how optimization models can be of support for the architectural decision-making process based on QoS tradeoffs along the whole software lifecycle. We have focused on the architectural design, the implementation/ deployment, and maintenance phases, and for each phase we have introduced an optimization model that supports the decisions on the basis of the available knowledge in the specific phase. We have merged the three models in the same approach, and we have shown the usefulness of our approach by applying it to the same example in the domain of medical information systems.

The work presented in this chapter is the result of our research effort in the last years. As we report here below, besides the models formulation, we have built software tools to support the automated model generation and solution. Basing on this experience we can assert that optimization modeling is a very promising approach to formulate certain problems in the field of software quality analysis. This is especially true in cases where decisions have to be made among different alternatives that may lead to different software costs.

The most evident limitation of such approaches nowadays is the necessity to express objective functions as well as constraints in closed mathematical formulas. This is not trivial for many non-functional properties and scenarios. In addition, with the increasing complexity of software systems based on components/services, the size of these models can sensibly grow. This aspect leads to prefer heuristic search-based techniques to exact optimization tools.

Therefore we devise for the near future the necessity to work in the definition of closed mathematical formulas for different quality attributes. Beside this, we also intend to work on relaxing the model assumptions that we have introduced throughout this chapter. In particular a quite relevant aspect to work on is represented by the dependencies among different quality attributes and among parameters within the same optimization model. In this direction, we also intend to investigate the use of search based techniques, such as metaheuristics, and the multi-objective optimization for solving large scale models.

For the model for architectural design phase we have already provided the tool, called CODER (Cost Optimization under DElivery and Reliability constraints) (Cortellessa et al. 2006), which generates and solves the model automatically. We are also designing an integrated tool, based on our optimization models that may assist software designers during the whole software life cycle. It would be interesting to embed such a tool into a CASE tool, for example the one presented in Cancian et al. (2007), for supporting and automating the development of a component-based system.

Acknowledgments This work has been partially supported by the VISION European Research Council Starting Grant (ERC-240555), and by European Commission funding under the 7th Framework Programme IAPP Marie Curie program for project ICEBERG no. 324356.

References

Aleti, A., Buhnova, B., Grunske, L., Koziolek, A., Meedeniya, I.: Software architecture optimization methods: a systematic literature review. IEEE Trans. Software Eng. **39**(5), 658–683 (2013)

Autili, M., Cortellessa, V., Ruscio, D.D., Inverardi, P., Pelliccione, P., Tivoli, M.: EAGLE: engineering software in the ubiquitous globe by leveraging uncErtainty. In: SIGSOFT FSE, pp. 488–491 (2011)

Bass, L., Klein, M., Bachmann, F: Quality attribute design primitives and the attribute driven design method. In: Software Product-Family Engineering, vol. 2290, Lecture Notes in Computer Science, pp. 169–186. Springer Berlin Heidelberg (2002)

Becha, H., Amyot, D.: Non-Functional properties in service oriented architecture – aconsumer's perspective. JSW **7**(3), 575–587 (2012)

Becker, S., Koziolek, H.: Transforming operational profiles of software components for quality of service predictions. In: Proceedings of the 10th Workshop on Component Oriented Programming (WCOP2005), 2005

Belotti, P., Lee, J., Liberti, L., Margot, F., Wächter, A.: Branching and bounds tightening techniques for non-convex MINLP. Optim. Methods Softw. **24**(4–5), 597–634 (2009)

Blum, C., Roli, A.: Metaheuristics in combinatorial optimization: overview and conceptual comparison. ACM Comput. Surv. **35**(3), 268–308 (2003)

Boehm, B., In, H.: Identifying quality-requirement conflicts. Softw. IEEE **13**(2), 25–35 (1996)

Breivold, H.P., Crnkovic, I., Larsson, M.: A systematic review of software architectureevolution research. Inf. Softw. Technol. **54**(1), 16–40 (2012)

Breivold, H.P., Larsson, M.: Component-based and service-oriented software engineering: key concepts and principles. In: EUROMICRO-SEAA, IEEE Computer Society, pp. 13–20 (2007)

Brereton, P., Budgen, D.: Component-based systems: a classification of issues. Computer **33**(11), 54–62 (2000)

Cancian, R.L., Stemmer, M.R., Schulter, A., Fröhlich, A.A.: A tool for supporting and automating the development of component-based embedded systems. J. Object Technol. **6**(9), 399–416 (2007)

Cardellini, V., Casalicchio, E., Grassi, V., Iannucci, S., Presti, F.L., Mirandola, R.: MOSES: A framework for QoS driven runtime adaptation of service-oriented systems. IEEE Trans. Softw. Eng. **38**(5), 1138–1159 (2012)

Chandran, S. K., Dimov, A., Punnekkat, S.: Modeling uncertainties in the estimation of software reliability. In: SSIRI, IEEE Computer Society, pp. 227–236 (2010)

Clements, P., Bachmann, F., Bass, L., Garlan, D., Ivers, J., Little, R., Merson, P., Nord, R., Stafford, J.: Documenting Software Architectures: Views and Beyond, 2nd edn. Addison Wesley (2011)

Cortellessa, V., Marinelli, F., Potena, P.: Automated Selection of Software Components Based on Cost/Reliability Tradeoff. In: EWSA, Lecture Notes in Computer Science, vol. 4344, pp. 66–81. Springer (2006)

Cortellessa, V., Potena, P.: Path-Based error propagation analysis in composition of software services. In: Software Composition, Lecture Notes in Computer Science, vol. 4829, pp. 97–112. Springer (2007)

Cortellessa, V., Crnkovic, I., Marinelli, F., Potena, P.: Experimenting the automated selection of COTS components based on cost and system requirements. J. Univers. Comput. Sci. **14**(8), 1228–1255 (2008)

Cortellessa, V., Potena, P.: How can optimization models support the maintenance of component-based software? In: 1st International Symposium on Search Based Software Engineering, pp. 97–100 (2009)

Cortellessa, V., Mirandola, R., Potena, P.: Selecting optimal maintenance plans based on cost/reliability tradeoffs for software subject to structural and behavioral changes. In: CSMR, IEEE, pp. 21–30 (2010)

Dai, Y.-S., Xie, M., Long, Q., Ng, S.-H.: Uncertainty analysis in software reliability modeling by bayesian analysis with maximum-entropy principle. Softw. Eng. IEEE Trans. **33**(11), 781–795 (2007)

Doran, D., Tran, M., Fiondella, L., Gokhale, S.S.: Architecture-based reliability analysis with uncertain parameters. In: SEKE, pp. 629–634 (2011)

Esfahani, N., Razavi, K., Malek, S.: Dealing with uncertainty in early software architecture. In: Proceedings of ACM SIGSOFT 2012/FSE-20 (New Ideas track) (2012)

Ghezzi, C., Pinto, L., Spoletini, P., Tamburelli, G.: Managing non-functional uncertainty via model-driven adaptivity. In: Proceedings of ICSE 2013 (2013)

Gokhale, S.: Architecture-based software reliability analysis: overview and limitations. Dependable Secure Comput. IEEE Trans. **4**(1), 32–40 (2007)

Goseva-Popstojanova, K., Trivedi, K.S.: Architecture-based approach to reliability assessment of software systems. Perform. Eval. **45**(2–3), 179–204 (2001)

Goseva-Popstojanova, K., Kamavaram, S.: Software reliability estimation under uncertainty: generalization of the method of moments. In: HASE, IEEE Computer Society, pp. 209–218 (2004)

Grunske, L.: Identifying "good" architectural design alternatives with multi-objective optimization strategies. In: ICSE, ACM, pp. 849–852 (2006)

Harman, M., Mansouri, S.A., Zhang, Y.: Search-based software engineering: Trends, techniques and applications. ACM Comput. Surv. **45**(1), 11:1–11:61 (2012)

http:\\www.lindo.com

Immonen, A., Niemelä, E.: Survey of reliability and availability prediction methods from the viewpoint of software architecture. Softw. Syst. Model. **7**(1), 49–65 (2008)

Jung, H.-W., Choi, B.: Optimization models for quality and cost of modular software systems. Eur. J. Oper. Res. **112**(3), 613–619 (1999)

Kazman, R., Klein, M., Barbacci, M., Longstaff, T., Lipson, H., Carrière, S.: The architecture tradeoff analysis method. In: ICECCS, pp. 68–78 (1998)

Kim, C.-K., Lee, D.H., Ko, I.-Y., Baik, J.: A Lightweight value-based software architecture evaluation. In: Eighth ACIS International Conference on Software Engineering, Artificial Intelligence, Networking, and Parallel/Distributed Computing, 2007. SNPD 2007, vol. 2, pp. 646–649, July 2007

Kontio, J.: A Case study in applying a systematic method for COTS selection. In: Proceedings of the 18th International Conference on Software Engineering, ICSE '96, IEEE Computer Society, pp. 201–209, Washington, DC, USA (1996)

Krka, I., Edwards, G., Cheung, L., Golubchik, L., Medvidovic, N.: A comprehensive exploration of challenges in architecture-based reliability estimation. In: Architecting Dependable Systems VI, vol. 5835, Lecture Notes in Computer Science, pp. 202–227 (2009)

Lee, J., Kang, S., Kim, C.-K.: Software architecture evaluation methods based on cost benefit analysis and quantitative decision making. Empir. Softw. Eng. **14**(4), 453–475 (2009)

Lundberg, L., Bosch, J., Häggander, D., Bengtsson, P.-O.: Quality attributes in software architecture design. In: Proceedings of the IASTED 3rd International Conference Software Engineering and Applications, pp. 353–362 (1999)

Malek, S., Medvidovic, N., Mikic-Rakic, M.: An Extensible framework for improving a distributed software system's deployment architecture. IEEE Trans. Softw. Eng. **38**(1), 73–100 (2012)

Meedeniya, I., Aleti, A., Grunske, L.: Architecture-driven reliability optimization with uncertain model parameters. J. Syst. Softw. **85**(10), 2340–2355 (2012)

Mirandola, R., Potena, P.: A QoS-based framework for the adaptation of service-based systems. Scalable Comput. Pract. Experience **12**(1) (2011)

Mohamed, A., Ruhe, G., Eberlein, A.: MiHOS: an approach to support handling the mismatches between system requirements and COTS products. Requir. Eng. **12**(3), 127–143 (2007)

Musa, J.: Operational profiles in software-reliability engineering. Softw. IEEE **10**(2), 14–32 (1993)

Neubauer, T., Stummer, C.: Interactive decision support for multiobjective COTS selection. In: 40th Annual Hawaii International Conference on System Sciences, 2007, HICSS 2007, pp. 283b–283b, Jan 2007

Park, J.: A high performance backoff protocol for fast execution of composite web services. Comput. Ind. Eng. **51**(1), 14–25 (2006)

Potena, P.: Optimization of adaptation plans for a service-oriented architecture with cost, reliability, availability and performance tradeoff. J. Syst. Softw. **86**(3), 624–648 (2013)

Potena, P., Crnkovic, I., Marinelli, F., Cortellessa, V.: Appendix of the chapter: software architecture quality of service analysis based on optimization models. Technical report, Dip. Informatica, Università de L'Aquila, [Online] (2016). http://www.di.univaq.it/cortelle/docs/TRChapter.pdf

Roshandel, R., Medvidovic, N., Golubchik, L.: A bayesian model for predicting reliability of software systems at the architectural level. In QoSA, vol. 4880, Lecture Notes in Computer Science, pp. 108–126. Springer (2007)

Sommerville, I.: Software Engineering (7th edn.). Pearson Addison Wesley (2004)

Szyperski, C.: Component Software: Beyond Object-Oriented Programming, 2nd edn. Addison-Wesley Longman Publishing Co., Inc. (2002)

Uchitel, S., Kramer, J., Magee, J.: Synthesis of behavioral models from scenarios. IEEE Trans. Softw. Eng. **29**(2), 99–115 (2003)

Vinek, E., Beran, P.P., Schikuta, E.: A dynamic multi-objective optimization framework for selecting distributed deployments in a heterogeneous environment. Procedia Comput. Sci. **4**, 166–175 (2011)

Wallnau, K., Stafford, J.A.: Dispelling the myth of component evaluation. In: Building Reliable Component-Based Software Systems (2002)

Wang, Y., Li, L., Huang, S., Chang, Q.: Reliability and covariance estimation of weighted k-out-of-n multi-state systems. Eur. J. Oper. Res. **221**(1), 138–147 (2012)

Wang DL, Zhu J, Li ZK, Paterson AH. Mapping QTLs with epistatic effects and QTL×environment interactions by mixed linear model approaches. Theor Appl Genet, 1999,99:1255–1264.

Wiesemann, W., Hochreiter, R., Kuhn, D.: A stochastic programming approach for QoS-aware service composition. In: CCGRID, pp. 226–233 (2008)

Yacoub, S., Cukic, B., Ammar, H.: A component-based approach to reliability analysis of distributed systems. In: Proceedings of the 18th IEEE Symposium on Reliable Distributed Systems, 1999, pp. 158–167 (1999)

Yang, J., Huang, G., Zhu, W., Cui, X., Mei, H.: Quality attribute tradeoff through adaptive architectures at runtime. J. Syst. Softw. **82**(2), 319–332 (2009)

Younas, M., Chao, K.-M., Laing, C.: Composition of mismatched web services in distributed service oriented design activities. Adv. Eng. Inform. **19**(2), 143–153 (2005)

Chapter 15
Key-Driver Analysis with Extended Back-Propagation Neural Network Based Importance-Performance Analysis (BPNN-IPA)

Josip Mikulić, Damir Krešić and Katarina Miličević

Abstract Importance-performance analysis (IPA) is a popular prioritization tool used to formulate effective and efficient quality improvement strategies for products and services. Since its introduction in 1977, IPA has undergone numerous enhancements and extensions, mostly with regard to the operationalization of attribute-importance. Recently, studies have promoted neural network-based IPA approaches to determine attribute-importance more reliably compared to traditional approaches. This chapter describes the application of back-propagation neural networks (BPNN) in an extended IPA framework with the goal of discovering key areas of quality improvements. The value of the extended BPNN-based IPA is demonstrated using an empirical case example of airport service quality.

Keywords Back-propagation · Neural network · Importance-performance analysis · Attribute importance · Service quality

15.1 Introduction

Originally introduced by Martilla and James in 1977 (Martilla and James 1977), the importance-performance analysis (IPA) has become one of the most popular analytical tools for prioritizing improvements of service attributes. According to the SCOPUS citation database, in July 2013 there were more than 300 papers bearing the name of the technique in the title, abstract or keywords, whereas the term appeared anywhere in the text in more than 1000 papers.

J. Mikulić (✉)
Faculty of Economics and Business, University of Zagreb, Zagreb, Croatia
e-mail: jmikulic@efzg.hr

D. Krešić · K. Miličević
Institute for Tourism, Zagreb, Croatia

© Springer International Publishing Switzerland 2016
C. Kahraman and S. Yanik (eds.), *Intelligent Decision Making in Quality Management*, Intelligent Systems Reference Library 97,
DOI 10.1007/978-3-319-24499-0_15

447

IPA usually departs from a formative multi-attribute model of customer satisfaction (CS). Put differently, the focal service is decomposed into key functional
and/or psychological attributes that significantly influence the customer experience
with the service. Such a model is then used to develop a questionnaire for gathering
the necessary IPA-input data. Following the original methodology, one set of
measurement items is used to measure perceived attribute-importance, and another
set to measure perceived attribute-performance. Arithmetic means of importance
and performance ratings are then plotted into a two-dimensional matrix. Grand
means of importance and performance ratings (or, alternatively, scale means), are
further taken to divide the matrix into four quadrants. Accordingly, four distinct
managerial recommendations can then be derived depending on the location of the
attributes within the matrix (Fig. 15.1).

Although the prioritization logic of the original IPA is intuitive and straightforward (priority rises with increasing importance and decreasing performance of
attributes), researchers have identified several shortcomings of the technique during
the past three decades. Whereas some authors were primarily concerned with
technical issues (e.g. the most appropriate way to divide the matrix into different
areas), a significantly larger number of scholars have raised conceptual issues that
mainly regard the importance-dimension in IPA. In order to enhance the reliability
(and validity) of the original methodology, researchers have thus proposed
numerous modifications with regard to both the conceptualization and the operationalization of attribute-importance in IPA. Most recently, IPA variants have been
introduced that utilize the power of the multilayer perceptron (MLP), a popular type
of back-propagation neural networks (BPNN), for assessing attribute-importance.
As several studies have shown, the integration of BPNNs into the IPA framework
can help to significantly increase the reliability of managerial implications (Deng
et al. 2008; Hu et al. 2009; Mikulić and Prebežac 2012; Mikulić et al. 2012).

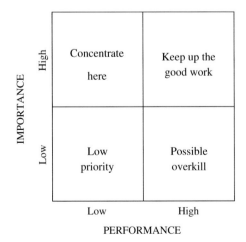

Fig. 15.1 Importance-performance matrix

In this chapter we present an extended BPNN-based IPA analytical framework which solves several significant shortcomings of traditional IPA. The value and application of the extended BPNN-IPA is demonstrated in an empirical case example of airport service quality. Before proceeding to the case study in Sect. 15.3, the following section reviews and summarizes recent advances regarding the IPA, particularly with regard to the integration of BPNNs into the analysis framework.

15.2 Literature Review

15.2.1 IPA and the Conceptualization of Attribute-Importance

The conceptualization and, subsequent, operationalization of attribute-importance is a controversial issue in IPA studies. The original methodology put forward the use of stated importance measures which assess the importance of attributes as perceived by the customer (Martilla and James 1977). This type of importance can be evaluated through rating-, ranking- or constant-sum scales. Contemporary IPA studies, however, employ increasingly derived measures of importance which are obtained by relating attribute-level performance to a measure of global service performance, like overall satisfaction or overall service quality (Grønholdt and Martensen 2005).

While several scholars have argued in favor of one of these two types of importance measures, recent IPA studies have revived the early ideas of Myers and Alpert (Myers and Alpert 1977) who stressed that these two types of measures should not be regarded as competing or conflicting measures. Rather they should be regarded as complementary measures because they assess different dimensions of the importance-construct (Van Ittersum et al. 2007). While stated measures assess an attributes general importance (referred to as *relevance*), derived measures assess an attributes actual influence in a particular study context (referred to as *determinance*). Most important, it is not reasonable to assume strong correlation between these two dimensions of importance. Aspects of a service which are perceived very important by the customer do not necessarily have to be those ones that will truly have the strongest influence on his satisfaction in a particular service transaction. If an attribute which is perceived very important by the customer performs according to the customer's expectations, then its actual influence on the customer's overall satisfaction might be smaller than the effect of an attribute which is perceived less important. This might occur in cases when the less important attribute performs below or above customer-expected levels, thus causing strong negative or positive customer reactions, respectively.

Following this line of thought, Mikulić and Prebežac (2011, 2012) have proposed a rather simple extension of IPA by integrating both stated and derived

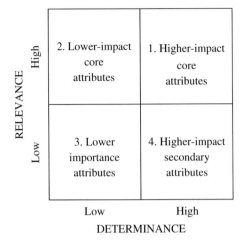

Fig. 15.2 Relevance-determinance matrix

measures of attribute-importance into a relevance-determinance matrix (RDM; Fig. 15.2). Since a three-dimensional representation of results might, however, be confusing (i.e. two importance dimensions and one performance dimension), the authors suggest marking attributes in the RDM that perform below and above average with a minus (−) and a plus (+), respectively.

The following recommendations apply to the four attribute categories (Mikulić and Prebežac 2012):

- *Higher-impact core attributes* (quadrant 1): These attributes are perceived very important by customers and they have a strong influence on overall satisfaction. The management should primarily focus on this category to strengthen the market position. Attributes from this attribute category that perform relatively low should be assigned highest priority in improvement strategies.

- *Lower-impact core attributes* (quadrant 2): These attributes are perceived very important, but they only have a relatively weak influence on overall satisfaction. Market-typical levels of performance should be ensured for these attributes. These attributes can turn into dissatisfiers with a strong influence on overall satisfaction when performance drops below a tolerated threshold.

- *Higher-impact secondary attributes* (quadrant 4): These attributes are perceived less important, but they have a strong influence on overall satisfaction. Attributes forming this category are likely part of the augmented product/service and can be used to differentiate from the competition. The importance of these attributes would be completely underestimated if using stated importance measures, only.

- *Lower-importance attributes/Lower-impact secondary attributes* (quadrant 3): These attributes should be assigned lower general priority in improvement strategies than the previous three categories.

15.2.2   IPA and the Problem of Multicollinearity

While stated importance is typically assessed through direct rating scales, derived importance is usually assessed by means of multiple regression or correlation analysis. A significant technical problem here, which limits the applicability of popular derived measures of attribute-importance, is strong correlation among the attributes which are used to predict overall CS. In particular, the problem is that a regression on correlated attributes violates a basic assumption of the technique, why it may produce invalid estimates of relative attribute-determinance. Typical consequences are (i) regression coefficients with reversed signs, although the zero-order correlation with the dependent variable is positive, (ii) significantly different weights for equally determinant variables, and (iii) exaggerated/suppressed regression coefficients (Johnson 2000). Although many research areas struggle with correlated variables, the problem can be characterized as a major 'plague' in CS research, as this area of research does not rely on metric measures of objective phenomena, but rather on limited scale-range measures of perceptions that frequently tend to be strongly correlated (Weiner and Tang 2005). Moreover, CS studies tend to analyze relatively large numbers of variables, which generally increases the risk of multicollinearity. Since the reliability and validity of derived importance measures directly affects the reliability and validity of attribute-prioritizations, ways need to be found to deal with this problem in IPA. Basically, there are three general options.

1. Bivariate approaches like zero-order correlation or bivariate regressions may be applied to circumvent the multicollinearity problem. However, these approaches are less than optimal because they fail to consider the influence of all other variables in estimations of relative attribute-determinance. Accordingly, these measures are generally not recommended for use with multi-attribute CS models.
2. The risk of high inter-correlations may be reduced by specifying attribute-models in which they are less likely to occur. Since the likelihood of occurrence is typically positively correlated with the number of explanatory variables in a regression model, researchers may, on the one hand, consider the use of hierarchical attribute-models to keep the number of predictors in a model at a reasonable level, but thereby preserving desired levels of detail. On the other hand, if the data are not based on hierarchical models, attributes may be factor analyzed in an exploratory manner to potentially obtain a decreased number of uncorrelated factors that enter the analysis. Similarly, but simpler, correlational matrices can be computed to identify highly correlated attributes that should be reconsidered for inclusion into the final model.
3. Researchers may use approaches that are capable of effectively dealing with correlated predictors. Several regression-based approaches have been proposed, involving measures of average variable contributions to R^2 across all possible sub-models (Kruskal 1987; Budescu 1993), variance-decomposition with

uncorrelated subsets of predictors (Genizi 1993), or heuristics based on predictor orthogonalization (Johnson 2000). However, in case of larger numbers of attributes, a severe limitation of these approaches is that they are either complicated to implement, or computationally very demanding. For example, 'all sub-set regression' procedures require $2^p - 1$ models for estimating the importances of p attributes—i.e.: 31 models for p = 5, 1023 models for p = 10, and even 32,767 models for p = 15. Since none of available statistical packages have built-in features for performing such analyses, these approaches are not very appealing to CS researchers.

15.2.3 IPA and the Application of Artificial Neural Networks

A valuable alternative to traditional statistical approaches that does not assume uncorrelated predictors is the multilayer perceptron (MLP), a popular class of back-propagation neural networks (BPNN) that has been applied in several IPA studies (Deng et al. 2008; Hu et al. 2009; Mikulić and Prebežac 2012; Mikulić et al. 2012). BPNNs are artificial neural networks with feed-forward architecture that use a supervised learning method. Back-propagation is the most widely used neural network architecture for classification and prediction. The idea of the BPNN goes back to 1974 with Werbos discussing the concept, while the algorithm was clearly defined in 1985 by Rumelhart and his colleagues who introduced the Propagation Learning Rule (Rumelhart et al. 1986). Nowadays, BPNNs are widely applied in numerous research areas, such as pattern recognition, medical diagnosis, sales forecasting or stock market returns, among others (Zong et al. 2014; Subbaiah et al. 2014; Kuo et al. 2014; Huo et al. 2014). A graphical presentation of a typical MLP is provided in Fig. 15.3.

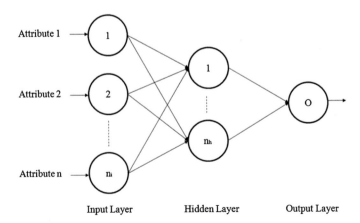

Fig. 15.3 Multilayer perceptron

An MLP consists of one input-layer, one or more hidden-layers, and one output-layer. Each layer comprises a number of neurons that process the data via nonlinear activation functions (e.g. sigmoid, hyperbolic-tangent). To draw an analogy to regression, the input-layer neurons can be referred to as predictors and the output-layer neurons as the dependent variable (typically this is one in regression-kind problems).

An important difference compared to regression is, however, that predictors are not directly related to the dependent variable, but via neurons in one or more hidden layers. These in turn determine the mapping relations which are stored as weights of connecting paths between the neurons. The nonlinear activation functions further enable the MLP to straightforwardly deal with indefinable nonlinearity, giving the MLP a significant technical advantage over regular linear regression (DeTienne et al. 2003). The most important difference towards regression is, however, that the MLP is a dynamic network model that uses a back-propagation algorithm to train and optimize the network. Errors between predicted and actual output values are iteratively fed back to the network in order to minimize this discrepancy according to some predefined rule or target (Haykin 1999). Put differently, the MLP *learns* from the data and dynamically updates the network weights. Sum-of-squares (SOS) error functions are typically used in combination with learning algorithms like the scaled conjugate gradient algorithm (Moller 1993), or the Broyden-Fletcher-Goldfarb-Shanno (BFGS) algorithm (Broyden et al. 1973).

Although MLPs are powerful prediction tools that can explain very large amounts of variance in dependent variables, MLPs do, however, not provide straightforward indicators of predictor determinance (i.e. derived predictor importance). Because of this, ANNs have been frequently termed as "black box" methodologies. Such indicators can, however, be obtained by using one of the following two approaches.

On the one hand, predictor determinance can be derived through connection-weight procedures—i.e. all weights connecting an input-layer neuron over hidden-layers to the output layer neuron are used to calculate a neuron's determinance (i.e. its influence on the dependent variable). The two most widespread, though conflicting, procedures are the algorithms proposed by Garson (Garson 1991) and Olden and Jackson (Olden and Jackson 2002). An empirical comparison using Monte-Carlo simulated data has, however, come to the conclusion that the latter approach performs significantly better, and thus it should be preferred (Olden and Jackson 2002).

On the other hand, predictor determinance can be derived through stepwise procedures. Here it is analyzed how the discrepancy between predicted and actual output values behaves when predictors are iteratively dropped from, or included into the network (Sung 1998). Analogously to analyzing changes in R^2 when dropping/including predictors in a regression model, a relatively larger increase of the network/model error, attributed to the omission of a particular predictor, can be interpreted as relatively larger predictor determinance. Conversely, a decrease of the network error would imply that the respective predictor should rather be omitted from the network, as it, in fact, decreases the overall model quality. Moreover, because the assumption of uncorrelated predictors is not made in MLPs, a noteworthy advantage over regression is that there is no need to average changes in model error over all predictor-orderings to ensure the reliability of determinance estimates. Since all-subset regressions become exponentially time-consuming with larger numbers of predictors (i.e. 2^{p-1} models are required to estimate the determinance of p attributes), this is a significant practical advantage of MLPs over similar regression-based approaches like e.g. dominance analysis, or Kruskal's averaging over orderings procedure.

15.3   An Application of the Extended BPNN-IPA

An overview of the extended BPNN-IPA methodology is given in Fig. 15.4.

The data used in this example were collected as part of a periodical survey on airline passenger satisfaction with services provided at a European international airport. The data were collected by means of a structured questionnaire in face-to-face interviews in the international departure area of the airport. Five-point direct rating scales were used to assess both the importance (1 = less important; 5 = very important) and performance (1 = very poor; 5 = excellent) of a series of airport attributes, as well as the level of overall satisfaction with the airport (1 = disappointed; 5 = delighted). Overall, 2025 fully completed questionnaires entered the subsequent data analysis.

In order to guide management efforts for improving the overall airport experience we will conduct an extended BPNN-based IPA. Following the approach proposed by Mikulić and Prebežac (2011, 2012), the traditional IPA framework is extended by using measures of both attribute-relevance and determinance. This facilitates a relative categorization of attributes according to their general importance, as perceived by passengers (i.e. attribute-relevance, AR), and their actual influence on overall passenger satisfaction with the airport services (i.e. attribute-determinance, AD).

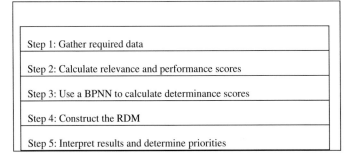

Step 1: Gather required data

Step 2: Calculate relevance and performance scores

Step 3: Use a BPNN to calculate determinance scores

Step 4: Construct the RDM

Step 5: Interpret results and determine priorities

Fig. 15.4 Methodology of the extended BPNN-IPA

To prepare the necessary input-data arithmetic means of attribute-performance ratings (AP) are first calculated:

$$AP_i = \frac{1}{n} \sum_{j=1}^{n} p_{i,j} \quad \forall i \in I; j = 1, \ldots, n \tag{15.1}$$

where $p_{i,j}$ is the performance rating for attribute i, $i \in I$ by respondent j, $j = 1, \ldots, n$, n the number of respondents, and I the set of analyzed attributes i.

Analogously, arithmetic means of importance ratings are calculated to obtain indicators of attribute-relevance (AR). To obtain indicators of AD an MLP-based sensitivity analysis is conducted. This analysis involves the following steps:

4. *Specification of MLP architecture*: AP ratings are specified as input-layer neurons and ratings of overall satisfaction with the airport as single output-layer neuron in a one-hidden layer MLP. The overall sample is partitioned into training, testing, and holdout samples (60, 20, and 20 % of the samples, respectively). The network training continues as long as the network error is decreasing in both the main dataset (i.e. training samples) and the testing samples. When the error between predicted and true output values starts increasing in the testing sample, training is stopped to prevent over-fitting. This stopping rule is necessary because over-fitted networks usually perform very well or perfect during training, but they also typically perform significantly weaker or badly on unseen data. The holdout samples are further used to cross-validate the performance of the MLP after the training is finished. The MLP can be considered reliable only if the network performs consistently well across all three independent samples.

5. *Network training*: The network is trained using a sum-of-squares error function and the BFGS learning algorithm. Network performance is assessed using the mean absolute percentage error (*MAPE*) and root mean squared error (*RMSE*). RMSE can be used to derive network goodness-of-fit (R^2):

$$\text{MAPE} = \frac{1}{n} \sum_{i=1}^{n} \left| \frac{y_i - a_i}{a_i} \right| * 100\% \; i = 1, \ldots, n \tag{15.2}$$

$$\text{RMSE} = \sqrt{\frac{\sum_{i=1}^{n} (y_i - a_i)^2}{n}} \tag{15.3}$$

$$R^2 = 1 - \frac{\text{RMSE}}{\sigma^2} \tag{15.4}$$

where y_i is the predicted output value for sample i, a_i the actual output value for sample i, n the number of samples, and σ^2 the variance of the actual output.

The following trial-and-error procedure is used to determine the best network configuration. In a first step several networks with varying activation functions and numbers of neurons in the hidden-layer are estimated. The correlations between true and predicted values are then checked to identify the best-performing networks. Here it is important that the network configurations provide consistent performance across the training, testing and holdout samples. After identifying the better-performing activation functions, these are then used to estimate another set of network configurations. The correlations between predicted and true output values are then checked again to identify the best performing activation functions and number of hidden-layer neurons. Using e.g. the *automated neural network* feature in newer versions of *Statsoft Statistica* (version 8.0 or higher) this whole trial-and-error procedure can easily be conducted, thereby using large numbers of network configurations to be estimated at a time (e.g. 5000 or higher).

6. *Estimation of attribute-determinance*: To obtain indicators of AD a global sensitivity analysis of the network error is conducted. While in a local sensitivity analysis the focus is on how sensitive the output is to a given domain of a predictor, global sensitivity focuses on how the output behaves when completely eliminating a predictor from the network. This is done by iteratively fixing the value of each particular predictor to its arithmetic mean before re-estimating the same network (with a particular predictor omitted). Accordingly, a larger increase of the network error can then be regarded as an indicator of larger influence of an attribute in explaining variations in the output (i.e. determinance). This type of indicator is very similar to changes in R^2 which are attributable to the omission of predictors from a regression model.

The results of the network performance assessment for our case example are provided in Table 15.1. The network we choose to estimate indicators of AD has 20 hidden-layer. neurons, exponential activation functions in the hidden layer, and identity functions in the output layer (in bold). For comparison, the coefficient of determination of the respective OLS regression model is $R^2 = 0.59887$.

Table 15.1 Assessment of network performances

Network configuration	R^2 (Training sample)	R^2 (Test sample)	R^2 (Holdout sample)	Activation functions (Hidden layer/Output layer)
13-20-1	**0.75994**	**0.74677**	**0.75358**	**Exponential/Identity**
13-5-1	0.75447	0.73327	0.74652	Exponential/Identity
13-6-1	0.80416	0.73053	0.72631	Exponential/Exponential

Table 15.2 Determinance, relevance and performance of airport attributes

Attribute	Determinance (%)	Relevance	Performance
1. Traffic connection between airport and city	93.63	4.27	3.91
2. Parking	56.86	3.57	3.70
3. Ease of way-finding	53.61	4.48	4.22
4. Information desk	71.10	4.22	4.21
5. Customs and body check procedure	73.71	4.29	4.19
6. Cafes and restaurants	42.75	3.60	3.66
7. Shopping possibilities	0.00	3.63	3.73
8. Availability of ATMs	50.14	4.16	4.16
9. Availability of Internet access	29.31	3.87	3.59
10. Availability of luggage carts	80.70	3.84	3.94
11. Comfort level and cleanliness	35.03	4.37	4.00
12. Staff politeness	67.29	4.62	4.20
13. Flight network	100.00	4.25	3.86
Grand mean	58.01	4.09	3.95

Min-max normalization is applied to the weights obtained from the sensitivity analysis for easier comparison across attributes (expressed as percentages). Final scores of AD, AR and AP are presented in Table 15.2.

Scores of AD and AR are then used to construct the two-dimensional RDM. The thresholds that divide the matrix into four quadrants are set at the values of the grand means of AD and AR (Fig. 15.5). The basic prioritization logic is to search for attributes that perform relatively lower (e.g. below the grand mean) starting from the first quadrant (higher-impact core attributes; highest general priority), over the fourth quadrant (higher-impact secondary attributes), the second quadrant (lower impact core attributes), to the third quadrant (lower importance attributes; lowest general priority).

The BPNN-IPA reveals that most attention should be paid to the quality of (13) the flight network and the (1) traffic connection between the airport and the city. These are the only two attributes located in the first quadrant that perform below average ($AP_{13} = 3.86$; $AP_1 = 3.91$). Improving the quality of these two attributes would be likely to significantly enhance overall passenger satisfaction with the airport.

If we move to the fourth quadrant, we see that the only higher-impact secondary attribute performs below average, i.e. (10) availability of luggage carts ($AP_{10} = 3.94$). Accordingly, this attribute should be considered next for improvement. It is noteworthy that the *importance* of this attribute would have been significantly underestimated if only stated importance measures had been used.

A look at the attributes in the second quadrant (lower-impact core attributes) reveals that no immediate action is needed here, because all attributes perform above average.

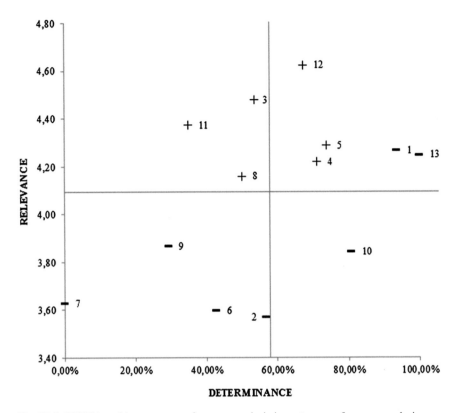

Fig. 15.5 BPNN-based importance-performance analysis importance-performance analysis

Finally, the focus is shifted to the attributes located in the third quadrant which have relatively lower general priority than attributes in the other three quadrants. Although these attributes have relatively lower relevance and determinance, the airport management should consider their improvement after having improved the previously mentioned attributes, because all the four attributes perform below average—i.e. (2) parking (AP_2 = 3.70), (6) cafes and restaurants (AP_6 = 3.66), (9) availability of Internet access (AP_9 = 3.59), and (7) shopping possibilities (AP_7 = 3.73).

15.4 Conclusion

This chapter described the application of back-propagation neural networks (BPNN) in an extended importance-performance analysis (IPA) framework with the goal of discovering and prioritizing key areas of quality improvements. The application of the extended BPNN-based IPA was demonstrated using an empirical case study of passenger satisfaction with services provided by an international airport. The extended BPNN-based IPA identified the most important key-drivers of passenger satisfaction and provided detailed improvement priorities of the various airport services.

From a methodological point of view, the applied framework solves two important shortcomings of traditional key-driver analyses, in particular of prevailing approaches to IPA:

First, by combining two different dimensions of attribute-importance into IPA (i.e. attribute-relevance and determinance), the general reliability of the analytical framework is significantly increased. With only few exceptions, IPA studies typically use a one-dimensional operationalization of importance, i.e. either they use relevance or determinance. Since these two measures do not necessarily have to converge, the reliability and validity of managerial implications from traditional IPA are at least questionable. That a one-dimensional operationalization of importance might mislead managers has also been demonstrated in the example used in this chapter. The importance of one attribute (availability of luggage carts) would have been significantly underestimated if only measures of relevance had been used. Here, relevance of the attribute was below average, while its determinance was significantly above average.

Second, by using the multilayer perceptron (MLP), a popular class of BPNNs for deriving attribute-determinance in IPA, the proposed framework provides more reliable determinance estimates compared to traditional regression-based analyses. This is because the MLP can effectively deal with correlated predictors, and it applies nonlinear rather than linear activation functions in modeling the data. The MLP can thus straightforwardly account for possible nonlinearities in the relationship between the performance of various service/product attributes and the level of global satisfaction. Application of the MLP is particularly valuable in customer satisfaction studies, as demonstrated in this chapter, because studies in this area typically analyze larger numbers of product or service attributes. Since there is usually a significant amount of correlation among these attributes, traditional regression-based analyses tend to provide distorted and, subsequently, unreliable determinance scores. With application of the MLP, reliability of determinance-scores is significantly improved. Moreover, since multicollinearity problems tend to increase with larger numbers of analyzed product/service attributes-predictors, application of the MLPdoes not force researchers to make large trade-offs between the desired level of detail of the attribute-model under study and the reliability of results.

For future IPA studies it is generally recommended to apply both relevance and determinance scores to determine an attribute's *importance*. With regard to the application of ANNs in assessing an attribute's influence on a dependent variable (like overall satisfaction), future IPA studies may consider the application of genetic algorithm for network optimization. Also, it would be useful to further investigate and compare different was of obtaining determinance weights from ANNs (e.g. connection-weights procedures *vs.* stepwise procedures), in order to provide some best practice guidelines for both practitioners and researchers in this area.

Key terms

- Back-propagation neural network: A feed-forward artificial neural network that uses supervised learning to map a set of input data onto a set of output data. The error (i.e. discrepancies between true and computed data) is back-propagated to the network until it is minimized according to some predefined rule.
- Importance-performance analysis: A widely applied analytical tool that is used to prioritize product/service attributes for improvement. The rationale is to compare the importance of product/service attributes with the attributes' performance using a two-dimensional matrix. The analysis is based on data from typical customer satisfaction surveys.
- Relevance: A dimension of the importance construct that could be referred to as *general importance*. The literature also uses the term *stated importance* to denote relevance. The relevance of a product/service attribute such refers to the attribute's importance without a particular performance context.
- Determinance: A dimension of the importance construct that could be referred to as *actual importance* or *impact*. The literature also uses the term derived importance to denote determinance. The determinance of a product/service attribute such refers to the attribute's actual influence on e.g. the customer's satisfaction given a particular context of attribute performances.
- Relevance-determinance asymmetry: The case when the relevance of a product/service attribute does not correspond with the attribute's determinance. E.g. the relevance of safety as an attribute of an airline flight certainly is very high. The attribute's actual importance or impact on a passenger's flight satisfaction (i.e. determinance), however, certainly depends on the attribute's level of performance. Such, it should not have a significant impact in case everything went fine on a flight.

References

Broyden, C.G., Dennis, J.E., More, J.J.: On the local and superlinear convergence of quasi-newton methods. IMA J. Appl. Math. **12**, 223–246 (1973)

Budescu, D.V.: Dominance analysis: a new approach to the problem of relative importance of predictors in multiple regression. Psychol. Bull. **114**, 542–551 (1993)

Deng, W.J., Chen, W.C., Pei, W.: Back-propagation neural network based importance-performance analysis for determining critical service attributes. Expert Syst. Appl. **34**, 1115–1125 (2008)

DeTienne, K.B., DeTienne, D.H., Joshi, S.A.: Neural networks as statistical tools for business researchers. Organ. Res. Methods **6**, 236–265 (2003)

Garson, G.D.: Interpreting neural-network connection weights. AI Expert **6**, 47–51 (1991)

Genizi, A.: Decomposition of R2 in multiple regression with correlated regressors. Stat. Sinica **3**, 407–420 (1993)

Grønholdt, L., Martensen, A.: Analysing customer satisfaction data: a comparison of regression and artificial neural networks. Int. J. Market Res. **47**, 121–130 (2005)

Haykin, S.: Neural networks: a comprehensive foundation. Prentice-Hall, Upper Saddle River (1999)

Hu, H.Y., Lee, Y.C., Yen, T.M., Tsai, C.H.: Using BPNN and DEMATEL to modify importance-performance analysis model: a study of the computer industry. Expert Syst. Appl. **36**, 9969–9979 (2009)

Huo, L., Jiang, B., Ning, T., Yin, B.: A BP neural network predictor model for stock price. In Intelligent Computing Methodologies, pp. 362–368. Springer International Publishing, New York (2014)

Johnson, J.W.: A heuristic method for estimating the relative weight of predictor variables in multiple regression. Multivar. Behav. Res. **35**, 1–19 (2000)

Kruskal, W.H.: Relative importance by averaging over orderings. Am. Stat. **41**, 6–10 (1987)

Kuo, R.J., Tseng, Y.S., Chen, Z.Y.: Integration of fuzzy neural network and artificial immune system-based back-propagation neural network for sales forecasting using qualitative and quantitative data. J. Intell. Manuf. 1–17 (2014)

Martilla, J.A., James, J.C.: Importance-performance analysis. J. Mark. **41**, 77–79 (1977)

Mikulić, J., Prebežac, D.: Rethinking the importance grid as a research tool for quality managers. Total Qual. Manag. **22**, 993–1006 (2011)

Mikulić, J., Prebežac, D.: Accounting for dynamics in attribute-importance and for competitor performance to enhance reliability of BPNN-based importance-performance analysis. Expert Syst. Appl. **39**, 5144–5153 (2012)

Mikulić, J., Paunović, Z., Prebežac, D.: An extended neural network-based importance-performance analysis for enhancing wine fair experience. J. Travel Tour. Mark. **29**, 744–759 (2012)

Moller, M.F.: A scaled conjugate gradient algorithm for fast supervised learning. Neural Netw. **6**, 525–533 (1993)

Myers, J.H., Alpert, M.I.: Semantic confusion in attitude research: salience vs Importance vs. Determinance. Adv. Consum. Res. **4**, 106–110 (1977)

Olden, J.D., Jackson, D.A.: Illuminating the "Black Box": a randomization approach for understanding variable contributions in artificial neural networks. Ecol. Model. **154**, 135–150 (2002)

Rumelhart, D.E., Hinton, G.E., Williams, R.J.: Learning internal representation by error propagation. Parallel Distrib. Proc. **1**, 318–362 (1986)

Subbaiah, R.M., Dey, P., Nijhawan, R.: Artificial neural network in breast lesions from fine-needle aspiration cytology smear. Diagn. Cytopathol. **42**, 218–224 (2014)

Sung, A.H.: Ranking importance of input parameters of neural networks. Expert Syst. Appl. **15**, 405–411 (1998)

Van Ittersum, K., Pennings, J.M.E., Wansink, B., van Trijp, H.C.M.: The validity of attribute-importance measurement: a review. J. Bus. Res. **60**, 1177–1190 (2007)

Weiner, J.L., Tang, J.: Multicollinearity in Customer Satisfaction Research. White paper, Ipsos Loyalty (2005)

Zong, R., Zhi, Y., Yao, B., Gao, J., Stec, A.A.: Classification and identification of soot source with principal component analysis and back-propagation neural network. Aust. J. Forensic Sci. **46**, 224–233 (2014)

Index

© Springer International Publishing Switzerland 2016
C. Kahraman and S. Yanik (eds.), *Intelligent Decision Making in Quality Management*, Intelligent Systems Reference Library 97,
DOI 10.1007/978-3-319-24499-0